ENCYCLOPEDIA OF
BIOLOGY

ENCYCLOPEDIA OF
BIOLOGY

DON RITTNER
AND
TIMOTHY L. McCABE, Ph.D.

Facts On File, Inc.

Encyclopedia of Biology

Facts On File, Inc.
132 West 31st Street
New York NY 10001

Library of Congress Cataloging-in-Publication Data

Rittner, Don.
Encyclopedia of biology / Don Rittner and Timothy L. McCabe.
p. cm.
Summary: Contains approximately 800 alphabetical entries, prose essays on important topics, line illustrations, and black-and-white photographs.
Includes bibliographical references (p.   ).
ISBN 0-8160-4859-2
1. Biology—Encyclopedias, Juvenile. [1. Biology—Encyclopedias.
2. Encyclopedias and dictionaries.] I. McCabe, Timothy Lee. II. Title.
QH309.2.R58 2004
570'.3—dc222003021279

Facts On File books are available at special discounts when purchased in bulk quantities for businesses, associations, institutions, or sales promotions. Please call our Special Sales Department in New York at (212) 967-8800 or (800) 322-8755.

You can find Facts On File on the World Wide Web at http://www.factsonfile.com

Text design by Joan M. Toro
Cover design by Cathy Rincon
Illustrations by Richard Garratt and Sholto Ainslie

Printed in the United States of America

VB FOF 10 9 8 7 6 5 4 3 2 1

This book is printed on acid-free paper.

Dedicated to

Louis F. Ismay and John F. Roach
Two superb teachers who taught me to always ask why

and to my family
Nancy, Christopher, Kevin, Jackson, Jennifer, & Jason

CONTENTS

ACKNOWLEDGMENTS

We would like to thank the following for their generosity in helping to make this book as complete as possible, especially in the use of images, biographies, essays, and encouragement: Darryl Leja, NHGRI, National Institutes of Health, for wonderful illustrations; Marissa Mills, Human Genome Management Information System, Oak Ridge National Laboratory, U.S. Department of Energy Genomes to Life Program; Celia Boyer, executive director, Health On the Net (HON) Foundation; Kristina Fallenias, Nobel Foundation; Fabienne Meyers, International Union of Pure and Applied Chemistry; Robert Dirig, Cornell University; Centers for Disease Control; William and Greta Wagle; John McConnell; John C. Brenner; Demetra Xythalis; Joseph Deuel, Petrified Sea Gardens, Inc.; Betty Harrison; James G. (Spider) Barbour; Hideki Horikami; Thomas Wittling; and to Nancy, Chris, Kevin, and Jack.

Finally, thanks to Frank K. Darmstadt, our very patient editor, and the rest of the staff at Facts On File for their contributions.

We apologize to anyone left out in error.

PREFACE

Despite the often extreme specialization and intimate knowledge required to make a contribution to science, most scientific disciplines are quick to adapt new technologies and advances developed from other fields. Inevitably, a new vocabulary follows these advances, the purpose of which is to convey meaning with a word that once required a descriptive paragraph or even a page.

The *Encyclopedia of Biology* pulls together the specialized terminology that has found its way into the language of the biologist. It addresses the often duplicitous meanings in an easily understood, succinct fashion. As each discipline has become more of a specialty, each has developed terms that serve as a shorthand for concepts within that discipline. On rare occasion, different disciplines develop the same term with radically different definitions. By indicating a discipline, the encyclopedia directs the reader to a definition relevant to the topic at hand. An example of this is the word *genotype*. Historically, this was a taxonomist's term meaning "the type of the genus." The genotype is important for classification and evolutionary studies. Subsequently, geneticists used genotype to refer to the genetic makeup of an organism. One needs to understand not only the meaning of words, but must also be able to put them in the context of the period in which they were written.

There will be new terms, new (and defunct) science Websites, new leaders, new disciplines, and even breathtaking new discoveries in science, but these will not detract from the utility of this encyclopedia. Bibliophiles need only pause to consider which books they consult most frequently. The reference book holds counsel over all others. Facts On File's *Encyclopedia of Biology* may not read like a novel, but it will help you read like a biologist.

—Tim McCabe, Ph.D.

INTRODUCTION

Facts On File's *Encyclopedia of Biology* is a reference to help in understanding the basic concepts in biology and its peripheral disciplines like ecology, botany, and even Earth science. Arranged in alphabetical order, the entries include biographies of individuals who have made major contributions as well as numerous line illustrations and photographs to help in visualizing technical concepts.

I have tried to include the more common terms you will likely encounter during your educational experience or even when you are out in the "real" world. There are literally thousands of biological terms. Many are so specific to major or minor subdisciplines of biology that you may never encounter them. You will not find those esoteric terms in this encyclopedia but, rather, a collection of terms that you should be familiar with to understand core biological principles and have a working knowledge of the field. You can also use this volume simply to increase your scientific vocabulary. A series of well-placed essays elaborate on some of the most important trends and issues in the field. One of these describes how the use of computer technology has revealed an artificial toe in a mummy that is thousands of years old. You will also learn how blood is used in forensic science to capture criminals and read about the latest trends in human cytogenetics. Other essays will make you think about your role in the world and explore some of the negative effects we humans have had on the biological world, in particular to the insect family.

The encyclopedia also includes appendixes with information about Internet Websites and biology-related software that is waiting for you to explore.

We humans are part of this immense biological world, and we interact with it in many ways. Some of those interactions have cost species their very existence. Some have helped us survive disease. In other cases, we have helped species come back from the brink of extinction. This complex interrelationship is not clearly understood even today, and that is why many who use this book are pursuing some aspect of biology as a career.

We have come a long way from Robert Hooke's first observation of a cell under a crude microscope to today's observations of atomic-level activity using electron microscopes. The use of computer science and technology has enabled huge leaps in our understanding of our biological world. The future hope of nanotechnology, using small robots to scurry through our bodies to fix organs or cure disease, is closer to becoming reality than it is to fiction. Other former sci-fi issues, like cloning humans, are on the forefront of discussion, and some have even claimed human cloning has happened. The mapping of our entire genetic makeup brings promise to thousands of people who have or carry genetically based disease. This has led at least one scientist to declare that we are moving into the "industrial revolution of biology," anticipating the exciting discoveries just around the corner from the analysis of all this genetic information. We are entering a world of molecular understanding of developmental biology all the way to the enigma of consciousness. But you cannot leap without first taking small steps. Use this book for the small steps, and heed the words of Cornell biologist James G. Needham (1888–1957), who once wrote:

> It is a monstrous abuse of the science of biology to teach it only in the laboratory—Life belongs in the fields, in the ponds, on the mountains, and by the seashore.

So, armed with this book, consider that your next assignment.

—Don Rittner
Schenectady, New York

ENTRIES A–Z

A

ABO blood groups Blood group antibodies (A, B, AB, O) that may destroy red blood cells bearing the antigen to which they are directed; also called "agglutinins." These red-cell antigens are the phenotypic expression of inherited genes, and the frequency of the four main groups varies in populations throughout the world. The antigens of the ABO system are an integral part of the red-cell membrane as well as all cells throughout the body and are the most important in transfusion practice.

See also LANDSTEINER, KARL.

abortion The termination of gestation before the fetus can survive on its own.

abscisic acid (ABA) A plant hormone ($C_{15}H_{20}O_4$) and weak acid that generally acts to inhibit growth, induces dormancy, and helps the plant tolerate stressful conditions by closing stomata. Abscisic acid was named based on a belief that the hormone caused the abscission (shedding) of leaves from deciduous trees during the fall.

At times when a plant needs to slow down growth and assume a resting (dormant) stage, abscisic acid is produced in the terminal bud, which slows down growth and directs the leaf primordia to develop scales that protect the dormant bud during winter. Because the hormone also inhibits cell division in the vascular cambium, both primary and secondary growth are put on hold during winter.

This hormone also acts as a stress agent that helps a plant deal with adverse conditions. For example, ABA accumulates on leaves and causes stomata to close, reducing the loss of water when a plant begins to wilt.

In 1963, abscisic acid was first identified and characterized by Frederick Addicott and colleagues. In 1965, the chemical structure of ABA was defined, and in 1967, it was formally called abscisic acid.

absorption spectrum Different pigments absorb light of different wavelengths. For example, chlorophyll effectively absorbs blue and red. The absorption spectrum of a pigment is produced by examining, through the pigment and an instrument called a spectroscope, a continuous spectrum of radiation. The energies removed from the continuous spectrum by the absorbing pigment show up as black lines or bands and can be graphed.

abyssal zone The portion of the ocean floor below 1,000–2,000 m (3,281–6,561 ft.), where light does not penetrate and where temperatures are cold and pressures are intense. It lies seaward of the continental slope and covers approximately 75 percent of the ocean floor. The temperature does not rise above 4°C. Because oxygen is present, a diverse community of invertebrates and fishes do exist, and some have adapted to harsh environments such as hydrothermal vents of volcanic creation. Food-producing organisms at this depth are chemoautotrophic prokaryotes and not photosynthetic producers.

See also OCEANIC ZONE.

acclimatization   Acclimatization is the progressive physiological adjustment or adaptation by an organism to a change in an environmental factor, such as temperature, or in conditions that would reduce the amount of oxygen to its cells. This adjustment can take place immediately or over a period of days or weeks. For example, the human body produces more erythrocytes (red blood cells) in response to low partial pressures of oxygen at high altitudes; short-term responses include shivering or sweating in warm-blooded animals.

accommodation   The automatic reflex adjustment that allows the focal length of the lens of an eye to change to focus on an object. The lens shape, more convex for near objects and less convex for distant objects, is caused by ciliary muscles acting on the elastic property of the lens.

acetylcholine   (ACh)   One of the most common neurotransmitters of the vertebrate nervous system, ACh is a chemical ($CH_3COOCH_2CH_2N+(CH_3)_3$) that transmits impulses between the ends of two adjacent nerves or neuromuscular junctions. Released by nerve stimulation (exciting or inhibiting), it is confined largely to the parasympathetic nervous system, where it diffuses across the gap of the synapse and stimulates the adjacent nerve or muscle fiber. It rapidly becomes inactive by the enzyme cholinesterase, allowing further impulses to occur.

acetyl CoA   A compound formed in the mitochondria when the thiol group (–SH) of coenzyme A combines with an acetyl group ($CH_3CO–$). It is important in the Krebs cycle in cellular respiration and plays a role in the synthesis and oxidation of fatty acids.

Fritz Albert Lipmann (1899–1986), a biochemist, is responsible for discovering coenzyme A and cofactor A, or CoA (A stands for *acetylation*), in 1947. He shared the 1953 Nobel Prize in physiology or medicine with HANS KREBS.

See also KREBS CYCLE.

achiral   *See* CHIRALITY.

acid   A chemical capable of donating a HYDRON (proton, H+) or capable of forming a covalent bond with an electron pair. An acid increases the hydrogen ion concentration in a solution, and it can react with certain metals, such as zinc, to form hydrogen gas. A strong acid is a relatively good conductor of electricity. Examples of strong acids are hydrochloric (muriatic), nitric, sulfuric, while examples of mild acids are sulfurous and acetic (vinegar). The strength of an acidic solution is usually measured in terms of its pH (a logarithmic function of the H+ ion concentration). Strong acid solutions have low pHs (typically around 0–3), while weak acid solutions have pHs in the range 3–6.

See also BASE; PH SCALE.

acidity constant   The equilibrium constant for splitting off a HYDRON from a BRØNSTED ACID.

acid-labile sulfide   Refers to sulfide LIGANDs, e.g., the BRIDGING LIGANDs in IRON–SULFUR PROTEINS, which are released as H_2S at acid pH.

See also FERREDOXIN.

acid precipitation   Because pure precipitation (e.g., rain) is slightly acidic (due to the reaction between water droplets and carbon dioxide, creating carbonic acid) with a potential pH of 5.6, acid precipitation refers to precipitation with a pH less than 5.6. Acid precipitation includes rain, fog, snow, and dry deposition. Anthropogenic (man-made) pollutants (carbon dioxide, carbon monoxide, ozone, nitrogen and sulfur oxides, and hydrocarbons) react with water vapor to produce acid precipitation. These pollutants come primarily from burning coal and other fossil fuels. Sulfur dioxide, which reacts readily with water vapor and droplets (i.e., has a short residence time in the atmosphere as a gas), has been linked to the weathering (eating away) of marble structures and the acidification of freshwater lakes (consequently killing fish). Natural interactions within the biosphere can also lead to acid precipitation.

acoelomate A solid-bodied animal lacking a body cavity, the space between the gut (digestive tract) and body wall. Simple animals do not have a body cavity as higher animals do; this body cavity is called a coelom in mammals and contains the gut (a cavity by itself), heart, and lungs, for example. Acoelomates are bilateral animals and are triploblastic (have three layers: ectoderm, endoderm, and mesoderm). They can move forward and have a degree of cephalization (centralization of neural and sensory organs in the head).

Representative phyla of acoelomates are the Platyhelminthes: flatworms that include the Turbellaria (nonconfined flatworms such as planarians), Monogenea (monogeneans), Trematoda (trematodes, or flukes), and Cestoidea (tapeworms). There are more than 20,000 species of flatworms living in wet environments such as marine or freshwater bodies and damp terrestrial areas.

See also COELOM.

aconitase A name for citrate (isocitrate) hydro-LYASE (aconitate hydratase), which catalyzes the interconversion of citrate, *cis*-aconitate ((Z)-prop-1-ene-1,2,3-tricarboxylate), and isocitrate. The active ENZYME contains a catalytic [4FE-4S] CLUSTER.

acrosome The acrosome is a special area or compartment that is located at the tip of the head of a sperm cell. It contains special digestive enzymes that on

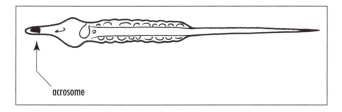

The acrosome is a special area or compartment that is located at the tip of the head of a sperm cell.

Mariposa lily from California. An example of an actinomorphic, radially symmetrical flower. *(Courtesy of Tim McCabe)*

contact with the egg help the sperm head penetrate the egg for fertilization. Directly behind the acrosome is the haploid nucleus (single set of unpaired chromosomes) that contains the genetic material.

See also FERTILIZATION.

actin A globular protein found in muscle tissue as thin filaments and in microfilaments that form portions of cell cytoskeletons. Actin links into chains, and paired chains twist helically around each other, forming microfilaments in muscle and other contractile elements in cells. Actin and myosin filaments interact to initiate muscle contraction.

Tropomyosin and troponin are two protein molecules associated with actin filaments in muscle. Tropomyosin runs along the length of the actin filament and covers the area of the actin molecule that interacts with myosin when at rest. On the other hand, when a muscle is contracted, tropomyosin is replaced with troponin as it binds to calcium ions. Troponin is

***Solanum* species as an example of an actinomorphic flower from Sierra Madre Oriental in Mexico.** *(Courtesy of Tim McCabe)*

located at regular intervals along the actin filament and allows actin to interact with myosin.

See also CYTOSKELETON.

actinomorphic   Symmetrical over more than one vertical plane; e.g., flowers that can be separated into symmetrical halves along any plane.

action potential   A localized rapid change in voltage that occurs across the membrane of a muscle or nerve cell when a nerve impulse is initiated. It is caused by a physicochemical change in the membrane during the interaction of the flow and exchange of sodium and potassium ions.

See also DEPOLARIZATION.

active center   The location in an ENZYME where the specific reaction takes place.

active site   The active site of an enzyme is the area—a depressed region comprising a few of the protein's amino acids—on a portion of an enzyme that binds to its substrate. The enzyme's specificity is based on the shape of the active area, which can alter itself to snugly fit the substrate to which it is binding by using weak chemical bonds.

See ACTIVE CENTER.

active transport   The movement of a substance across a biological membrane, such as living cells, against a concentration (diffusion) gradient with the help of metabolic energy, usually provided by ATP (adenosine triphosphate). Active transport serves to maintain the normal balance of ions in cells, in particular ions of sodium and potassium, which play a vital role in nerve and muscle cells. Because a molecule is "pumped" across the membrane against its gradient with the help of metabolic energy, it is referred to as "active" transport.

The sodium–potassium "pump" that exchanges sodium (Na+) for potassium (K+) across the plasma membrane of animal cells is an example of the active transport mechanism.

It is the carriage of a solute across a biological membrane from low to high concentration that requires the expenditure of metabolic energy.

adaptive radiation   The process where a population of plants or animals evolves into a number of different ones over time, usually as a response to multiplying and living under different environmental conditions. Subpopulations from the common ancestor develop as a response to adapting to the new environmental conditions, and new species evolve from this original parent stock.

Impressive rapid adaptive radiations have occurred over time after mass extinctions caused by cataclysmic episodes on the Earth. Plate tectonics, volcanism, and possible Earth-comet-asteroid collisions all have wiped the landscape clean, allowing survivors and new species to rapidly fill the voids of these new adaptive zones.

address-message concept   Refers to compounds in which part of the molecule is required for binding (address) and part for the biological action (message).

adenosine 5'-triphosphate (ATP)   Key NUCLEOTIDE in energy-dependent cellular reactions, in combination with Mg(II). The reaction: ATP + water → ADP + phosphate is used to supply the necessary energy.

See also ATP.

adenylyl cyclase An enzyme, embedded in the plasma membrane, that converts ATP to cyclic adenosine monophosphate (cyclic AMP, or cAMP) in response to a chemical signal. It is activated when a signal molecule binds to a membrane receptor. Cyclic AMP acts as a second messenger, relaying the signal from the membrane to the metabolic machinery of the cytoplasm.

adrenal glands A pair of small triangular endocrine glands (one above each kidney in animals) that are ductless and secrete hormones into the blood. The glands are composed of two portions. The adrenal cortex, which forms an outer shell on each and is controlled by the pituitary gland, responds to endocrine signals in reacting to stress and homeostatic conditions by (a) secreting steroid hormones (corticosteroid, cortisol, and aldosterone) that deal with carbohydrate metabolism and with salt and water balance (electrolyte metabolism), such as the reabsorption of water by the kidneys, and (b) releasing androgens (male sex hormone) and estrogens (female sex hormone).

The adrenal cortex surrounds the central medulla, is controlled by the nervous system, and responds to nervous inputs resulting from stress and produces adrenaline and noradrenaline, hormones that increase blood sugar level and reduce body fat. The adrenal glands are also known as the suprarenal glands.

See also GLAND.

adrenodoxin A [2FE-2S] FERREDOXIN involved in electron transfer from NADPH+ (the reduced form of NADP [nicotinamide adenine dinucleotide phosphate, a coenzyme]), via a REDUCTASE, to CYTOCHROME P-450 in the adrenal gland.

Adrian, Edgar Douglas (1889–1977) British *Physiologist* Edgar Douglas Adrian was born on November 30, 1889, in London to Alfred Douglas Adrian, a legal adviser to the British Local Government Board. He attended the Westminster School, London, and in 1908 enrolled at Trinity College, Cambridge. At Cambridge University, he studied physiology, receiving a bachelor's degree in 1911.

In 1913 he entered Trinity College, studied medicine, did his clinical work at St. Bartholomew's Hospital, London, and received his M.D. in 1915.

In 1929 he was elected Foulerton professor of the Royal Society and in 1937 became professor of physiology at the University of Cambridge until 1951, when he was elected master of Trinity College, Cambridge. He was chancellor of the university from 1968 until two years before his death.

He spent most of his research studying the physiology of the human nervous system, particularly the brain, and how neurons send messages. In 1932 he shared the Nobel Prize in physiology or medicine for his work on the function of the neuron. He is considered one of the founders of modern neurophysiology.

He wrote three books, *The Basis of Sensation* (1927), *The Mechanism of Nervous Action* (1932), and *The Physical Basis of Perception* (1947), and was knighted baron of Cambridge in 1955. He died on August 4, 1977, and is buried at Trinity College.

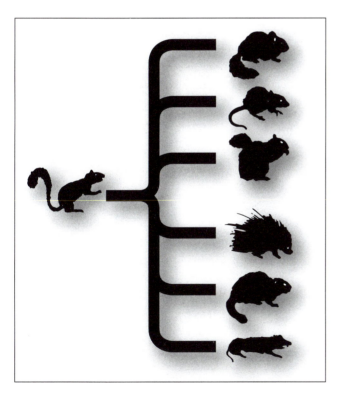

Adaptive radiation is the process where a population of plants or animals evolves into a number of different ones over time, usually as a response to multiplying and living under different environmental conditions.

aerobic Any organism, environmental condition, or cellular process that requires atmospheric oxygen. Aerobic microorganisms, called aerobes, require the presence of oxygen for growth. An aerobe is capable of using oxygen as a terminal electron acceptor and can tolerate oxygen levels higher than that present in the air (21 percent oxygen). They have a respiratory type of metabolism, and some aerobes are also capable of growing anaerobically with electron accepters other than oxygen.

See also ANAEROBIC.

affinity The tendency of a molecule to associate with another. The affinity of a DRUG is its ability to bind to its biological target (RECEPTOR, ENZYME, transport system, etc.). For pharmacological receptors, it can be thought of as the frequency with which the drug, when brought into the proximity of a receptor by diffusion, will reside at a position of minimum free energy within the force field of that receptor.

For an AGONIST (or for an ANTAGONIST), the numerical representation of affinity is the reciprocal of the equilibrium dissociation constant of the ligand–receptor complex, denoted K_A, calculated as the rate constant for offset (k_{-1}) divided by the rate constant for onset (k_1).

age structure The relative number of individuals of each age in a population, or the composition of a country by age groups. Since generations coexist over a time period, an age structure develops and is important in foreseeing the growth rate of an entire population. Except for the "baby boom" generation of the late 1940s–50s, the United States has a pretty even age distribution.

Agnatha A superclass family of jawless vertebrate fish that probably originated during the late Precambrian or early Cambrian. This superclass is the source of the oldest vertebrate fossils, dating some 465 million years ago during the Paleozoic era. Early agnathans, such as the now extinct ostracoderms, were encased in bony plates. While most were small, no larger than 20 in., they lacked paired fins and had circular mouths or slits with no jaws, although there were exceptions that had paired fins. Agnathans were most likely bottom-dwelling mud suckers or suspension feeders, taking in food through the mouth and then trapped in the gill, which also functioned as the area for gas exchange.

Only about 60 species comprising two classes of agnathans exist today, while the rest declined and disappeared during the Devonian period. Two classes, the Myxini (hagfishes) and Cephalaspidormorphi (lampreys), are all that remain.

Both hagfishes and lampreys lack paired appendages and lack body armor. Hagfishes are scavengers, living only in salt water and feeding on dead fish or marine worms, and lack a larval stage. Lampreys use their round mouth and a rasping tongue to latch on the side of a fish, penetrate the skin, and ingest its blood. As larvae, they live in freshwater streams, are suspension feeders, and migrate to the sea and lakes when they become adults. Some species only feed while in the larval stage, reproduce, and die. The agnathans are considered the most primitive living vertebrates known today.

agonist An endogenous substance or a DRUG that can interact with a RECEPTOR and initiate a physiological or a pharmacological response characteristic of that RECEPTOR (contraction, relaxation, secretion, ENZYME activation, etc.).

agonistic behavior This behavior usually involves two animals in a competitive contest, which can be in the form of combat, threat, or ritual, for food, a sexual partner, or other need. The end result is one becoming a victor while the other surrenders or becomes submissive, both exhibiting different traits. When one surrenders, it stops the combat because the continued battle could end up injuring both. Likewise, any future combat between the two individuals will likely end with the same result as the first and will not last as long.

Many animal social groups are maintained by agonistic behavior where one individual becomes dominant, others become subdominant, and so on down the line, each controlling the others in a dominance hierarchy or "pecking order." This dominant behavior can be used to control access to food or mates. Chickens, gorillas, and wolves are good examples of social groups maintained by dominance.

Agonistic behavior is used to defend territories, areas that a dominant individual will defend for feeding, mating, rearing, or any combination of these activities.

AIDS (acquired immunodeficiency syndrome) AIDS is the name given to the late stages of HIV infection, first discovered in 1981 in Los Angeles, California. By 1983 the retrovirus responsible for it, the human immunodeficiency virus (HIV), was first described, and since then millions around the world have died from contracting the disease. It is thought to have originated in central Africa from monkeys or to have developed from contaminated vaccines used in the world's first mass immunization for polio.

AIDS is acquired mostly by sexual contact either through homo- or heterosexual practice by having unprotected sex via vaginal or anal intercourse. The routes of infection include infected blood, semen, and vaginal fluid. The virus can also be transmitted by blood by-products, through maternofetal infection (where the virus is transmitted by an infected mother to the unborn child in the uterus), or by maternal blood during parturition, or by breast milk consumption upon birth. Intravenous drug abuse also is a cause.

The virus destroys a subgroup of lymphocytes, essential for combating infections, known as the helper T cells, or CD4 lymphocytes, and suppresses the body's immune system, leaving it prone to infection.

Infection by the virus produces antibodies, but not all those exposed develop chronic infection. For those that do, AIDS or AIDS-related complex (ARC) bring on a variety of ailments involving the lymph nodes, intermittent fever, loss of weight, diarrheas, fatigue, pneumonia, and tumors. A person infected, known as HIV-positive, can remain disease-free for up to 10 years, as the virus can remain dormant before full-blown AIDS develops.

While HIV has been isolated from bodily fluids such as semen to breast milk, the virus does not survive outside the body, and it is considered highly unlikely that ordinary social contact can spread the disease. However, the medical profession has developed high standards to deal with handling blood, blood products, and body fluids from HIV-infected people.

In the early discovery stage of the disease, AIDS was almost certainly fatal, but the development of antiviral drugs, such as zidovudine (AZT), didanosine (ddI), zalcitabine (ddc), lamivudine (3TC), stavudine (DAT), and protease inhibitors used in combination with the others, has showed promise in slowing or eradicating the disease. Initial problems with finding a cure have to do with the fact that glycoproteins encasing the virus display a great deal of variability in their amino acid sequences, making it difficult to prepare a specific AIDS vaccine.

During the 1980s and 1990s, an AIDS epidemic brought considerable media coverage to the disease, especially as well-known celebrities such as actors Rock Hudson and Anthony Perkins, Liberace, and others died from it. Hudson was the first to admit having the disease in 1985. During the 1980s and 1990s, the homosexual community became active in lobbying for funds to study the disease, as it early on was considered simply a "gay" disease. ACT UP, acronym for the AIDS Coalition to Unleash Power, began as a grassroots AIDS organization associated with nonviolent civil disobedience in 1987. ACT UP became the standard-bearer for protest against governmental and

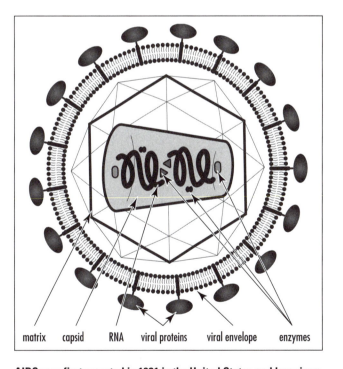

| matrix | capsid | RNA | viral proteins | viral envelope | enzymes |

AIDS was first reported in 1981 in the United States and has since become a major epidemic, killing nearly 12 million people and infecting more than 30 million others worldwide. The disease is caused by HIV, a virus that destroys the body's ability to fight infections and certain cancers. (Courtesy of Darryl Leja, NHGRI, National Institutes of Health)

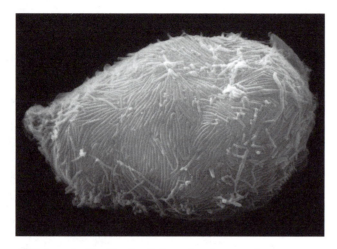

HIV-infected white blood cell. Scanning electron micrograph (SEM) of the abnormal surface of a white blood cell infected with HIV virus that causes AIDS. Glycosaminoglycan (GAG) gene expression for cell surface proteins is abnormal. The HIV virus's genetic material controls the cell, forcing it to express abnormal structural proteins. Normally these surface GAG proteins are tiny spheres, but here the cell's outer surface is formed from large irregular rods. HIV causes AIDS (acquired immunodeficiency syndrome). AIDS destroys white blood cells, leading to a weakened immune system. Magnification: × 6600 at 4.5 × 5.5 inch size. *(Courtesy © NIBSC/Photo Researchers, Inc.)*

societal indifference to the AIDS epidemic. Public attitude changed when heterosexuals became infected, and greater education on the causes of the disease became more widespread, initiated by celebrities such as Elizabeth Taylor and the American Foundation for AIDS Research, where fundraising activities made national news coverage.

There have been significant advances in the treatment for HIV/AIDS by attacking the virus itself, strengthening the immune system, and controlling AIDS-related cancers and opportunistic infections. At present, there is still no cure or vaccine.

albumin   A type of protein, especially a protein of blood PLASMA, that transports various substances, including metal ions, drugs, and XENOBIOTICS.

aldehydes   Aldehydes are organic chemicals that contain the –CHO (aldehyde) group, a carbonyl group (C=O) that has the carbon and hydrogen atom bound. They are the result of the oxidation of alcohols and, when further oxidized, form carboxylic acids. Methanal (formaldehyde) and ethanal (acetaldehyde) are common examples.

aldosterone   An adrenal steroid hormone, derived from cholesterol, that is secreted by the adrenal cortex and acts on the distal tubules of the kidney to control the elimination of salts and water by the kidneys. The presence of the pituitary hormone ACTH, changing levels of sodium and potassium, and variations in blood volume stimulate the production of aldosterone by the cortex.

Aldosterone is a component of the rennin-angiotensin-aldosterone system (RAAS) that is a complex feedback system that functions in homeostasis.

See also GLAND.

algae (singular, alga)   A large and diverse group of photosynthetic organisms formerly called simple plants but now members of their own phyla, the Protoctista, that also includes the slime molds and protozoa. Algae, some 17,000 species or more, live in aquatic and moist inland regions. They do not have roots, stems, or leaves and have no vascular water-conducting systems. They reproduce by spores, and in some species the spores are mobile with the use of flagella. They range from simple single cells (e.g., *Euglena*) to "plants" many feet long (e.g., kelps such as *Macrocytis*) and make up marine seaweed and much of the plankton that provide food for other species.

Unusual growth outbursts result in "algal blooms" or "red tides" and occur when there is an increase in nutrient levels in a body of water.

Cyanophyta, or blue-green algae, are now classified as cyanobacteria. The phylum Cyanophyta also includes chloroxybacteria. Cyanophytes contain phycocyanin, a photosynthetic pigment giving them a blue color. The red pigment phycoerythrin is also almost always present. They are diverse and can live as single cells or as colonies or large filaments. Some are nitrogen fixers in soil and others, like lichens, display symbiosis with a fungus, usually with a member of Ascomycota. The fungus provides the host plant for the algae cells that are distributed throughout and provide food to the fungus while the fungus protects the algae cells. While many lichen have a different fungi component, they often have the same algae species. Some common species of lichens are British soldiers (*Cladonia cristatella*), pixie cups (*Cladonia grayi*), cedar lichen (*Vulpicidia viridis*), wrinkled shield lichen (*Flavoparmelia caperata*), and reindeer lichen (*Cladina subtenuis*).

The chloroxybacteria, or green-grass bacteria, contain both chlorophylls a and b but do not contain the red or blue pigments of the blue-green algae. They are nonmotile, aerobic organisms.

Other phyla of algae include the Bacillariophyceae, comprising the diatoms; Charophyceae, fresh- or brackish- water algae that resemble bryophytes; Chlorophyceae, or green algae believed to be the progenitor of plants; Chrysophyceae, or yellow-green algae; Dinophyceae, unicellular algae with two flagella; Phaeophyceae, or brown algae; and Rhodophyceae, or red algae.

The Bacillariophyceae or diatoms are unicellular algae that are found in single, colonial, or filamentous states. Under the microscope they often are beautifully symmetrical, as their cell walls, or frustules, are composed of silica and are bivalved, one of which overlaps the other, and the frustule is often punctated and ornamented. The two orders, Centrales and Pennales, occupy two different environments. The centric diatoms (Centrales) are circular in shape with radial symmetry and live mostly in marine environments. The pennate diatoms (Pennales) are elliptical in shape, have bilateral symmetry, and are found in freshwater environments.

Deposits of fossil diatoms known as diatomaceous earth have been mined and used for years in paints, abrasives, and other products such as chalk. The famous White Cliffs of Dover in England (rising to 300 feet) are composed of massive amounts of diatoms—coccoliths—that were laid down some 790 million years ago when Great Britain was submerged by a shallow sea.

The Charophyceae, also known as stoneworts, and which resemble bryophytes, live in fresh and brackish water especially rich in calcium, where they become stiff and lime encrusted. The stoneworts consist of a complex branched thallus with an erect stemlike structure and many whorls of short branches.

The Chlorophyceae (or green algae) are the closest to plants in pigment composition and structure and are related based on a common ancestor. More than 7,000 species live in freshwater and marine environments as unicellular parts of plankton, in damp soil and even snow as colonies or filaments, symbiotically with other eukaryotes, or mutually with fungi as lichens.

The Chrysophyceae or golden algae, named because of their yellow and brown carotene and xanthophyll pigments, are typically cells with two flagella at one end of the cell, and many live among freshwater and marine plankton. While most are unicellular, a few are colonial.

The Dinophyceae or dinoflagellates are algae that are unicellular with two flagella of unequal length contained in channels on the cell surface. They can change shape with different water temperatures and are very tolerant of chemical and physical conditions.

The Phaeophyceae or brown algae are the largest and most complex of the algae. All members are multicellular, and the majority live in marine environments, especially common in cool water along temperate coastlines. Many of the marine seaweeds are brown algae.

The Rhodophyceae or red algae are more recent and have lost their flagellate stages in their life cycle. Some species are actually black and not red, as those that live deeper in waters, because of different levels of the pigment phycoerythrin. While most are found in warm coastal waters of the tropical oceans, some also live in freshwater and on land in the soil. Most rhodophytes are multicellular, sharing seaweed status with brown algae.

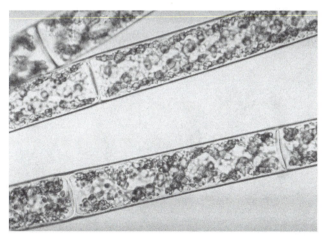

Spirogyra is a filamentous algae that can be found in almost every pond or ditch. *(Courtesy of Hideki Horikami)*

Because the various forms of algae are now assigned to different phyla, the words *alga* and *algae* are used informally and have no taxonomic status.

allantois   During the embryonic stage of mammals, birds, and reptiles, the allantois, a small sac, is one of four extra-embryonic membranes (along with amnion, yolk sac, chorion) and serves several functions, such as a repository for the embryo's nitrogenous waste (chiefly uric acid) in reptiles and birds (in the egg). The allantois provides oxygen to mammals, birds, and reptiles, as well as food in mammals (via the placenta).

The membrane of the allantois works with the chorion in respiratory functions, allowing the exchange of gases between the embryo and surrounding air. In humans, it is involved in the development of the urinary bladder.

See also EMBRYO.

allele   An allele is one of two or more alternative forms of a gene that can exist at a single locus. Each allele therefore has a unique nucleotide sequence and may lead to different phenotypes for a given trait. If the alleles for a gene are identical, the organism is called homozygous. If the alleles are different, the organism is heterozygous.

If two alleles are different, one becomes dominant and is fully expressed in appearance in the organism, while the other is recessive and has no noticeable effect on the appearance of the organism.

This is shown in the color of your eyes, determined by the genes inherited from your parents. The gene for brown eyes is dominant and overrides genes for other eye colors. The gene for blue eyes is recessive and will appear when there are no genes for other eye colors. A person with brown eyes may have a recessive, or "hidden," gene for blue eyes. Therefore, two brown-eyed parents may each give a recessive gene for blue eyes to their child, who would then have blue eyes. Gray, green, and other eye colors result from a complex mixing of different eye color genes.

See also GENE.

Allen's Rule   In warm-blooded animals, the warmer the climate, the longer the appendages (ears, legs, wings) as compared with closely related taxa from colder areas.

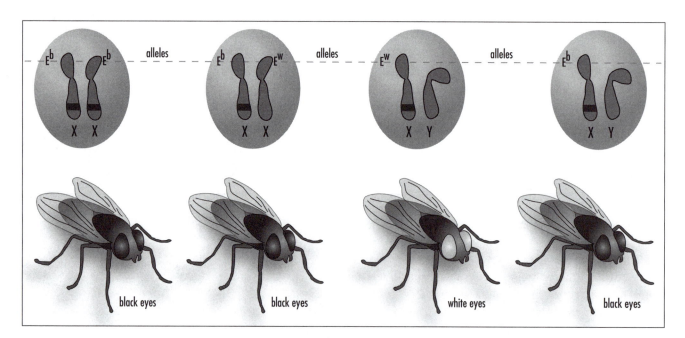

An allele is one of the variant forms of a gene at a particular locus, or location, on a chromosome. Different alleles produce variation in inherited characteristics such as hair color or blood type. In an individual, one form of the allele (the dominant one) may be expressed more than another form (the recessive one). *(Courtesy of Darryl Leja, NHGRI, National Institutes of Health)*

allochronic speciation Speciation that takes place related to time rather than space; populations that are reproductively isolated due to mating at different times.

See also SPECIATION.

allometric growth The variation in the relative rates of growth of various parts of the body, which helps shape the organism. In other words, it is the pattern of growth whereby different parts of the body grow at different rates with respect to each other. Allometry is the study of relative growth and of changes in proportion with increase in size. For example, human arms and legs grow at a faster rate than the body and head, making adult proportions strikingly different from those of infants. Another striking example is the male fiddler crab *Uca pugnax*. In small males, the two claws are of equal weight, each weighing about 8 percent of the total crab weight. However, as the crab enlarges, its large crushing claw grows more rapidly, eventually constituting about 38 percent of the crab's weight.

In 1932, Sir Julian Huxley described a simple mathematical method for the detection and measurement of the allometric growth. In order to compare the relative growth of two components (one of which may be the whole body), they are plotted logarithmically on *x*- and *y*-axes:

$$\log y = \log b + k \log x$$

The slope of the resulting regression is called the allometric growth ratio, often designated as *k*.

k = 1, both components are growing at the same rate.

k < 1, the component represented on the *y*-axis is growing more slowly than the component on the *x*-axis.

k > 1, the *y*-axis component is growing faster than the *x*-axis component.

Another formula for measuring allometric growth is $Y = bx^a$, where *Y* is equal to the mass of the organ, *x* = mass of the organism, *a* = growth coefficient of the organ, and *b* = a constant.

Yet another formula for measuring allometric growth is $Y = bx^{a/c}$, where *a* and *c* are the growth rates for two body parts.

Allometric growth studies have also been applied to animal husbandry, archaeology, and urban systems studies.

allometry The study of relative growth and of changes in proportion with increase in size.

allopatric speciation One of two methods of speciation (the other is sympatric), allopatric speciation happens when the ancestral population becomes segregated by a geographical barrier. The Karner blue butterfly (*Lycaeides melissa samualis*) became allopatric from its parent the Melissa blue butterfly (*Lycaeides melissa melissa*) when the climate changed and restricted various populations along its range to northeastern pine barrens environments several thousand years ago. As populations become isolated, the isolated gene pools accumulate different genetic traits by microevolution. Small populations are more likely to evolve into separate species than larger isolated populations. Several populations of the Karner blue butterfly are now separated from each other by human-made development and may be evolving into separate subspecies or species, even though geographically they are isolated by only a few miles in some cases.

Conditions that favor allopatric speciation are when one population becomes isolated at the fringe of the parent population's range. This splinter population, called a peripheral isolate, is likely to become allopatric because the gene pool of the isolate may already be different, since living on the border of the range encourages the expression of the extremes of any genotypic CLINEs that existed in the original population. Furthermore, if the population is small enough, a FOUNDER EFFECT will occur, giving rise to a gene pool that is not that of the parent population.

Genetic drift will also occur until the peripheral isolate becomes a larger population and will continue to change the gene pool at random until the population grows. New mutations or combinations of existing alleles that are neutral in adaptive value now may become fixed in the population by chance, causing genotypic or phenotypic divergence from the parent population. For example, the Karner blue butterfly has a row of orange spots on the top of the hindwing, whereas, the ancestral parent, the Melissa blue butterfly, has orange spots on the top of both front and hindwing, a phenotypic variation.

Another factor in causing allopatric speciation is that evolution via natural selection may take a different road in the peripheral population. The isolate will encounter selection factors that are different from and perhaps more severe than that experienced by the parent because the isolate is living in an environment slightly, or completely, different from that of the parent. These small isolated populations are not guaranteed to become new species, as they are more often likely to become extinct, yet it is clear in evolutionary history that allopatric speciation does occur.

See also SPECIATION.

allopolyploid A type of polyploid (having a nucleus that contains more than two sets of chromosomes) species, often a plant, resulting from two different species interbreeding and combining their chromosomes. Hybrids are often sterile because they do not have sets of homologous chromosomes, making pairing nonexistent unless two diploid hybrids double the chromosome numbers, resulting in a fertile allotetraploid that now contains two sets of homologous chromosomes. Plant breeders find that this is beneficial, since it is possible to breed the advantages of different species into one. Triticale (a "new" grain created by crossing rye and durum wheat) is an allopolyploid that was developed from wheat and rye. Some crops are naturally allopolyploid, such as cotton, oats, tall fescue, potatoes, wheat, and tobacco. It is estimated that half of all angiosperms (flowering plants) are polyploid.

allosteric binding sites A type of binding site contained in many ENZYMEs and RECEPTORs. As a consequence of the binding to allosteric binding sites, the interaction with the normal ligand (ligands are molecules that bind to proteins) may be either enhanced or reduced. Ligand binding can change the shape of a protein.

allosteric effector Specific small molecules that bind to a protein at a site other than a catalytic site and that modulate (by activation or INHIBITION) the biological activity.

allosteric enzyme An ENZYME that contains a region, separate from the region that binds the SUB-STRATE for catalysis, where a small regulatory molecule binds and affects that catalytic activity. This effector molecule may be structurally unrelated to the substrate, or it may be a second molecule of substrate. If the catalytic activity is enhanced by binding, the effector is called an activator; if it is diminished, the effector is called an INHIBITOR.

allosteric regulation The regulation of the activity of allosteric ENZYMEs.

See also ALLOSTERIC BINDING SITES; ALLOSTERIC ENZYME.

allosteric site A specific receptor site on an enzyme molecule not on the active site (the site on the surface of an enzyme molecule that binds the substrate molecule). Molecules bind to the allosteric site and change the shape of the active site, either enabling the substrate to bind to the active site or prevent the binding of the substrate.

The molecule that binds to the allosteric site is an inhibitor because it causes a change in the three-dimensional structure of the enzyme that prevents the substrate from binding to the active site.

allozyme An enzyme form, a variant of the same enzyme (protein) that is coded for by different alleles at a single locus.

See also ENZYME.

alpha helix Most proteins contain one or more stretches of amino acids that take on a particular shape in three-dimensional space. The most common forms are alpha helix and beta sheet.

Alpha helix is spiral shaped, constituting one form of the secondary structure of proteins, arising from a specific hydrogen-bonding structure; the carbonyl group (–C=O) of each peptide bond extends parallel to the axis of the helix and points directly at the –N–H group of the peptide bond four amino acids below it in the helix. A hydrogen bond forms between them [–N–H . . . . . O=C–] and plays a role in stabilizing the helix conformation. The alpha helix is right-handed and twists clockwise, like a corkscrew, and makes a complete turn every 3.6 amino acids. The distance between two turns is 0.54 nm. However, an alpha helix can also be left-handed. Most enzymes contain sections of alpha helix.

The alpha helix was discovered by Linus Pauling in 1948.

See also HELIX.

alternation of generations   A life cycle in plants where there is both a multicellular diploid form (the sporophyte generation) and a multicellular haploid form (the gametophyte generation).

Gametophytes produce haploid gametes that fuse zygotes that are forming. These zygotes then develop into diploid sporophytes. Meiosis in the sporophytes produces haploid spores, with division by meiosis giving rise to the next generation of gametophytes.

Alternation of generations occurs in plants and certain species of ALGAE. Ferns and fern allies (such as the club moss) are common examples that display alternation of generations. The above ground parent fern plant (the diploid sporophyte, or spore-bearing plant) has two full sets of chromosomes (two of each kind of chromosome). It sheds its single-celled haploid spores, having one set of chromosomes (one of each kind), which fall to the ground, and these in turn grow into a different plant, the gametophyte or prothallus, also haploid. The gametophyte has special bodies within the plant called archegonia (female cells) and antheridia (male cells). Here sexual fertilization takes place, and a new diploid sporophyte then grows.

There are four main groups of plants considered to be "fern allies," a diverse group of vascular plants that are neither flowering plants nor ferns and that reproduce by shedding spore to initiate an alternation of generations. These are the Lycophyta (Lycopsida, the club mosses; Selaginellopsida, the spike mosses; and Isoëtopsida, the quillworts); the Archeophyta (Sphenopsida, the horsetails and scouring-rushes; Psilopsida, the whiskbrooms; and Ophioglossopsida, the adder's-tongues and grape-ferns); the Pteridophyta (ferns); and Spermatophyta (flowering plants).

In some examples of alternation of generations—for example, in certain algae species such as in some green or brown forms—the alternation of generations takes on two different approaches. Where the sporophytes and gametophytes are structurally different, the two generations are heteromorphic. If the sporophytes and gametophytes look the same and have different chromosome pairs, the generations are said to be isomorphic.

altruistic behavior   The aiding of another individual at one's own risk or expense. This can be in the form of one animal sending out a distress call to warn others of impending trouble, although putting itself in danger by giving out its location. Strangers coming to the rescue of other strangers, such as victims in an accident, hurricane, or earthquake, is another example of altruistic behavior.

alveolus (plural, alveoli)   Latin for "hollow cavity." There are several definitions for alveolus. It is a thin, multilobed air sac that exchanges gases in the lungs of mammals and reptiles at the end of each bronchiole, a very fine respiratory tube in the lungs. An alveolus is lined with many blood capillaries where the exchange of carbon dioxide and oxygen takes place.

It is also the name given to the socket in the jawbone in which a tooth is rooted by means of the periodontal membrane, the connective tissue that surrounds the root and anchors it.

Furthermore, it is the term used to describe a single hexagonal beehive cell found in a honeycomb. It is also the term that refers to the milk-secreting sacs of the mammary gland.

ambidentate   LIGANDs, such as $(NCS)^-$, that can bond to a CENTRAL ATOM through either of two or more donor atoms.

amicyanin   An ELECTRON TRANSFER PROTEIN containing a TYPE 1 COPPER site, isolated from certain bacteria.

amino acid   An organic molecule possessing both acidic carboxylic acid (–COOH) and basic amino (–NH$_2$) groups attached to the same tetrahedral carbon atom.

Amino acids are the principal building blocks of proteins and enzymes. They are incorporated into

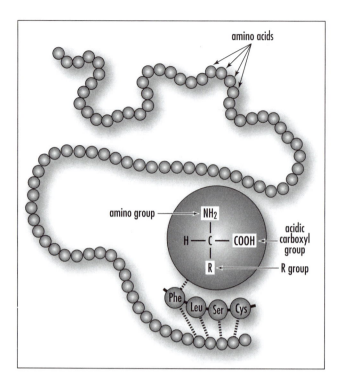

Amino acids comprise a group of 20 different kinds of small molecules that link together in long chains to form proteins. Often referred to as the "building blocks" of proteins. *(Courtesy of Darryl Leja, NHGRI, National Institutes of Health)*

proteins by transfer RNA according to the genetic code while messenger RNA is being decoded by ribosomes. The amino acid content dictates the spatial and biochemical properties of the protein or enzyme during and after the final assembly of a protein. Amino acids have an average molecular weight of about 135 daltons. While more than 50 have been discovered, 20 are essential for making proteins, long chains of bonded amino acids.

Some naturally occurring amino acids are alanine, arginine, asparagine, aspartic acid, cysteine, glutamine, glutamic acid, glycine, histidine, isoleucine, leucine, lysine, methionine, phenylalanine, proline, serine, threonine, tryptophan, tyrosine, and valine.

The two classes of amino acids that exist are based on whether the R-group is hydrophobic or hydrophilic. Hydrophobic or nonpolar amino acids tend to repel the aqueous environment and are located mostly in the interior of proteins. They do not ionize or participate in the formation of hydrogen bonds. On the other hand, the hydrophilic or polar amino acids tend to interact with the aqueous environment, are usually involved in the formation of hydrogen bonds, and are usually found on the exterior surfaces of proteins or in their reactive centers. It is for this reason that certain amino acid R-groups allow enzyme reactions to occur.

The hydrophilic amino acids can be further subdivided into polar with no charge, polar with negatively charged side chains (acidic), and polar with positively charged side chains (basic).

While all amino acids share some structural similarities, it is the side groups, or "R"-groups as they are called, that make the various amino acids chemically and physically different from each other so that they react differently with the environment. These groupings, found among the 20 naturally occurring amino acids, are ionic (aspartic acid, arginine, glutamic acid, lysine, and histidine), polar (asparagine, serine, threonine, cysteine, tyrosine, and glutamine), and nonpolar amino acids (alanine, glycine, valine, leucine, isoleucine, methionine, phenylalanine, tryptophan, and proline).

Amino acids are also referred to as amphoteric, meaning they can react with both acids and alkali, which makes them effective buffers in biological systems. A buffer is a solution where the pH usually stays constant when an acid or base is added.

In 1986 scientists found a 21st amino acid, selenocysteine. In 2002 two teams of researchers from Ohio State University identified the 22nd genetically encoded amino acid, called pyrrolysine, a discovery that is the biological equivalent of physicists finding a new fundamental particle or chemists discovering a new element.

Amino acid supplements are widely used in exercise and dietary programs.

See also PROTEIN.

amino acid residue (in a polypeptide) When two or more amino acids combine to form a peptide, the elements of water are removed, and what remains of each amino acid is called amino acid residue. Amino acid residues are therefore structures that lack a hydrogen atom of the amino group (–NH–CHR–COOH), or the hydroxy moiety of the carboxy group (NH$_2$–CHR–CO–), or both (–NH–CHR–CO–); all units of a peptide chain are therefore amino acid residues. (Residues of amino acids that contain two amino groups or two carboxy groups may be joined by isopeptide bonds, and so may not have the formulas shown.) The residue in a peptide that has an amino group that is free, or at least not acylated by another amino acid residue (it may, for example, be acylated or formylated), is called N-terminal; it is the N-terminus. The residue that has a free carboxy group, or at least does not acylate another amino acid residue (it may, for example, acylate ammonia to give –NH–CHR–CO–NH$_2$), is called C-terminal.

The following is a list of symbols for amino acids (use of the one-letter symbols should be restricted to the comparison of long sequences):

A	Ala	Alanine
B	Asx	Asparagine or aspartic acid
C	Cys	Cysteine
D	Asp	Aspartic acid
E	Glu	Glutamic acid
F	Phe	Phenylalanine
G	Gly	Glycine
H	His	Histidine
I	Ile	Isoleucine
K	Lys	Lysine
L	Leu	Leucine
M	Met	Methionine
N	Asn	Asparagine
P	Pro	Proline
Q	Gln	Glutamine
R	Arg	Arginine
S	Ser	Serine
T	Thr	Threonine
V	Val	Valine
W	Trp	Tryptophan
Y	Tyr	Tyrosine
Z	Glx	Glutamine or glutamic acid

aminoacyl-tRNA synthetases (aaRSs) When ribosomes pair a tRNA (transfer ribonucleic acid) with a codon (three bases in a DNA or RNA sequence), an amino acid is expected to be carried by the tRNA. Since each tRNA is matched with its amino acid before it meets the ribosome, the ribosome has no way of knowing if the match was made. The match is made by a family of enzymes called aminoacyl-tRNA synthetases. These enzymes charge each tRNA with the proper amino acid via a covalent ester bond, allowing each tRNA to make the proper translation from the genetic code of DNA into the amino acid code of proteins. Cells make at least 20 different aminoacyl-tRNS synthetases, one for each of the amino acids.

Aminoacyl-tRNA synthetases belong to two classes, depending on which amino acid they specify. Class I enzymes usually are monomeric and attach to the carobxyl of their specific amino acid to the 2' OH of adenosine 76 in the tRNA molecule. Class II enzymes are either dimeric or tetrameric and attach to their amino acids at the 3' OH. These enzymes catalyze first by activating the amino acid by forming an aminoacyl-adenylate. Here the carboxyl of the amino acid is linked to the alpha-phosphate of ATP, displacing pyrophosphate. After the corrected tRNA is bound, the aminoacyl group of the aminoacyl-adenylate is transferred to the 2' or 3' terminal OH of the tRNA.

Recent studies have shown that aminoacyl-tRNA synthetases can tell the difference between the right and the wrong tRNA before they ever start catalysis, and if the enzyme binds aminoacyl-adenylate first, it is even more specific during tRNA binding. Previous studies have also proved that aminoacyl-tRNA synthetases reject wrong tRNAs during catalysis. Other research has shown that specific aaRSs play roles in cellular fidelity, tRNA processing, RNA splicing, RNA trafficking, apoptosis, and transcriptional and translational regulation. These new revelations may present new evolutionary models for the development of cells and perhaps opportunities for pharmaceutical advancements.

amino group (–NH$_2$) A functional group (group of atoms within a molecule that is responsible for certain properties of the molecule and reactions in which it takes part), common to all amino acids, that consists of a nitrogen atom bonded covalently to two hydrogen atoms, leaving a lone valence electron on the nitrogen atom capable of bonding to another atom. It can act as a base in solution by accepting a hydrogen ion and carrying a charge of +1. Any organic compound that has an amino group is called an amine and is a derivative of the inorganic compound ammonia, NH$_3$. A primary amine has one hydrogen atom replaced, such as in the amino group. A secondary amine has two hydrogens replaced. A tertiary amine has all three hydrogens replaced. Amines are created by decomposing organic matter.

amniocentesis Amniocentesis is the removal of about two tablespoons of amniotic fluid via a needle inserted through the maternal abdomen into the uterus and amniotic sac. This is done to gain information about the condition, and even the sex, of the fetus. The fluid contains cells from the fetus and placenta.

Some women have a greater chance of giving birth to a baby with a chromosome problem, and amniocentesis can provide the answers if performed at about 16 weeks gestation (second trimester) or later. Chromosome analysis and alpha-fetoprotein (AFP) tests are two such tests, and these check for chromosome abnormalities such as Down's syndrome and whether there are any openings in the fetal skin, such as in the spine, that could lead to neural-tube defects like spina bifida or anencephaly, or inherited disorders such as cystic fibrosis.

While the procedure is relatively safe, some problems that can occur are miscarriage (1 in 200, or 0.5 percent chance), cramping, and infections (less than 1 in 1,000).

Amniocentesis can also be performed during the second and third trimesters to determine fetal lung maturity, to verify the health of the fetus in cases of Rh sensitivity, and to identify any infections.

First used in 1882 to remove excess amniotic fluid, it is often used in late pregnancy to test for anemia in fetuses with Rh disease and to check if the fetal lungs are advanced enough for delivery to occur.

amnion The amnion is a thin, but tough, transparent membranous sac and innermost of the four extra embryonic membranes (allantois, yolk sac, chorion) that encloses the embryo of reptiles, birds, and mammals. These membranes hold the amniotic fluid and form a protective layer for the fetus, insulating it from bacteria and infection.

See also EMBRYO.

amniotes Any of the vertebrates such as reptiles, birds, and mammals that have an amnion surrounding the embryo.

amniotic egg A calcium based or leathery shelled water-retaining egg that enables reptiles, birds, and egg-laying mammals, such as the monotremes (duck-billed platypus and two species of echidna, spiny anteaters), to complete their life cycles on dry land.

amoebic dysentery Dysentery caused by a protozoan parasite (*Entamoeba histolytica*), mostly caused by poor sanitary conditions and transmitted by contaminated food or water.

amphibian Cold-blooded, or ectothermic, vertebrates in the class Amphibia. These include the frogs and toads (order Anura, or Salientia), salamanders and newts (order Urodela, or Caudata), and the caecilians, limbless amphibians (order Apoda, or Gymnophiona). There are more than 11,000 species of amphibians, and they are believed to be the first vertebrate species to live on land.

Located between the fish and reptiles on the evolutionary scale, they are the most primitive of the terrestrial vertebrates and undergo a metamorphosis from water-breathing limbless larva (tadpole) to land-loving, or partly terrestrial, air-breathing four-legged adult.

Eggs are typically deposited in water or a wet protected place, although some do lay eggs in dry places. The eggs are not shelled and do not possess the membranes that are common in reptiles or higher vertebrates. Adults have moist skins with no scales or small scales, and they are specialized in living habitats. Each has its own evolutionary adaptations from the jumping ability (over 17 feet in some cases) of frogs and toads, to the limbless caecilians, to the long tails of the salamanders and newts. For example, frogs can enter aestivation, a period of dormancy similar to hibernation, when experiencing long periods of heat or drought conditions, and they can breathe through their skin in a process called cutaneous gas exchange. The most poisonous frog known, *Phyllobates terribilis*, only needs 0.00000007 ounce of skin secretion to kill a predator, while an antibiotic secreted from the African clawed frog (*Xenopus laevis*) may someday be used to treat burns and cystic fibrosis.

Over the last 50 years, many species of amphibians around the world have declined markedly in numbers; some species have become extinct. In many instances, these declines are attributable to adverse human influences acting locally, such as deforestation, draining of wetlands, and pollution.

However, in 1988, herpetologists (scientists who study amphibians) from many parts of the world reported declines in amphibian populations in protected, or pristine, habitats such as national parks and nature reserves, where such local effects could not be blamed. This suggested that there may be one or more global factors that are affecting climatic and atmospheric changes and adversely affecting amphibians, such as increased UV-B radiation, widespread pollution, acid rain, and disease. In effect, the decline could be the result of human-induced changes to the global ecosystem and could have far-reaching consequences for human survival.

amphipathic molecule   A molecule that has both a hydrophilic (water soluble, polar) region and a hydrophobic (water hating, nonpolar) region. The hydrophilic part is called the head, while the hydrophobic part is called the tail. Lipids (phospholipids, cholesterol and other sterols, glycolipids [lipids with sugars attached], and sphingolipids) are examples of amphipathic molecules.

Amphipathic molecules act as surfactants, materials that can reduce the surface tension of a liquid at low concentrations, and are used in wetting agents, demisters, foaming agents, and emulsifiers.

anabolism   The processes of metabolism that result in the synthesis of cellular components from precursors of low molecular weight.

See also METABOLISM.

anaerobic   Any organism or environmental or cellular process that does not require the use of free oxygen. Certain bacteria such as *Actinomyces israeli, Bacteroides fragilis, Prevotella melaninogenica, Clostridium difficile,* and *Peptostreptococcus* are anaerobes.

In effect, an anaerobic organism does not need oxygen for growth. Many anaerobes are even sensitive to oxygen. Obligate (strict) anaerobes grow only in the absence of oxygen. Facultative anaerobes can grow either in the presence or in the absence of oxygen.

See also AEROBIC.

anagenesis   A pattern of evolutionary change along a single, unbranching lineage involving the transformation of an entire population, sometimes so different from the ancestral population that it can be called a separate species. Examples would be one taxon replacing another or the transformation of a single ancestral species into a single descendant species. Anagenesis is also known as phyletic evolution and is the opposite of CLADOGENESIS.

In medicine, it refers to the regeneration of tissue or structure.

analog   A DRUG whose structure is related to that of another drug but whose chemical and biological properties may be quite different.

See also CONGENER.

analogy   The similarity of structure between two species that are not closely related; usually attributed to convergent evolution. Structures that resemble each other due to a similarity in function without any simi-

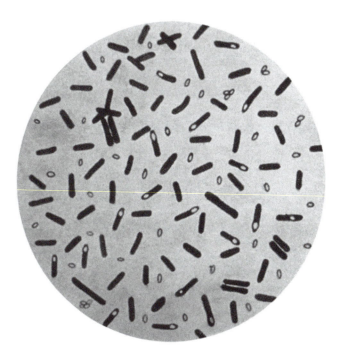

A photomicrograph of *Clostridium botulinum,* a strictly anaerobic bacterium, stained with Gentian violet. The bacterium *C. botulinum* produces a nerve toxin that causes the rare but serious paralytic illness botulism. *(Courtesy of Centers for Disease Control and Prevention)*

larity in underlying structure (or origin) are called analogous structures. For example, birds and bats each have their forelimbs modified as wings. They are analogous because they evolved independently after the earliest birds and bats diverged from their common ancestor, who did not have wings. However, the details of their structures are quite different.

anation Replacement of the LIGAND water by an anion in a COORDINATION entity.

androgens Steroid sex hormones, such as testosterone secreted by the testes in males, and others secreted by the adrenal cortex in humans and higher animals, as well as by the adrenal glands and ovaries in mammals. Androgens stimulate the development and maintenance of the male reproductive system such as sperm production, sexual behavior, and muscle development. Secondary sex characteristics such as the growth of pubic hair in females is also a product of androgens, as is the deepening of the voice at puberty.

Testosterone is present in a number of forms, such as free testosterone, as testosterone bound to a protein, sex hormone binding globulin (SHBG), and as dihydrotestosterone. Testosterone and synthetic androgens (anabolic steroids) have been used for infertility, athletic enhancement, erectile dysfunction, and libido problems, but their use can cause side effects such as muscle weakness, muscle atrophy, little facial and body hair, and even changes in the size of the genitalia. Prolonged use can damage the liver, and their use is banned in many sports.

Other androgens are androsterone (excreted in urine), which reinforces masculine characteristics; dihydrotestosterone, which is a metabolite synthesized mainly in the liver from free testosterone by the enzyme 5-alpha-reductase and which levels are proportionally correlated to sex drive as well as erectile capabilities; and dehydroepiandrosterone, which are adrenal androgens that have been linked to puberty and aging.

See also HORMONE.

androgynous Term applied to flowering plants that have both staminate and pistillate flowers, or to cryptograms (ferns, mosses, fungi, algae) where the antheridia and archegonia are together.

anemia Condition in which there is a reduction in the number of red blood cells or amount of HEMOGLOBIN per unit volume of blood below the reference interval for a similar individual of the species under consideration, often causing pallor and fatigue.

See also HEMOGLOBIN.

aneuploidy Aneuploidy is the gain or loss of individual chromosomes from the normal diploid set of 46 and is the most common cytogenetic abnormality caused when homologous chromosomes fail to separate during the first division of meiosis.

When a loss of a chromosome occurs, it is called monosomy and is rarely seen in live births, since most monosomic embryos and fetuses are lost to spontaneous abortion at very early stages of pregnancy. One exception to this is the loss of an X chromosome, which produces Turner syndrome in about one out of every 5,000 female births.

The more common gain of a single chromosome is called trisomy and has been associated with various cancers. A common autosomal trisomy is Down's syndrome in humans.

Another form of aneuploidy is called nullisomy, which is the loss of both pairs of homologous chromosomes and is almost always fatal to humans, since humans have no extra disposable chromosomes in the genome.

Tetrasomy is the gain of an extra pair of homologous chromosomes and is a rare chromosomal aberration. It can cause metopic craniosynostosis, facial anomalies, cranial asymmetry, atrioseptal defects, hydronephrosis, flexion contractures of the lower limbs, sensorineural hearing loss, and mental retardation.

angiosperm A flowering plant. There are close to 250,000 species of flowering plants, second in abundance only to insects. All have three basic organs (roots, stems, and leaves) and represent the most abundant and advanced terrestrial plants, which include trees, herbaceous plants, herbs, shrubs, all grasses, and some aquatic plants. Angiosperms are the source of most of the food on which human beings and other mammals rely and of many raw materials and natural products that provide the infrastructure for modern civilizations.

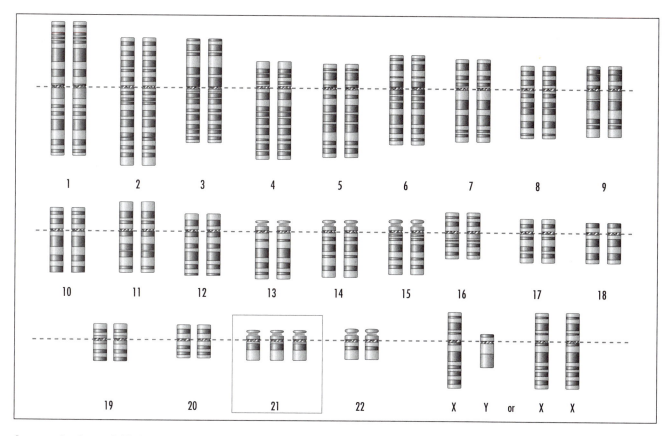

An example of aneuploidy for an individual possessing three copies of a particular chromosome instead of the normal two copies. *(Courtesy of Darryl Leja, NHGRI, National Institutes of Health)*

Angiosperms are divided into two large groups. The dicotyledonea, or dicotyledons (also called magnoliopsida), the larger of the two groups, includes trees and shrubs and herbaceous plants. Dicots have two seed leaves (cotyledons) in the embryo. The smaller of the two groups is the monocotyledoneae, or monocotyledons (also called liliopsida), that include rice, corn, palms, bananas, coconuts, grasses, lilies, orchids, and garden plants. Monocots have a single seed leaf in the embryo.

The life cycles of the angiosperms have several advantages over those of conifers, or gymnosperms, the only other group of seed-bearing plants, and from which scientists believe the angiosperms evolved during the Cretaceous era some 145 million years ago. They reproduce via flowers instead of cones; their ovules are embedded in female sporophylls instead of being exposed on a bare ground surface (e.g., apple); the gametophyte is reduced; and seeds are enclosed in fruits that develop from the ovary or related structures.

Angiosperms have a true flower that is either a highly modified shoot with modified stem and leaves or a condensed and reduced compound strobilus (conelike structure) or inflorescence (flower cluster). Floral parts are in the form of sepals, petals, stamens, and carpels, while the ovules—the structure that develops in the plant ovary and contains the female gametophyte—are contained within the megasporophylls that are sealed in most angiosperm families. Pollination is facilitated by wind, water, or many animals. Self-pollination as well as parthenogenesis, a process by which embryonic development is initiated directly from an unfertilized cell, are common. Double fertilization occurs in all members of the phylum to produce the unusual stored food tissue called endosperm. Sexual reproduction in flowering plants occurs by this process of double fertilization in which one fertilization event forms an

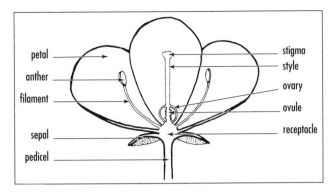

A schematic of a typical angiosperm flower.

embryo, and a second fertilization event produces endosperm, a polyploid embryo-nourishing tissue found only in the angiosperms. Seeds are dispersed through a variety of forms such as fruits, follicles, capsules, berries, drupes, samaras, nuts, and achenes. *Angiosperm* is a combination of the Latin word *angi-* (enclosed) and the Greek word *sperma* (seed).

anion An atom or molecule that has a negative charge; a negatively charged ion.
See also ION.

anisotropy The property of molecules and materials to exhibit variations in physical properties along different molecular axes of the substance.

annual A plant that completes its entire life cycle—germinates, grows, flowers, and seeds—in a single year or growing season.
See also BIENNIAL; PERENNIAL.

antagonist A DRUG or a compound that opposes the physiological effects of another. At the RECEPTOR level, it is a chemical entity that opposes the RECEPTOR-associated responses normally induced by another bioactive agent.

anterior Referring to the head end of a bilaterally symmetrical animal; the front of an animal.

anther In angiosperms, it is the terminal pollen sac of the stamen. The pollen grains with male gametes form inside the anther. It is the pollen that fertilizes the ovules. The anther is the primary male reproductive structure at the apex of the flower's stamen, the male sexual organ.
See also STAMEN.

antheridium The multicellular male sex organ or gametangium where motile male gametes (sperm) are formed and protected in algae, fungi, bryophytes (mosses, liverworts, etc.), and pteridophytes (ferns).

anthrax Bacterial disease of animals and humans caused by contamination with spores from *Bacillus anthracis* through inhalation or skin entry (cutaneous); can be used as an agent of bioterrorism.
See also BACTERIA.

anti In the representation of STEREOCHEMICAL relationships, *anti* means "on opposite sides" of a reference plane, in contrast to *syn*, which means "on the same side."

antibiotic A chemical agent that is produced synthetically or by an organism that is harmful to another organism. It is used to combat disease, either topically or by ingestion, in humans, animals, and plants. It can be made from a mold or bacterium and kills or slows the growth of other microbes, in particular bacteria. Penicillin, one of the most famous antibiotics, was accidentally discovered by the British bacteriologist SIR ALEXANDER FLEMING in 1928.

Antibiotic resistance can occur when antibiotics are used repetitively. While most of the targeted bacteria are killed by a dose of antibiotics, some escape death, and these remaining bacteria have or develop a genetic resistance to the antibiotic. Unfortunately, this resistance trait can be passed on to their offspring.
See also DRUG.

antibody A soluble immunoglobulin blood protein produced by the B cells, white blood cells, that develop in the bone marrow (also known as B lymphocytes, plasma cells) in response to an antigen (a foreign substance). Antibodies are produced in response to disease and help the body fight against a particular disease by binding to the antigen and killing it, or making it more vulnerable to action by white blood cells. They help the body develop an immunity to diseases.

Each antibody has two light (L) and two heavy (H) immunoglobulin polypeptide chains linked together by disulfide bonds, with two antigen-binding sites. There

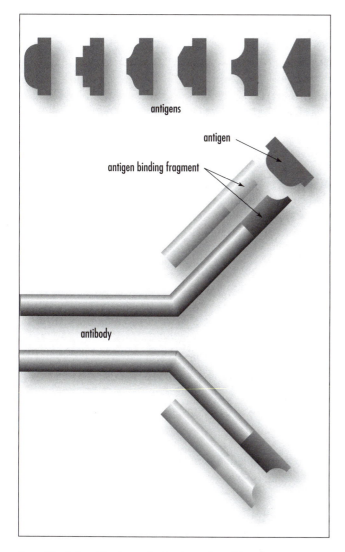

An antibody is a blood protein that is produced in response to and that counteracts an antigen. Antibodies are produced in response to disease and help the body fight against particular diseases. In this way, antibodies help the body develop an immunity to disease. *(Courtesy of Darryl Leja, NHGRI, National Institutes of Health)*

are more than 1,000 possible variations, yet each antibody recognizes only one specific antigen. Antibodies are normally bound to a B cell, but when an antibody encounters an antigen, the B cell produces copies of the antibody with the assistance of helper T cells (a lymphocyte that undergoes a developmental stage in the thymus). The released antibodies then go after and bind to the antigen, either killing it or marking it for destruction by phagocytes.

There are five immunoglobulins: IgC, IgA, IgM, IgD, and IgE.

IgA, or immunoglobulin A, comprises about 10–15 percent of the body's total immunoglobulins and is found in external secretions such as saliva, tears, breast milk, and mucous, both intestinal and bronchial. They are secreted on the surface of the body as a first defense against bacteria and viral antigens in an attempt to prevent them from entering the body.

IgM or immunoglobulin M antibodies are produced in response to new or repeat infections and stay in the body for a short time after infection. They make up from 5 to 10 percent of the total immunoglobulins and are the first to show up in the serum after an antigen enters. IgM is produced during the primary immune response. It is the IgMs that capture and bind antigens to form large insoluble complexes that are cleared from the blood.

IgG or immunoglobulin G (gamma globulin) antibodies remain in the body for long periods of time after infection and are the most common type, comprising about 80 percent of the body's total immunoglobulins. They are in the serum and are produced in substantial quantities during the secondary immune response, and along with IgM activate the complement system, which results in the destruction of the membrane of pathogens. The IgGs act by agglutinating, by opsonising, by activating complement-mediated reactions against cellular pathogens, and by neutralizing toxins.

IgE or immunoglobulin E is associated with mast cells, which are basophils, a type of granular white blood cell that has left the bloodstream and entered a tissue. Mast cells release histamine and heparin, chemicals that mediate allergic reactions. Not surprisingly, IgE is responsible for immediate hypersensitivity (allergic) reactions and immune defense against parasites.

IgD or immunoglobulin D is a specialized immunoglobulin, but its function is currently unknown. It is found in small amounts in the serum.

anticodon   A specialized sequence of three nucleotides on a tRNA (transfer ribonucleic acid) molecule. The anticodon associates with a complementary triplet of bases—the codon—on an mRNA (messenger RNA) molecule during protein synthesis.

The tRNA molecule acts like a "ferry" whose job is to "pick up a passenger" (read the code from the mRNA) and then "shuttle it" (dock to the corresponding amino acid) into place. The other end of the tRNA molecule has an acceptor site where the tRNA's specific amino acid will bind.

The 20 amino acids in the table below can create 64 different tRNA molecules, 61 for tRNA coding and three codes for chain termination (pairing up with "stop codons" that end the mRNA message), and each amino acid can create more than one set of codons.

See also CODON.

Amino Acid:

A = Adeninez
C = Cytosine
G = Guanine
U = Uracil

Alanine	GCC, GCA, GCG, GCU
Arginine	AGA, AGG, CGU, CGA, CGC, CGG
Asparagine	AAC, AAU
Aspartic Acid	GAC, GAU
Cysteine	UGC, UGU
Glutamic Acid	GAA, GAG
Glutamine	CAA, CAG
Glycine	GGA, GGC, GGG, GGU
Histidine	CAC, CAU
Isoleucine	AUA, AUC, AUU
Leucine	UUA, UUG, CUA, CUC, CUG, CUU
Lycine	AAA, AAG
Methionine (initiation)	AUG
Phenylalanine	UUC, UUU
Proline	CCA, CCC, CCG, CCU
Serine	UCA, UCC, UCG, UCU, AGC, AGU
Threonine	ACA, ACC, ACG, ACU
Tryptophan	UGG
Tyrosine	UAC, UAU
Valine	GUA, GUC, GUG, GUU
"Stop"	UAA, UAG, UGA

antidiuretic hormone (ADH)   Also known as vasopressin, ADH is a nine–amino acid peptide secreted from the posterior pituitary gland. The hormone is packaged in secretory vesicles with a carrier protein called neurophysin within hypothalamic neurons, and both are released upon hormone secretion. The single most important effect of antidiuretic hormone is to conserve body water by reducing the output of urine. It binds to receptors in the distal or collecting tubules of the kidney and promotes reabsorption of water back into the circulation.

The release of ADH is based on plasma osmolarity, the concentration of solutes in the blood. For example, loss of water (e.g., sweating) results in a concentration of blood solutes, so plasma osmolarity increases. Osmoreceptors, neurons in the hypothalamus, stimulate secretion from the neurons that produce ADH. If the plasma osmolarity falls below a certain threshold, the osmoreceptors do nothing and no ADH is released. However, when osmolarity increases above the threshold, the osmoreceptors stimulate the neurons and ADH is released.

antiferromagnetic   *See* FERROMAGNETIC.

antigen   A foreign substance, a macromolecule, that is not indigenous to the host organism and therefore elicits an immune response.

antimetabolite   A structural ANALOG of an intermediate (substrate or COENZYME) in a physiologically occurring metabolic pathway that acts by replacing the natural substrate, thus blocking or diverting the biosynthesis of physiologically important substances.

antisense molecule   An OLIGONUCLEOTIDE or ANALOG thereof that is complementary to a segment of RNA (ribonucleic acid) or DNA (deoxyribonucleic acid) and that binds to it and inhibits its normal function.

aphotic zone   The deeper part of the ocean beneath the photic zone, where light does not penetrate sufficiently for photosynthesis to occur.

See also OCEANIC ZONE.

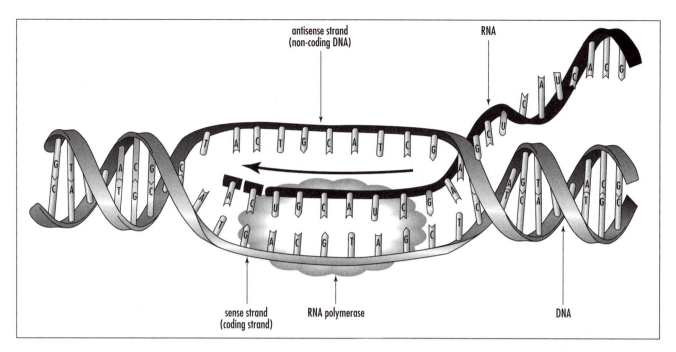

An antisense molecule is the noncoding strand in double-stranded DNA. The antisense strand serves as the template for mRNA synthesis. *(Courtesy of Darryl Leja, NHGRI, National Institutes of Health)*

apical dominance Concentration of growth at the tip of a plant shoot, where a terminal bud partially inhibits axillary bud growth. It is thought to be caused by the apical bud producing a great deal of IAA (auxin), which is transported from the apical bud to the surrounding area and causes lateral buds to stay dormant.

apical meristem Embryonic plant tissue (meristematic cells) in the tips of roots and in the buds of shoots that supplies cells via mitosis for the plant to grow in length.

apomixis The ability of certain plants to reproduce clones of themselves, i.e., the scaly male fern group, *Dryopteris affinis* (Lowe) Fraser-Jenkins.

apomorphic character A phenotypic character, or homology, in which the similarity of characters found in different species is the result of common descent, i.e., the species evolved after a branch diverged from a phylogenetic tree.

Two characters in two taxa are homologues if they are the same as the character that is found in the ancestry of the two taxa or they are different characters that have an ancestor/descendant relationship described as preexisting or novel. The ancestral character is termed the plesiomorphic character, and the descendant character is termed the apomorphic character. Examples are the flippers of whales and human arms.

See also PHENOTYPE.

apoplast The cell-wall continuum of an organ or a plant; in a plant it includes the xylem. The movement of substances via cell walls is called apoplastic transport.

apoprotein A protein without its characteristic PROSTHETIC GROUP or metal.

apoptosis Cells die by injury or commit "suicide." Apoptosis is a programmed cell death (PCD) brought about by signals that trigger the activation of a flood

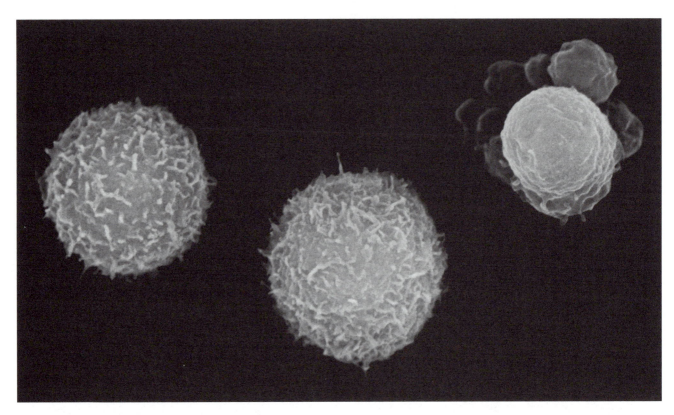

Scanning Electron Micrograph (SEM) of human white blood cells (leucocytes) showing one cell undergoing apoptosis. Apoptosis is the process of "genetically programmed cell death." At upper right, an apoptotic white blood cell has shrunk and its cytoplasm has developed blebs (grapelike clusters). Normal white blood cells are seen beside it. These white blood cells are myeloid leucocytes, originating from bone marrow. The human myeloid cell line depends on growth factors to survive, and cells undergo apoptosis when deprived of growth factors. Research on apoptosis may provide genetic treatments for diseases such as cancer. Magnification: ×7,500 at 8×10-in. size. *(Courtesy © Dr. Gopal Murti/Photo Researchers, Inc.)*

of "suicide" proteins in the cells destined to die. The destined cells then go through a number of molecular and morphological changes until they finally die. PCD is important in proper development in mitosis and cells that may be threatening to the host organism. It can be induced by a variety of stimuli, such as ligation of cell surface receptors, starvation, growth factor/survival factor deprivation, heat shock, hypoxia, DNA damage, viral infection, and cytotoxic/chemotherapeutical agents. *Apoptosis* is a word of Greek origin meaning "falling off or dropping off." There is a Web site devoted to the topic at http://www.celldeath.de/main-fram.htm.

aposematic coloration   The bright coloration of animals with effective physical or chemical defenses that

A New York tiger moth (*Grammia virginiensis*) exuding a toxic yellow froth from prothoracic glands. This is an example of lepidoptera showing chemical defense and aposematic coloration. *(Courtesy of Tim McCabe)*

acts as a warning to experienced predators. The larvae of the monarch butterfly and *Phymateus morbillosus,* a foaming grasshopper from South Africa, are two examples. The warning coloration alerts the predator, who may have eaten a similar-looking animal and was sickened by it, to avoid it. This also helps those species that mimic others in appearance, such as the viceroy butterfly and the monarch butterfly.

See also MIMICRY.

aquaporins (AQPs)   The aquaporins are a family of proteins known for facilitating water transport. An aquaporin is a transport protein in the plasma membranes of a plant or animal cell that specifically facilitates the diffusion of water across the membrane (osmosis).

Aquaporin-1, or CHIP-28, discovered in 1992 by Peter Agre, is the major water channel of the red blood cells. In the kidneys, it is involved in the reabsorption of most of the waste filtered through the glomeruli. It is also thought to influence the movement of CO_2 across the cell membrane, since it is present in most cells that have high levels of CO_2. Aquaporin-2, or WCH-CD, is a water channel that makes the principal cells of the medullary collecting duct in the kidneys more permeable to water. Lack of a functional aquaporin-2 gene leads to a rare form of nephrogenic diabetes insipidus. There are many more aquaporins that have been discovered in the more water-permeable parts of the body, such as the moist surface tissues of the alveoli in the lung, the kidney tubules, the choroid plexus of the brain where cerebrospinal fluid is produced, the ciliary epithelium of the eye where aqueous humor is formed, and the salivary and lacrimal tear glands. Aquaporins are believed to be involved in mechanisms defending against brain edema, congestive heart failure, and many other clinical entities.

aquation   The incorporation of one or more integral molecules of water into another chemical species with or without displacement of one or more atoms or groups.

See also HYDRATION.

aqueous solution   A solution in which water is the solvent or dissolving medium, such as salt water, rain, or soda.

Archaea   One of two prokaryotic (no nucleus) domains, the other being the Bacteria. Archaeans include organisms that live in some of the most extreme environments on the planet and resemble bacteria. They are single-cell organisms that, with bacteria, are called prokaryotes. Their DNA is not enclosed in a nucleus. Bacteria and archaea are the only prokaryotes; all other life forms are eukaryotes. Archaeans are among the earliest forms of life that appeared on Earth billions of years ago, and it is believed that the archaea and bacteria developed separately from a common ancestor nearly 4 billion years ago.

Some archaeans are "extremophiles," that is, they live near rift vents in the deep sea at temperatures well over 100°C (212°F). Others live in hot springs (such as the hot springs of Yellowstone National Park, where some of archaea were first discovered) or in extremely alkaline or acid waters. They have been found inside the digestive tracts of cows, termites, and marine life, where they produce methane. They also live in the anoxic muds of marshes and at the bottom of the ocean and in petroleum deposits deep underground. They are also quite abundant in the plankton of the open sea and even have been found in the Antarctic. They survive in these harsh conditions by using a variety of protective molecules and enzymes.

Three groups of archaeans are known and include the Crenarchaeota, those that are extremophiles; the Euryarchaeota, methane producers and salt lovers; and the Korarchaeota, an all-inclusive group that contains a number of types that are little understood today.

Archaeans produce energy by feeding on hydrogen gas, carbon dioxide, and sulfur and can even create energy from the sun by using a pigment around the membrane called a bacteriorhodopsin that reacts with light and produces ATP.

The archaeans were not discovered as a separate group until the late 1970s.

archaezoa   This group is believed to be the first to diverge from the prokaryotes. They lack mitochondria (converts foods into usable energy), though some archaezoans have genes for mitochondrial. They also lack an endoplasmic reticulum (important for protein synthesis) and golgi apparatus (important for glycosylation, secretion), have no peroxisomes (use oxygen to carry out catabolic reactions), and have small ribosomes similar to bacteria.

Archaezoa has three known subgroups: diplomonads, microspoidians, and trichomonads. They are usually found with flagellas in moist/damp environments such as streams, lakes, underground water deposits, and in damp soil.

Some members have been found in harsh environments and can exist in bodies of water that can drop below –20° Fahrenheit and around ocean floor vents that exceed 320°F. These organisms can survive in a variety of environments as long as they are in water.

Many archaezoans are parasites and feed off their host. The species *Giardia,* which causes abdominal cramps and severe diarrhea, uses a ventral suction cup to attach to the human intestinal epithelium. Some species have chloroplasts that allow them to take in light energy and use it when needed. Some species contain hydrogenosomes, organelles that are similar to mitochondria but do not respire with oxygen. They convert pyruvate into acetate, CO_2, and H_2, allowing extra ATP synthesis without respiration.

Since they have no mitochondria or plastid, it is believed that they are the intermediate stage between prokaryotes and eukaryotes and are also used as evidence for the evolution of the nucleus before the organelles.

archegonium   In plants, the multicellular flask-shaped female gametangium (a moist chamber in which gametes develop in bryophytes, ferns, and gymnosperms).

archenteron   The endoderm-lined gut (enteron) hollow cavity formed during the gastrulation process in metazoan embryos. The archenteron is formed by the infolding of part of the outer surface of the BLASTULA and opening to the exterior via the BLASTOPORE. Also called the primitive gut, or gastrocoel in early embryonic development, it is the digestive cavity. The term is Greek for "primitive intestine."

archipelago   A group or chain of islands clustered in a body of water, e.g., the African Bazaruto Archipela-go, consisting of five islands: Bazaruto, Magaruque, Santa Carolina, Benguera (Benguerra), and Bangue.

Aristotle (384 B.C.E.–322 B.C.E.) Greek *Philosopher* Aristotle, a Greek philosopher and scientist, has had more influence on the field of science than anyone. His influence, which lasted more than 2,000 years, was due to the fact that he was the first to depart from the old Platonic school of thinking by reasoning that accurate observation, description, inductive reasoning, and interpretation was the way to understand the natural world. Since he was the first to use this method, he is often called the "Father of Natural History."

Born in 384 B.C.E. in the Ionian colony of Stagirus (now Macedonia), Aristotle was the son of Nicomachus, a physician and grandfather of Alexander the Great. At 17, he became a student in Plato's academy in Athens and stayed there for more than 20 years as a student and teacher. In 347 B.C.E., he moved to the princedom of Atarneus in Mysia (northwestern Asia Minor), ruled by Hermias, and who presided over a small circle of Plato followers in the town of Assos. Aristotle befriended Hermias, joined the group, and eventually married Hermias's niece and adopted daughter Pythias.

Around 342 B.C.E., he moved to Mieza, near the Macedonian capital Pella, to supervise the education of 13-year-old Alexander the Great. Aristotle returned to Athens in 335 B.C.E. to teach, promote research projects, and organize a library in the Lyceum. His school was known as the Peripatetic School. After Alexander's death in 323 B.C.E., Aristotle was prosecuted and had to leave Athens, leaving his school to Theophrastus. He died shortly after at Chalcis in Euboea in 322 B.C.E.

While his writings were immense, one of his works particularly influenced the field of meteorology for over 2,000 years. *Meteorologica* (meteorology) was written in 350 B.C.E. and comprised four books, although there are doubts about the authenticity of the last one. They deal mainly with atmospheric phenomena, oceans, meteors and comets, and the fields of astronomy, chemistry, and geography.

Aristotle attempted to explain the atmosphere in a philosophical way and discussed all forms of "meteors," a term then used to explain anything suspended in the atmosphere. Aristotle discussed the philosophical nature of clouds and mist, snow, rain and hail, wind, lightning and thunder, rivers, rainbows, and climatic changes. His ideas posited the existence of four elements (earth, wind, fire, and water), each arranged in separate layers but capable of mingling.

Aristotle's observations in the biological sciences had some validity, but many of his observations and conclusions regarding weather and climate were wrong, and it was not until the 17th century—with the invention of meteorological instruments such as the hygrometer, thermometer, and barometer—that his ideas were disproved scientifically. However, he correctly reasoned that the earth was a sphere, recorded information regarding the bathymetry of seas, correctly interpreted dolphins and whales as mammals, separated vertebrates into oviparous and viviparous, and described and named many organisms, including crustaceans and worms, mollusks, echinoderms, and fish from the Aegean Sea.

Arrhenius, Svante August (1859–1927) Swedish Chemist, Physicist

Svante August Arrhenius was born in Vik (or Wijk), near Uppsala, Sweden, on February 19, 1859. He was the second son of Svante Gustav Arrhenius and Carolina Christina (née Thunberg). Svante's father was a surveyor and an administrator of his family's estate at Vik. In 1860, a year after Arrhenius was born, his family moved to Uppsala, where his father became a supervisor at the university. He was reading by the age of three.

Arrhenius received his early education at the cathedral school in Uppsala, excelling in biology, physics, and mathematics. In 1876, he entered the University of Uppsala and studied physics, chemistry, and mathematics, receiving his B.S. two years later. While he continued graduate classes for three years in physics at Uppsala, his studies were not completed there. Instead, Arrhenius transferred to the Swedish Academy of Sciences in Stockholm in 1881 to work under Erick Edlund to conduct research in the field of electrical theory.

Arrhenius studied electrical conductivity of dilute solutions by passing electric current through a variety of solutions. His research determined that molecules in some of the substances split apart, or dissociated from each other, into two or more ions when they were dissolved in a liquid. He found that while each intact molecule was electrically balanced, the split particles carried a small positive or negative electrical charge when dissolved in water. The charged atoms permitted the passage of electricity, and the electrical current directed the active components toward the electrodes. His thesis on the theory of ionic dissociation was barely accepted by the University of Uppsala in 1884, since the faculty believed that oppositely charged particles could not coexist in solution. He received a grade that prohibited him from being able to teach.

Arrhenius published his theories ("Investigations on the Galvanic Conductivity of Electrolytes") and sent copies of his thesis to a number of leading European scientists. Russian-German chemist Wilhelm Ostwald, one of the leading European scientists of the day and one of the principal founders of physical chemistry, was impressed and visited him in Uppsala, offering him a teaching position, which he declined. However, Ostwald's support was enough for Uppsala to give him a lecturing position, which he kept for two years.

The Stockholm Academy of Sciences awarded Arrhenius a traveling scholarship in 1886. As a result, he worked with Ostwald in Riga with physicist Friedrich Kohlrausch at the University of Wurzburg, with physicist Ludwig Boltzmann at the University of Graz, and with chemist Jacobus Van't Hoff at the University of Amsterdam. In 1889, he formulated his rate equation that is used for many chemical transformations and processes, in which the rate is exponentially related to temperature, known as the "Arrhenius equation."

He returned to Stockholm in 1891 and became a lecturer in physics at Stockholm's Hogskola (high school) and was appointed physics professor in 1895 and rector in 1897. Arrhenius married Sofia Rudbeck in 1894 and had one son. The marriage lasted a short two years. Arrhenius continued his work on electrolytic dissociation and added the study of osmotic pressure.

In 1896, he made the first quantitative link between changes in carbon dioxide concentration and climate. He calculated the absorption coefficients of carbon dioxide and water based on the emission spectrum of the moon, and he also calculated the amount of total heat absorption and corresponding temperature change in the atmosphere for various concentrations of carbon dioxide. His prediction of a doubling of carbon dioxide from a temperate rise of 5–6°C is close to modern predictions. He predicted that increasing reliance on fossil fuel combustion to drive the world's increasing industrialization would, in the end, lead to increases in the concentration of CO_2 in the atmosphere, thereby giving rise to a warming of the Earth.

In 1900, he published his *Textbook of Theoretical Electrochemistry*. In 1901 he and others confirmed the Scottish physicist James Clerk Maxwell's hypothesis that cosmic radiation exerts pressure on particles. Arrhenius went on to use this phenomenon in an effort to explain the aurora borealis and solar corona. He supported the Norwegian physicist Kristian Birkeland's explanation of the origin of auroras that he proposed in 1896. He also suggested that radiation pressure could carry spores and other living seeds through space and believed that life on earth was brought here under those conditions. He likewise believed that spores might have populated many other planets, resulting in life throughout the universe.

In 1902, he received the Davy Medal of the Royal Society and proposed a theory of immunology. The following year he was awarded the Nobel Prize for chemistry for his work that originally had been perceived as improbable by his Uppsala professors. He also published his *Textbook of Cosmic Physics*.

He became director of the Nobel Institute of Physical Chemistry in Stockholm in 1905 (a post he held until a few months before his death). He married Maria Johansson and had one son and two daughters. The following year he also had time to publish three books, *Theories of Chemistry, Immunochemistry,* and *Worlds in the Making*.

He was elected a foreign member of the Royal Society in 1911, the same year he received the Willard Gibbs Medal of the American Chemical Society. Three years later he was awarded the Faraday Medal of the British Chemical Society. He was also a member of the Swedish Academy of Sciences and the German Chemical Society.

During the latter part of his life his interests included the chemistry of living matter and astrophysics, especially the origins and fate of stars and planets. He continued to write books such as *Smallpox and Its Combating* (1913), *Destiny of the Stars* (1915), *Quantitative Laws in Biological Chemistry* (1915), and *Chemistry and Modern Life* (1919). He also received honorary degrees from the universities of Birmingham, Edinburgh, Heidelberg, and Leipzig and from Oxford and Cambridge Universities. He died in Stockholm on October 2, 1927, after a brief illness, and is buried at Uppsala.

arteriosclerosis   Also known as "hardening of the arteries." It is a disease whereby the arteries thicken and the inner surfaces accumulate deposits of hard plaques of cholesterol, calcium, fibrin, and other cellular debris. The arteries become inelastic and narrowed, which increases the stress on the heart as it pumps blood through, and complete obstruction with loss of blood supply can occur. This is a common cause for high blood pressure. There are hereditary links that are associated with increased risk of heart attack and stroke. When arteriosclerosis occurs in large arteries, such as the aorta, it is often referred to as atherosclerosis.

See also ARTERY.

artery   A blood vessel that carries oxygenated (except the pulmonary artery) blood away from the heart via the right and left ventricles to organs throughout the body. The main trunk of the arterial system in the body is called the aorta. The aortic divisions are the abdominal aorta, thoracic aorta, aortic artery, and ascending aorta. The pulmonary artery carries unoxygenated blood from the heart to the lungs for oxygenation.

See also VEIN.

arthritis   Inflammation of one or more of the joints in the body.

Arthropoda   An animal phylum where individuals have a segmented body, exoskeleton, and jointed legs.

artificial selection   Artificial selection is the conscious attempt by human beings to alter the environments or traits of other organisms (including their own environment) so as to alter the evolution of these

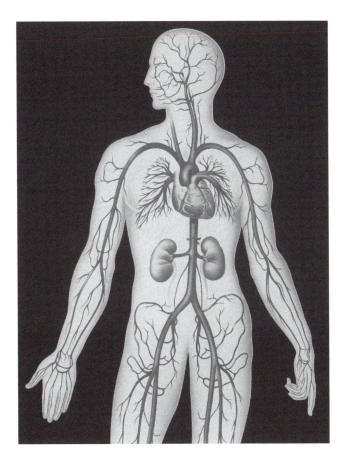

Illustration of the arterial system in the human body, shown in a standing figure. The heart and kidneys are also shown. Note the feathery network of blood vessels in the left and right lungs (next to the heart). Arteries are the blood vessels that carry oxygen-rich blood to the body's tissues. Veins (not shown) carry blood back to the heart. The average adult has about five liters of blood. At rest, this volume of blood passes through the heart each minute. *(Courtesy © John Bavosi/Photo Researchers, Inc.)*

which ascopores are found and in which karyogamy is performed, i.e., two (dikaryotic) nuclei fuse (karyogamy) to form diploid nuclei. Asci vary in shape from narrow and elongate to nearly round. While the number of ascospores per ascus is usually eight, numerous other counts of ascospores per ascus are also known.

In medicine ASCUS stands for atypical squamous cells of undetermined significance and means that irregular cells have shown up on a Pap smear.

asexual reproduction   A type of reproduction, without meiosis or syngamy (the fusion of two gametes in fertilization), involving only one parent that produces genetically identical offspring by budding, by the division of a single cell, or by the entire organism breaking into two or more parts. The offspring has the identical genes and chromosomes as the parent. Most plants are capable of asexual reproduction by means of specialized organs called propagules, such as tubers, stolons, gemma cups, and rhizomes.

Asexual reproduction is also known as vegetative reproduction. Examples of organisms that reproduce by asexual reproduction include aspens, dandelions, strawberries, walking fern (*Asplenium rhizophyllum*), and yeast. While asexual reproduction guarantees reproduction (no dependence on others), it does not allow genetic variation.

See also SEXUAL REPRODUCTION.

organism's species. It is used in the selective breeding of domesticated plants and animals to encourage the occurrence of desirable traits or new breeds. Chickens are artificially selected to produce better eggs, and pet fish are selectively bred to produce vibrant colors and other desirable traits.

ascus (plural, asci)   In Ascomycota (blue, green, and red molds), a saclike spore capsule located at the tip of the fruiting body, called the ascocarp in dikaryotic (containing two differing haploid nuclei) hyphae, in

A tailless whip scorpion (arthropod) from a cave in the Bahamas is an example of a troglodyte, an animal that lives underground. *(Courtesy of Tim McCabe)*

assimilation To transform food and other nutrients into a part of the living organism.

assimilative *See* ASSIMILATION.

assimilator *See* ASSIMILATION.

associative learning The acquired ability to associate one stimulus with another, such as one linked to a reward or punishment; also called classical conditioning and trial-and-error learning.

assortative mating A type of nonrandom or preference mating in which mating partners resemble each other in certain phenotypic characteristics. It can be a preference or avoidance of certain individuals as mates based on physical or social traits.

astigmatism Distorted vision, especially at close distances, resulting from an irregularly shaped cornea.

asymmetric carbon A carbon atom covalently bonded to four different atoms or groups of atoms.

asymmetric synthesis A traditional term for stereoselective synthesis. A chemical reaction or reaction sequence in which one or more new elements of CHIRALITY are formed in a SUBSTRATE molecule and which produces the STEREOISOMERiC (ENANTIOMERiC or DIASTEREOISOMERiC) products in unequal amounts.

asymmetry parameter In nuclear quadrupole resonance spectroscopy, the parameter, η, is used for describing nonsymmetric fields. It is defined as $\eta = (q_{xx} - q_{yy})/q_{zz}$ in which q_{xx}, q_{yy}, and q_{zz} are the components of the field gradient q (which is the second derivative of the time-averaged electric potential) along the $x-$, $y-$ and $z-$axes. By convention q_{zz} refers to the largest field gradient, q_{yy} to the next largest, and q_{xx} to the smallest when all three values are different.

atomic number The atomic number is equal to the number of positively charged protons in an atom's nucleus and determines which element an atom is. The atomic number is unique for each element and is designated by a subscript to the left of the elemental symbol. The atomic number for hydrogen is 1; it has one proton. Elements are substances made up of atoms with the same atomic number. Most of the elements are metals (75 percent) and the others are nonmetals.

atomic weight or mass The total atomic mass (the weighted average of the naturally occurring isotopes), which is the mass in grams of one MOLE of the atom. The atomic weight is calculated by adding the number of protons and neutrons together. The atomic weight of hydrogen is 1.0079 grams per mole.

ATP (adenosine triphosphate) An adenine (purine base), ribose, and three phosphate units containing nucleoside triphosphate that (a) releases free energy when its phosphate bonds are hydrolyzed and (b) produces adenosine diphosphate (ADP) and inorganic phosphorous. This energy is used to drive ENDERGONIC REACTIONS in cells (chemical reactions that require energy input to begin). ATP is produced in the cristae of mitochondria and chloroplasts in plants and is the driving force in muscle contraction and protein synthesis in animals. It is the major energy source within cells.

ATP synthase (proton translocating ATPase) A protein complex (a chemiosmotic enzyme) that synthesizes adenosine triphosphate (ATP) from adenosine diphosphate (ADP) and enables phosphate coupling with an electrochemical ion gradient across the membrane. It is found in cellular membranes and the inner membrane of mitochondria, the thylakoid membrane of chloroplasts, and the plasma membrane of prokaryotes. The protein consists of two portions: a soluble fraction that contains three catalytic sites and a membrane-bound portion that contains anion channels. It functions in chemiosmosis, the use of ion gradients across membranes, with adjacent electron transport chains, and it uses the energy stored across the photosynthetic membrane (a hydrogen-ion concentration gradient) to add inorganic phosphate to ADP, thereby creating ATP. This allows hydrogen ions (H+) to diffuse into the mitochondrion.

atrioventricular valve   A valve in the heart between each atrium and ventricle. It prevents a backflow of blood when the ventricles contract.

atrium (plural, atria)   An upper chamber that receives blood from the veins returning to the vertebrate heart and then pushes the blood to the ventricles, the lower chambers. There is a left and right atrium. Oxygenated blood returns from the lungs into the left atrium and gets pushed down to the left ventricle. The left ventricle pumps the blood out to the rest of the body, transporting the oxygen to parts of the body that need it. Blood returning from its voyage through the body arrives in the right atrium. It then goes into the right ventricle from which it goes through the lungs again to get more oxygen, and the cycle continuously repeats itself.

auranofin   *See* GOLD DRUGS.

autacoid   A biological substance secreted by various cells whose physiological activity is restricted to the vicinity of its release; it is often referred to as local HORMONE.

autogenesis model   According to autogenesis ("self-generating"), eukaryotic cells evolved by the specialization of internal membranes originally derived from prokaryotic plasma membranes. This is another word for spontaneous generation or abiogenesis.

autoimmune disease   An immunological disorder in which the immune system turns against itself. Autoimmunity can be the cause of a broad spectrum of human illnesses. Autoimmune diseases were not accepted into the mainstream of medicine until the 1950s and 1960s.

They are diseases in which the progression from benign autoimmunity to pathogenic autoimmunity happens over a period of time and is determined by both genetic influences and environmental triggers. Examples of autoimmune diseases are idiopathic thrombocytopenic purpura, Graves' disease, myasthenia gravis, pemphigus vulgaris (cause of pemphigus), and bullous pemphigoid (a blistering disease).

autonomic nervous system (ANS)   A division of the nervous system of vertebrates. The nervous system consists of two major subdivisions: the central nervous system (CNS), made up of the brain and spinal cord, and the peripheral nervous system (PNS), which comprises ganglia and peripheral nerves outside the brain and the spinal cord. The peripheral nervous system is divided into two parts: the somatic, which is concerned with sensory information about the environment outside the body as well as muscle and limb position; and the autonomic nervous system that regulates the internal environment of vertebrates. It consists of the sympathetic (fight/flight), parasympathetic (rest/rebuild), and enteric nervous systems. The ANS is involved in the function of virtually every organ system.

The parasympathetic nervous system takes care of essential background operations such as heart/lungs and digestion, while the sympathetic nervous system provides stress-response and procreation strategies and functions. The enteric nervous system takes care of controlling the function of the gut.

The sympathetic nerves form part of the nerve network connecting the organ systems with the central nervous system. The sympathetic nerves permit an animal to respond to stressful situations and helps control the reaction of the body to stress. Examples of the sympathetic reactions are increase in heart rate, decrease in secretion of salivary and digestive glands, and dilation of pupils. The parasympathetic nerves connect both somatic and visceral organs to the central nervous system, and their primary action is to keep body functions normalized. The ANS works to conserve the body's resources and to restore equilibrium to the resting state.

autophytic   The process whereby an organism uses photosynthesis to make complex foods from inorganic substances.

autopolyploid   A type of polyploid species resulting from one species doubling its chromosome number to become tetraploid, which may self-fertilize or mate with other tetraploids. This can result in sympatric speciation, where a new species can evolve in the geographical midst of its parent species because of reproductive isolation.

See also POLYPLOIDY.

autoreceptor   Present at a nerve ending, a RECEPTOR that regulates, via positive or negative feedback processes, the synthesis and/or release of its own physiological ligand.

See also HETERORECEPTOR.

autosome   A chromosome that is not directly involved in determining sex, as opposed to the sex chromosomes or the mitochondrial chromosome. Human cells have 22 pairs of autosomes.

autotroph   Any organism capable of making its own food. It synthesizes its own organic food substances from inorganic compounds using sources such as carbon dioxide, ammonia, and nitrates. Most plants and many protists and bacteria are autotrophs. Photoautotrophs can use light energy to make their food (photosynthesis). Chemoautotrophs use chemical energy to make their food by oxidizing compounds such as hydrogen sulfide (H_2S). Heterotrophs are organisms that must obtain their energy from organic compounds.

See also TROPHIC LEVEL.

auxins   A group of plant hormones that produce a number of effects, including plant growth, phototropic response through the stimulation of cell elongation (photopropism), stimulation of secondary growth,

apical dominance, and the development of leaf traces and fruit. An important plant auxin is indole-3-acetic acid (IAA). (IAA and synthetic auxins such as 2,4-D and 2,4,5-T are used as common weed killers.)

auxotroph   A nutritionally mutant organism that is unable to synthesize certain essential molecules (e.g., mineral salts and glucose) and that cannot grow on media lacking these molecules normally synthesized by wild-type strains of the same species without the addition of a specific supplement like an amino acid.

Aves   The vertebrate class of birds, characterized by feathers and other flight adaptations, such as an active metabolism, and distinguished by having the body more or less completely covered with feathers and the forelimbs modified as wings. Birds are a monophyletic lineage that evolved once from a common ancestor, and all birds are related through that common origin. There are about 30 orders of birds, about 180 families, and about 2,000 genera with 10,000 species.

axillary bud   An embryonic shoot present in the angle formed by a leaf and stem. Also called the lateral bud.

axon   A process from a neuron, usually covered with a myelin sheath, that carries nerve impulses away from the cell body and to the synapse in contact with a target cell. The end of the axon contains vesicles (hollow spheres), in which transmitters are stored, and specialized structures forming the synapse.

See also NEURON.

azurin   An ELECTRON TRANSFER PROTEIN, containing a TYPE 1 COPPER site, that is isolated from certain bacteria.

B

bacteria One of two prokaryotic (no nucleus) domains, the other being the ARCHAEA. Bacteria are microscopic, simple, single-cell organisms. Some bacteria are harmless and often beneficial, playing a major role in the cycling of nutrients in ecosystems via aerobic and anaerobic decomposition (saprophytic), while others are pathogenic, causing disease and even death. Some species form symbiotic relationships with other organisms, such as legumes, and help them survive in the environment by fixing atmospheric nitrogen. Many different species exist as single cells or colonies, and they fall into four shapes based on the shape of their rigid cell wall: coccal (spherical), bacillary (rod-shaped), spirochetal (spiral/helical or corkscrew), and vibro (comma-shaped). Bacteria are also classified on the basis of oxygen requirement (aerobic vs. anaerobic).

In the laboratory, bacteria are classified as gram-positive (blue) or gram-negative (pink) following a laboratory procedure called a Gram's stain. Gram-negative bacteria, such as those that cause the plague, cholera, typhoid fever, and salmonella, for example, have two outer membranes, which make them more resistant to conventional treatment. They can also easily mutate and transfer these genetic changes to other strains, making them more resistant to antibiotics. Gram-positive bacteria, such as those that cause anthrax and listeriosis, are more rare and are treatable with penicillin but can cause severe damage by either releasing toxic chemicals (e.g., clostridium botulinum) or by penetrating deep into tissue (e.g., streptococci). Bacteria are often called germs.

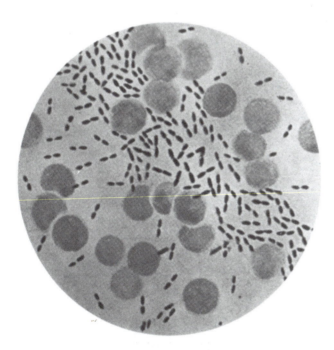

Photomicrograph of *Streptococcus (Diplococcus) pneumoniae* bacteria, using Gram's stain technique. *Streptococcus pneumoniae* is one of the most common organisms causing respiratory infections such as pneumonia and sinusitis, as well as bacteremia, otitis media, meningitis, peritonitis, and arthritis. *(Courtesy of Centers for Disease Control and Prevention, 1979)*

bacteriochlorin (7,8,17,18-tetrahydroporphyrin) A reduced PORPHYRIN with two pairs of nonfused saturated carbon atoms (C-7, C-8 and C-17, C-18) in two of the pyrrole rings.

See also ISOBACTERIOCHLORIN.

bacteriochlorophyll   *See* CHLOROPHYLL.

bacterium (plural, bacteria)   A single-celled prokaryotic microorganism in the bacteria domain.

See also BACTERIA.

balanced polymorphism   The maintenance of two or more alleles in a population due to the selective advantage of the heterozygote. A heterozygote is a genotype consisting of two different alleles of a gene for a particular trait (Aa). Balanced polymorphism is a type of polymorphism where the frequencies of the coexisting forms do not change noticeably over many generations. Polymorphism is a genetic trait controlled by more than one allele, each of which has a frequency of 1 percent or greater in the population gene pool. Polymorphism can also be defined as two or more phenotypes maintained in the same breeding population.

See also POLYMORPHISM.

Banting, Frederick Grant (1891–1941) Canadian *Physician* Frederick Grant Banting was born on November 14, 1891, at Alliston, Ontario, Canada, to William Thompson Banting and Margaret Grant.

He went to secondary school at Alliston and then to the University of Toronto to study divinity before changing to the study of medicine. In 1916 he took his M.B. degree and joined the Canadian Army Medical Corps and served in France during World War I. In 1918 he was wounded at the battle of Cambrai, and the following year he was awarded the Military Cross for heroism under fire.

In 1922 he was awarded his M.D. degree and was appointed senior demonstrator in medicine at the University of Toronto. In 1923 he was elected to the Banting and Best Chair of Medical Research, which had been endowed by the legislature of the Province of Ontario.

Also in 1922, while working at the University of Toronto in the laboratory of the Scottish physiologist John James Richard MACLEOD, and with the assistance of the Canadian physiologist Charles Best, Banting discovered insulin after extracting it from the pancreas. The following year he received the Nobel Prize in medicine along with Macleod. Angered that Macleod, rather than Best, had received the Nobel Prize, Banting divided his share of the award equally with Best. It was Canada's first Nobel Prize. He was knighted in 1934. The word *banting* was associated with dieting for many years.

In February 1941 he was killed in an air disaster in Newfoundland.

Bárány, Robert (1876–1936) Austrian *Physician* Robert Bárány was born on April 22, 1876, in Vienna, the eldest son of the manager of a farm estate. His mother, Maria Hock, was the daughter of a well-known Prague scientist. The young Bárány contracted tuberculosis, which resulted in permanent knee problems.

He completed medical studies at Vienna University in 1900, and in 1903, he accepted a post as demonstrator at the otological clinic.

Bárány developed a rotational method for testing the middle ear, known as the vestibular system, that commands physical balance by integrating an array of neurological, biological, visual, and cognitive processes to maintain balance. The middle ear's vestibular system is made up of three semicircular canals and an otolith. Inside the canals are fluid and hairlike cilia that register movement. As the head moves, so does the fluid, which in turn moves the cilia that send signals to the brain and nervous system. The function of the otolith, a series of calcium fibers that remain oriented to gravity, is similar. Both help the body to stay upright. Bárány's contributions in this area won him the Nobel Prize in physiology in 1914. To receive his award, he had to be released from a Russian prisoner of war camp in 1916 at the request of the prince of Sweden.

After the war he accepted the post of principal and professor of the Otological Institute in Uppsala, where he remained for the remainder of his life.

During the latter part of his life, Bárány studied the causes of muscular rheumatism. Although he suffered a stroke, this did not prevent him from writing on the subject. He died at Uppsala on April 8, 1936. An elite organization called the Bárány Society is named after him and is devoted to vestibular research.

barchan   A crescent-shaped dune with wings, or horns, pointing downwind.

bark   The outer layer or "skin" of stems and trunks that forms a protective layer. It is composed of all the tissues outside the vascular cambium in a plant growing in thickness. Bark consists of phloem, phelloderm, cork cambium, and cork.

Barr body   One of the two X chromosomes in each somatic cell of a female is genetically inactivated. The Barr body is a dense object or mass of condensed sex chromatin lying along the inside of the nuclear envelope in female mammalian cells; it represents the inactivated X chromosome. X inactivation occurs around the 16th day of embryonic development. Mary Lyon, a British cytogeneticist, introduced the term *Barr body*.

basal body (kinetosome)   A eukaryotic cell organelle within the cell body where a flagellum arises, which is usually composed of nine longitudinally oriented, equally spaced sets of three microtubules. They usually occur in pairs and are structurally identical to a centriole.

Not to be confused with basal body temperature (BBT), which is the lowest body temperature of the day, usually the temperature upon awakening in the morning. BBT is usually charted daily and is used to determine fertility or to achieve pregnancy.

basal metabolic rate (BMR)   BMR is the number of calories your body burns at rest to maintain normal body functions and changes with age, weight, height, gender, diet, and exercise.

base   A substance that reduces the hydrogen ion concentration in a solution. A base has less free hydrogen ions (H+) than hydroxyl ions (OH−) and has a pH of more than 7 on a scale of 0–14. A base is created when positively charged ions (base cations) such as magnesium, sodium, potassium, and calcium increase the pH

An example of a deflation zone (low ground behind fore dunes) and an example of barchan dunes in Morro Bay, California. *(Courtesy of Tim McCabe)*

of water when released to solution. They have a slippery feel in water and a bitter taste. A base will turn red litmus paper blue (acids turn blue litmus red). The three types of bases are: Arrhenius, any chemical that increases the number of free hydroxide ions (OH⁻) when added to a water-based solution; Bronsted or Bronsted-Lowry, any chemical that acts as a proton acceptor in a chemical reaction; and Lewis, any chemical that donates two electrons to form a covalent bond during a chemical reaction. Bases are also known as alkali or alkaline substances, and when added to acids they form salts. Some common examples of bases are soap, ammonia, and lye.

See also ACID; BRØNSTED BASE; HARD BASE; LEWIS BASE.

basement membrane The thin extracellular layer composed of fibrous elements, proteins, and space-filling molecules that attaches the epithelium tissue (which forms the superficial layer of skin and some organs and the inner lining of blood vessels, ducts, body cavities, and the interior of the respiratory, digestive, urinary, and reproductive systems) to the underlying connective tissue. It is made up of a superficial basal lamina produced by the overlying epithelial tissue, and an underlying reticular lamina, which is the deeper of two layers and produced by the underlying connective tissue. It is the layer of tissue that cells "sit" or rest on.

base pairing The specific association between two complementary strands of nucleic acids that results from the formation of hydrogen bonds between the base components (adenine [A], guanine [G], thymine [T], cytosine [C], uracil [U] of the NUCLEOTIDES of each strand (the lines indicate the number of hydrogen bonds):

A=T and G

C in DNA, A=U and G

C (and in some cases GU) in RNA

Single-stranded nucleic acid molecules can adopt a partially double-stranded structure through intrastrand base pairing.

See also NUCLEOSIDES.

base-pair substitution There are two main types of mutations within a gene: base-pair substitutions and base-pair insertions or deletions. A base-pair substitution is a point mutation; it is the replacement of one nucleotide and its partner from the complementary deoxyribonucleic acid (DNA) strand with another pair of nucleotides. Bases are one of five compounds—adenine, guanine, cytosine, thymine, and uracil—that form the genetic code in DNA and ribonucleic acid (RNA).

basicity constant *See* ACIDITY CONSTANT.

basidiomycetes A group of fungi whose sexual spores (basidiospores) are borne in a basidium, a club-shaped reproductive cell. Includes the orders Agaricales (mushrooms) and Aphyllophorales.

See also FUNGI.

basidium (plural, basidia) A specialized club-shaped sexual reproductive cell found in the fertile area of the hymenium, the fertile sexual spore-bearing tissues of all basidiomycetes, and that produces sexual spores on the gills of mushrooms. Shaped like a baseball bat, it possesses four slightly inwardly curved horns or spikes called sterigma on which the basidiospores are attached.

Batesian mimicry A type of mimicry described by H. W. Bates in 1861 that describes the condition where a harmless species, the mimic, looks like a different species that is poisonous or otherwise harmful to predators, the model, and in this way gains security and protection by counterfeiting its appearance. Since many predators have become sick from eating a poisonous animal, they will avoid any similar looking animals in the future. Examples of Batesian mimicry include the Viceroy mimicking the Monarch butterfly and the clearwing moth that resembles a bee by having yellow and black coloring.

See also MIMICRY.

bathyal zone The deepest part of the ocean where light does not penetrate.

See also PELAGIC ZONE.

B cell or lymphocyte A type of white blood cell, or lymphocyte, that makes up 25 percent or more of the white blood cells in the body. The other class of lymphocyte is T cells. B cells develop in the bone marrow and spleen, and during infections they are transformed into plasma cells that produce large quantities of antibody (immunoglobulin) directed at specific pathogens. A cancer of the B lymphocytes is called a B-cell lymphoma.

behavioral ecology A subdiscipline that seeks to understand the functions, or fitness consequences, of behavior in which animals interact with their environment.

Békésy, Georg von (1899–1972) Hungarian *Physicist* Georg von Békésy was born in Budapest, Hungary, on June 3, 1899, to Alexander von Békésy, a diplomat, and his wife Paula. He received his early education in Munich, Constantinople, Budapest, and in a private school in Zurich. He received a Ph.D. in physics in 1923 from the University of Budapest for a method he developed for determining molecular weight. He began working for the Hungarian Telephone and Post Office Laboratory in Budapest until 1946. During the years 1939–46 he was also professor of experimental physics at the University of Budapest.

While his research was concerned mainly with problems of long-distance telephone transmission, he conducted the study of the ear as a main component of the transmission system. He designed a telephone earphone and developed techniques for rapid, nondestructive dissection of the cochlea.

In 1946 he moved to Sweden as a guest of the Karolinska Institute and did research at the Technical Institute in Stockholm. Here he developed a new type of audiometer. The following year he moved to the United States to work at Harvard University in the Psycho-Acoustic Laboratory and developed a mechanical model of the inner ear. He received the Nobel Prize in 1961 for his discoveries concerning the physical mechanisms of stimulation within the cochlea. He moved on to the University of Hawaii in 1966, where a special laboratory was built for him.

He received numerous honors during his lifetime. He died on June 13, 1972, in Honolulu.

benthic zone A lower region of a freshwater or marine body. It is below the pelagic zone and above the abyssal zone, which is the benthic zone below 9,000 m. Organisms that live on or in the sediment in these environments are called benthos.

See also OCEANIC ZONE.

beringia All of the unglaciated area that encompassed northwestern North America and northeastern Asia, including the Bering Strait, during the last ice age.

berry A pulpy and stoneless fruit containing one or more seeds, e.g., strawberry.

beta sheet Preferentially called a beta pleated sheet; a regular structure in an extended polypeptide chain,

The Monarch butterfly is a chemically protected species that is mimicked by the Viceroy. This is known as Batesian mimicry. *(Courtesy of Tim McCabe)*

stabilized in the form of a sheet by hydrogen bonds between CO and NH groups of adjacent (parallel or antiparallel) chains.

beta strand Element of a BETA SHEET. One of the strands that is hydrogen bonded to a parallel or antiparallel strand to form a beta sheet.

beta turn A hairpin structure in a polypeptide chain reversing its direction by forming a hydrogen bond between the CO group of AMINO ACID RESIDUE n with the NH group of residue (n+3).
    See also HELIX.

biennial A plant that requires two years or at least more than one season to complete its life cycle. In the first year, plants form vegetative growth, and in the second year they flower. (Latin *biennialis*, from *biennis*; *bis*, twice, and *annus*, year)

bifunctional ligand A LIGAND that is capable of simultaneous use of two of its donor atoms to bind to one or more CENTRAL ATOMS.
    See also AMBIDENTATE.

bilateral symmetry Characterizing a body form having two similar sides—one side of an object is the mirror image of its other half—with definite upper and lower surfaces and anterior and posterior ends. Also called symmetry across an axis.

   In plants, the term applies to flowers that can be divided into two equal halves by only one line through the middle. Most leaves are bilaterally symmetrical.

bilateria Members of the branch of eumetazoans possessing bilateral symmetry. Many bilaterian animals exhibit cephalization, an evolutionary trend toward concentration of sensory structures, mouth, and nerve ganglia at the anterior end of the body. All bilaterally symmetrical animals are triploblastic, that is, having three germ layers: ectoderm, endoderm, and mesoderm.

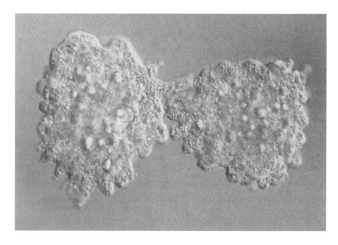

Amoeba proteus **showing cell division via binary fission. Amoebas are protozoans, the simplest form of animal life. (Courtesy of Hideki Horikami)**

binary fission A type of asexual reproduction in prokaryotes (cells or organisms lacking a membrane-bound, structurally discrete nucleus and other subcellular compartments) in which a cell divides or splits into two "daughter" cells, each containing a complete copy of the genetic material of the parent. Examples of organisms that reproduce this way are bacteria, paramecium, and *Schizosaccharomyces pombe* (an ascomycetous species of yeast). Also known as transverse fission.

binding constant *See* STABILITY CONSTANT.

binding site A specific region (or atom) in a molecular entity that is capable of entering into a stabilizing interaction with another molecular entity. An example of such an interaction is that of an ACTIVE SITE in an enzyme with its SUBSTRATE. Typical forms of interaction are by hydrogen bonding, COORDINATION, and ion-pair formation. Two binding sites in different molecular entities are said to be complementary if their interaction is stabilizing.

binomial (binomial name) Each organism is named using a Latin-based code consisting of a combination of two names, the first being a generic (genus) name and the second a specific trivial name, which, together, constitute the scientific name of a species. *Lupinus perennis,* or wild blue lupine, is an example. Both names are italicized, and both names used together constitute the species name. This is an example of the binomial nomenclature, critical to the system of classification of plants and animals. Linnaeus, a Swedish naturalist, developed the system in the 18th century. The hierarchy lists the smallest group to largest group: species, genus, family, order, class, division, and kingdom. The first person to formally describe a species is often included, sometimes as an abbreviation, when the species is first mentioned in a research article (e.g., *Lupinus perennis* L., where L. = Linnaeus, who first produced this binomial name and provided an original description of this plant).

binuclear   Less frequently used term for the IUPAC recommended term *dinuclear.*
   See also NUCLEARITY.

bioassay   A procedure for determining the concentration, purity, and/or biological activity of a substance (e.g., vitamin, hormone, plant growth factor, antibiotic, enzyme) by measuring its effect on an organism, tissue, cell, enzyme, or receptor preparation compared with a standard preparation.

bioavailability   The availability of a food component or a XENOBIOTIC to an organ or organism.

biocatalyst   A catalyst of biological origin, typically an ENZYME.

bioconjugate   A molecular species produced by living systems of biological origin when it is composed of two parts of different origins, e.g., a conjugate of a xenobiotic with some groups, such as glutathione, sulfate, or glucuronic acid, to make it soluble in water or compartmentalized within the cell.

bioconversion   The conversion of one substance to another by biological means. The fermentation of sug-

ars to alcohols, catalyzed by yeasts, is an example of bioconversion.
   See also BIOTRANSFORMATION.

biodiversity (**biological diversity**)   The totality of genes, species, and ecosystems in a particular environment, region, or the entire world. Usually refers to the variety and variability of living organisms and the ecological relationships in which they occur. It can be the number of different species and their relative frequencies in a particular area, and it can be organized on several levels, from specific species complexes to entire ecosystems or even molecular-level heredity studies.

bioenergetics   The study of the energy transfers in and between organisms and their environments and the regulation of those pathways. The term is also used for a form of psychotherapy that works through the body to engage the emotions and is based on the work of Wilhelm Reich and psychiatrist Alexander Lowen in the 1950s.

biofacies   A characteristic set of fossil fauna. Facies is a geological term that means "aspect" and is used for defining subdivisions based on an aspect or characteristic of a rock formation, such as lithofacies, based on physical characteristics, or biofacies, based on the fossil content.

biogeochemical cycles   Both energy and inorganic nutrients flow through ecosystems. However, energy is a one-way process that drives the movement of nutrients and is then lost, whereas nutrients are cycled back into the system between organisms and their environments by way of molecules, ions, or elements. These various nutrient circuits, which involve both biotic and abiotic components of ecosystems, are called biogeochemical cycles. Major biogeochemical cycles include the water cycle, carbon cycle, oxygen cycle, nitrogen cycle, phosphorus cycle, sulfur cycle, and calcium cycle. Biogeochemical cycles can take place on a cellular level (absorption of carbon dioxide by a cell) all the way to global levels (atmosphere and ocean interactions).

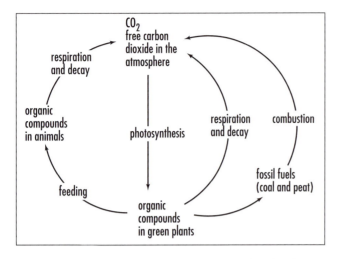

The carbon cycle, one of the main biogeochemical cycles that processes and transfers nutrients from organisms to their environment.

These cycles take place through the biosphere, lithosphere, hydrosphere, and atmosphere.

biogeographic boundary (zoogeographical region) Six to nine regions that contain broadly similar fauna. Consists of Nearctic, Palearctic, Neotropical, Aethiopian, Oriental, and Australian, and some include Holarctic, Palaeotropical, and Oceana.

biogeography The study of the past and present distribution of life.

bioisostere (nonclassical isostere) A compound resulting from the exchange of an atom or of a group of atoms with another broadly similar atom or group of atoms. The objective of a bioisosteric replacement is to create a new compound with similar biological properties to the parent compound. The bioisosteric replacement can be physicochemically or topologically based.
See also ISOSTERE.

bioleaching Extraction of metals from ores or soil by biological processes, mostly by microorganisms.

biological clock The internal timekeeping that drives or coordinates a circadian rhythm.

biological control (integrated pest management) Using living organisms to control other living organisms (pests), e.g., aphids eaten by lady beetles.

biological half-life The time at which the amount of a biomolecule in a living organism has been reduced by one half.
See also HALF-LIFE.

biological magnification (bioaccumulation) The increase in the concentration of heavy metals (e.g., mercury) or organic contaminants (e.g., chlorinated hydrocarbons [CBCs]) in organisms as a result of their consumption within a food chain/web. An excellent example is the process by which contaminants such as polychlorinated biphenyls (PCBs) accumulate or magnify as they move up the food chain. For example, PCBs concentrate in tissue and internal organs, and as big fish eat little fish, they accumulate all the PCBs that have been eaten by everyone below them in the food chain.

biological species A population or group of populations whose members can interbreed or have the potential to interbreed.
See also SPECIES.

bioluminescence The process of producing light by a chemical reaction by a living organism, e.g., glowworms, fireflies, and jellyfish. Usually produced in organs called photopores or light organs, bioluminescence can be used for luring prey or as a courting behavior.

biomass The dry weight of organic matter in unit area or volume, usually expressed as mass or weight of a group of organisms in a particular habitat. The term also refers to organic matter that is available on a renewable basis such as forests, agricultural crops, wood and wood wastes, animals, and plants.

biome A large-scale recognizable grouping, a distinct ecosystem, that includes many communities of a similar nature that have adapted to a particular environment. Deserts, forests, grasslands, tundra, and the oceans are biomes. Biomes have changed naturally and moved many times during the history of life on Earth. In more recent times, change has been the result of human-induced activity.

biomembrane Organized sheetlike assemblies, consisting mainly of proteins and lipids (bilayers), that act as highly selective permeability barriers. Biomembranes contain specific molecular pumps and gates, receptors, and enzymes.

biomimetic Refers to a laboratory procedure designed to imitate a natural chemical process. Also refers to a compound that mimics a biological material in its structure or function.

biomineralization The synthesis of inorganic crystalline or amorphous mineral-like materials by living organisms. Among the minerals synthesized biologically in various forms of life are fluorapatite ($Ca_5(PO_4)_3F$), hydroxyapatite, magnetite (Fe_3O_4), and calcium carbonate ($CaCO_3$).

biopolymers Macromolecules, including proteins, nucleic acids, and polysaccharides, formed by living organisms.

bioprecursor prodrug A PRODRUG that does not imply the linkage to a carrier group, but results from a

Cuatraciénagas dunes in Mexico showing an example of the biodiversity of plants found in the xeric conditions of gypsum sands and deserts. *(Courtesy of Tim McCabe)*

molecular modification of the active principle itself. This modification generates a new compound, able to be transformed metabolically or chemically, the resulting compound being the active principle.

biosensor   A device that uses specific biochemical reactions mediated by isolated enzymes, immunosystems, tissues, organelles, or whole cells to detect chemical compounds, usually by electrical, thermal, or optical signals.

biosphere   The entire portion of the Earth between the outer portion of the geosphere (the physical elements of the Earth's surface crust and interior) and the inner portion of the atmosphere that is inhabited by life; it is the sum of all the planet's communities and ecosystems.

biotechnology   The industrial or commercial manipulation and use of living organisms or their components to improve human health and food production, either on the molecular level (genetics, gene splicing, or use of recombinant deoxyribonucleic acid [DNA]) or in more visible areas such as cattle breeding.

biotic   Pertains to the living organisms in the environment, including entire populations and ecosystems.

biotransformation   A chemical transformation mediated by living organisms or ENZYME preparations. The chemical conversion of substances by living organisms or enzyme preparations.
See also BIOCONVERSION.

bivalve   A mollusk having two valves or shells that are hinged together, e.g., mussels and clams.

blastocoel   The fluid-filled cavity that forms in the center of the blastula embryo. The blastula is an early stage in the development of an ovum, consisting of a hollow sphere of cells enclosing the blastocoel.
See also BLASTULA.

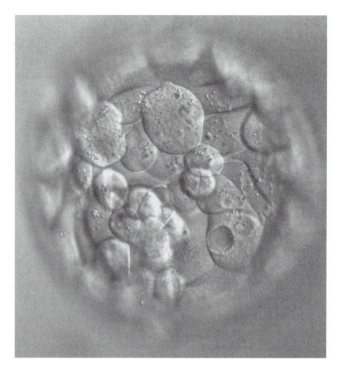

Light micrograph of a human embryo at the blastocyst stage. This early blastocyst is about four days old, appearing as a hollow ball of cells. On around day six, the embryo will begin to implant into the wall of the uterus. Magnification unknown. *(Courtesy © Pascal Goetgheluck/Photo Researchers, Inc.)*

blastocyst   An embryonic stage in mammals; a hollow ball of 30–150 cells produced one week after FERTILIZATION in humans. It is a sphere made up of an outer layer of cells called the trophectoderm, a fluid-filled cavity called the BLASTOCOEL, and a cluster of cells on the interior called the INNER CELL MASS. It is the inner cell mass that becomes what is known as the FETUS.
See also EMBRYO.

blastopore   The opening of the ARCHENTERON (primitive gut) in the gastrula that develops into the mouth in protostomes (metazoans such as the nematodes, flatworms, and mollusks that exhibit determinate, spiral cleavage and develop a mouth from the blastopore) and the anus in deuterostomes. (Animals such as the chordates and echinoderms in which the first opening in the embryo becomes the anus, while the mouth appears at the other end of the digestive system.)

blastula   Early stage of animal development of an embryo, where a ball forms consisting of a single layer of cells that surrounds the fluid-filled cavity called the blastocoel. The term *blastula* is often used interchangeably with the term *blastocyst*.

See also BLASTOCOEL; BLASTOCYST.

bleomycin (BLM)   A glycopeptide molecule that can serve as a metal-chelating ligand. The Fe(III) complex of bleomycin is an antitumor agent, and its activity is associated with DNA cleavage.

BLM   *See* BLEOMYCIN.

blood   Blood is an animal fluid that transports oxygen from the lungs to body tissues and returns carbon dioxide from body tissues to the lungs through a network of vessels such as veins, arteries, and capillaries. It transports nourishment from digestion, hormones from glands, carries disease-fighting substances to tissues, as well as wastes to the kidneys. Blood contains red and white blood cells and platelets that are responsible for a variety of functions, from transporting substances to fighting invasion from foreign substances. Some 55 percent of blood is a clear liquid called plasma. The average adult has about five liters of blood.

See also ARTERY; CAPILLARY; VEIN.

blood-brain barrier (BBB)   The blood-brain barrier is a collection of cells that press together to block many substances from entering the brain while allowing others to pass. It is a specialized arrangement of brain capillaries that restricts the passage of most substances into the brain, thereby preventing dramatic fluctuations in the brain's environment. It maintains the chemical environment for neuron functions and protects the brain from the entry of foreign and harmful substances. It allows substances in the brain such as glucose, certain ions, and oxygen and others to enter, while unwanted ones are carried out by the endothelial cells. It is a defensive system to protect the central nervous system.

What is little understood is how the blood-brain barrier is regulated, or why certain diseases are able to manipulate and pass through the barrier.

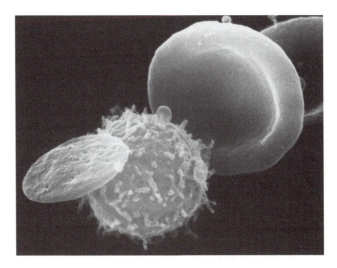

Scanning electron micrograph (SEM) showing three types of cells found in human blood. At right is a red blood cell (erythrocyte), a biconcave disc that transports oxygen around the body. A white blood cell (center) is roughly spherical with microvilli projecting from its surface. Different types of white cells are active in the body's immune response to infection. The waferlike cell at left is a blood platelet, which functions to control clotting and thus prevents bleeding from damaged vessels. Each cubic millimeter of blood contains approximately 5 million red cells, 7,000 white cells, and 250,000 platelets. Magnification: ×3,850 at 35-mm size, ×27,000 at 8 × 10-in. size. *(Courtesy © NIBSC/Photo Researchers, Inc.)*

There is evidence that multiple sclerosis attacks occur during breakdowns of the blood-brain barrier. A study in rats showed that flavinoids, such as those found in blueberries and grape seeds among others, can inhibit blood-brain barrier breakdown under conditions that normally lead to such breakdown.

Researchers at the University of Maryland School of Medicine in Baltimore have identified a receptor in the human brain that regulates the interface between the bloodstream and the blood-brain barrier and could lead to a new understanding of this nearly impenetrable barrier and to treatment of diseases that affect the brain. They found that two proteins, zonulin and zot, unlock the cell barrier in the intestine, attach themselves to receptors in the intestine to open the junctions between the cells, and allow substances to be absorbed. The new research indicates that zonulin and zot also react with similar receptors in the brain, suggesting that it may become feasible to develop a new generation of drugs able to cross the blood-brain barrier.

Blood Identification through the Ages

by John C. Brenner and Demetra Xythalis

Blood is a fluid that circulates throughout the body, transporting oxygen, nutrients and waste materials. Blood is composed of various formed elements such as red blood cells (erythrocytes), white blood cells (leukocytes), platelets (thrombocytes), and a liquid fraction called plasma, each containing a vast array of biochemical constituents. Red blood cells comprise the majority of the formed elements in the blood. Hemoglobin is a chemical that is found in red blood cells, consisting of an iron-containing pigment, heme, and a protein component, globin. The components of blood are controlled genetically and have the potential of being a highly distinctive feature for personal identification.

The field of forensic science is the study and practice of the application of natural sciences for the purpose of the law. One of the disciplines in forensic science is forensic serology, which involves the identification and characterization of blood and body fluids, either in a liquid or dried state, in association with a criminal or civil investigation. Blood and dried bloodstains are two of the most important and most frequently encountered types of evidence in criminal investigation of crimes such as homicides, assaults, and rapes.

Since the 1900s, forensic serologists have attempted to identify blood and/or bloodstains found at crime scenes. When serology was in its early stages, stains at crime scenes were identified just as blood. Now that forensic serologists can individualize human blood by identifying all of its known factors, the result could be evidence of the strongest kind for linking a suspect to the crime scene or finding a lost victim.

When examining dried bloodstains, the forensic serologists are trying to determine the following: (1) Is the stain blood? (2) If the stain is blood, is it human or animal? (3) If

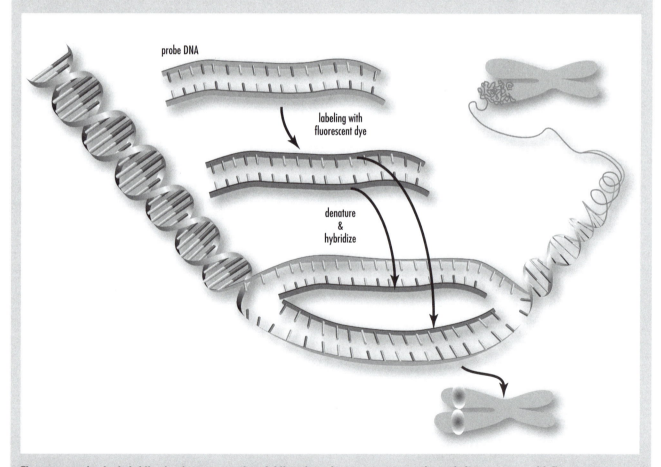

probe DNA

labeling with
fluorescent dye

denature
&
hybridize

Fluorescence in situ hybridization is a process that vividly paints chromosomes or portions of chromosomes with fluorescent molecules. This technique is useful for identifying chromosomal abnormalities and gene mapping. (Courtesy of Darryl Leja, NHGRI, National Institutes of Health)

the stain is human blood, can it be associated with a particular individual? The forensic serologist, in an attempt to do blood identification, uses two categories of tests: the presumptive test, which is nonspecific for blood, and the confirmatory, which is specific for blood species.

The first step in determining whether a crime scene stain is blood involves the use of a chemical screening test or presumptive test. Some presumptive tests used by forensic serologists include benzidine (introduced in 1904), phenolphthalein (1901), leucomalachite green (1904), and luminol (1928).

The identification of benzidine as a carcinogen led to its discontinuance as a screening test for blood. Another chemical used for screening stains for blood is phenolphthalein. Both tests consist of a two-step procedure. The first step is to moisten a white filter paper with distilled water. Apply the filter paper to the suspected bloodstain. A portion of the stain will transfer onto the moistened paper. Add the leucomalachite green reagent to the paper. The second step is to add hydrogen peroxide to the filter paper and look for a color change on the paper. A positive result will yeild a bluish-green color. When phenolphthalein is mixed with a dried bloodstain, with the addition of hydrogen peroxide, the hemoglobin in the blood causes the formation of a deep pink color. The leucomalachite green test is a presumptive test for blood that is used by many laboratories today. The heme group in hemoglobin catalyzes the oxidation by peroxide of the malachite green to produce a bluish-green reaction when the suspected stain is blood.

Luminol is a chemical reagent that is very useful in locating small traces of blood at a crime scene. Unlike the above-mentioned chemical screening test, the reaction of luminol with blood results in the production of light rather than color. The one requirement for the use of luminol is that the scene be completely dark. Luminol is a chemical that can be used as a spray at crime scenes and will react with any blood present, causing a luminescence.

The second step in blood identification is the use of confirmatory tests. Confirmatory blood identification tests are specific for the heme component of hemoglobin. A positive confirmatory test result is taken as positive proof of the presence of blood in a questioned stain. Some of the confirmatory tests include: microcrystalline tests, Teichmann and Takayama tests, ring precipitin test, gel diffusion method, and electrophoresis (1907). The Teichmann (1853) and the Takayama (1905) confirmatory tests are based on the observation that heme, in the presence of certain chemicals, will form characteristic crystals that can be seen using a microscope.

The precipitin test is based on antibody molecules interacting with antigens to form a precipitate that can be visualized under the proper light conditions or with a stain. Serologists use this test to determine whether the origin of the bloodstain is human or animal. The ring precipitin test involves layering a dilute saline extract of the bloodstain on top of the antihuman serum in a capillary tube. Because of the density of the antihuman serum, the bloodstain extract will layer on top, and the two solutions will not mix, thus forming a cloudy ring or band at the interface between the two solutions.

The gel diffusion method is based on the fact that antigens and antibodies will diffuse, or move toward each other, on an agar-gel-coated plate, such as the Ouchterlony plate. The extracted bloodstain and the human antiserum are placed in separate holes opposite each other on the gel. A white precipitate line will form where the antigens and antibodies meet if the bloodstain is of human origin.

The electrophoretic method, or crossover electrophoresis, is a sensitive method using an electric current that is passed through a gel plate. A line of precipitation formed between the hole containing the bloodstain extract and the hole containing the human antiserum denotes a specific antigen antibody reaction.

The next generation of confirmatory test, which in some forensic laboratories has replaced the electrophoretic or crossover electrophoresis is called the One Step ABAcard HemaTrace. This test utilizes a combination of monoclonal and polyclonal antibody reagents to selectively detect human hemoglobin. Adding a portion of the prepared elution from the bloodstain to the sample well and observing the development of indicative colored lines conduct the test. The species specificity of the reaction is based on the recognition by antibodies of antigens displayed on human hemoglobin. A positive result is visualized because gold-conjugated, monoclonal antibody-Hb immune complexes are captured and condensed on the test mambrane by stationary phase polyclonal anti-human hemoglobin antibodies, causing the gold particles to condense. This produces a pink-colored line at the test area. Absence of this colored line indicates a negative result.

There are many different substances in human blood that can be grouped to individualize the blood. Blood contains many inherited factors referred to as genetic markers. Determination of factors in a person's blood is called blood grouping or blood typing. True individualization of a specimen of blood would mean that a sufficiently large number of genetic markers could be typed so that nobody else in the

(continues)

Blood Identification through the Ages
(continued)

world would have that particular combination of genetic markers.

In 1900, Dr. Karl Landsteiner announced one of the most significant discoveries of the 20th century—the typing of human blood. Out of Landsteiner's work came the classification system that is called the ABO system. Blood group systems are the most well known and widely recognized class of genetic markers.

Bloodstains can be typed for ABO using two different procedures: (1) by detecting the antibodies of the serum or (2) by detecting the antigens of the red cells. Detection of the antibodies is the older method. This procedure was first extensively employed by Lattes in Italy in 1913. The procedure has been modified and improved with the development of new antisera. In this test, which detects antibodies in dried bloodstains, two portions of the stains are placed onto microscope slides. Type A red cells are added to one glass slide, and Type B red cells to the other slide. If the bloodstain on the slide contains anti-A antibodies, the A red cells that were added will agglutinate, which looks like crosslinked cells under the microscope. Agglutination of the A cells indicates that anti-A is present, and therefore, that the bloodstain is of blood group B.

The other approach to typing dried bloodstains is the detection of the antigens that are on the surface of the red blood cells. When blood dries, the red cells break apart, but the red blood cell antigens are still present in the dried stains. The two major methods that have been used are absorption-inhibition and absorption-elution. The absorption-inhibition method depends on the ability to estimate the amount of antibody present in an antiserum before and after exposure to a stain extract containing a possible antigen.

The absorption-elution method is based on the theory that blood-group antibodies can bind to their specific red-cell surface antigens in bloodstains. The antigen-antibody complex can then be dissociated and the antibodies recovered. The breaking of the antigen-antibody bond can be done by increasing the temperature. Removing specific antibodies from complexes with their antigens in this way is called elution.

Another main class of blood constituents used as genetic markers is the polymorphic enzymes. The enzymes of interest to the forensic serologists are primarily located within the red blood cell and are commonly referred to as isoenzymes. These enzyme forms can be grouped from a bloodstain to further individualize the blood. Red-cell isoenzymes are frequently typed by a procedure called elec-trophoresis. This procedure brings about the separation of different proteins based primarily upon differences in net charge, and it is usually done with some kind of starch-gel or on a cellulose acetate support. The most important enzyme systems used in forensic serology are phosphoglu-comutase (PGM), erythrocyte acid phosphatase (EAP), esterase D (ESD), adenylate kinase (AK), adenosine deami-nase (ADA), and glyoxalase I (GLO).

The main features of the molecular architecture of deoxyribonucleic acid (DNA) were first formulated by Watson and Crick in 1953, who at the same time pointed out how the proposed structure would account for the three basic attributes of genetic material: gene specificity, gene replication, and gene mutation. It was not until 1985 that forensic scientists discovered that portions of the DNA structure of certain genes are as unique to an individual as their fingerprints. Alec Jeffreys and his colleagues at Leicester University, England, who were responsible for these revelations, named the process for isolating and reading these DNA markers "DNA fingerprinting." The DNA typing of biological fluid and stains finally gives the forensic scientist the ability to link crime scene evidence such as bloodstains to a single individual.

The separation of DNA fragments of different sizes usually can be efficiently accomplished by agarose gel electrophoresis. The agarose gels are thick, which makes them difficult to process in terms of hybridization, washing, and autoradiography. To overcome these problems, a transfer technique was developed that transferred the DNA fragments from the agarose gel onto a nylon membrane. This technique was first described by E. Southern in 1975, and it is called Southern blotting. If a specific recognition base sequence is present, the restriction enzyme recognizing that site will cleave the DNA molecule, resulting in fragments of specific base-pair lengths. Restriction fragment length polymorphism (RFLPs) generates different DNA fragment lengths by the action of specific endonucleases. To visualize the separate RFLPs, a nylon sheet is treated with radioactively labeled probes containing a base sequence complementary to the RFLPs being identified, a process called hybridization. Once the radioactive sequences are on the nylon membrane, the membrane is exposed to a piece of X-ray film. The developed X-ray film shows DNA fragments that combined with radioactive probe. The size of the bloodstain (i.e., the amount of blood required) on forensic evidence and the time required to obtain the DNA information from the evidence were two drawbacks to this procedure.

As the push to individualize forensic bloodstain proceeded, the next advancement came in 1983, with molecular biologist Kary Mullis's development of a process called polymerase chain reaction (PCR). PCR has revolutionized the approach to the recovery of DNA from a variety of sources. Availability of oligonucleotide primers is the key to the amplification process. PCR consists of three steps, beginning with the denaturing of the double-strand DNA, separated by heating to 90–96°C. The second step involves hybridization or annealing, in which one primer is annealed to the flanking end of each DNA target sequence complementary strand. The third step uses a thermally stable Taq polymerase to mediate the extension of the primers. The result is two new helices in place of the first, each one composed of the original strands plus its newly assembled complementary strand.

All eukaryotic genomes contain regions of simple repetitive DNA, called short tandem repeats (STR) or microsatellites, which consist of variable numbers of tandem repeats (VNTRs). The number of repeats at an STR locus can be highly variable among individuals, resulting in different-length polymorphisms that can be detected by relatively simple use of the PCR-based assays. STR loci are useful to forensic science because of their small range of alleles, their high sensitivity, and suitability even if the DNA

is degraded. Today the forensic laboratory using the PCR/STR analysis can individualize bloodstains obtained from forensic evidence with a very high probability of identifying a single individual.

In the past, forensic scientists who handled biological forensic evidence were only able to tell the investigating official whether the dried stain at the crime scene was blood. Today the forensic laboratory reports contain information about dried stains at crime scenes that can be related to one individual. This information has been tremendously helpful in the investigation of crimes. With the establishment of a DNA database, physical evidence collected from the crime scene that contains biological stains can be analyzed even if there is no suspect. The DNA profile developed from forensic bloodstain evidence can be compared with various DNA databases to develop a match, which could lead to identification of an individual.

—**John C. Brenner**, M.S., is a forensic scientist, and **Demetra Xythalis** is a senior lab technician. Both work at the New York State Police Forensic Investigation Center in Albany, New York.

blood pressure The hydrostatic force that blood exerts against the wall of a blood vessel. This pressure is greatest during the contraction of the ventricles of the heart (systolic pressure), which forces blood into the arterial system. Pressure falls to its lowest level when the heart is filling with blood while at rest (diastolic pressure). Blood pressure varies depending on the energy of the heart action, the elasticity of the walls of the arteries, and the volume and viscosity (resistance) of the blood. Blood pressure rises and falls throughout the day.

When the blood flows through the vessels at a greater than normal force, reading consistently above 140/90 mm Hg (millimeters of mercury), it is called hypertension or high blood pressure. High blood pressure strains the heart; harms the arteries; and increases the risk of heart attack, stroke, and kidney problems. About one in every five adults in the United States has high blood pressure. Elevated blood pressure occurs more often in men than in women, and in African Americans it occurs almost twice as often as in Caucasians. Essential hypertension (hypertension with no

known cause) is not fully understood, but it accounts for about 90 percent of all hypertension cases in people over 45 years of age.

Low blood pressure is called hypotension and is an abnormal condition in which the blood pressure is lower than 90/60 mm Hg. When the blood pressure is too low, there is inadequate blood flow to the heart, brain, and other vital organs.

An optimal blood pressure is less than 120/80 mm Hg.

blotting A technique used for transferring DNA, RNA, or protein from gels to a suitable binding matrix, such as nitrocellulose or nylon paper, while maintaining the same physical separation.

blue copper protein An ELECTRON TRANSFER PROTEIN containing a TYPE 1 COPPER site. Characterized by a strong absorption in the visible region and an EPR (ELECTRON PARAMAGNETIC RESONANCE SPECTROSCOPY) signal with an unusually small HYPERFINE coupling to the copper nucleus. Both characteristics are attributed to COORDINATION of the copper by a cysteine sulfur.

bond energy (bond dissociation energy)   Atoms in a molecule are held together by covalent bonds, and to break these bonds atoms need bond energy. The source of energy to break the bonds can be in the form of heat, electricity, or mechanical means. Bond energy is the quantity of energy that must be absorbed to break a particular kind of chemical bond. It is equal to the quantity of energy the bond releases when it forms. It can also be defined as the amount of energy necessary to break one mole of bonds of a given kind (in gas phase).

bone imaging   The construction of bone tissue images from the radiation emitted by RADIONUCLIDES that have been absorbed by the bone. Radionuclides such as ^{18}F, ^{85}Sr, and ^{99m}Tc are introduced as complexes with specific LIGANDs (very often phosphonate ligands) and are absorbed in the bones by metabolic activity.

See also IMAGING.

book lungs   The respiratory pouches or organs of gas exchange in spiders (arachnids), consisting of closely packed blood-filled plates, sheets, or folds for maximum surface aeration and contained in an internal chamber on the underside of the abdomen. They look like the pages of a book.

Bordet,   Jules (1870–1961) Belgian *Bacteriologist, Immunologist* Jules Bordet was born in Soignies, Belgium, on June 13, 1870. He was educated in Brussels and graduated with a doctor of medicine in 1892. Two years later he went to Paris and began work at the Pasteur Institute, where he worked on the destruction of bacteria and explored red blood cells in blood serum, contributing to the founding of serology, the study of immune reactions in bodily fluids. In 1901 he returned to Belgium to found the Pasteur Institute of Brabant, Brussels, where he served until 1940. He was director of the Belgian Institute and professor of bacteriology at the University of Brussels (1907–35).

His work in immunology included finding two components of blood serum responsible for bacteriolysis (rupturing of bacterial cell walls) and the process of hemolysis (rupturing of foreign red blood cells in blood serum). Working with his colleague Octave Gengou, Bordet developed several serological tests for diseases such as typhoid fever, tuberculosis, and syphilis. The bacteria responsible for whooping cough, *Bordetella pertussis,* was named for him after he and Gengou discovered it in 1906. In 1919, he received the Nobel Prize in physiology and medicine for his immunological discoveries.

He was the author of *Traité de l'immunité dans les maladies infectieuses* (Treatise on immunity in infectious diseases) and numerous medical publications.

Bordet was a permanent member of the administrative council of Brussels University, president of the First International Congress of Microbiology (Paris, 1930), and member of numerous scientific societies. He died on April 6, 1961.

bottleneck effect   A dramatic reduction in genetic diversity of a population or species when the population number is severely depleted by natural disaster, by disease, or by changed environmental conditions. This limits genetic diversity, since the few survivors are the resulting genetic pool from which all future generations are based.

Bovet,   Daniels (1907–1992) Swiss *Physiologist* Daniel Bovet was born in Neuchâtel, Switzerland, on March 23, 1907, to Pierre Bovet, professor of pedagogy at the University of Geneva, and Amy Babut. He graduated from the University of Geneva in 1927 and then worked on a doctorate in zoology and comparative anatomy, which he received in 1929.

During the years 1929 until 1947 he worked at the Pasteur Institute in Paris, starting as an assistant and later as chief of the institute's Laboratory of Therapeutic Chemistry. Here he discovered the first synthetic antihistamine, pyrilamine (meplyramine). In 1947 he went to Rome to organize a laboratory of therapeutic chemistry and became an Italian citizen. He became the laboratory's chief at the Istituto Superiore di Sanità, Rome. Seeking a substitute for curare, a muscle relaxant, for anesthesia, he discovered gallamine (trade name Flaxedil), a neuromuscular blocking agent used today as a muscle relaxant in the administration of anesthesia.

He and his wife Filomena Nitti published two important books, *Structure chimique et activité pharmacodynamique des médicaments du système nerveux végétatif* (The chemical structure and pharmacodynamic activity of drugs of the vegetative nervous system) in 1948 and, with G. B. Marini-Bettòlo, *Curare and Curare-like Agents* (1959). In 1957 he was awarded the Nobel Prize for physiology or medicine for his discovery relating to synthetic compounds for the blocking of the effects of certain substances occurring in the body, especially in its blood vessels and skeletal muscles.

Bovet published more than 300 papers and received numerous awards. He served as the head of the psychobiology and psychopharmacology laboratory of the National Research Council (Rome) from 1969 until 1971, when he became professor of psychobiology at the University of Rome (1971–82). He died on April 8, 1992, in Rome.

Bowman's capsule   A cup-shaped receptacle in the kidney that contains the glomerulus, a semipermeable twisted mass of tiny tubes through which the blood passes and is the primary filtering device of the nephron, a tiny structure that produces urine during the process of removing wastes. Each kidney is made up of about 1 million nephrons. Blood is transported into the Bowman's capsule from the afferent arteriole that branches off of the interlobular artery. The blood is filtered out within the capsule, through the glomerulus, and then passes out by way of the efferent arteriole. The filtered water and aqueous wastes are passed out of the Bowman's capsule into the proximal convoluted tubule, where it passes through the loop of Heinle and into the distal convoluted tubule. Eventually the urine passes and filters through the tiny ducts of the calyces, the smallest part of the kidney collecting system, where it begins to be collected and passes down into the pelvis of the kidney before it makes its way to the ureter and to the bladder for elimination.

brachyptery   A condition where wings are disproportionately small in relation to the body.

brain imaging   In addition to MAGNETIC RESONANCE IMAGING, which is based on the absorption by the brain of electromagnetic radiation, brain images can be acquired by scintillation counting (scintigraphy) of radiation emitted from radioactive nuclei that have crossed the blood-brain barrier. The introduction of radionuclides into brain tissue is accomplished with the use of specific $^{99m}Tc(V)$ complexes with lipophilic ligands.

See also IMAGING.

brain stem (brainstem)   The oldest and inferior portion of the brain that consists of the midbrain, pons, reticular formation, thalamus, and medulla oblongata,

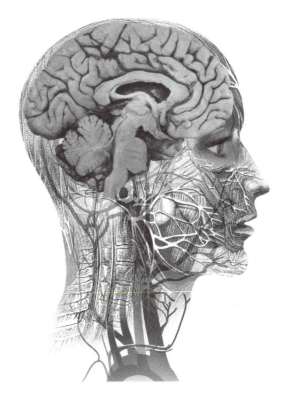

Artwork combining profiles of brain and head anatomy. The brain is seen sliced in half to show internal anatomy. The brain's major area, the cerebrum, includes the folded outer layer (cerebral cortex) that produces memory, language, and conscious movement. The central space is a brain ventricle. The brain stem, at the base of the brain, controls subconscious functions like breathing. It extends downwards and connects to the spinal cord in the neck. The cerebellum (round area, at left of the brainstem) controls balance as well as muscle coordination. The head and neck blood vessels branch from the major chest vessels at bottom. *(Courtesy © Mehau Kulyk/Photo Researchers, Inc.)*

and forms a cap on the anterior end of the spinal cord. The brain stem is the base of the brain and connects the brain's cerebrum to the spinal cord. It shares several features in common with the brain of reptiles and controls automatic and motor basic functions such as heart rate and respiration and also is the main channel for sensory and motor signals.

bridging ligand   A bridging ligand binds to two or more CENTRAL ATOMS, usually metals, thereby linking them together to produce polynuclear coordination entities. Bridging is indicated by the Greek letter μ appearing before the ligand name and separated by a hyphen. For an example, see FEMO-COFACTOR.

bronchiole   A series of small tubes or airway passages that branch from the larger tertiary bronchi within each lung. At the end of the bronchiole are the alveoli, thousands of small saclike structures that make up the bulk of the lung and where used blood gets reoxygenated before routing back through the heart.

See also LUNG.

Brønsted acid   A molecular entity capable of donating a hydron to a base (i.e., a "hydron donor") or the corresponding chemical species.

See also ACID.

Brønsted base   A molecular entity capable of accepting a hydron from an acid (i.e., a "hydron acceptor") or the corresponding chemical species.

See also BASE.

Brownian movement   The rapid but random motion of particles colliding with molecules of a gas or liquid in which they are suspended.

bryophytes   The mosses (Bryophyta), liverworts (Hepatophyta), and hornworts (Anthocerophyta); a group of small, rootless, thalloid (single cell, colony, filament of cells, or a large branching multicellular structure) or leafy nonvascular plants with life cycles dominated by the gametophyte phase. These plants inhabit the land but lack many of the terrestrial adaptations of vascular plants, such as specialized vascular or transporting tissues (e.g., xylem and phloem).

Terrestrial bryophytes are important for soil fixation and humus buildup. In pioneer vegetation, they provide a suitable habitat for seedlings of early pioneering plants. Bryophytes are also early colonizers after fire and contribute to nutrient cycles.

bubonic plague   A bacterial disease marked by chills, fever, and inflammatory swelling of lymphatic glands found in rodents and humans. It is caused by *Pasteurella pestis* and transmitted by the oriental rat flea. The famous Black Death that devastated the population of Europe and Asia in the 1300s was a form of bubonic plague.

budding   An asexual means of propagation in which a group of self-supportive outgrowths (buds) from the parent form and detach to live independently, or else remain attached to eventually form extensive colonies. The propagation of yeast is a good example of budding.

Also a type of grafting that consists of inserting a single bud into a stock.

buffer   A molecule or chemical used to control the pH of a solution. It consists of acid and base forms and minimizes changes in pH when extraneous acids or bases are added to the solution. It prevents large changes in pH by either combining with H^+ or by releasing H^+ into solution.

See also PH SCALE.

bulk flow (pressure flow)   Movement of water due to a difference in pressure between two locations. The movement of solutes in plant phloem tissue is an example.

C

C3 plant   The majority of photosynthetic plants that produce, as the initial steps of CO_2 incorporation, a three-carbon compound, phosphoglyceric acid (PGA), as the first stable intermediate (CALVIN CYCLE). The PGA molecules are further phosphorylated (by ATP) and are reduced by NADPH to form phosphoglyceraldehyde (PGAL), which then serves as the starting material for the synthesis of glucose and fructose, which, when combined, make sucrose that travels through the plant. Velvetleaf (*Abutilon theophrasti*) is an example of a C3 plant.

C4 plant   A small number of plants that incorporate CO_2 using a carboxylase for the CO_2 capture, producing a four-carbon compound (carboxylic acid) as a stable intermediary in the first step of photosynthesis. C4 plants (e.g., corn) supply CO_2 for the CALVIN CYCLE.

CADD   *See* COMPUTER-ASSISTED DRUG DESIGN.

cage   An aggregate of molecules, generally in the condensed phase, that surround the fragments formed by thermal or photochemical dissociation of a species.

calcitonin   Calcitonin is a hormone produced by the thyroid gland that acts primarily on bone. It inhibits bone removal by osteoclasts and promotes bone formation by osteoblasts; lowers blood calcium levels.

calmodulin   A Ca^{2+} binding protein involved in muscular contraction.

calorie   An energy measurement unit; the amount of energy required to raise the temperature of 1 g of water by 1°C. A kilocalorie (1,000 calories) is used in food science to describe the energy content of food products.

calpain   A calcium-activated neutral protease.

Calvin cycle   The second major stage in photosynthesis after light reactions—discovered by chemist Melvin Calvin (1911–97)—whereby carbon molecules from CO_2 are fixed into sugar (glucose) and mediated by the enzyme rubisco (ribulose-1-5-biphosphate carboxylase). It occurs in the stroma of chloroplasts. The Calvin cycle is also known as the dark reaction, as opposed to the first-stage light reactions.

CAM (crassulacean acid metabolism)   A metabolic adaptation of certain plants, particularly xerophytes (desert loving, e.g., succulents), in arid areas that allows them to take up CO_2 at night, not during the day, store it as organic acid (malic), and release CO_2 by decarboxylation of the acids for fixing into sugar. This reduces transpirational water loss during photosynthesis. The CALVIN CYCLE occurs during the day.

Cambrian explosion A period about 530 million years ago (Cambrian age) when a large explosion of species, both in number and diversity, appeared on Earth. It lasted about 10 million years, and it is the first recorded evidence through the fossil record of larger and more complex life forms appearing.

Canadian shield A geographic area of Canada centered around Hudson Bay and composed of 2- to 3-billion-year-old igneous and metamorphic shield rock. It covers much of northern Canada.

cancer Diseases in which abnormal cells divide and grow unchecked and can spread from the original site to other parts of the body; often fatal.

capillary The smallest blood vessels in the circulatory system. Capillaries have thin walls that facilitate the transfer of oxygen and glucose into a cell and the removal of waste products such as carbon dioxide back out into the blood stream, to be carried away and taken out of the body via the lungs. They act as the bridge between the arteries, which carry blood away from the heart, and the veins, which carry blood back to the heart.

See also BLOOD.

capsid The outer protein coat or shell of a virus surrounding its genetic material. Also capsid bugs (capsidae), which number over 6,000 species and live on plants, sucking juice and damaging cultivated plants.

carbohydrate A large class of compounds that contain carbon, hydrogen, and oxygen in a general formula of $Cn(H_2O)n$. Classified from simple to complex, they form mono-, di-, tri-, poly-, and heterosaccharides. Examples include sugars (monosaccharide, di-and polysaccharides), starches, and cellulose. Carbohydrates are used as an energy source by organisms, and most are formed by green plants and are obtained by animals via food intake.

carbon dioxide (CO_2) A colorless, odorless gas that makes up the fourth most abundant gas in the atmosphere. Used by plants in carbon fixation. Atmospheric CO_2 has increased about 25 percent since the early 1800s due to burning fossil fuels and deforestation. Increased amounts of CO_2 in the atmosphere enhance the greenhouse effect, blocking heat from escaping into space and contributing to the warming of Earth's lower atmosphere and affecting the world's biota. This is a major issue currently being debated by scientists around the world.

See also GREENHOUSE EFFECT.

carbon fixation The process by which carbon atoms from CO_2 gas are incorporated into sugars. Carbon fixation occurs in the chloroplasts of green plants or any photosynthetic or chemoautotrophic organism.

carbonic anhydrase A zinc-containing ENZYME (carbonate hydrolyase, carbonate dehydratase) that catalyzes the reversible decomposition of carbonic acid to carbon dioxide and water.

Carboniferous period A geological time period (360 to 280 millions of years ago) during the middle-to-late Paleozoic era. It is divided into the Pennsylvanian period (325 to 280 millions of years ago) and the Mississippian period (360 to 325 millions of years ago).

See also GEOLOGIC TIME.

carbon monoxide (CO) A colorless, odorless gas that is toxic.

carbon monoxide dehydrogenases ENZYMES that catalyze the oxidation of carbon monoxide to carbon dioxide. They contain IRON-SULFUR CLUSTERS and either nickel and zinc, or MOLYBDOPTERIN. Some nickel-containing enzymes are also involved in the synthesis of acetyl coenzyme A from CO_2 and H_2.

carbonyl group A functional group with an oxygen atom double-bonded to a carbon atom, e.g., aldehydes (joined to at least one hydrogen atom) and ketones (carbonyl group is joined to alkyl groups or aryl groups).

carboplatin A "second generation" platinum drug effective in cancer chemotherapy, named *cis*-diammine (cyclobutane-1,1-dicarboxylato)platinum(II). Carboplatin is less toxic than the "first generation" antitumor drug, CISPLATIN.

carboxyl group A functional group that consists of a carbon atom joined to an oxygen atom by a double bond and to a hydroxyl group; present in all carboxylic acids.

carcinogen Any substance that can produce cancer.

cardiac muscle One of the three muscle types (the others are skeletal and smooth); found in the walls of the heart, each rectangular heart muscle cell has one central nucleuslike smooth muscle, but it is striated like skeletal muscle. These cells are joined by intercalated discs, physical connections between the fibers of the myocardium, that relay each heartbeat through gap junctions (electrical synapses). Each strong and rhythmical contraction of the cardiac muscle is controlled by the autonomic nervous system and is involuntary.

cardiac output The amount of blood that is pumped each minute from the left ventricle into the aorta or from the right ventricle into the pulmonary trunk.

cardiotech A species radiolabeled with ^{99m}Tc with the formula $[Tc(CNR)_6]^+$ (R=*tert*-butyl) known for IMAGING the heart after a heart attack.

cardiovascular system The human circulatory system; the heart and all the vessels that transport blood to and from the heart.

carnivore Any animal that eats the meat of other animals.
See also HERBIVORE.

carotenoids A large family of natural phytochemicals, accessory pigments found in plants (in chloroplasts) and animals that are composed of two small six-carbon rings connected by a carbon chain that must be attached to cell membranes. Their variety of colors absorb wavelengths that are not available to chlorophyll and so serve

The ribbon snake, a type of carnivorous tree snake in Venezuela, will eat birds and serves as a model for mimicry among caterpillars as well as a Batesian model. *(Courtesy of Tim McCabe)*

to transfer their captured energy from the sun to help in photosynthesis. Carotenoids color fruits and vegetables and give them their characteristic red, orange, and yellow colors and serve as antioxidants in human nutrition. Over 600 carotenoids are known.

carpal bones Hand bones. The carpal bones include the navicular, lunate, pisiform, capitate, trapezium, trapezoid, hamate, and the triquetrum. They are arranged in two rows, the proximal (near the body) and the distal (near the fingers).

See also SKELETON.

carpal tunnel A small passage located below the wrist at the heel of the hand where the median nerve, the major nerve to the hand, as well as tendons that bend the fingers pass through.

carpel The female reproductive part of the flower, including the ovary, style, and stigma.

Carrel, Alexis (1873–1944) French *Surgeon* Alexis Carrel was born in Lyons, France, on June 28, 1873, to a businessman, also named Alexis Carrel, who died when his son was very young. Carrel was educated at home by his mother Anne Ricard and at St. Joseph School, in Lyons. He received a bachelor of letters degree in 1889 from the University of Lyons, a bachelor of science the following year, and, in 1900, his Ph.D. at the same university. He worked as prosector at the Lyons Hospital and taught anatomy and operative surgery at the university. By 1906, he was at the Rockefeller Institute for Medical Research, where he carried out most of his landmark experiments.

Influenced by the assassination by knife of the president of France in 1894, he dedicated himself to develop a way to suture blood vessels, which ironically he developed after he studied with a French embroidress who showed him how to do embroidery. His first attempt was made in France in 1902. He subsequently developed the triangulation technique of vascular suture. He won the Nobel Prize in physiology or medicine in 1912 for his work on vascular suture and the transplantation of blood vessels and organs.

During World War I, Carrel served as a major in the French army medical corps and helped devise a widely used method of treating war wounds, called the Carrel-Dakin method, a method of wound irrigation in which the wound is intermittently irrigated with Dakin's solution, a germicidal fluid (no longer used).

Carrel's researches were mainly concerned with experimental surgery and the transplantation of tissues and whole organs. As early as 1902, he published a technique for the end-to-end anastomosis (union) of blood vessels, and during the next few years he did every conceivable form of anastomosis, although many were not accepted until the 1950s. In 1908, he devised methods for the transplantation of whole organs and had tested kidney and heart transplantations as early as 1905. In 1910 he demonstrated that blood vessels could be kept for long periods in cold storage before they were used as transplants in surgery, and he also conducted aortocoronary bypass surgery, before the advent of anticoagulants.

In 1935, in collaboration with Charles Lindbergh, Carrel devised a machine for supplying a sterile respiratory system to organs removed from the body. Carrel was able to perform surgeries that showed that circulation, even in such vital organs as the kidneys, could be interrupted for as long as two hours without causing permanent damage. The cover of the June 13, 1938, *Time* magazine showed Charles Lindbergh and Alexis Carrel with the new perfusion pump.

His books, such as *The Culture of Organs* and *Man, the Unknown, Treatment of Infected Wounds* (with Georges Debelly), were important works. He died in Paris on November 5, 1944.

carrier An individual who is heterozygous for a recessive disease-causing trait but who does not necessarily show any symptoms and can pass the mutant gene to offspring. If both parents are homozygous for the trait, the chance that a newborn child will be affected is one out of four.

carrier-linked prodrug (**carrier prodrug**) A PRODRUG that contains a temporary linkage of a given active substance with a transient carrier group that provides improved physicochemical or pharmacokinetic properties and that can be easily removed in vivo, usually by a hydrolytic cleavage.

carrying capacity A population's maximum capacity within a habitat that a single species can maintain before the habitat degrades or becomes destroyed. While a species may go over the carrying capacity, the long-term viability of the habitat is destined to lessen or be depleted.

cartilage A rubbery but firm and flexible shock-absorber tissue that cushions bones at the joints and can be found in other areas like the spine, throat, ears, and nose. Made up of cells called chondroblasts that secrete a cartilage matrix (containing cells called chondrocytes when surrounded by a matrix), an intracellular substance. Cartilage is covered by a membrane called perichondrium that serves for nutrition and growth of the cartilage. Osteoarthritis can occur when cartilage is worn away faster than it is replaced. The underlying bones then start to rub against each other.

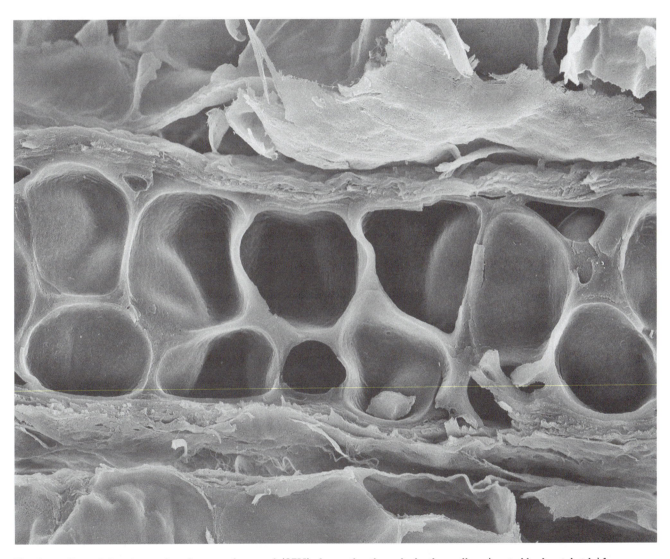

Elastic cartilage. Colored scanning electron micrograph (SEM) of a section through elastic cartilage (central horizontal strip) from a pinna (external ear). Elastic cartilage maintains the semirigid, flexible shape of the ear. It contains fibers of the protein elastin. Cartilage is a firm and flexible connective tissue. It is composed of chondrocyte cells embedded in holes (black) in an extracellular matrix (brown). The layers of skin (top and bottom) enclosing the cartilage are also seen. The pinna channels sound waves into the internal ear. Magnification unknown. *(Courtesy © Science Photo Library/Photo Researchers, Inc.)*

cascade prodrug   A PRODRUG for which the cleavage of the carrier group becomes effective only after unmasking an activating group.

casparian strip   A band of suberin, a waxy substance that waterproofs the walls of each plant's root cells; prevents and controls passive water and mineral uptake into the central vascular tube of roots (steles).

caste   Morphologically distinct individuals within a colony, e.g., ants, that are also behaviorally specialized such as queens, workers, soldiers, etc.

catabolic pathway   The process for taking large complex organic molecules and breaking them down into smaller ones, which release energy that can be used for metabolic processes.

catabolism   Reactions involving the breaking down of organic SUBSTRATES, typically by oxidative breakdown, to provide chemically available energy (e.g., adenosine triphosphate [ATP]) and/or to generate metabolic intermediates.

catabolite   A naturally occurring METABOLITE.

catabolite activator protein (CAP)   A protein that binds cyclic adenosine monophosphate (cAMP), a regulatory molecule, to DNA in organisms. When this interaction takes place, the gene promoter is made accessible to the enzyme RNA polymerase, and transcription of the gene can begin.

catalase   A HEME protein that catalyzes the DISPROPORTIONATION of dihydrogen peroxide to O_2 and water. It also catalyzes the oxidation of other compounds, such as ethanol, by dihydrogen peroxide. A nonheme protein containing a dinuclear manganese CLUSTER with catalase activity is often called pseudocatalase.

catalyst   Any substance that speeds up a chemical reaction without itself being consumed by the reaction.

catalytic antibody (abzyme)   An ANTIBODY that catalyzes a chemical reaction analogous to an enzymatic reaction, such as an ester hydrolysis. It is obtained by using a hapten that mimics the transition state of the reaction.
See also ENZYME.

cataract   The clouding of the natural lens of the eye or surrounding membrane, making it difficult to see.

catecholamine   A class of hormones, two of which are known to be important in a medical emergency. These are epinephrine and norepinephrine. Dopamine and dopa are also catecholamines. All the catecholamines stimulate high blood pressure and can trigger symptoms usually associated with threatening situations leading to a panic attack.

Epinephrine (adrenaline). A hormone released by the adrenal gland, which is the drug of choice for the treatment of anaphylaxis. Indeed, those who are allergic to insect stings and certain foods should always carry a self-injecting syringe of epinephrine.

Epinephrine increases the speed and force of heartbeats and, therefore, the work that can be done by the heart. It dilates the airways to improve breathing and narrows blood vessels in the skin and intestine so that an increased flow of blood reaches the muscles and allows them to cope with the demands of exercise. Usually treatment with this hormone stops an anaphylactic reaction. Epinephrine has been produced synthetically as a drug since 1900.

Norepinephrine (noradrenaline). A hormone released by the adrenal gland. Norepinephrine is released, along with epinephrine, from the adrenals and from nerves when heart failure takes place. These hormones are the first line of defense during any sudden stress. The release of these hormones cause the heart to pump faster, making up for the pumping problem caused by heart failure.

cation   A positively charged ION.

cation exchange The ability of soils to attract and exchange cations with cations of soil solutions; high for clays and humus and low for sand.

catkin A hanging group of unisexual flowers (either male or female) without petals, e.g., willows.

Caucasian A member of the human race that is white skinned.

caudal A fin, or direction toward a tail.

CBS (**colloidal bismuth subcitrate**) *See* DE-NOL.

CD *See* CIRCULAR DICHROISM.

celiac disease (**celiac sprue**) Celiac disease is a malabsorption disorder characterized by a permanent gluten-sensitive enteropathy resulting in malabsorption, failure to thrive, and other gastrointestinal manifestations. However, it should not be confused with a food allergy or hypersensitivity to food products.

Celiac disease is an inherited cell-mediated hypersensitivity involving a tissue-bound immune cell, often resulting in delayed reaction to a food allergen such as wheat, rye, oats, or barley. Gluten, a protein in these grains, is thought to be the offending agent. The disease has also been referred to as gluten enteropathy, gluten intolerance, gluten intolerant enteropathy, gluten-sensitive enteropathy, nontropical sprue, and wheat allergy.

The onset of the disease has no age restriction, but there are many hypotheses related to possible causative factors. In some adults, symptoms leading to a diagnosis of celiac sprue have been observed to appear following severe emotional stress, pregnancy, an operation, or a viral infection.

cell The basic unit of life, capable of growing and multiplying. All living things are either single, independent cells or aggregates of cells. A cell is usually composed of cytoplasm and a nucleus, and it is surrounded by a membrane or wall. Cells can be categorized by the presence of specific cell surface markers called clusters of differentiation.

cell center (**centrosome**) The organelle centrally located near the nucleus where the microtubules are organized and the location of the spindle pole during mitosis. A pair of centrioles, arrays of microtubules, are found in the center in the cells of animals.

cell cycle The reproductive cycle of the eukaryotic cell: the orderly sequence of events (M, G1, S, and G2 phases) when a cell duplicates its contents and divides into two.

See also CELL-CYCLE PHASES; MITOSIS.

cell-cycle phases The sequence of events that cells go through between mitotic divisions. The cycle is divided into gap 0 (G0), gap 1 (G1), synthesis phase (S) when DNA is replicated, Gap 2 (G2), and mitosis (M).

G0 phase Period of time when the cell pauses in cell division between M (mitosis) and S (synthesis) phases. Normal cells in this phase have exactly one set of chromosome pairs.

G1 phase Period of time after mitosis but before S phase of the cell-division cycle; the cell is making preparations for DNA synthesis.

G2 phase Period of time after the S phase and before mitosis of the cell-division cycle. In this phase, the cells have duplicated their DNA and formed two sets of chromosome pairs, in preparation for division. G2 follows the S phase and precedes the M (mitosis) phase.

See also MITOSIS.

cell division When two daughter cells are created from one cell.

cell fractionation Separation of a cell's individual subcellular components (membranes, nucleus, cytoplasm, mitochondria) by the use of centrifuges, which allows closer study of individual cellular components.

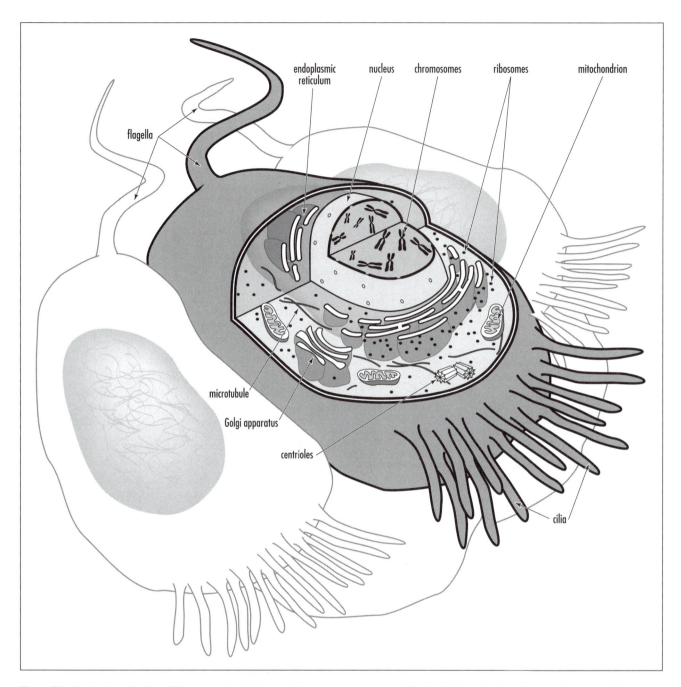

The cell is the basic unit of any living organism. It is a small, watery compartment filled with chemicals and a complete copy of the organism's genome. *(Courtesy of Darryl Leja, NHGRI, National Institutes of Health)*

cell-mediated immunity (CMI; cellular immunity) The branch of the immune system in which the reaction to foreign material is performed by specific defense cells (killer cells, macrophages, and other white blood cells) rather than antibodies.

cell membrane A two-layered structure of material surrounding living cells. Most cell membranes have proteins, such as receptors and enzymes, embedded in them. The membrane holds the cell together, controls which substances go in or out, and maintains homeostasis.

cell plate A membrane that forms in an area of the cytoplasm from the fusion of vesicles (which flatten) of a dividing plant cell during cytokinesis and which will develop into a new cell wall.

cellular differentiation The process of embryonic cells developing into their destined specific forms and functions as an organism develops; a result of gene expression. The process by which different cells, all sharing the same DNA, are capable of performing different tasks.

cellular respiration The process in which adenosine triphosphate (ATP) is created by metabolizing glucose and oxygen with the release of carbon dioxide. Occurs in the mitochondria of eukaryotes and in the cytoplasm of prokaryotes.

See also ATP.

cellulose A polysaccharide (polymer of glucose) that is found in the cell walls of plants. A fiber that is used in many commercial products like paper.

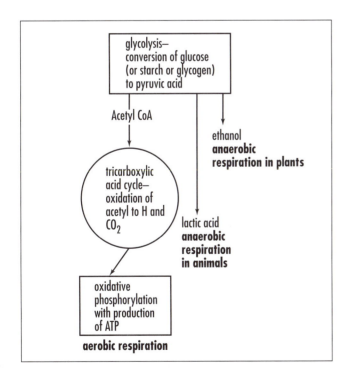

Cellular respiration is the process in which ATP is created by metabolizing glucose and oxygen with the release of carbon dioxide.

cell wall A tough surrounding layer of a cell. In plant cells, it is formed of cellulose embedded in a polysaccharide-protein matrix and is composed of primary and secondary cell walls: the primary is flexible, while the secondary is more rigid. The cell wall provides structural support and protection.

Celsius, Anders (1701–1744) Swedish *Astronomer, Physicist* Anders Celsius was a Swedish astronomer, physicist, and mathematician who introduced the Celsius temperature scale that is used today by scientists in most countries. He was born in Uppsala, Sweden, a city that has produced six Nobel Prize winners. Celsius was born into a family of scientists, all originating from the province of Hälsingland. His father Nils Celsius was a professor of astronomy, as was his grandfather Anders Spole. His other grandfather, Magnus Celsius, was a professor of mathematics. Both grandfathers were at the university in Uppsala. Several of his uncles were also scientists.

Celsius's important contributions include determining the shape and size of the Earth; gauging the magnitude of the stars in the constellation Aries; publication of a catalog of 300 stars and their magnitudes; observations on eclipses and other astronomical events; and a study revealing that the Nordic countries were slowly rising above the sea level of the Baltic. His most famous contribution falls in the area of temperature, and the one he is remembered most for is the creation of the Celsius temperature scale.

In 1742 he presented to the Swedish Academy of Sciences his paper, "Observations on Two Persistent Degrees on a Thermometer," in which he presented his observations that all thermometers should be made on a fixed scale of 100 divisions (centigrade) based on two points: 0° for boiling water, and 100° for freezing water. He presented his argument on the inaccuracies of existing scales and calibration methods and correctly presented the influence of air pressure on the boiling point of water.

After his death, the scale that he designed was reversed, giving rise to the existing 0° for freezing and 100° for boiling water, instead of the reverse. It is not known if this reversal was done by his student Martin Stromer; or by botanist Carolus Linnaeus, who in 1745 reportedly showed the senate at Uppsala University a thermometer so calibrated; or if it was done by Daniel Ekström, who manufactured most of the thermometers

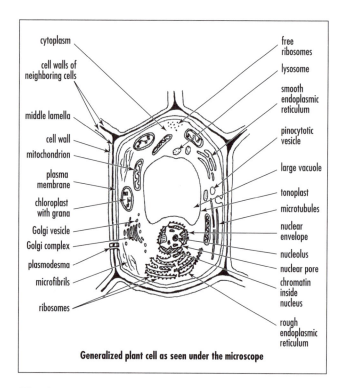

cytoplasm

cell walls of neighboring cells

middle lamella

cell wall

mitochondrion

plasma membrane

chloroplast with grana

Golgi vesicle

Golgi complex

plasmodesma

microfibrils

ribosomes

free ribosomes

lysosome

smooth endoplasmic reticulum

pinocytotic vesicle

large vacuole

tonoplast

microtubules

nuclear envelope

nucleolus

nuclear pore

chromatin inside nucleus

rough endoplasmic reticulum

Generalized plant cell as seen under the microscope

All cells have cell walls that provide a tough surrounding layer for a cell.

used by both Celsius and Linneaus. However, Jean Christin from France made a centigrade thermometer with the current calibrations (0° freezing, 100° boiling) a year after Celsius and independent of him, and so he may therefore equally claim credit for the existing "Celsius" thermometers.

For years Celsius thermometers were referred to as "centigrade" thermometers. However, in 1948, the Ninth General Conference of Weights and Measures ruled that "degrees centigrade" would be referred to as "degrees Celsius" in his honor. The Celsius scale is still used today by most scientists.

Anders Celsius was secretary of the oldest Swedish scientific society, the Royal Society of Sciences in Uppsala, between 1725–44 and published much of his work through that organization, including a math book for youth in 1741. He died of tuberculosis on April 25, 1744, in Uppsala.

Celsius scale (centigrade scale)   A temperature scale with the range denoted by °C. The normal freezing point of water is 0°C, and the normal boiling point of water is 100°C. The scale was named after Anders Celsius, who proposed it in 1742 but designated the freezing point to be 100 and the boiling point to be 0 (reversed after his death).
See also CELSIUS, ANDERS.

Cenozoic era   Age of the mammals. The present geological era, beginning directly after the end of the Mesozoic era, 65 million years ago, and divided into the Quaternary and Tertiary periods.
See also GEOLOGIC TIME.

central atom   The atom in a COORDINATION entity that binds other atoms or group of atoms (LIGANDs) to itself, thereby occupying a central position in the coordination entity.

central nervous system   That part of the nervous system that includes the brain and spinal cord. The brain receives and processes signals delivered through the spinal cord, where all signals are sent and received from all parts of the body, and in turn the brain then sends directions (signals) to the body.

centriole   A pair of short, cylindrical structures composed of nine triplet microtubules in a ring; found at the center of a centrosome; divides and organizes spindle fibers during MITOSIS and MEIOSIS.
See also CENTROSOME.

centromere   A specialized area, the constricted region, near the center of a chromosome to which spindle fibers attach during cell division; the location where the two sister chromatids are joined to one another.

centrosome (microtubule organizing center)   The structural organizing center in cell cytoplasm, near the nucleus, where all microtubules originate; if folded it can become a centriole or a basal body for cilia and flagella.

cephalic Pertains to the head.

cephalochordate A chordate with no backbone (subphylum Cephalochordata), eg., lancelets.

cerebellum A part of the vertebrate hindbrain; controls muscular coordination in both locomotion and balance.

cerebral cortex The outer surface (3–5 mm) of the cerebrum and the sensory and motor nerves. It controls most of the functions that are controlled by the cerebrum (consciousness, the senses, the body's motor skills, reasoning, and language) and is the largest and

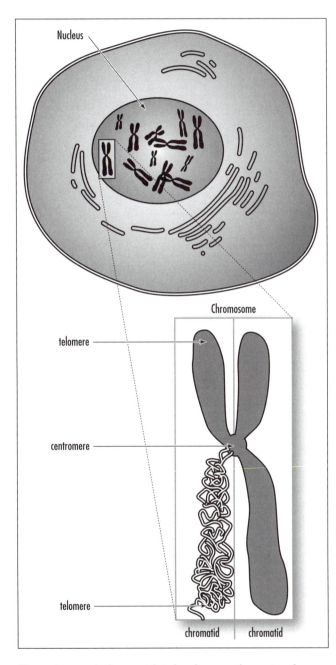

The centromere is the constricted region near the center of a human chromosome. This is the region of the chromosome where the two sister chromatids are joined to one another. *(Courtesy of Darryl Leja, NHGRI, National Institutes of Health)*

Scanning electron micrograph (SEM) of neurons (nerve cells) from the human cerebral cortex (the outer, heavily folded, grey matter of the brain). Neurons exist in varying sizes and shapes throughout the nervous system, but all have a similar basic structure: a large central cell body (large, light-gray bodies) containing a nucleus and two types of processes. These are a single axon (a nerve fiber), which is the effector part of the cell that terminates on other neurons (or organs), and one or more dendrites, smaller processes that act as sensory receptors. Similar types of neurons are arranged in layers within the cerebral cortex. Magnification: ×3,000 at 8 × 8 in. ×890 at 6 × 6 cm size. *(Courtesy © CNRI/Photo Researchers, Inc.)*

most complex part of the mammalian brain. The cortex is broken up into five lobes, each separated by an indentation called a fissure: the frontal lobe, the parietal lobe, the temporal lobe, the occipital lobe, and the insula. It is composed of six layers that have different densities and neuron types from the outermost to innermost: molecular layer, external granular layer, external pyramidal layer, internal granular layer, internal pyramidal layer, and the multiform layer. Vertical columns of neurons run through the layers.

See also BRAIN STEM.

cerebrum The largest part of the brain; divided into two hemispheres (right and left) that are connected by nerve cells called the corpus callosum. It is the most recognized part of the brain and comprises 85 percent of its total weight. The cerebrum is where consciousness, the senses, the body's motor skills, reasoning, and language take place.

See also BRAIN STEM.

ceruloplasmin A copper protein present in blood plasma, containing TYPE 1, TYPE 2, and TYPE 3 COPPER centers, where the type 2 and type 3 are close together, forming a trinuclear copper CLUSTER.

See also MULTICOPPER OXIDASES.

Chain, Ernst Boris (1906–1979) German *Biochemist* Ernst Boris Chain was born on June 19, 1906, in Berlin, to Dr. Michael Chain, a chemist and industrialist. He was educated at the Luisen gymnasium, Berlin, with an interest in chemistry. He attended the Friedrich-Wilhelm University, Berlin, and graduated in chemistry in 1930. After graduation he worked for three years at the Charité Hospital, Berlin, on enzyme research. In 1933, after the rise of the Nazi regime in Germany, he left for England.

In 1935 he was invited to Oxford University, and in 1936 he became a demonstrator and lecturer in chemical pathology. In 1948 he was appointed scientific director of the International Research Centre for Chemical Microbiology at the Istituto Superiore di Sanità, Rome. He became professor of biochemistry at Imperial College, University of London, in 1961, serving in that position until 1973. Later, he became a pro-

fessor emeritus and a senior research fellow (1973–76) and a fellow (1978–79).

From 1935 to 1939 he worked on snake venoms, tumor metabolism, the mechanism of lysozyme action, and the invention and development of methods for biochemical microanalysis. In 1939 he began a systematic study of antibacterial substances produced by microorganisms and the reinvestigation of penicillin. Later he worked on the isolation and elucidation of the chemical structure of penicillin and other natural antibiotics.

With pathologist Howard Walter FLOREY (later Baron Florey), he isolated and purified penicillin and performed the first clinical trials of the antibiotic. For their pioneering work on penicillin Chain, Florey, and FLEMING shared the 1945 Nobel Prize in physiology or medicine.

Later his research topics included the carbohydrate–amino acid relationship in nervous tissue, a study of the mode of action of insulin, fermentation technology, 6-aminopenicillanic acid and penicillinase-stable penicillins, lysergic acid production in submerged culture, and the isolation of new fungal metabolites.

Chain was the author of many scientific papers and a contributor to important monographs on penicillin and antibiotics, and was the recipient of many awards including being knighted in 1969. He died on August 12, 1979.

channels Transport proteins that act as gates to control the movement of sodium and potassium ions across the plasma membrane of a nerve cell.

See also ACTIVE TRANSPORT.

chaparral Dense vegetation of fire-adapted thick shrubs and low trees living in areas of little water and extreme summer heat in the coastal and mountainous regions of California. Similar community types exist in the coastal and mountainous regions of South Africa (fynbos), Chile (matorral), Spain (maquis), Italy (macchia), and Western Australia (kwongan). Also referred to as coastal sagebrush.

chaperonin A member of the set of molecular chaperones, located in different organelles of the cell and involved either in transport of proteins through BIOMEMBRANEs (by unfolding and refolding the proteins) or in assembling newly formed POLYPEPTIDEs.

character A synonym for a trait in TAXONOMY.

character displacement The process whereby two closely related species interact in such a way, such as intense competition between species, as to cause one or both to diverge still further. This is most often apparent when the two species are found together in the same environment, e.g., large and small mouth bass.

charge-transfer complex An aggregate of two or more molecules in which charge is transferred from a donor to an acceptor.

charge-transfer transition An electronic transition in which a large fraction of an electronic charge is transferred from one region of a molecular entity, called the electron donor, to another, called the electron acceptor (intramolecular charge transfer), or from one molecular entity to another (intermolecular charge transfer).

chelation Chelation involves COORDINATION of more than one sigma-electron pair donor group from the same LIGAND to the same CENTRAL ATOM. The number of coordinating groups in a single chelating ligand is indicated by the adjectives didentate, tridentate, tetradentate, etc.

chelation therapy The judicious use of chelating (metal binding) agents for the removal of toxic amounts of metal ions from living organisms. The metal ions are sequestered by the chelating agents and are rendered harmless or excreted. Chelating agents such as 2,3-dimercaptopropan-1-ol, ethylenediaminetetraacetic acid, DESFERRIOXAMINE, and D-penicillamine have been used effectively in chelation therapy for arsenic, lead, iron, and copper, respectively.
See also CHELATION.

chemical equilibrium The condition when the forward and reverse reaction rates are equal and the concentrations of the products remain constant. Called the law of chemical equilibrium.

chemical shift *See* NUCLEAR MAGNETIC RESONANCE SPECTROSCOPY.

chemiosmosis A method of making ATP that uses the electron transport chain and a proton pump to transfer hydrogen protons across certain membranes and then utilize the energy created to add a phosphate group (phosphorylate) to ADP, creating ATP as the end product.

chemoautotroph (chemolithotroph) An organism that uses carbon dioxide as its carbon source and obtains energy by oxidizing inorganic substances.

chemoheterotroph Any organism that derives its energy by oxidizing organic substances for both a carbon source and energy.

chemoreceptor A sense organ, cell, or structure that detects and responds to chemicals in the air or in solution.

chemotherapy The treatment of killing cancer cells by using chemicals.
See also CANCER.

chiasma The x-shaped point or region where homologous chromatids have exchanged genetic material through crossing over during MEIOSIS. The term is also applied to the site where some optic nerves from each eye cross over to the opposite side of the brain, forming the optic tract.

chigger Red, hairy, very small mites (arachnids) of the family Trombiculidae, such as *Trombicula alfreddu-*

gesi and *T. splendens* Ewing. Also called jiggers and redbugs. Chiggers cause skin irritation in humans and dermatitis in animals.

chigoe flea   A flea (*Tunga penetrans*) that attacks bare feet and causes nodular swellings and ulcers around toenails, as well as between the toes and the sole.

chirality   A term describing the geometric property of a rigid object (or spatial arrangement of points or atoms) that is nonsuperimposable on its mirror image; such an object has no symmetry elements of the second kind (a mirror plane, a center of inversion, a rotation reflection axis). If the object is superimposable on its mirror image, the object is described as being achiral.

chi-square test   A statistical exercise that compares the frequencies of various kinds or categories of items in a random sample with the frequencies that are expected if the population frequencies are as hypothesized by the researcher.

chitin   The long-chained structural polysaccharide found in the exoskeleton of invertebrates such as crustaceans, insects, and spiders and in some cell walls of fungi. A beta-1,4-linked homopolymer of N-acetyl-D-glucosamine.

chlamydospore   A thick-walled asexual resting spore of certain fungi assuming a role for survival in soil or in decaying crop debris from year to year.

chlorin   In organic chemistry, it is an unsubstituted, reduced PORPHYRIN with two nonfused saturated carbon atoms (C-2, C-3) in one of the pyrrole rings.

chlorophyll   Part of the photosynthetic systems in green plants. Generally speaking, it can be considered as a magnesium complex of a PORPHYRIN in which a double bond in one of the pyrrole rings (17-18) has been reduced. A fused cyclopentanone ring is also pre-

sent (positions 13-14-15). In the case of chlorophyll *a*, the substituted porphyrin ligand further contains four methyl groups in positions 2, 7, 12, and 18, a vinyl group in position 3, an ethyl group in position 8, and a $-(CH_2)_2CO_2R$ group (R=phytyl, (2*E*)-(7*R*, 11*R*)-3,7, 11,15-tetramethylhexadec-2-en-1-yl) in position 17. In chlorophyll *b*, the group in position 7 is a –CHO group. In bacteriochlorophyll *a*, the porphyrin ring is further reduced (7-8), and the group in position 3 is now a $-COCH_3$ group. In addition, in bacteriochlorophyll *b*, the group in position 8 is a $=CHCH_3$ group.
See also PHOTOSYNTHESIS.

chloroplast   The double membrane organelle of eukaryotic photosynthesis; contains enzymes and pigments that perform photosynthesis.
See also EUKARYOTES; PHOTOSYNTHESIS.

cholera   An acute infection of the small intestines by *Vibrio cholerae* that is transmitted by ingesting fecal-contaminated water or food, or raw or undercooked seafood. Symptoms include diarrhea, abdominal cramps, nausea, vomiting, and severe dehydration. Endemic to India, Africa, the Mediterranean, South and Central America, Mexico, and the United States. Highly infectious disease that can be fatal, but there is a vaccine against it.

cholesterol   A soft, waxy, fat-soluble steroid formed by the liver and a natural component of fats in the bloodstream (as lipoproteins); most common steroid in the human body and used by all cells in permeability of their membranes. It is used in the formation of many products such as bile acids, vitamin D, progesterone, estrogens, and androgens. In relation to human health, there is the high-density "good" cholesterol (HDL), which protects the heart, and the low-density "bad" cholesterol (LDL), which causes heart disease and other problems.

chondrichthyes   Cartilaginous fishes; internal skeletons are made of cartilage and reinforced by small bony plates, while the external body is covered with hooklike scales. There are over 900 species in two subclasses, Elasmobranchii (sharks, skates, and rays) and Holocephali (chimaeras).

chondrin A substance that forms the matrix of cartilage, along with collagen; formed by chondrocytes.

chordate One of the most diverse and successful animal groups. The phylum Chordata includes fish, amphibians, reptiles, birds, mammals, and two invertebrates (tunicates and lancelets). Characterized by having at various times of their life a notochord (primitive spine, skeletal rod), pharyngeal slits, and hollow nerve cord ending in the brain area; usually have a head, a tail, and a digestive system, with an opening at both ends of the body. Their bodies are elongate and bilaterally symmetrical. Includes the hemichordates (vertebrates), cephalochordates (e.g., amphioxus), and urochordates (e.g., sea squirts).

chorion One of the four extraembryonic membranes, along with AMNION, YOLK SAC, and ALLANTOIS. It contributes to the formation of the PLACENTA in mammals; outermost membrane.

chorionic villus sampling A prenatal diagnosis technique that takes a small sample of tissue from the placenta and tests for certain birth defects. This is an early detection test, as it can be performed 10 to 12 weeks after a woman's last menstrual cycle.

chromatin The combination of DNA and proteins that make up the chromosomes of eukaryotes. Exists as long, thin fibers when cells are not dividing; not visible until cell division takes place.

chromatography A method of chemical analysis where a compound is separated by allowing it to migrate over an absorbent material, revealing each of the constituent chemicals as separate layers.

Chromista Brown algae, diatoms, and golden algae, placed together under a new proposed kingdom name.

chromophore That part of a molecular entity consisting of an atom or group of atoms in which the electronic transition responsible for a given spectral band is approximately localized.

chromosome The self-replicating gene-carrying member found in the cell nucleus and composed of a DNA molecule and proteins (chromatin). Prokaryote organisms contain only one chromosome (circular DNA), while eukaryotes contain numerous chromosomes that comprise a genome. Chromosomes are divided into functional units called genes, each of which contains the genetic code (instructions) for making a specific protein.
See also GENE.

chronic Long lasting and severe; the opposite of acute. Examples of chronic diseases include: chronic atopic dermatitis; chronic bronchitis; chronic cough; chronic rhinitis; chronic ulcerative colitis.

chytrid A group of fungi not completely understood by science. Not visible to the eye, they are small, with a mycelium and central sporophore, looking like a miniature octopus. They reproduce by means of self-propelling spores.

cilium (plural, cilia) A hair-like oscillating structure that is used for locomotion or for moving particles. It projects from a cell surface and is composed of nine

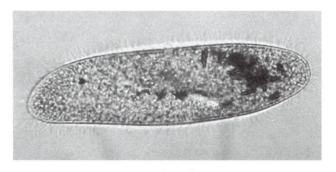

A single-celled organism, *Paramecium caudatum*, that uses cilia for locomotion. (*Courtesy Hideki Horikami*)

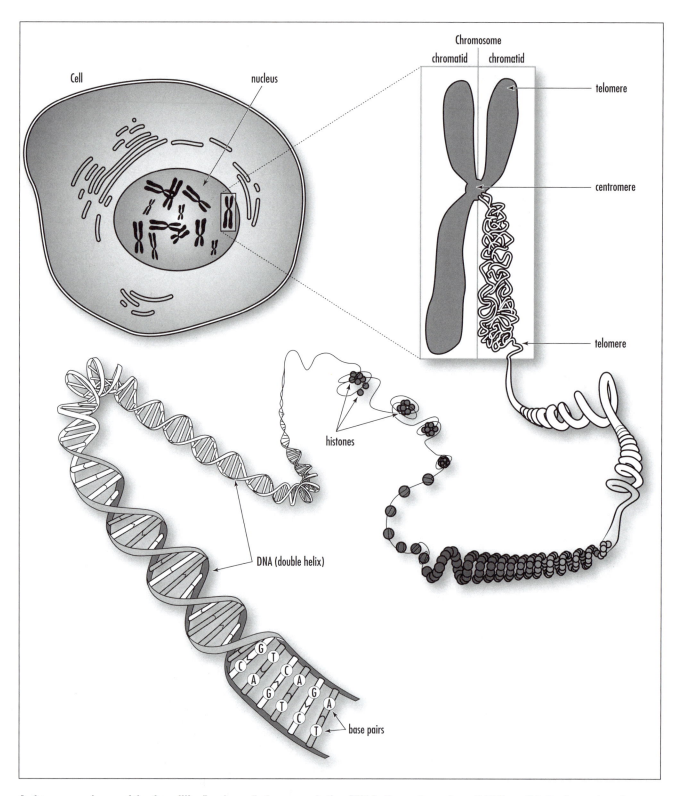

Cell

nucleus

Chromosome

chromatid   chromatid

telomere

centromere

telomere

histones

DNA (double helix)

G
C
A
G
T
C
T

T
A
C
A
G
A
T

base pairs

A chromosome is one of the threadlike "packages" of genes and other DNA in the nucleus of a cell. Different kinds of organisms have different numbers of chromosomes. Humans have 23 pairs of chromosomes, 46 in all: 44 autosomes and two sex chromosomes. Each parent contributes one chromosome to each pair, so children get half of their chromosomes from their mothers and half from their fathers. *(Courtesy of Darryl Leja, NHGRI, National Institutes of Health)*

Human Cytogenetics: Historical Overview and Latest Developments

by Betty Harrison

Cytogenetics, the study of chromosomes, was revolutionized by the discovery that quinicrine staining under ultraviolet light produces a unique banding pattern. In 1970, Dr. Torbjorn O. Caspersson and his group discovered that human chromosomes fluoresce when stained with quinicrine mustard, giving a distinct banding pattern to each chromosome pair. It was later found that chromosomes show a similar banding pattern with the use of stain. Giemsa/Trypsin or Wright's/Trypsin stain is now preferable to quinicrine, as they allow the use of the light microscope and also provide stable preparations.

Banding of chromosomes allows pairing of, rather than grouping of, homologous chromosomes of similar size. The number of bands identified is routinely between 450 and 550, and high resolution is 550 and higher bands in a haploid set. A band is a region that is distinguishable from a neighboring region by the difference in its staining intensity. Banding permits a more detailed analysis of chromosome rearrangements such as translocations, deletions, duplications, insertions, and inversions.

Several types of banding procedures have been developed since the early 1970s in addition to Giemsa banding. R-banding or reverse banding is the opposite of Giemsa (Wright's) banding pattern. There are staining techniques for specific regions of the chromosomes: silver staining, which stains the nucleolus-organizing regions (NORs), and C-banding (constitutive heterochromatin), which stains the centromere of all chromosomes and the distal portion of the Y chromosome. The size of the C-band on a given chromosome is usually constant in all cells of an individual but is highly variable from person to person.

The development of banding techniques in the early 1970s was followed by the development of "high resolution banding" in the late 1970s. This technique divides the landmark bands into sub-bands of contrasting shades of light and dark regions. High-resolution or extended banding is produced by a combination of the induction of cell synchronization followed by the precise timing of harvesting. The cells are then examined in prophase, prometaphase, and early metaphase. At these stages of cell division, very small chromosome changes can be detected. This is useful clinically to find previously undetected chromosome aberrations not found at lower banding levels, to localize breakpoints in rearranged chromosomes, and to help establish phenotype–genotype relationships at a more precise level.

The normal chromosome complement of 46 is called diploid. If there are 46 chromosomes with structural abnormalities, it is referred to as pseudodiploid. Numerical abnormalities are called aneuploidy. When there are more than 46, it is called hyperdiploid, and when fewer, it is called hypodiploid. Chromosome loss in whole or part is a monosomy, while the gain of a single chromosome when they are paired is a trisomy. The presence of two or more cell lines in an individual is known as mosaicism. When, for example, the abnormal line is a trisomy with a normal line, the overall phenotypic effect of the extra chromosome is generally decreased.

The types of cells examined are usually from peripheral blood, bone marrow, amniotic fluid, chorionic villi, and solid-tissue biopsies. These tissues are analyzed in diagnostic procedures in prenatal diagnosis, multiple miscarriages, newborns and children with abnormal phenotypes and abnormal sexual development, hematological disorders, and solid tumors. Autosomal chromosome abnormalities generally have more serious consequences than sex chromosome abnormalities. Chromosome abnormalities are a major cause of fetal loss. These numbers decrease by birth, and since some trisomies result in early death, their frequency is lower in children and even lower in adults.

Numerical changes are the most common chromosome abnormalities. Most numerical changes are the result of nondisjunction in the first meiotic division. Mitotic nondisjunction typically results in mosaicism. Chromosome structural changes can be balanced or unbalanced. Structural abnormalities may be losses, rearrangements, or gains, while numerical abnormalities are losses or gains. Both numerical and structural changes may result in phenotypic abnormalities.

Chromosome abnormalities may be constitutional or acquired. Constitutional abnormalities may be associated with phenotypic anomalies (i.e., Down's syndrome, Turner syndrome) or result in a normal phenotype (i.e., balanced familial translocation). Acquired abnormalities are usually those associated with malignant transformation, such as cancers.

A significant advance in the past decade has been the addition of fluorescent in situ hybridization (FISH) or molecular cytogenetics. FISH has become both a diagnostic and a research tool in cytogenetic laboratories. These procedures involve the denaturation of DNA followed by hybridization with a specific probe that has been tagged with a fluorochrome and stained with a counterstain. These preparations are viewed through a fluorescent microscope with a 100-W mercury lamp and appropriate filter sets.

(continues)

Human Cytogenetics: Historical Overview and Latest Developments
(continued)

FISH, for example, provides rapid results in prenatal diagnosis. Chromosome analysis using classical chromosome analysis on cultured cells from amniotic fluid takes 1–2 weeks to complete. FISH is performed on the uncultured amniocytes and is complete in 24–48 hours. FISH screens for only numerical abnormalities in chromosomes 13, 18, 21, X, and Y. However, these chromosomes make up approximately 90 percent of the total chromosome abnormalities that result in birth defects.

FISH can be used for identification of marker chromosomes, microdeletion syndromes, rearrangements and deletions, detection of abnormalities in leukemia, myeloproliferative disorders, and solid tumors. For example, a chromosome deletion not clearly visible with standard chromosome analysis may be detected with a specific probe for the region (i.e., Di George syndrome).

Cytogenetic analysis of malignant cells proved valuable in the diagnosis and in some cases prognosis of hematological malignancies. FISH has increased the importance of chromosome analysis in the diagnosis of these patients. Many nonrandom cytogenetic abnormalities associated with a specific hematological malignancy have been found. These findings have contributed to the understanding and in some cases treatment of these malignancies. The use of FISH techniques has greatly improved the diagnostic accuracy because, in addition to specific translocation probes, FISH can detect abnormalities in both interphase and metaphase cells.

A current FISH test that may be used for the identification of subtle rearrangements or to characterize complex translocations involving more than two chromosomes is spectral karyotyping (M-FISH, or multiplex in-situ hybridization). This test simultaneously identifies entire chromosomes using 24 different colors.

Recent research using telomere probes in cases of unexplained mental retardation has revealed that approximately 6 percent may be due to subtelomere rearrangements.

An additional research tool, especially in cancer cytogenetics, is comparative genomic hybridization (CGH). This technique reveals whole chromosome gains and losses as well as deletions and amplifications of very small chromosome segments. Cytogenetic analysis has been an extremely valuable tool for screening and for diagnosing genetic disorders, and it will continue to play a vital role in medical service and research.

—**Betty Harrison,** M.S., is director of the
Cytogenetics Laboratory in the department
of Obstetrics, Gynecology and Reproductive
Sciences at Albany Medical College,
Albany, New York.

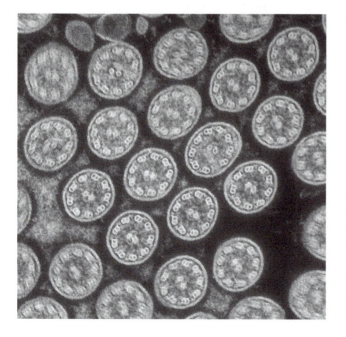

outer double microtubules with two inner single microtubules, and it is anchored by a basal body. Single-cell organisms, such as protozoa, use them for locomotion. The human female fallopian tubes use cilia to transport the egg from the ovary to the uterus.

Colored transmission electron micrograph (TEM) of a cross section through cilia (circles), from the lining (epithelium) of the human trachea, or windpipe. Cilia in the trachea are hairlike projections that beat rhythmically to move mucus away from the gas-exchanging parts of the lungs, up toward the throat where it can be swallowed or coughed up. They project in parallel rows, with 300 on each cell, measuring up to 10 μm in length. Each cillium contains a central core (axoneme), which consists of 20 microtubules arranged as a central pair, surrounded by nine peripheral doubtlets (as seen). Magnification: ×40,000 at 6 × 7 cm size. *(Courtesy © Science Photo Library/Photo Researchers, Inc.)*

circadian rhythm A biological process that oscillates with an approximate 24-hour periodicity, even if there are no external timing cues; an internal daily biological clock present in all eukaryotes.

circular dichroism (CD) A spectroscopic method that measures the difference in absorbance of left- and right-handed circularly polarized light by a material as a function of the wavelength. Most biological molecules, including proteins and NUCLEIC ACIDS, are CHIRAL and show circular dichroism in their ultraviolet absorption bands, which can be used as an indication of SECONDARY STRUCTURE. Metal centers that are bound to such molecules, even if they have no inherent chirality, usually exhibit CD in absorption bands associated with LIGAND-based or ligand-metal CHARGE-TRANSFER TRANSITIONS. CD is frequently used in combination with absorption and MCD studies to assign electronic transitions.

cis In inorganic nomenclature, *cis* is a structural prefix designating two groups occupying adjacent positions. (The term is not generally recommended for precise nomenclature purposes of complicated systems.)
  See also TRANS.

cisplatin *cis*-Diamminedichloroplatinum(II). An antitumor drug highly effective in the chemotherapy of many forms of cancer. Of major importance in the antitumor activity of this drug is its interaction with the NUCLEIC ACID bases of DNA.

cistron A segment of DNA that codes for a single polypeptide domain; another name for a gene.

cladistics A way to classify organisms by common ancestry, based on the branching of the evolutionary family tree. Organisms that share a common ancestor and have similar features are put into groups called clades. At each diverging line, there are two branching lines of descendants, and evolution plays a role in future changes in characteristics.

cladogenesis The evolutionary splitting of lineages; one or more new species comes from an existing parent species, i.e., speciation. Also called branching evolution.

cladogram A pictorial representation of a branching tree that depicts species divergence from a common ancestry.

class The taxonomic ranking of plants and animals that is between phylum and order.
  See also TAXON.

classical conditioning The presentation of two stimuli at the same point in time: a neutral stimulus and a conditioned stimulus; the changes in behavior arising from the presentation of one stimulus in the presence of another. The pairing leads to the neutral stimulus associating with the properties of the conditioned.

cleavage The process of cell division in an early embryo. Initial stages in embryonic development where the zygote converts to a ball of cells through divisions of clearly marked blastomeres, usually from a succession from first through sixth cleavages (2–64 cells). Each species of organism displays a characteristic cleavage pattern that can be observed. Cleavage divides the embryo without increasing its mass.

cleavage furrow A groove composed of actin-rich contractile microfilaments that draws in tight to separate daughter cells during cytokinesis. Also called the contractile ring.

cleistogamous A flower that does not open and is self-pollinated. Pollen is transferred directly from the anthers to the stigma of the same flower.

cleptoparasite The parasitic relation in which a female seeks out the prey or stored food of another female, usually of another species, and appropriates it for the rearing of her own offspring.

climax The final stage in succession where the constituent species populations fluctuate normally instead of acting as replacements of other species. The constituent species will self-perpetuate as long as all natural conditions are favorable and continue.

cline The establishment of plant populations over a specific geographic range that have adapted to different locations and have become slightly different from one another. The plants show a gradient of change over the range, with the frequency of a particular gene either increasing or decreasing over the range. Under the right conditions, speciation may occur over time.

cloaca An all-purpose opening that serves as a digestive, excretory, and reproductive tract for most vertebrates, with the exception of the majority of mammals.

clonal deletion A mechanism whereby the loss of lymphocytes of a particular specificity is due to contact with either "self" or an artificially introduced antigen.

The faces of identical "Megan" and "Morag," the world's first cloned sheep aged 9 months. These Welsh mountain sheep were the product of research by Dr. Ian Wilmut and colleagues at the Roslin Institute in Edinburgh, Scotland. The research involved culturing identical embryonic cells from sheep to produce a "cell line." Next, a sheep egg cell had its DNA removed, and one of these embryonic cells was implanted into the egg. A spark of electricity then stimulated the egg to grow into a lamb, nourished in the womb of a surrogate sheep. The ability to clone farm animals, first achieved in 1996, may provide benefits to agriculture and biotechnology. *(Courtesy © James King-Holmes/Photo Researchers, Inc.)*

clonal selection theory (Burnett theory) Clonal selection theory states that the specificity and diversity of an immune response are the result of selection by an antigen of specifically reactive clones from a large repertoire of preformed lymphocytes, each with individual specificities.

clone A population of organisms, cells, viruses, or DNA molecules that is derived from the replication of a single genetic progenitor. In the case of B cells, each B cell has a typical Ig, and so all the cells that descend from one B cell (the clone) have the same Ig. Typically, a cancer is a clone of cells. Sometimes, *clone* is also used to refer to a number of recombinant DNA molecules all carrying the same inserted SEQUENCE.

cloning vector Any organism or agent (virus, plasmid) that is used to introduce foreign DNA into host cells.

closed circulatory system A type of circulatory system where the blood is contained within a system of vessels and the heart; blood vessels carry blood through all the organs.

cluster A number of metal centers grouped closely together that can have direct metal-bonding interactions or interactions through a BRIDGING LIGAND, but are not necessarily held together by these interactions. Examples can be found under the entries [2FE-2S], [4FE-4S], FERREDOXIN, HIPIP, IRON-SULFUR CLUSTER, FEMO-COFACTOR, FERRITIN, METALLOTHIONEIN, NITROGENASE, and RIESKE IRON-SULFUR PROTEIN.

clusters of differentiation (CD) Cluster of antigens, with which antibodies react, that characterize a cell surface marker.

Lymphocytes can be divided into subsets either by their functions or by surface markers. The availability of monoclonal Abs raised against lymphocytes has allowed for the demonstration of several lymphocyte subsets, which express a combination of certain molecules on their surfaces. These surface markers have been designated clusters of differentiation (CD). Already, 78 CDs have been identified as well as the monoclonal Abs used to define them, their molecular weights, and cellular distribution. For example, CD23, the receptor for the FC portion of IgE and CD8 (T8), a protein embedded in the cell surface of suppressor T lymphocytes.

CD4 One of the most infamous CDs due to its importance in AIDS. CD4 (T4) is the protein embedded on the surface of T helper and other white blood cells to which HIV attaches itself. It is also found to a lesser degree on the surface of monocytes/ macrophages, Langerhans cells, astrocytes, ker-

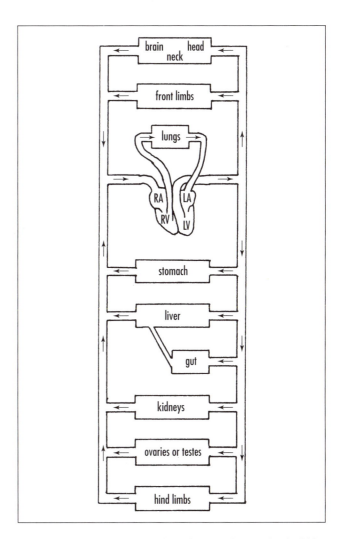

A closed circulatory system, where the blood is contained within a system of vessels and the heart.

atinocytes, and glial cells. HIV invades cells by attaching itself to the CD4 molecule (CD4 receptor). The number of T4 cells in a blood sample is used to measure the health of the immune system in people with HIV.

Helper T cell (CD4 cell, helper, helper cell, T helper cell, T helper lymphocyte, T4 cell) A subset of T cells that carry the T4 marker and are essential for turning on antibody production, activating cytotoxic T cells, and initiating other immune responses. The number of T4 cells in a blood sample is used to measure the health of the immune system in people with HIV. T helper lymphocytes contain two subsets, TH1 and TH2 cells.

CD8 cell (T suppressor cell, T8 cell) The existence of these cells is a relatively recent discovery, and hence their functioning is still somewhat debated. The basic concept of suppressor T cells is a cell type that specifically suppresses the action of other cells in the immune system, notably B cells and T cells, thereby preventing the establishment of an immune response. How this is done is not known with certainty, but it seems that certain specific antigens can stimulate the activation of the suppressor T cells. Discrete epitopes have been found that display suppressor activity on killer T cells, T helper cells, and B cells. This suppressor effect is thought to be mediated by some inhibitory factor secreted by suppressor T cells. It is not any of the known lymphokines. A fact that renders the study of this cell type difficult is the lack of a specific surface marker. Most suppressor T cells are CD8 positive, as are cytotoxic T cells.

clutch The eggs laid in a nest by an individual bird.

cobalamin (vitamin B$_{12}$) A vitamin synthesized by microorganisms and conserved in animals in the liver. Deficiency of vitamin B$_{12}$ leads to pernicious ANEMIA. Cobalamin is a substituted CORRIN-Co(III) complex in which the cobalt atom is bound to the four nitrogen atoms of the corrin ring, an axial group R and 5,6-dimethylbenzimidazole. The latter is linked to the cobalt by the N-3 nitrogen atom and is bound to the C-1 carbon of a ribose molecule by the N-1 nitrogen atom. Various forms of the vitamin are known with different R groups, such as R=CN, cyanocobalamin;

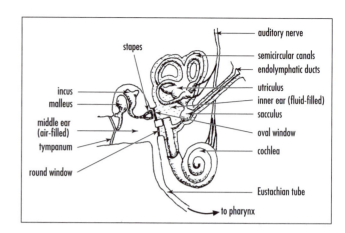

Sound signals pass from the cochlea via the oscillating hair cells, which transform them into electrical signals along the auditory nerve to the brain stem, where they activate other nerves in the brain.

R=OH, hydroxocobalamin; R=CH$_3$, methylcobalamin; R=adenosyl, COENZYME B12.

cochlea The inner ear; a circular or coiled snaillike shell that contains a system of liquid-filled tubes with tiny hair cells. Sound signals pass from the cochlea via the oscillating hair cells, which transform them into electrical signals along the auditory nerve to the brain stem, where they activate other nerves in the brain.

cockroach The order Blattodea that contains the insects also known as the "cucaracha," black beetle, water bug, Yankee settler, shiner, and a host of other names; it is one of the most hated insects known to man. There are 4,000 species, but only about 12 are commonly associated with humans. Common species include: *Blatella germanica, Blatta orientalis, Periplaneta americana, Periplaneta australasiae, Periplaneta brunnea, Periplaneta fuliginosa,* and *Supella longipalpa.*

codominance When both alleles in a heterozygote are expressed phenotypically.

codon A sequence of three consecutive NUCLEOTIDES that occurs in mRNA and (a) directs the incorporation

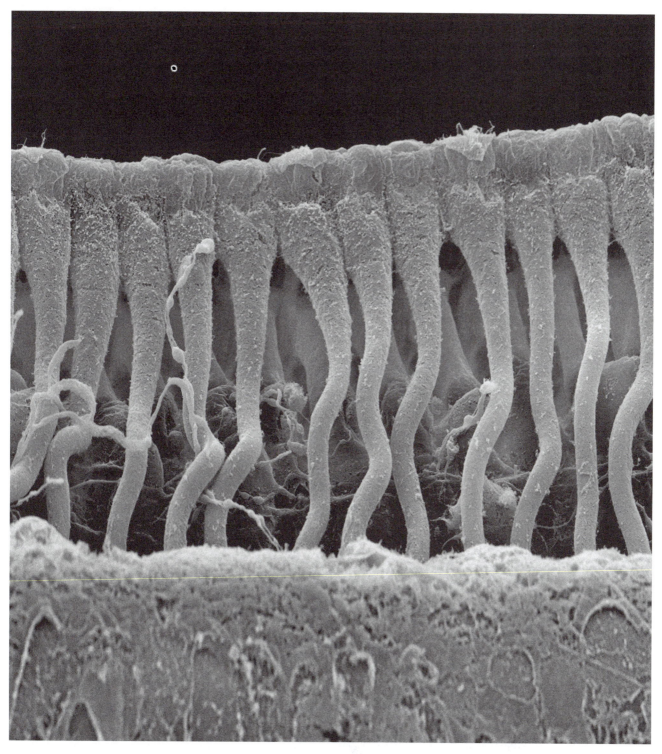

Scanning electron micrograph (SEM) of a vertical section through part of the cochlea inside a human ear. The section shows part of the row of columnar outer pillar cells that runs along the organ of Corti, the auditory sense organ. The outer pillar cells arise from the basilar membrane (across bottom), and their upper surfaces (across top) form part of the surface of the organ of Corti. This organ lies on the basilar membrane, an internal surface of the cochlear duct. The organ of Corti also contains hairlike cilia (not seen) and an overlying tectorial membrane (removed). Sound waves deform hairlike cilia and trigger auditory nerve impulses. Magnification: ×600 at 6 × 7 cm size.
(Courtesy © Science Photo Library/Photo Researchers, Inc.)

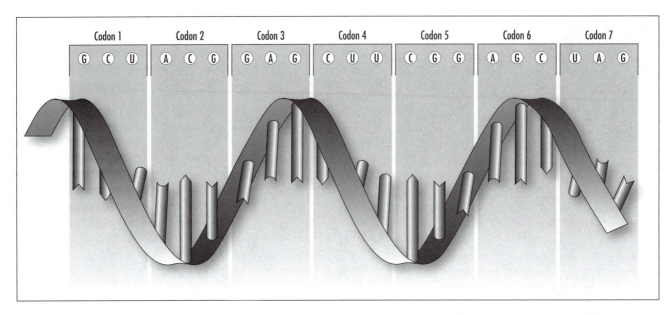

| Codon 1 | Codon 2 | Codon 3 | Codon 4 | Codon 5 | Codon 6 | Codon 7 |
| G C U | A C G | G A G | C U U | C G G | A G C | U A G |

A codon consists of three bases in a DNA or RNA sequence that specify a single amino acid. *(Courtesy of Darryl Leja, NHGRI, National Institutes of Health)*

of a specific amino acid into a protein, or (b) represents the starting or termination signal of protein synthesis.

See also MESSENGER RNA.

coefficient of kinship   The kinship coefficient expresses the chance of finding common genes on the same locus. It also expresses the probability that alleles drawn randomly from each of two individuals are identical by descent. It is also the relationship between a pair of individuals.

coefficient of variation   A measure of dispersion around the mean (average).

coelacanth   A 400-million-year-old "living fossil," believed to have gone extinct 65 million years ago during the Cretaceous period and predating the dinosaurs by millions of years. The coelacanth was found alive in 1938 in South Africa. Today, *Latimeria chalumnae* and *Latimeria menadoensis* represent a once widespread family of sarcopterygian (fleshy-finned) coelacanth fishes. There are more than 120 species known from fossils.

coelom   A fluid-filled body cavity lined with mesoderm where organs can develop.

coelomate   Any organism whose body cavity is lined by mesoderm; animals possessing a coelom. These included the phylas: Entoprocta, Ectoprocta, Phoronida, Brachiopoda, Mollusca, Priapulida, Sipuncula, Echiura, Annelida, Tardigrada, Pentastoma, Onychophora, Arthropoda, Pogonophora, Echinodermata, Chaetognatha, Hemichordata, and Chordata.

See also ACOELOMATE.

coenocytic   Having multiple nuclei embedded in cytoplasm without cross walls; the nuclei lie in a common matrix. Also denotes a mycelium where the hyphae lack septa, i.e., members of the Oomycota and Chytridiomycota.

coenzyme   A low-molecular-weight, nonprotein organic compound (often a NUCLEOTIDE) participating in enzymatic reactions as a dissociable acceptor or donor of chemical groups or electrons.

See also ENZYME.

coevolution The evolution of two species where the evolutionary changes in one of the species influences the evolution of the other. A classic example is the long, narrow bill of the hummingbird. It has coevolved with tubular flowers, and the adaptation of its bill allows it to feed on plants with long, tubular flowers. These flowers in turn have adapted (coevolved) for fertilization by the hummingbirds when they take their nectar.

See also EVOLUTION.

cofactor An organic molecule or ion (usually a metal ion) that is required by an ENZYME for its activity. It may be attached either loosely (COENZYME) or tightly (PROSTHETIC GROUP).

cohesion The force of attraction between molecules of the same substance that allows them to bind.

coitus An alternative term for intercourse.

Coleoptera The taxonomic order that comprises the insect group of beetles, one of the most adaptable and numerous insect groups. Beetles go through complete metamorphosis, have their hind wings covered by their fore wings, and are found in a variety of habitats.

collagen The most abundant fibrous protein in the human body (about 30 percent) and in the animal kingdom; shapes the structure of tendons, bones, and connective tissues. There are several types (I, II, III, IV) that are found in bone, skin, tendons, cartilage, embryonic tissues, and basement membranes.

collecting duct The area in the kidneys where urine is collected. Distal tubules of several nephrons join to form the collecting duct, which consists of the arcuate renal tubule, straight collecting tubule, and the papillary duct. Also known as the tubulus renalis colligens, or renal collecting tubule.

Colored transmission electron micrograph (TEM) of a section through healthy collagen fibers from human skin. Collagen is the major structural protein in the body, forming a large part of bones, tendons, and tissues. Magnification unknown. *(Courtesy © Science Photo Library/Photo Researchers, Inc.)*

A blister beetle (chemically protected) located in the Kelso Dunes, California, that feeds on flowers and pollinates them in the process. *(Courtesy of Tim McCabe)*

Collembola An arthropod order resembling small insects that are wingless and can jump remarkable distances.

collenchyma cell One of the three major plant cell types (dermal, ground, and vascular). Collenchyma cells are part of the ground tissue (ground tissues include parenchyma, collenchyma, and sclerenchyma cells) and are elongated and thick, with uneven cell walls and arranged in strands to provide support in areas of the plant that are growing.

colloidal bismuth subcitrate (CBS) *See* DE-NOL.

colony-stimulating factor (CSF) The category includes granulocyte-colony stimulating factor (G-CSF), macrophage-colony stimulating factor (M-CSF), and granulocyte-macrophage-colony stimulating factor (GM-CSF). These are all cytokine proteins that stimulate growth and reproduction of certain kinds of blood cells in the bone marrow. Also referred to as growth factors. The production of white blood cells is controlled by colony-stimulating factors. Cancer chemotherapy and inherited disorders are among the causes of low white-cell counts, which lower resistance to infection. Thus, CSFs are being investigated not only as a way to counteract low white-cell counts but also as a way to produce specific types of white blood cells. In addition, there is hope that CSFs can stimulate the body to produce additional bone marrow as well as cause some cancer cells to stop dividing.

combinatorial library A set of compounds prepared by combinatorial synthesis.

combinatorial synthesis A process to prepare large sets of organic compounds by combining sets of building blocks.

CoMFA *See* COMPARATIVE MOLECULAR FIELD ANALYSIS.

commensalism One of the forms of symbiosis. In this case, one organism benefits and the other is not affected.

community All of the organisms, plant and animal, that inhabit a specific geographic area.

companion cell A type of plant cell that is connected to a sieve-tube member, making up the phloem tissue. It retains the nucleus and dense cytoplasm to service adjacent sieve tube members, and it helps pump sugars into the phloem.

comparative molecular field analysis (CoMFA) A three-dimensional quantitative structure-activity relationship (3D-QSAR) method that uses statistical correlation techniques for analysis of the quantitative relationship between (a) the biological activity of a set of compounds with a specified alignment and (b) their three-dimensional electronic and steric properties. Other properties such as hydrophobicity and hydrogen bonding can also be incorporated into the analysis.
See also THREE-DIMENSIONAL QUANTITATIVE STRUCTURE-ACTIVITY RELATIONSHIP.

competitive exclusion principle (Gause's law) The condition where one species is driven out of a community by extinction due to interspecific competition; one species will dominate the use of resources and have a reproductive advantage, forcing the other to disappear.

competitive inhibitor A substance that resembles the substrate for an enzyme, both in shape and size, and competes with the substrate for the substrate binding site on the enzyme, thereby reducing the rate of reaction by reducing the number of enzyme molecules that successfully bind.

complementary DNA (cDNA) A laboratory-produced DNA section that is created by extracting a single-stranded RNA from an organism as a template and transcribing it back into a double-stranded DNA using the enzyme reverse transcriptase. However, the cDNA does not include introns, those portions of the DNA that were spliced out while still in the cell. Used for research purposes and can be cloned into plasmids for storage.

complement fixation The consumption of complement, a complex of nine blood serum proteins that interact sequentially with specific antibodies (and concentrates in inflamed regions), by an antibody-antigen reaction containing complement-fixing antibodies. Used as a test to detect antibodies that react against a particular antigen such as a virus.

complement system A set of 30 glycoproteins in the blood serum in the form of components, factors, or other regulators that work at the surface of cells as receptors. Inactive until activated by immune responses, the system acts to dissolve and remove immune complexes and kill foreign cells.

complete digestive tract (**alimentary canal**) A tube that has an opening and end (mouth and anus) that is used in digestion. The complete digestive tract is one where food is ingested at one end of the tract, the mouth, and wastes from digestion are passed out of the tract at the other end, the anus. An incomplete digestive tract has just one opening used both to take in food and to eliminate wastes.

complete flower Any flower that has all four major parts: SEPALS, PETALS, STAMENS, and CARPELS.

Compositae (**Asteraceae**) The composites (also known as the daisy or sunflower family), Compositae or Asteraceae, are one of the largest plant families, containing almost 20,000 species. Most of these species are herbs, but there are also some shrubs, trees, and vines. The family includes many edible salad plants (e.g., lettuce, endive, chicory, and artichoke); cultivated species such as the marigolds, daisies, sun-

flowers, and chrysanthemums; as well as many common weeds and wildflowers. It is primarily the latter, for example, ragweed and mugwort, that are involved in pollen-induced seasonal allergies.

Ragweed (**Ambrosia**)
Ragweed refers to the group of approximately 15 species of weed plants, belonging to the Compositae family. Most ragweed species are native to North America, although they are also found in Eastern Europe and the French Rhône Valley. The ragweeds are annuals characterized by their rough, hairy stems and mostly lobed or divided leaves. The ragweed flowers are greenish and inconspicuously concealed in small heads on the leaves. The ragweed species, whose copious pollen is the main cause of seasonal allergic rhinitis (hay fever) in eastern and middle North America, are the common ragweed (*A. artemisiifolia*) and the great, or giant, ragweed (*A. trifida*). The common ragweed grows to about 1 meter (3.5 feet); is common all across North America; and is also commonly referred to as Roman wormwood, hogweed, hogbrake or bitterweed. The giant ragweed, meanwhile, can reach anywhere up to 5 meters

An example of pollinators (moths) and a plant in the Compositae family (thistle flower) from Painted Rock, Colorado. *(Courtesy of Tim McCabe)*

(17 feet) in height and is native from Quebec to British Columbia in Canada and southward to Florida, Arkansas, and California in the United States.

Due to the fact that ragweeds are annuals, they can be eradicated simply by mowing them before they release their pollen in late summer.

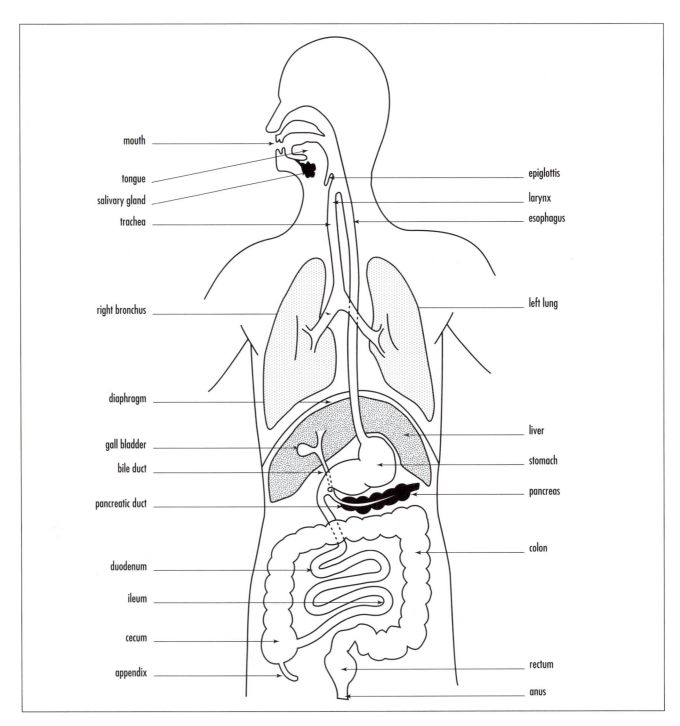

mouth

tongue

salivary gland

trachea

right bronchus

diaphragm

gall bladder

bile duct

pancreatic duct

duodenum

ileum

cecum

appendix

epiglottis

larynx

esophagus

left lung

liver

stomach

pancreas

colon

rectum

anus

The complete digestive tract is one where food is ingested at one end of the tract, the mouth, and wastes from digestion are passed out of the tract at the other end, the anus.

Mugwort *(Artemisia vulgaris, A. campestris, A. dracunculus, A. rupestris, A. mutellina, A. absinthium, A. maritima, A. austriaca, A. pontica, A. laciniata, A. abrotanum, A. annua, A. tilessii)* A shrubby weed most commonly found on wasteland, mugwort can reach heights of up to 2 meters (7 feet) and is characterized by quite small, yellow to reddish brown flowers and a woody stem. The mugwort pollen season (in central Europe) is generally late July to September, with a peak around mid-August. Mugwort is known to cross-react with almost all members of the Compositae family, especially the ragweeds, as well as dandelions, sunflowers, chamomille, and all daisylike flowers. Mugwort also displays an important cross-reaction in the context of food allergies to celery.

compound The combination of two or more different elements, held together by chemical bonds. The elements in a given compound are always combined in the same proportion by mass (law of definite proportion).

compound eye A multifaceted eye found in most invertebrates. The eye is composed of many separate cylinder-shaped (hexagonal) units called ommatidia. Each ommatidium has its own surface area, lens (crystalline cone), light receptors (retinulae), and optic nerve fiber. The images from the collection of ommatidia are then processed.

comproportionation Describes a chemical reaction when a mixture of species in different oxidation states reacts to produce a product that is in a different but more stable intermediate oxidation state. A type of redox reaction. For example, when iodide ions and

Light micrograph of hexagonal facets, called ommatidia, that form the compound eye of a dragonfly. The compound eye is a characteristic of insects, although the size, shape, and number of facets vary among species. The dragonfly, with 30,000 facets, has the largest insect eye. Each ommatidia is a light-sensitive unit consisting of a lens immediately behind the cuticular surface and light-sensitive cells. Light stimulating these cells is converted into electrical signals that are passed to the brain. The insect sees a mosaic image made up from separate bits of information entering each ommatidia. Magnification: ×55 at 35-mm size, ×110 at 6 × 7-cm size. *(Courtesy © John Walsh/Photo Researchers, Inc.)*

iodate ions react together, they form elemental iodine. The reverse of DISPROPORTIONATION.

computational chemistry   A discipline using mathematical methods for the calculation of molecular properties or for the simulation of molecular behavior.

computer-assisted drug design (CADD)   Involves all computer-assisted techniques used to discover, design, and optimize biologically active compounds with a putative use as DRUGS.

concanavalin A   A protein from jack beans, containing calcium and manganese, that agglutinates red blood cells and stimulates T lymphocytes to undergo mitosis.

condensation reaction (dehydration reaction)   A (usually stepwise) reaction in which two or more reactants (or remote reactive sites within the same molecular entity) yield a single main product with accompanying formation of water or of some other small molecule, e.g., ammonia, ethanol, acetic acid, hydrogen sulfide.

The mechanism of many condensation reactions has been shown to comprise consecutive addition and elimination reactions, as in the base-catalyzed formation of (E)-but-2-enal (crotonaldehyde) from acetaldehyde via 3-hydroxybutanal (aldol). The overall reaction in this example is known as the aldol condensation.

The term is sometimes also applied to cases where the formation of water or another simple molecule does not occur.

cone cell   A photoreceptor cell of the eye that is found in the retina and densely populates the central portion, called the macula. It is responsible for seeing color and fine visual detail.

See also ROD CELL.

confidence limits   A statistical parameter defining the lower and upper boundaries/values of a confidence interval. A range of values that is estimated from a sample group that is highly likely to include the true, although unknown, value.

configuration   In the context of stereochemistry, the term is restricted to the arrangement of atoms of a molecular entity in space that distinguishes the entity as a STEREOISOMER, the isomerism of which is not due to CONFORMATION differences.

conformation   A spatial three-dimensional arrangement of atoms in a molecule that can rotate without breaking any bonds.

congener   A substance—literally con- (with) generated—synthesized by essentially the same synthetic chemical reactions and the same procedures. An ANALOG is a substance that is analogous in some respect to the prototype agent in chemical structure.

The term congener, while most often a synonym for homologue, has become somewhat more diffuse in meaning, so that the terms congener and analog are frequently used interchangeably in the literature.

conidium   A nonmotile asexual spore borne at the tip of a special hyphal branch called a conidiophore. It is diverse in form: single or multicelled; simple or complex; round, elongated, or spiral shaped. It is found in ascomycetes and basidiomycetes only.

conifer   A seed-bearing evergreen tree or shrub, a gymnosperm, that reproduces by the use of cones. Conifers inhabit cool temperate regions and have leaves in the form of needles or scales. Examples include pines, fir, spruce, and hemlock. The gymnosperms are the plant order of nonflowering plants, which are characterized by the fact that their seeds are exposed to the air during all stages of development. The name gymnosperm means "naked seeds." Gymnosperms are woody plants and are pollinated by wind, hence their potential for inducing seasonal allergy. The seed-bearing structure is typically a cone. Gymnosperm members include the cycads (e.g., sago palm); ginkgoes; conifers (order Pinales) (e.g., monkey-puzzle, nutmeg);

An example of a bristlecone pine (gymnosperm) reflecting climax conditions and ancient life in the White Mountains of California. *(Courtesy of Tim McCabe)*

family Cupressaceae (e.g., cedar); family Taxaceae (e.g., yew); family Taxodiaceae (e.g., redwood); and family Pinaceae (e.g., pine).

See also DECIDUOUS.

conjugation The process of transferring genetic material between two organisms that are temporarily joined.

connective tissue (**myofascial matrix; fascia**) A very strong tissue that is the main support system for the body, an important component of muscles, tendons, ligaments, cartilage, and bones. It wraps around various systems as a tough fibrous sheath, giving shape and strength. Composed of cells and extracellular matrix.

consensus sequence A SEQUENCE of DNA, RNA, protein, or carbohydrate—derived from a number of similar molecules—that comprises the essential features for a particular function.

conservation biology A branch of biology concerned with the loss of world biodiversity.

conspecific Refers to animals (individuals or populations) of the same species.

continental drift Two hundred million years ago the Earth's continents were joined together to form one gigantic supercontinent called Pangaea. As the rock

The meeting of the North American tectonic plate and the European tectonic plate can be clearly seen near Pingvellir, Iceland, where ravines and cliffs mark the line of the Atlantic Fault. To the left of the picture is the eastern edge of the North American continental plate; to the right is the western edge of Europe. The Atlantic Fault cuts across Iceland from northeast (in the distance) to southwest. The two plates are slowly moving apart by the process of continental drift as the Atlantic Ocean widens. This process causes Iceland's intense seismic and volcanic activity. The last major earthquake occurred at Pingvellir in 1789, when part of the land sank by 50 centimeters in 10 days. *(Courtesy © Simon Fraser/Photo Researchers, Inc.)*

plates that the continents sit on moved, the supercontinent broke up and began to move apart. This process is known as continental drift.

See also GONDWANALAND; PANGAEA.

contraception The conscious and deliberate act of preventing pregnancy.

contrast agent Paramagnetic (or FERROMAGNETIC) metal complex or particle causing a decrease in the relaxation times (increase in relaxivity) of nuclei detected in an image, usually made of water.

See also IMAGING.

convection Fluid or air circulation driven by temperature gradients; the rising of warm air and the sinking of cool air. The transfer of heat by circulation or movement of heated liquid or gas.

convergent evolution When two unrelated species share similar traits arising from each species independently adapting to a similar environmental condition.

cooperativity The phenomenon that binding of an effector molecule to a biological system either enhances or diminishes the binding of a successive molecule, of the same or different kind, to the same system. The system may be an ENZYME or a protein that specifically binds another molecule such as oxygen or DNA. The effector molecule may be an enzyme SUBSTRATE or an ALLOSTERIC EFFECTOR. The enzyme or protein exists in different CONFORMATIONS, with different catalytic rates or binding affinities, and binding of the effector molecule changes the proportion of these conformations. Enhanced binding is named positive cooperativity; diminished binding is named negative cooperativity. A well-known example of positive cooperativity is in HEMOGLOBIN. In biocatalysis it was originally proposed that only multiSUBUNIT enzymes could respond in this way. However, single-subunit enzymes may give such a response (so-called mnemonic enzymes).

See also BIOCATALYST.

coordination A coordination entity is composed of a CENTRAL ATOM, usually that of a metal, to which is attached a surrounding array of other atoms or group of atoms, each of which is called a LIGAND. A coordination entity can be a neutral molecule, a cation, or an anion. The ligands can be viewed as neutral or ionic entities that are bonded to an appropriately charged central atom. It is standard practice to think of the ligand atoms that are directly attached to the central atom as defining a coordination polyhedron (tetrahedron, square plane, octahedron, etc.) about the central atom. The coordination number is defined as being equal to the number of sigma-bonds between ligands and the central atom. This definition is not necessarily appropriate in all areas of (coordination) chemistry. In a coordination formula, the central atom is listed first. The formally anionic ligands appear next, and they are listed in alphabetic order according to the first symbols of their formulas. The neutral ligands follow, also in alphabetic order, according to the same principle. The formula for the entire coordination entity, whether charged or not, is enclosed in square brackets. In a coordination name, the ligands are listed in alphabetic order, without regard to charge, before the name of the central atom. Numerical prefixes indicating the number of ligands are not considered in determining that order. All anionic coordination entities take the ending -ate, whereas no distinguishing termination is used for cationic or neutral coordination entities.

cordilleran A system of parallel mountain ranges forming the spine of continents (e.g., Andes in South America, Rocky Mountains in North America). Spanish for mountain range.

Cori, Carl Ferdinand (1896–1984) Austrian *Biochemist* Carl Ferdinand Cori was born in Prague on December 5, 1896, to Carl I. Cori, director of the Marine Biological Station in Trieste. He studied at the gymnasium in Trieste and graduated in 1914, when he entered the German University of Prague to study medicine. During World War I, he served as a lieutenant in the sanitary corps of the Austrian army on the Italian front; he returned to the university to graduate as a doctor of medicine in 1920. He spent a year at the University of Vienna and a year as assistant in pharmacology at the University of Graz until, in 1922, he accepted a position as biochemist at the State Institute for the Study of Malignant Diseases in Buffalo, New York. In 1931, he was appointed professor of pharmacology at the Washington University Medical School in St. Louis, where he later became professor of biochemistry.

He married Gerty Theresa CORI (née Radnitz) in 1920. They worked together in Buffalo. When he moved to St. Louis, she joined him as a research associate. Gerty Cori was made professor of biochemistry in 1947.

Jointly, they researched the biochemical pathway by which glycogen, the storage form of sugar in liver and muscle, is broken down into glucose. They also determined the molecular defects underlying a number of genetically determined glycogen-storage diseases. For these discoveries the Coris received the 1947 Nobel Prize in physiology or medicine.

They became naturalized Americans in 1928. He died on October 20, 1984, in Cambridge, Massachusetts. His wife died earlier, in 1957.

See also CORI, GERTY THERESA (NÉE RADNITZ).

Cori, Gerty Theresa (née Radnitz) (1896–1957) Austrian *Biochemist* Gerty Theresa Cori (née Radnitz) was born in Prague on August 15, 1896, and received her primary education at home before entering a lyceum for girls in 1906. She entered the medical school of the German University of Prague and received the doctorate in medicine in 1920. She then spent two years at the Carolinian Children's Hospital before emigrating to America with her husband, Carl, whom she married in 1920. They worked together in Buffalo, and when he moved to St. Louis, she joined him as a research associate. She was made professor of biochemistry in 1947.

Jointly, they researched the biochemical pathway by which glycogen, the storage form of sugar in liver and muscle, is broken down into glucose. They also determined the molecular defects underlying a number of genetically determined glycogen-storage diseases. For these discoveries the Coris received the 1947 Nobel Prize for physiology or medicine. She died on October 26, 1957.

See also CORI, CARL FERDINAND.

cork cambium A narrow cylindrical sheath of plant tissue (meristematic) that produces cork cells that replace the epidermis during secondary growth. The resulting cork is impregnated with suberin, a waterproof, waxy fatty acid derivative.

corphin The F-430 cofactor found in methyl-coenzyme M reductase, a nickel-containing ENZYME that catalyzes one step in the conversion of CO_2 to methane in methanogenic bacteria. The Ni ion in F-430 is coordinated by the tetrahydrocorphin LIGAND. This ligand combines the structural elements of both PORPHYRINS and CORRINS.

See also COORDINATION; METHANOGENS; OXIDOREDUCTASE.

corpus luteum A secreting tissue in the ovary, formed from a collapsed follicle, that produces increasing levels of estrogen as well as progesterone after ovulation. These hormones prepare the endometrium for the implantation of a fertilized egg. However, if pregnancy does not occur, the corpus luteum regresses and these hormone levels decline. This results in the breakdown of the endometrium and initiates menstrual bleeding. If pregnancy does occur, the corpus luteum begins to produce human chorionic gonadotropin (HCG).

corrin A ring-contracted PORPHYRIN derivative that is missing a carbon from one of the mesopositions (C-20). It constitutes the skeleton $C_{19}H_{22}N_4$ upon which various B12 vitamins, COFACTORs, and derivatives are based.

cortex Generic term for the outer layer of an organ. Also the region of parenchyma cells in the root between the stele and epidermis filled with ground tissue.

coterie The basic society of prairie dogs, or a small, close group.

cotransport A simultaneous transporting of two solutes across a membrane by a transporter going one way (symport) or in opposite directions (antiport).

cotyledons Leaflike structures (seed leaves) produced by the embryo of flowering plants, the dicots (Magnoliopsida), and the monocots (Liliopsida). They serve to absorb nutrients in the seed until the seedling is able to produce true leaves and begin photosynthesis. In monocots, the embryo has a single cotyledon, while in dicots, the embryo has two cotyledons.

See also DICOT; MONOCOT.

countercurrent exchange The effect caused when two fluids move past each other in opposite directions and facilitate the efficient exchange of heat, gas, or substance. For example: the passage of heat from one blood vessel to another; rete mirabile, the countercurrent exchange structure of capillaries that allows gas uptake in a fish swim bladder; the kidney nephron loop, a tubular section of nephron between the proximal and distal convoluted tubules where water is conserved and urine concentrates by a countercurrent exchange system; and the upper airway where, upon expiration, heat and moisture are retained and given up to relatively cool and dry inspired gases.

Cournand, André-Frédéric (1895–1988) French *Physiologist* André-Frédéric Cournand was born in Paris on September 24, 1895, to Jules Cournand, a stomatologist, and his wife Marguérite Weber. He received his early education at the Lycée Condorcet, received a bachelor's degree at the Faculté des Lettres of the Sorbonne in 1913, and received a diploma of physics, chemistry, and biology of the Faculté des Sciences the following year.

He began medical studies in 1914, but served in the French Army from 1915 to 1918, returning to medical studies at the Interne des Hôpitaux de Paris in 1925. He received an M.D. from the Faculté de Médecine de Paris in May 1930 and secured a residency in the Tuberculosis (later Chest) Service of the Columbia University Division at Bellevue Hospital, New York. He became chief resident of this service and conducted research on the physiology and physiopathology of respiration under the guidance of Dickinson W. RICHARDS. He became an American citizen in 1941 and retired from Columbia in 1964.

Together, Cournand and Richards collaborated in clinical lung and heart research. They perfected a procedure introduced by Werner Forssmann and called Forssmann's procedure, now called cardiac catheterization (a tube is passed into the heart from a vein at the elbow). This made it possible to study the functioning of the diseased human heart and to make more accurate diagnoses of the underlying anatomic defects. They also used the catheter to examine the pulmonary artery, improving the diagnosis of lung diseases as well. He shared the Nobel Prize in physiology or medicine with Dickinson W. Richards and Werner FORSSMANN for their discoveries concerning heart catheterization and circulatory changes in 1956.

Cournand served on the editorial boards of many medical and physiological publications, including *Circulation, Physiological Reviews, The American Journal of Physiology,* and also *Journal de Physiologie* and *Revue Française d'Etúdes Cliniques et Biologiques.* He was a member of numerous scientific organizations and received awards for his work. He died on February 19, 1988, at Great Barrington, Massachusetts.

court (lek) The area defended by individual males within an area where birds gather for display and courtship.

covalent bond A region of relatively high electron density between nuclei that arises at least partly from sharing of electrons and gives rise to an attractive force and characteristic internuclear distance.

crista The inner membrane of mitochondrion, where respiration takes place; location of the electron transport chain and enzymes that catalyze ATP synthesis. Also the term applied to sensory cells within an ear's semicircular canal that detect fluid movement. Also means crest, such as the *crista galli,* the comb on a rooster.

Croll, James (1821–1890) British *Carpenter, Physicist* James Croll was born in Cargill, Perthshire, Scotland, on January 2, 1821. He was the son of David Croll, a stonemason from Little Whitefield, Perthshire, and Janet Ellis of Elgrin. He received an elementary school education until he was 13 years old. His knowledge of science was the result of vigilance, since he was self-taught. On September 11, 1848, he married Isabella MacDonald, daughter of John MacDonald.

Croll started his career as a carpenter apprenticed to a wheelwright when he was young; then he became a joiner at Banchory and opened a shop in Elgin. In 1852, he opened a temperance hotel in Blairgowrie, and later, in 1853, became an insurance agent for the Safety Light Assurance Company ending up in Leicester.

His first book, *The Philosophy of Theism,* was published in 1857 and based on the influence of the metaphysics of Jonathan Edward. However due to an injury, he ended up obtaining a job as a janitor at Anderson's College and Museum in Glasgow in 1859. Being a janitor gave him enough free time after his daily chores to utilize the museum's extensive library. There, he would spend the night reading books on physics, including the works of Joseph A. Adhémar, the French mathematician, who noted in 1842 that the Earth's orbit is elliptical rather than spherical. Adhémar proposed in his book *Revolutions de la Mer, Deluges Periodics* (Revolutions of the sea, periodic floods) that the precession of the equinoxes produced variations in the amount of solar radiation striking the planet's two hemispheres during the winter time (insolation), and this, along with gravity effects from the sun and moon on the ice caps, is what produced ice ages alternately in each hemisphere during a 26,000-year cycle. Precession is the slow gyration of Earth's axis around the pole of the ecliptic, caused mainly by the gravitational pull of the sun, moon, and other planets on Earth's equatorial bulge. Croll also read about the new calculations of the Earth's orbit by French astronomer Urbain Jean Joseph Leverrier (a discoverer of the planet Neptune).

Croll decided to work on the origins of the ice ages, since he did not agree with the prevailing attitude that they were leftover relics from the biblical Great Flood, and additionally he found errors in Adhémar's work. Croll came to the conclusion that the overriding force changing climate and creating the ice ages on Earth was due to variations in insolation, which is the rate of delivery of solar radiation per unit of horizontal surface, i.e., the sunlight hitting the Earth.

Croll first realized that Adhémar did not take into account the shape of the Earth's orbit that varied over time and its effect on precession, so he calculated the eccentricity over several million years. This eccentricity (the distance between the center of an eccentric and its axis), in this case the degree of Earth's elliptic orbit, he proposed, varied on a time scale of about 100,000 years. Since variations in eccentricity only produced small changes in the annual radiation budget of Earth, and not enough to force an ice age, Croll developed the idea of climatic feedbacks, such as changes in surface albedo (reflection). He predicted that the last ice age was over about 80,000 years ago.

During the 1860s, he published his theories in a number of papers: "On the Physical Cause of Changes of Climate during Geological Epochs" (1864); "The Eccentricity of the Earth's Orbit" (1866, 1867); "Geological Time and Date of Glacial and Miocene Periods" (1868); "The Physical Cause of the Motion of Glaciers" (1869, 1870); "The Supposed Greater Loss of Heat by the Southern Hemisphere" (1869); "Evolution by Force Impossible: A New Argument against Materialism" (1877). During this time he was the keeper of maps and correspondence at the Scottish Geological Survey starting in 1867, where he mingled with some of the best geologists of the time until he retired in 1880.

In 1875 he published *Climate & Time in Their Geological Relations,* where he summed up his research on the ancient condition of the Earth. On January 6, 1876, he was elected a fellow of the Royal Society of London. Charles Darwin was among the many supporters of his nomination. He received an LL.D. (law degree) that year from St. Andrews College. While his main interests were in the field of paleoclimate change, he also put forth theories about ocean currents and their effects on climate during modern times.

However, some of his thoughts and ideas were wrong. For example, Croll believed that ice ages varied in the hemispheres, and his estimated age for the last ice advance ending 80,000 to 100,000 years ago was wrong. It ended between 14,000 and 10,000 years ago, as research currently shows. Because of these errors, Croll fell out of vogue until 1912, when Yugoslav geologist Milutin Milankovitch revised Croll's theories in his book, *Canon of Insolation.*

Croll published close to 90 papers on a variety of subjects, such as "Ocean Currents" (1870, 1871, 1874); "Change of Obliquity of Ecliptic: Its Effect on Climate" (1867); "Physical Cause of Submergence during Glacial Epoch" (1866, 1874); "Boulder Clay of Caithness & Glaciation of North Sea" (1870); "Method of Determining Mean Thickness of Sedimentary Rocks" (1871);

and "What Determines Molecular Motion? The Fundamental Problem of Nature" (1872).

A famous debate on the nature of deep-sea circulation between Croll and Irish scientist William Carpenter during the 1860s to 1880s was well discussed in the literature and around scientific circles via correspondence. In 1885, he published *Climate and Cosmology* to answer critics of his earlier work *Climate & Time in Their Geological Relation*. Five years later, plagued by ill health his whole life, he died in Perth on December 15, at age 69, shortly after publishing a small book called *The Philosophical Basis of Evolution*.

Cro-Magnon An early group of *Homo sapiens* (humans) that lived in Europe around 40,000 years ago.

crossing over A process during meiosis when alleles on homologous chromosomes (chromosomes that pair with each other at meiosis) switch places, increasing the possible combinations of alleles and thus increasing the variability of the whole genome. Also called recombination.

See also RECOMBINANT.

cross-pollination When pollen from the anther of a flower of one plant is transferred to the flowers (stigma) of a different plant.

See also POLLINATION.

cross-reactivity The ability of an immunoglobulin, specific for one antigen, to react with a second antigen. A measure of relatedness between two different antigenic substances.

Crustacea All crustaceans have two pairs of antennae, a pair of mandibles, a pair of compound eyes (usually on stalks), two pair of maxillae on their heads, and a pair of appendages on each body segment (head, thorax, and abdomen). There are about 30,000 species of this subphylum within five classes (Remipedia, Cephalocarida, Branchiopoda, Maxillopoda, and Malacostraca). Includes lobsters, crabs, crayfish, shrimp, copepods, isopods, barnacles, and others.

Many of them are important economic species for human consumption.

cryptic Describes the ability to conceal or camouflage.

cryptic coloration A camouflage technique whereby an organism matches its background, concealing itself from predators or prey, e.g., the peppered moth.

See also MIMICRY.

crystal field Crystal field theory is the theory that interprets the properties of COORDINATION entities on the basis that the interaction of the LIGANDS and the CENTRAL ATOM is a strictly ionic or ion-dipole interaction resulting from electrostatic attractions between the central atom and the ligands. The ligands are regarded as point negative (or partially negative) charges surrounding a central atom; covalent bonding is completely neglected. The splitting or separation of energy levels of the five degenerate d-orbitals in a transition metal, when the metal is surrounded by ligands arranged in a particular geometry with respect to the metal center, is called the crystal field splitting.

C-terminal amino acid residue *See* AMINO ACID RESIDUE.

Curie relation *See* MAGNETIC SUSCEPTIBILITY.

cuticle A protective impermeable waxy substance formed from the polymer cutan that covers the outside of leaves, stems, and fruits and forms the protective layer of arthropods.

cyanobacteria Bacteria, formerly known as blue-green algae; aquatic and photosynthetic organisms that live in water and manufacture their own food. Their fossils go back more than 3.5 billion years, making them the oldest known species, and they are the contributors to the origin of plants.

See also ALGAE.

cybernetics The science that studies the methods to control behavior and communication in animals (and machines).

cyclic AMP (cAMP; 3',5'-AMP) Cyclic adenosine monophosphate. A compound synthesized from ATP (by the enzyme adenylyl cyclase) in living cells that acts as an intercellular and extracellular second messenger mediating peptide and amine hormones.

cyclic electron flow Two photosystems are present in the thylakoid membrane of chloroplasts: photosystem I and photosystem II. The two photosystems work together during the light reactions of photosynthesis. The light-induced flow of electrons beginning with and returning to photosystem I to produce ATP without production of NADPH (nicotine adenine dinucleotide phosphate with hydrogen) is cyclic electron flow. The generation of ATP by this process is called noncyclic photophosphorylation.

cyclin A protein found in dividing cells that activates protein kinases (cyclin-dependent protein kinases), a type of enzyme that adds or removes a phosphate group from a target protein and controls the progression of one phase of the cell cycle to the next. The concentration of the cyclin increases and decreases during the cell cycle.

cyclin-dependent kinase A protein kinase, an enzyme involved in regulating cell growth and division, that must be attached to cyclin to become activated.

cytochrome A HEME protein that transfers electrons and exhibits intense absorption bands (the α and β bands, the α band having the longer wavelength) between 510 and 615 nm in the reduced form. Cytochromes are designated types *a, b, c,* or *d,* depending on the position of the α band, which depends on the type of heme. The iron undergoes oxidation–reduction between oxidation states Fe(II) and Fe(III). Most cytochromes are hemochromes, in which the fifth and sixth COORDINATION sites in the iron are occupied by strong field LIGANDs, regardless of the oxidation state of iron. Cytochromes can be distinguished by the wavelength of the α band, such as cytochrome *c*-550. Certain specific cytochromes with particular functions are designated with suffixes, such as cytochrome a_1, b_2, etc.

cytochrome-c oxidase An ENZYME, ferrocytochrome-c: dioxygen OXIDOREDUCTASE, CYTOCHROME aa_3. The major respiratory protein of animal and plant MITOCHONDRIA, it catalyzes the oxidation of Fe(II)-cytochrome *c*, and the reduction of dioxygen to water. Contains two HEMEs and three copper atoms, arranged in three centers. Heme a_3 and copper-B form a center that reacts with dioxygen; the second heme is cytochrome *a*; the third site, copper-A, is a dinuclear center.
See also NUCLEARITY.

cytochrome P-450 General term for a group of HEME-containing MONOOXYGENASES. Named from the prominent absorption band of the Fe(II)-carbonyl complex. The heme comprises PROTOPORPHYRIN IX, and the proximal LIGAND to iron is a cysteine sulfur. Cytochromes P-450 of microsomes in tissues such as liver are responsible for METABOLISM of many XENO-BIOTICs, including drugs. Others, such as the mitochondrial ENZYMEs from adrenal glands, are involved in biosynthetic pathways such as those of steroids. The reaction with dioxygen appears to involve higher oxidation states of iron, such as Fe(IV)=O.
See also MITOCHONDRIA.

cytokines Cytokines are soluble glycoproteins released by cells of the immune system (secreted primarily from leukocytes) that act nonenzymatically through specific receptors to regulate immune responses. Cytokines resemble hormones in that they act at low concentrations bound with high affinity to a specific receptor.

cytokinesis The final stage of mitosis, when the parent cell divides equally by cell-wall formation into two daughter cells by way of a constriction and drawing in of an actin/myosin ring around the center of the cell.

cytoplasm The part of protoplasm in a cell outside of and surrounding the nucleus. The contents of a cell other than the nucleus. Cytoplasm consists of a fluid containing numerous structures, known as organelles, that carry out essential cell functions.

See also CELL.

cytoplasmic determinants Substances distributed in an embryo, but present in an unfertilized egg, that appear in different blastomeres at the initial cleavage stage and influence their development fate.

cytoplasmic streaming The movement and flow of cytoplasm, the living part of a cell outside the nuclear membrane. The primary method of movement of mate-rials within cells, e.g., chloroplasts moving up to the surface of the leaf and then down, which appear to help in photosynthesis.

See also CELL.

cytoskeleton The internal support system and framework of a cell, comprising numerous microfila-ments and tubules that branch throughout the cell. The cytoskeleton serves not only as mechanical support but also in transport functions.

See also SKELETON.

cytosol The semifluid portion of the cytoplasm, not including organelles.

cytoxic T cells (**T killer cells**) Cells that kill target cells bearing appropriate antigen within the groove of an MHC (major histocompatibility complex) class I molecule that is identical to that of the T cell.

D

Dale, Henry Hallett (1875–1968) British *Physiologist* Sir Henry Hallett Dale was born in London on June 9, 1875, to Charles James Dale, a businessman, and Frances Ann Hallett. He attended Tollington Park College in London, Leys School, Cambridge, and in 1894 he entered Trinity College with a scholarship. He graduated through the Natural Sciences Tripos in 1898, specializing in physiology and zoology.

In 1900 he gained a scholarship and entered St. Bartholomew's Hospital, London, for the clinical part of the medical course. He received a B.Ch. at Cambridge in 1903 and became an M.D. in 1909.

He took an appointment as pharmacologist at the Wellcome Physiological Research Laboratories in 1904 and became director of these laboratories in 1906, working for some six years. In 1914, he was appointed director of the department of biochemistry and pharmacology at the National Institute for Medical Research in London, and in 1928 he became the director of this institute, serving until his retirement in 1942, when he became professor of chemistry and a director of the Davy-Faraday Laboratory at the Royal Institution, London.

In 1911, he was the first to identify the compound histamine in animal tissues, and he studied its physiological effects, concluding that it was responsible for some allergic and anaphylactic reactions. After successfully isolating acetylcholine in 1914, he established that it was found in animal tissue, and in the 1930s he showed that it is released at nerve endings in the parasympathetic nervous system, thus establishing acetylcholine's role as a chemical transmitter of nerve impulses.

In 1936 he shared the Nobel Prize in physiology or medicine with his friend German pharmacologist Otto LOEWI for their discoveries in the chemical transmission of nerve impulses.

He was knighted in 1932 and appointed to the Order of Merit in 1944. In addition to numerous articles in medical and scientific journals that record his work, he was the author of *Adventures in Physiology* (1953), and *An Autumn Gleaning* (1954).

Sir Henry was president of the Royal Society (1940–45) and others, and he received many awards. He married his first cousin Ellen Harriet Hallett in 1904. He died on July 23, 1968, in Cambridge.

dalton A unit of measurement of molecular weight based on the mass of one-twelfth the mass of ^{12}C, i.e., 1.656×10^{-24}. A dalton is also called an atomic mass unit, or amu, and is used to measure atomic mass. Protein molecules are express in kilodaltons (kDa). The dalton was named in honor of John Dalton (1766–1844), an English chemist and physicist.

Dam, Henrik (1895–1976) Danish *Biochemist* Carl Peter Henrik Dam was born in Copenhagen on February 21, 1895, to druggist Emil Dam and his wife Emilie (née Peterson), a teacher. He attended the Polytechnic Institute, Copenhagen, and graduated with a degree in chemistry in 1920. The same year he was appointed assistant instructor in chemistry at the School of Agriculture and Veterinary Medicine, advancing to full instructor in biochemistry at the Physiological Laboratory of the University of Copenhagen in 1923.

In 1925 Dam became assistant professor at the Institute of Biochemistry, Copenhagen University, and three years later was promoted to associate professor until 1941. On submitting a thesis *Nogle Undersøgelser over Sterinernes Biologiske Betydning* (Some investigations on the biological significance of the sterines) to the University of Copenhagen in 1934, he received a Ph.D. in biochemistry.

He discovered vitamin K and its anticoagulant effects while studying the sterol metabolism of chicks in Copenhagen and was awarded the Nobel Prize in physiology or medicine in 1943 for this work.

He conducted research at Woods Hole Marine Biological Laboratories in Massachusetts during the summer and autumn of 1941 and was a senior research associate at the University of Rochester, New York, between 1942 and 1945, and he was an associate member at the Rockefeller Institute for Medical Research in 1945.

Dam was appointed professor of biochemistry at the Polytechnic Institute, Copenhagen, in 1941, though the designation of his chair at the Polytechnic Institute was changed to professor of biochemistry and nutrition in 1950.

After his return to Denmark in 1946, he concentrated his research on vitamin K, vitamin E, fats, cholesterol, and nutritional studies in relation to gallstone formation.

He published over 300 articles in biochemistry and was a member of numerous scientific organizations. Dam died on April 17, 1976.

darling effect The stimulation of reproductive activity by the activity of other members of the species in addition to the mating pair. Also called the Fraser darling effect.

Darwinian fitness The measure of an individual's relative genetic contribution to the gene pool of the next generation; the longer an individual survives and the more it reproduces, the higher the fitness and the higher the chance that a hereditary characteristic will be reproduced.

Darwinism The evolutionary theory advanced by Charles Darwin during the mid-19th century suggesting that present-day species have evolved from simpler ancestors. A newer version of the theory is incorporated as "neo-Darwinism" or "modern synthesis."

day-neutral plant A plant where the length of day is not an influence on development; the plant will flower regardless of day length (photoperiodism). It is now known that it is not the length of the light period, but the length of uninterrupted darkness, that is critical to floral development. Examples are tomato, corn, cucumber, and strawberries.
See also PHOTOPERIODISM.

decapods Crustaceans that have five pairs of walking legs and a well-developed carapace, e.g., shrimps, lobsters, hermit crabs, and crabs.

deciduous A plant, tree, or shrub that sheds its leaves at the end of the growing season.
See also CONIFER.

decomposers A trophic level or group of organisms such as fungi, bacteria, insects, and others that as a group digest or break down organic matter (dead animals, plants, or other organic waste) by ingesting, secreting enzymes or other chemicals, and turning them into simpler inorganic molecules or compounds that are released back into the environment.

decumbent A plant or part of the plant that is reclining or lying on the ground but with the tip or apex ascending or pointing up. Also called prostrate, as in a prostrate shrub.

dehiscent A fruit or seed capsule that splits open.

dehydration reaction (condensation reaction) A chemical reaction in which two organic molecules become linked to each via COVALENT BONDs with the removal of a molecule of water; common in synthesis reactions of organic chemicals.

dehydrogenase An OXIDOREDUCTASE that catalyzes the removal of hydrogen atoms from a SUBSTRATE.

deletion A type of mutation where an alteration or loss of a segment of DNA occurs from a chromosome as the consequence of transposition, i.e., when DNA is being moved from one position from one genome to another. Such mutations can lead to disease or genetic abnormality. A terminal deletion refers to breakage and loss off the end of a chromosome, while an interstitial deletion is the loss of material from within the chromosome, but between the ends. Examples of deletion are

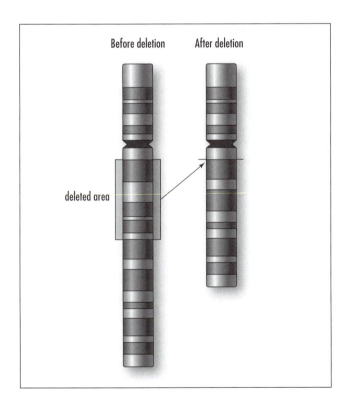

A particular kind of mutation, i.e., the loss of a piece of DNA from a chromosome. Deletion of a gene or part of a gene can lead to a disease or abnormality. *(Courtesy of Darryl Leja, NHGRI, National Institutes of Health)*

Angelman syndrome, a combination of birth defects caused by inheriting both copies of the No. 15 chromosome from the father, and Prader-Willi syndrome, a combination of birth defects caused by inheriting both copies of the No. 15 chromosome from the mother or by inheriting a deletion of a region in the proximal long arm of chromosome No. 15 from the father. Hypopigmentation, which is unusually lighter hair, eyes, and skin color in relation to other family members, is common in Prader-Willi syndrome, especially in individuals with a partial deletion of the long arm of chromosome 15. Cri du Chat syndrome is a rare combination of birth defects caused by a deletion of chromosome 5p.

deme One or more local populations of a taxon that can interbreed.

demography The scientific inquiry into the vital statistics of populations that includes sizes, age–sex compositions, ratios, distributions, densities, growth, natality, mortality, migration, and other characteristics as well as the causes, characteristics, and consequences of changes in these factors. Because the study seeks those relationships that can be expressed precisely and quantitatively, demographers use quantitative analytical methods, but they turn to other disciplines such as anthropology, sociology, and others to explain them. U.S. Census data are a major source of demography for human studies.

denaturation In DNA denaturation, two strands of DNA are separated as a result of the disruption of the hydrogen bonds following exposure to extreme conditions such as high temperature, chemical treatment, pH changes, salt concentration, and others. Denaturation in proteins by heat, acids, bases, or other means results in a change in the three-dimensional structure of the protein so that it cannot perform its function and becomes biologically inactive.

dendrite The thin extension of a neuron that forms synapses by producing or responding to neurotransmitters. A dendrite forms connections with the axons of other neurons and transmits nerve impulses toward the cell body. A dendrite is also a branch or treelike figure produced on or in a rock or mineral or a lichen form.

See also NEURON.

dendrochronology Tree-ring dating. The process of determining the age of a tree or wood by counting the number of annual growth rings.

dendrogram A treelike or graphical diagram that summarizes the process of hierarchical clustering showing evolutionary change.

See also CLADISTICS.

dengue (dandy fever) An epidemic disease found in tropical and subtropical regions. Caused by the dengue virus (genus *Flavivirus* [family Flaviviridae]), which is carried by a mosquito of the genus *Aedes* (*Aedes aegypti* or *Aedes albopictus*). First described in 1827 in Zanzibar; an outbreak occurred in Philadelphia in 1780, then called breakbone fever.

denitrification The reduction of nitrates to nitrites, nitrogen monoxide (nitric oxide), dinitrogen oxide (nitrous oxide), and ultimately dinitrogen catalyzed by microorganisms, e.g., facultative AEROBIC soil bacteria under ANAEROBIC conditions.

De-nol Trade name for the potassium salt or mixed ammonium potassium salt of a bismuth citrate complex, used in the treatment of ulcers.

de novo design The design of bioactive compounds by incremental construction of a ligand model within a model of the RECEPTOR or ENZYME-active site, the structure of which is known from X-ray or nuclear magnetic resonance (NMR) data.

density In biology, the number of individuals per unit area or volume.

density-dependent factor An external process or biological factor, such as disease, competition, or predation, that has a greater effect on a population as the population density increases. A dense population living closely together is more likely to have more of its individuals afflicted and affected by disease than a population that is less dense, with individuals living farther apart from one another. The term also refers to a population regulation factor in ecosystems where the communities have many species and where many biological interactions are taking place. The term can also refer to limiting factors that have an increasing effect on a population as the population increases in size.

density-dependent inhibition A process where most normal animal cells stop dividing when they come into contact with each other.

density-independent factor An external process or set of physical factors (weather, flooding, fire, pollution, etc.) that reduces a population, regardless of size. This can occur in areas with few species with few biological interactions. The term applies to limiting factors that affect all populations, regardless of their density.

denticity The number of donor groups from a given LIGAND attached to the same CENTRAL ATOM.

deoxyribonucleic acid (DNA) A high-molecular-mass linear polymer, composed of NUCLEOTIDES containing 2-deoxyribose and linked between positions 3' and 5' by phosphodiester groups. DNA contains the genetic information of organisms. The double-stranded form consists of a DOUBLE HELIX of two complementary chains that run in opposite directions and are held together by hydrogen bonds between pairs of the complementary NUCLEOTIDES. The way the helices are constructed may differ and is usually designated as A, B, Z, etc. Occasionally, alternative structures are found, such as those with Hoogsteen BASE PAIRING.

See also GENETICS.

deoxyribose   A five-carbon sugar ($C_5H_{10}O_4$) component of DNA. Joins with a phosphate group and base to form a deoxyribose nucleotide, the subunit of nucleic acids.

depolarization   A process where a neuron's electrical charge becomes less negative as the membrane potential moves from resting potential (70 mV) toward 0 mV; a decrease in voltage. The loss of membrane polarity is caused by the inside of the cell membrane becoming less negative in comparison to the outside. Depolarization is caused by an influx of NA+ ions through voltage-gated Na+ channels in axons.

Depolarization is a reduction in potential that usually ends with more positive and less negative charge. Hyperpolarization is the opposite, an increase in potential that ends with more negative and less positive charge. Repolarization is when the state returns to resting potential. Action potentials are caused by depolarization in nerve cells. An action potential is a one-way, self-renewing wave of membrane depolarization that propagates at rapid speed (up to 120 m/sec) along the length of a nerve axon. Julius Bernstein first proposed the concept of depolarization in 1868.

deposit-feeder   A land organism (e.g., earthworm) that eats sediment and processes it through a digestive tract or an aquatic organism (e.g., marine annelid) that ingests bottom sediments such as the sand and mud of a water body. Both digest the microorganisms and other organic matter, with the rest of the material passing through the gut. Examples of deposit feeders are most oligochaetes (earthworms [family Lumbricidae] and small freshwater forms like *Tubifex*), which includes about 3,500 species. Polychaetes (nonselective or selective deposit feeders), such as *Nereis* (common clamworm) and other marine worms such as bloodworms, lugworms, fanworms, and scaleworms, number about 8,000 species. Forms include the sedentary *Arenicola* (lugworm), which is a subsurface (burrow dwelling) deposit feeder, and *Amphrites* (terrebellid), which is a tube-dwelling selective deposit feeder.

dermal tissue system   The outside protective covering (skin) of young plants consisting of a waxy type (cuticle) epidermis, a tightly packed single outer layer of cells that protects and reduces water loss. Stomata regulate gases passing in and out of the plant and are usually located on the underside of leaves, and guard cells regulate the opening by changing water pressure within the cell to swell or shrink. Also the outer tissue

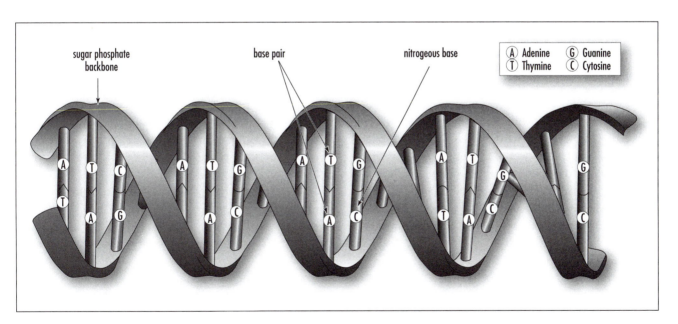

DNA is the chemical inside the nucleus of a cell that carries the genetic instructions for making living organisms. *(Courtesy of Darryl Leja, NHGRI, National Institutes of Health)*

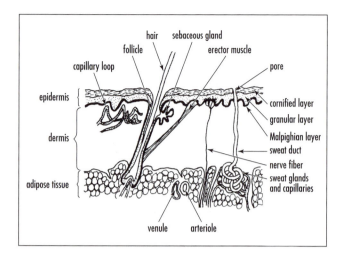

The epidermis serves as a protective layer against invasion of foreign substances, both chemical and animal (parasites).

of a secondary plant body, a periderm consisting of cork that serves as a protective packed-cell arrangement for woody stems.

desferal *See* DESFERRIOXAMINE.

desferrioxamine (dfo) Chelating agent used worldwide in the treatment of iron overload conditions, such as HEMOCHROMATOSIS and THALASSEMIA.
    See also CHELATION.

desmosome At certain points along adjacent surfaces of cells there are intercellular attachments: *zonula occluden* (tight junction), *zonula adherens* (belt desmosome), and *macula adherens* (spot desmosome). Belt desmosomes are a specialized lateral cell-to-cell adhesion, or anchoring junction, that anchors cells, usually epithelial cells, to each other or to extracellular matter. Consisting of dense protein plaques, they form tight attachments to other cells with intermediate filaments consisting of cytoskeleton material that serves as support and structure between adjacent cells, between cells, and as an extracellular matrix. Found in tissues that have been stretched or are subjected to friction, e.g., heart muscle. "Spot" desmosomes are found in all epithelial cells and other tissues such as smooth muscle and are buttonlike

contact points between cells and "spot welds" between cells and adjacent plasma membranes.

determinate cleavage (**mosaic cleavage**) A type of cleavage in protostomes, usually spiral, in which the fates of the cells (blastomeres) are fixed and cannot be changed very early in development. Determinate cleavage was first discovered in nematodes.

determinate growth A growth characteristic of specific duration in which an organism stops growing after it reaches a certain size or achieves a specific goal. As examples: a crop that stops growing and dries after producing grain; a human adult that stops growing after achieving final height; a plant that ripens all of its seeds at the same time; or an apical meristem that differentiates into flowers, terminating the production of additional leaves and stems.

determination In many organisms, the fates of the earliest embryonic cells are not determined and have the potential to develop into many different cell types. Determination is the process whereby cells are committed to a particular development fate as the embryo grows. Portions of the gene are selected for expression in different embryonic cells, which gradually restricts cell fate. Cells can progress from being capable of forming any cell type (totipotent), such as the zygote; to being capable of forming most tissues of an organism (pluripotent); to being fully determined.

    As an embryo develops, its cells become determined and committed to developing into particular parts of the embryo and later adult structures. Following determination, cells eventually differentiate into their final, and often specialized, forms.

    Determination is a slow process in which a cell's potency is progressively restricted as it develops, and the determined state is heritable (a type of cell memory) via somatic cell division. It is irreversible most of the time, but there have been examples of a cell reverting back to an undetermined state.

detritus Accumulated organic debris from dead organisms, often an important source of nutrients in a food web. A detrivore is any organism that obtains most of its nutrients from detritus.

deuterostomes One of the two groups of coelomates, animals that have a coelom or body cavity lined with mesoderm. The deuterostomes, which includes echinoderms and chordates, are animals where the first opening in the embryo during gastrulation becomes the anus, while the mouth appears at the other end of the digestive system; opposite of the protostomes (mollusks, annelids, arthropods), where the mouth forms first during gastrulation and before the future anus. The blastopore, which is the opening of the archenteron in the gastrula, is the site of both mouth or anus development in both groups.

Devonian period A geological period that existed during the middle of the Paleozoic era. The Devonian period existed between 406 and 360 million years ago and is separated into the Early Devonian period (406 to 387 million years ago), the Middle Devonian period (387 to 374 million years ago), and the Late Devonian period (374 to 360 million years ago). It is also called the age of fishes.

See also GEOLOGICAL TIME.

dfo *See* DESFERRIOXAMINE.

diabetes mellitus An ailment characterized by hyperglycemia resulting from the body's inability to use blood glucose for energy. There are two types. In type 1 diabetes, the pancreas ceases to make insulin, and blood glucose does not enter the cells to be used for energy. In type 2 diabetes, the pancreas fails to make sufficient insulin, or the body is unable to use insulin correctly. It is estimated to affect some 17 million people in the United States and is the sixth leading cause of death.

diamagnetic Substances having a negative MAGNETIC SUSCEPTIBILITY are diamagnetic. They are repelled by a magnetic field.

diaphragm A dome-shaped sheet of thin skeletal muscle that separates the lungs and heart from the abdomen and assists in breathing. With inspiration, the diaphragm contracts and flattens downward, while the volume of the thoracic cavity increases, allowing air to enter the respiratory tract. After expiration, the diaphragm relaxes back to its dome shape until the next inspiration.

A diaphragm is also a modern contraceptive devices that prevents sperm from reaching and entering the egg.

diastereoisomers STEREOISOMERs not related as mirror images.

diastole One of two phases of the beating heart. Diastole is the three-step phase when the heart muscle (ventricles) relaxes, causing blood to fill the heart chambers. SYSTOLE is when the ventricles contract. Diastolic blood pressure is the blood pressure measured during diastole, when the chambers fill with blood. It is the force exerted by blood on arterial walls. The lowest blood pressure measured in the large arteries is about 80 mm Hg under normal conditions for a young adult male.

The three phases—early, mid, and late diastole—deal with the conditions of the ventricle, semilunar valve, and atrioventricula valve, while filling, and the reactions of the SA (sinoatrial) node (the heart's "pacemaker") and AV (atrioventricular) node, which regulate contractions.

See also BLOOD PRESSURE.

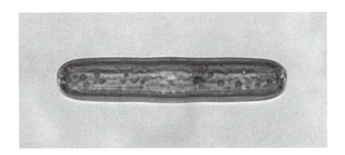

The Bacillariophyceae or diatoms are unicellar algae that are found in single, colonial, or filamentous states. *(Courtesy of Hideki Horikami)*

diatoms *See* ALGAE; CHROMISTA.

dichotomous Either halving or branching by pairs.

dicot (dicotyledon) A member of a subclass of the angiosperms (division Anthophyta) characterized by the presence of two cotyledons in the seed, a reticulated netlike system of veins in the leaves, flower petals in fours or fives, vascular system arranged in a ring in the cortex, root development from the radicle with a fibrous root, and a three-pore pollen structure.

There are about 250,000 species of angiosperms around the world. They make up the largest classification of plants. It is the development of flowers for reproduction that sets them apart from other plant types.

Many of our food and economic staples are from angiosperms, e.g., peanuts, flax, spinach, rice, corn, cotton.

The class Angiospermae (division Anthophyta) is the largest classification of plants. Its distinctive characteristic is the development of flowers, which are used for reproduction. The oldest angiosperm fossils came from the Cretaceous period of the Mesozoic era.

The other angiosperm class is the monocotyledons (monocots), which evolved from dicot ancestors early in the development of flowering plants. There are many more dicots than monocots, and the two groups differ radically in many ways. Dicots are woody or herbaceous; monocots are only herbaceous. The monocots's pollen (single pore), vascular arrangement in stems and roots (scattered bundles), seed leaf (one), flower parts (multiples of three), and root structure (adventitious with taproot) are also different.

See also MONOCOT.

differentiation A process in embryonic development where unspecialized cells take on their individual traits, reach mature form, and progressively become specialized for specific functions such as tissues and organs.

diffusion The random dispersion or spreading out of molecules from a region of high concentration to one of low concentration, stopping when the concentration is equally dispersed.

digestion The process by which living organisms break down ingested food in the alimentary tract into more easily absorbed and assimilated products using enzymes and other chemicals. Digestion can occur in aerobic conditions, where waste is decomposed by microbial action in the presence of oxygen, or under anaerobic conditions when waste is decomposed under microbial action in the absence of oxygen. In anaerobic conditions such as in a large animal facility (e.g., a dairy farm), the by-product, a low-energy biogas that is made with the combination of methane and carbon dioxide, can be used as an energy source.

dihybrid cross The inheritance of two characteristics from two parents at the same time (e.g., leaf shape and stem color). If one trait inherited does not affect the other, the dihybrid cross is two monohybrid crosses operating concurrently. Traits that do not influence the inheritance of each other are said to assort independently. Demonstrates that Mendel's Principle of Independent Assortment allows each trait to be considered separately, since each trait is inherited independently of the other.

dihydrofolate An oxidation product of TETRAHYDROFOLATE that appears during DNA synthesis and other reactions. It must be reduced to tetrahydrofolate to be of further use.

See also FOLATE COENZYMES.

dikaryon The occurrence of two separate haploid nuclei in each cell in the mycelium of some forms of fungi such as basidiomycetes. If the nuclei are both the same genotype, it is said to be homokaryotic; if nuclei are of different genotypes, it is said to be heterokaryotic.

dikaryotic Mycelium or spores containing two sexually compatible nuclei per cell. Common in the BASIDIOMYCETES.

dimorphism The ability of a species to have more than one color or body form, such as differences between sexes.

See also SEXUAL DIMORPHISM.

dinoflagellates *See* ALGAE.

dinosaurs Animals that arose from a common reptilian ancestor that had developed a hole in the bone of its hip socket, was erect, was bipedal or quadrupedal, lived on land, and only lived during the Mesozoic era between 245–144 million years ago. Birds are the direct descendents of dinosaurs.

dinuclear *See* NUCLEARITY.

dioecious A term describing individual organisms that produce only one type of gamete. Asparagus is a dioecious plant species, having a male (staminate) and female (pistillate) flowers on separate plants. The term applies to species where sexes are always separate, e.g., humans. It is the opposite of MONOECIOUS.

dioxygenase An ENZYME that catalyzes the INSERTION of two oxygen atoms into a SUBSTRATE, both oxygens being derived from O_2.

diploid cell A cell with two sets of chromosomes, with one set inherited from one parent and the other set inherited from the other parent.

diplopods A subclass of Myriapoda, the common millipede. They are cylindrical and bear two pairs of legs on each segment and were one of the first animals to venture onto dry land, in the Silurian period (438 million to 408 million years ago).

Diptera Insects in the order Diptera include mosquitoes, flies, gnats, and midges. Dipterons have two wings. Some dipterons are disease vectors (e.g., malaria, yellow fever), while others are important pollinators, pest predators, and parasites. Dipterons undergo complete metamorphosis and usually have more than one generation per year.

directional selection When natural selection favors a phenotype at one extreme of the phenotypic range, giving it an advantage over other individuals in the population, then that particular trait becomes more common in that population as other traits are reduced or eliminated from the population. Over geologic time, directional selection can lead to major changes in morphology and behavior of a population or species.

The classic example is England's peppered moth (*Biston betularia*) and industrial melanism, the gradual darkening of the wings of many species of moths and butterflies living in woodlands darkened by industrial pollution. The light form of the moth was camouflaged among light-colored lichens on London's darker-colored trees before the Industrial Revolution. The dark form of the moth was not observed until 1848, since it was eaten by predators that easily spotted it. After the lichens were killed off by the effects of soot pollution from the Industrial Revolution, the dark form increased in number, and by 1948, 90 percent of the peppered moths were dark colored. The light form of the moth continues to dominate populations in unpolluted areas outside London. Antibiotic resistance in bacteria and pesticide resistance in insects are other examples of directional selection.

disaccharide A class of sugar, a carbohydrate, created by linking together a pair of monosaccharides, which are simple sugars. An example of a disaccharide is sucrose, which is glucose joined to fructose. Other examples include lactose, which is glucose joined with galactose, and maltose, which is two glucoses joined together. While disaccharides can be decomposed into monosaccharides, monosaccharides cannot be degraded by hydrolysis. However, disaccharides can be degraded by hydrolysis into monosaccharides.

dismutase An ENZYME that catalyzes a DISPROPORTIONATION reaction.

dismutation   *See* DISPROPORTIONATION.

dispersion   The distribution pattern, or spacing apart, of individuals from each other within geographic population boundaries. Dispersion can be an aggregated clump, where individuals are concentrated in specific locations of their habitat, the most common example. This is usually because of unequal distribution of available resources or due to social or reproduction associations. Uniform dispersion, where everyone is evenly spaced, is based on individual interactions such as competition. Random dispersion is when individuals are spaced randomly in an unpredictable manner.

disposition   *See* DRUG DISPOSITION.

disproportionation (**dismutation**)   Any chemical reaction of the type A + A ➜ m' + A" where A, A', and A" are different chemical species. The reverse of disproportionation is called COMPROPORTIONATION.

disruptive selection (**diversifying selection**)   Acts against individuals in the middle of the range of phenotypes, instead favoring both ends of extreme or unusual traits and working against common traits in a population. Species may evolve into separate ecotypes, that is, types of individuals within the same species that have adapted to the special conditions they occupy. BATESIAN MIMICRY is often used as an example of disruptive selection. Diversifying selection results in an overall increase in genetic diversity.

dissimilatory   Related to the conversion of food or other nutrients into products plus energy-containing compounds.

dissociation constant   *See* STABILITY CONSTANT.

distomer   The enantiomer of a chiral compound that is the less potent for a particular action. This definition does not exclude the possibility of other effects or side effects of the distomer.
   See also EUTOMER.

DNA   *See* DEOXYRIBONUCLEIC ACID.

DNA ligase   A linking enzyme involved in replicating and repairing DNA molecules. It seals "nicks" in the backbone of a single strand of a double-stranded DNA molecule; connects Okazaki fragments—short, single-stranded DNA fragments on the lagging strand—during DNA replication, producing a complementary strand of DNA; and links two DNA molecules together by catalyzing the formation of a (phosphodiester) bond between the 5' and 3' ends of the nicked DNA backbone.

DNA methylation   A biochemical event that adds a methyl group (–CH3) to DNA, usually at the base cytosine or adenosine, and may be a signal for a gene or part of a chromosome to turn off gene expression and become inactive.

DNA polymerase   An enzyme that catalyzes the synthesis of new complementary DNA molecules from single-stranded DNA templates and primers. Different DNA polymerases are responsible for replication and repair of DNA, and they extend the chain by adding nucleotides to the 3' end of the growing DNA. DNA polymerase catalyzes the formation of covalent bonds between the 3' end of a new DNA fragment and the 5' end of the growing strand.

DNA probe   A single strand of DNA that is labeled or tagged with a fluorescent or radioactive substance and binds specifically to a complementary DNA sequence. The probe is used to detect its incorporation through hybridization with another DNA sample. DNA probes can provide rapid identification of certain species like mycobacterium.
   See also NUCLEIC ACID PROBE.

docking studies Molecular modeling studies aiming at finding a proper fit between a ligand and its binding site.

dodo A large and plump bird (*Raphus cucullatus*) that is now extinct. Dutch sailors began using the Indian Ocean island of Mauritius as a stopover in 1598; the last dodo was killed in 1681. The bird's extinction was due to destruction of its habitat and the importation of animals such as pigs, rats, and monkeys that ate its eggs and cut off its food supply. Overhunting also contributed to the bird's demise. Recently scientists have determined, through DNA analysis, that the long-extinct dodo belongs in the dove and pigeon family.

Doisy, Edward Adelbert (1893–1986) American *Biochemist* Edward Adelbert Doisy was born in Hume, Illinois, on November 3, 1893, to Edward Perez and his wife Ada (née Alley). Doisy was educated at the University of Illinois, receiving a B.A. degree in 1914 and an M.S. degree in 1916. He received a Ph.D. in 1920 from Harvard University.

From 1915 until 1917 he was assistant in biochemistry at Harvard Medical School, and the following two years he served in war in the sanitary corps of the U.S. Army. From 1919 until 1923 he was an instructor, associate, and associate professor at Washington University School of Medicine. In 1923 he became professor of biochemistry at St. Louis University School of Medicine, and the following year he was appointed director of the department of biochemistry retiring in 1965 (emeritus 1965–86).

Doisy and his associates isolated the sex hormones estrone (1929), estriol (1930), and estradiol (1935). He also isolated two forms of vitamin K and synthesized it

Laboratory worker reviewing DNA band patterns. *(Courtesy of Centers for Disease Control and Prevention)*

in 1936–39. For his work on vitamin K, Doisy was awarded the Nobel Prize in physiology or medicine for 1943.

Later, Doisy improved the methods used for the isolation and identification of insulin and contributed to the knowledge of antibiotics, blood buffer systems, and bile acid metabolism.

In 1936 he published *Sex Hormones* and in 1939 published, in collaboration with Edgar Allen and C. H. Danforth, a book entitled *Sex and Internal Secretions*. He died on October 23, 1986, in St. Louis.

Domagk, Gerhard Johannes Paul (1895–1964)

German *Biochemist* Gerhard Johannes Paul Domagk was born on October 30, 1895, in Lagow, a small town in the Brandenburg Marches. He attended school in Sommerfeld, where his father was assistant headmaster, until age 14. His mother, Martha Reimer, came from farming stock in the Marches, where she lived in Sommerfeld until 1945, when she was expelled from her home and died from starvation in a refugee camp.

Domagk became a medical student at Kiel and served in the army during World War I. After being wounded in 1914, he worked in the cholera hospitals in Russia. He noticed that medicine of the time had little success, and he was moved by the helplessness of the medical men of that time in treating cholera, typhus, diarrhea infections, and other infectious diseases. He recognized that surgery had little value in the treatment of these diseases, and also noticed that amputations and other radical treatments were often followed by severe bacterial infections.

In 1918 he resumed his medical studies at Kiel and graduated in 1921. In 1923 he moved to Greifswald and a year later became a university lecturer in pathological anatomy. In 1925 he held the same post at the University of Münster and in 1958 became a professor.

During the years 1927–29 he was given a leave of absence from the University of Münster to do research in the laboratories of the I. G. Farbenindustrie, at Wuppertal. In 1932 he tested a red dye, Prontosil rubrum. While the dye itself had no antibacterial properties, when he slightly changed its chemical makeup, it showed a remarkable ability to stop infections caused by streptococcal bacteria in mice. He had discovered the sulfa drugs that have since revolutionized medicine and saved many thousands of lives. He was awarded

the 1939 Nobel Prize in physiology or medicine for his discovery. He died on April 24, 1964.

domain An independently folded unit within a protein, often joined by a flexible segment of the polypeptide chain. Domain is also the highest taxonomic rank in the animal kingdom, which consists of three domains: Eukarya, Bacteria, and Archaea. The Archaea are commonly known as extremophiles, occurring in the deep sea vents and hot sulfur springs, whereas the Eukarya comprise the higher life forms, including humans.

dominance hierarchy A social order, or ranking, developed by a group of individuals that live together by which certain individuals gain status and exert power over others. Every individual in the group is ranked relative to all other community members of the same sex. Female rank is usually determined by the relative rank of their mothers, and male ranking may also be determined by the mother's rank, or by competition with other males. Individuals who are higher in the dominance hierarchy usually have greater access to food, sex, and other resources. Those males or females at the highest level of ranking are called alpha male and alpha female.

dominant allele An allele that controls the phenotype produced and blocks the phenotype expression of another allele of the same gene, whether or not that gene is dominant or recessive. This is in contrast to a recessive allele, which is expressed only when its counterpart allele on the matching chromosome is recessive.

donor atom symbol A polydentate LIGAND possesses more than one donor site, some or all of which may be involved in COORDINATION. To indicate the points of ligation, a system is needed. The general and systematic system for doing this is called the kappa convention: single ligating atom attachments of a polyatomic ligand to a coordination center are indicated by the italic element symbol preceded by a Greek kappa, κ. In earlier practice, the different donors of the ligand were denoted by adding to the end of the name of the ligand the italicized symbol(s) for the atom or atoms through which attachment to the metal occurs.

double-blind study A clinical study of potential and marketed DRUGS, where neither the investigators nor the subjects know which subjects will be treated with the active principle and which ones will receive a placebo.

double circulation A transportation system for the blood that has separate pulmonary and systemic systems. The heart pumps blood to the lungs and back, then to the body and back via a network of blood vessels. In humans (but not all animals) the blood travels through the heart twice on each complete journey around the body. There is no mixing of the two kinds of blood (oxygen-rich blood is completely separated from oxygen-poor blood). A double circulation system maintains the high blood pressure needed for efficient transport of materials around the body.

double fertilization Restricted to angiosperms, the flowering plants, it is a process where one male sperm cell pollinates an egg to form a zygote, a diploid embryo, while another male sperm joins with two other polar nuclei to form a triploid cell, becoming the endosperm in the ovule. Corn is an example.

double helix Two strands of DNA coiled about a central axis, usually a right-handed HELIX. The two sugar–phosphate backbones wind around the outside of the bases (A = adenine, G = guanine, T = thymine, C = cytosine). The strands are antiparallel, thus the phosphodiester bonds run in opposite directions. As a result, the structure has major and minor grooves at the surface. Each adenine in one strand of DNA is hydrogen bonded to a thymine in the second strand; each guanine is hydrogen bonded to a cytosine.

See also DEOXYRIBONUCLEIC ACID.

double prodrug (pro-prodrug) A biologically inactive molecule that is transformed in vivo in two steps (enzymatically and/or chemically) to the active species.

Down's syndrome The most common and readily identifiable chromosomal abnormality associated with mental retardation. There are 47 instead of the usual 46 chromosomes, and the extra chromosome, chromosome 21, changes the orderly development of the body and brain, showing several symptoms including a characteristic body type, mental retardation, increased susceptibility to infections, and various heart and other organ abnormalities. It is caused by one of the parent's gametes not dividing properly or

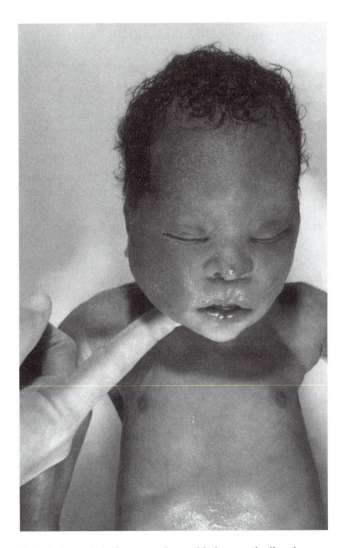

This photograph depicts a newborn with the genetic disorder Down's syndrome due to the presence of an extra 21st chromosome. The estimated incidence of Down's syndrome is between 1:1,000 to 1:1,1000 live births. Each year approximately 3,000 to 5,000 children are born with this chromosomal disorder. *(Courtesy of Centers for Disease Control and Prevention)*

where one of the parents has chromosome 14 and 21 merge.

Approximately 4,000 children with a Down's syndrome are born in the United States each year, or about one in every 800 to 1,000 live births. The incidence is higher for women over age 35, but the condition can occur at any age for women.

drug Any substance presented for treating, curing, or preventing disease in human beings or in animals. A drug can also be used for making a medical diagnosis or for restoring, correcting, or modifying physiological functions (e.g., the contraceptive pill).

drug disposition Refers to all processes involved in the absorption, distribution, METABOLISM, and excretion of DRUGs in a living organism.

drug latentiation The chemical modification of a biologically active compound to form a new compound, which in vivo will liberate the parent compound. Drug latentiation is synonymous with PRODRUG design.

drug targeting A strategy aiming at the delivery of a compound to a particular tissue of the body.

drumlin An oval or elongated hill of glacial drift that looks like an overturned canoe from the air.

drupe A fleshy or pulpy fruit with a single seed enclosed in a pit.

dual-action drug A compound that combines two desired different pharmacological actions at a similarly efficacious dose.

duodenum The first part of the small intestine; short, wide, U-shaped, about 12 inches long, and closest to the stomach. The bile duct (gallbladder) and pancreatic duct (pancreas) both open into the duodenum.

After food mixes with stomach acid, it moves into the duodenum and mixes with digestive juices from the pancreas, liver, and gallbladder, which starts the process of breaking down food into its constituent parts.

duplication A specific kind of mutation: production of one or more copies of any piece of DNA, including a gene or even an entire chromosome. A chromosome structural aberration from an error in meiosis. Duplication of a portion of a chromosome resulting from fusion with a fragment from a homologous chromosome.

dwarfism Animals that evolve on islands are affected by gigantism or dwarfism, the evolution of body form

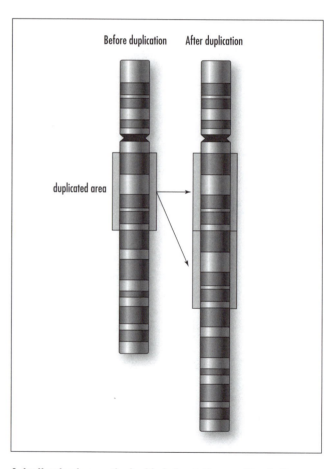

A duplication is a particular kind of mutation resulting in the production of one or more copies of any piece of DNA, including a gene or even an entire chromosome. (Courtesy of Darryl Leja, NHGRI, National Institutes of Health)

as either large (e.g., Komodo dragon weighs up to 365 pounds) or small (e.g., Island fox in the Channel Islands). Island animal populations tend to acquire different sizes than their mainland counterparts. Dwarfism may be due to limited food supply, but the reasons for both gigantism and dwarfism are not known fully.

See also GIGANTISM.

dynein   A molecular motor, a complex believed to be made of 12 distinct protein parts, that performs basic transportation tasks critical to the cell. Converts chemical energy stored in an ATP molecule into mechanical energy that moves material though the cell along slender filaments called microtubules. One of the most important functions occurs during cell division, when it helps move chromosomes into proper position. It also plays a part in the movement of eukaryotic flagella and cilia.

Molecular motors play a critical role in a host of cell functions, such as membrane trafficking and cell movement during interphase, and for cell asymmetry development. During cell division, they are responsible for establishing the mitotic or meiotic spindle, as well as segregating chromosomes and dividing the cell at cytokinesis. It is the last part of the mitotic cycle during which the two daughter cells separate. Motors either move along actin tracks (members of the myosin superfamily) or microtubules (the dynein and kinesin superfamilies). Based on the Greek *dunamis*, meaning "power."

See also ATP.

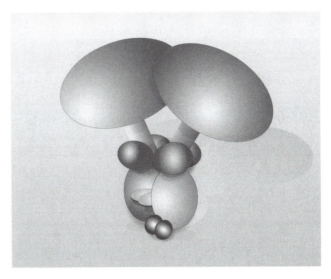

The dynein motor, a cellular complex believed to be composed of 12 distinct protein parts, performs fundamental transportation tasks critical to the cell. Defects in its structure can prove fatal. This machine converts chemical energy stored in an ATP molecule into mechanical energy that moves material though the cell along slender filaments called microtubules. One of the dynein motor's most important functions occurs during cell division, when it helps move chromosomes into proper position. *(Courtesy of U.S. Department of Energy Genomes to Life program: www.DOEGenomesToLife.org)*

dysentery   Sickness that usually involves the abdomen causing cramps, vomiting, and swelling. Caused by a bacterium, *Shigella bacillus*, or a protozoon, *Entamoeba histolytica*.

ecdysone A juvenile steroid hormone that affects arthropods and that belongs to the larger class of ecdysteroids, sterol derivatives that as a whole affect a variety of conditions relating to molting and metamorphosis, including eliciting molting, regulating growth of motor neurons, controlling choriogenesis, stimulating growth and development of imaginal discs, initiating breakdown of larval structures during metamorphosis, and eliciting the deposition of cuticle by the epidermis. In insects, ecdysone primarily elicits and stimulates molting. It acts on specific genes, stimulating the synthesis of proteins involved in these bodily changes, and is produced by prothoracic glands in insects and Y-organs (a gland near the external adductor muscles) in crustaceans. Ecdysone—formerly called alpha ecdysone, and beta ecdysone, or ecdysterone (now called 20-hydroxyecdysone [20-HE])—is believed to be the active form. Ecdysone is not the active molting hormone. Various tissues, including the fat body, convert ecdysone to 20-hydroxyecdysone, the active form of molting hormone.

echinoderm Diversified marine animals (phylum Echinodermata) that include the classes Crinoidea (sea lilies), Asteroidea (starfish), Ophiuroidea (brittle stars or snake stars), Echinoidea (sea urchins and sand dollars), and Holothuroidea (sea cucumbers).

echolocation A form of sensory perception used by animals like bats to orient themselves in flight, detect objects, seek food, and communicate. Bats, for example, send out a series of short, high-pitched sounds, called echoes, that travel, hit an object, and bounce back, giving the bat the ability to judge distance, size, shape, and motion.

eclosion The emergence of an adult insect from the pupa case or, less commonly, the hatching of an egg.

EC nomenclature for enzymes A classification of ENZYMES according to the Enzyme Commission of the International Union of Biochemistry and Molecular Biology. Enzymes are allocated four numbers, the first of which defines the type of reaction catalyzed; the next two define the SUBSTRATES; and the fourth is a catalog number. Categories of enzymes are EC 1, OXIDOREDUCTASES; EC 2, TRANSFERASES; EC 3, HYDROLASES; EC 4, LYASES; EC 5, ISOMERASES; EC 6, LIGASES (synthetases).

ecological efficiency Each transfer of energy from one trophic level to another has an ecological efficiency associated with it. Ecological efficiency refers to the transfer of energy up trophic levels; it is the ratio of secondary productivity to primary productivity consumed.

Ecological efficiency goes down as you move up the trophic levels. Ecological efficiencies generally range from 5 to 20 percent, meaning that this percentage of primary-producer biomass consumed is converted into new consumer biomass.

Ecological efficiency depends on assimilation efficiency (that portion of the consumed energy assimilated) and on net production efficacy (that portion of the consumed energy converted into biomass). Assimilation efficiencies are greater for carnivores (50–90 percent) than for herbivores (20–60 percent).

ecological niche The totality of biotic and abiotic resources an organism interacts with while living in its environment.

ecological succession A transitional change in the biological community, where a group of plant and/or animal species gives way to another set of species over time, in response to a sequence of events such as fire, storms, human activities, or other natural or human-made occurrences. The term also refers to the normal evolution of a community from pioneer stage to climax community when equilibrium between species and its environment occurs.

ecology The study of all life forms and their interactions with their environment.

ecosystem Any natural system—including biotic and abiotic parts—that interacts as a unit to produce a stable functioning system through cyclical exchange of materials.

ectoderm The outer layer of an embryo's three primary germ layers (endoderm, mesoderm, ectoderm) that gives rise to the nervous system and epidermis in vertebrates.

ectoparasite A parasite that feeds from the exterior of its host.

ectotherm A cold-blooded organism that relies on obtaining its heat from certain behavior techniques and from the external environment (sun); e.g., snakes, alligators, lizards, fish, or amphibians.

EDRF *See* ENDOTHELIUM-DERIVED RELAXING FACTOR.

effector cell A cell that performs a specific function in response to a stimulus. A gland or muscle cell that responds to stimuli from the body. Cells with full immune functions capable of participating in the immune response by destroying foreign cells or tissues; effector lymphocytes can mediate the removal of pathogens from the body.

efficacy Describes the relative intensity with which AGONISTs vary in the response they produce, even when they occupy the same number of RECEPTORs and with the same AFFINITY. Efficacy is not synonymous with INTRINSIC ACTIVITY.

Efficacy is the property that enables DRUGs to produce responses. It is convenient to differentiate the properties of drugs into two groups: those that cause them to associate with the receptors (affinity) and those that produce stimulus (efficacy). This term is often used to characterize the level of maximal responses induced by agonists. In fact, not all agonists of a receptor are capable of inducing identical levels of maximal response. Maximal response depends on the efficiency of receptor coupling, i.e., from the cascade of events that, from the binding of the drug to the receptor, leads to the observed biological effect.

EF-hand A common structure to bind Ca^{2+} in CALMODULIN and other Ca^{2+}-binding proteins consisting of a HELIX (E), a loop, and another helix (F).

egg The mature female reproductive cell.

Ehrlich, Paul (1854–1915) *German/Polish Immunologist* Paul Ehrlich was born on March 14, 1854, near Breslau, Germany (now Wroclaw, Poland), to Ismar Ehrlich and his wife Rosa Weigert, whose nephew was the great bacteriologist Karl Weigert.

Ehrlich was educated at the Breslau Gymnasium and then at the Universities of Breslau, Strassburg, Freiburg-im-Breisgau, and Leipzig. He received his doctorate in medicine in 1878 for his dissertation on the theory and practice of staining animal tissues based on the work of aniline dyes discovered by W. H. Perkin in 1853.

In 1878 Ehrlich was appointed assistant professor at the Berlin Medical Clinic, where he continued his work with dyes used for staining tissues, classifying them as being basic, acid, or neutral. His work on staining granules in blood cells laid the foundation for future work on hematology (the study of blood and blood-forming tissues) and in the field of staining of tissues.

In 1882 Ehrlich published his method of staining the tubercle bacillus that Robert Koch had discovered, and it was this technique that later became the precursor for the currently used Gram method of staining bacteria. Ehrlich himself had a bout of tuberculosis. Ehrlich also discovered the blood–brain barrier when he noticed that the dyes injected into an animal brain would not stain.

In 1899 he became director of the newly created Royal Institute of Experimental Therapy in Frankfurt and of the Georg-Speyer Haus, founded by Frau Franziska Speyer for chemotherapy studies, which was built next door to Ehrlich's institute. It is here that he began work on serum antitoxins and chemotherapy and came up with the concept of the "magic bullet," a compound that could be made to selectively target a disease-causing organism, killing only that organism. His research programs were guided by his theory that the germicidal capability of a molecule depended on its structure, especially its side chains, which could bind to the disease-causing organism. After many trials searching hundreds of agents with the help of the nearby Cassella chemical works, which donated samples of new compounds produced in their laboratory, in 1909 he found a cure for syphilis. The agent he identified was arsphenamine, trade name Salvarsan (the 606th substance tested) and later Neosalvaran (the 914th substance tested). Ehrlich became one of the founders of chemotherapy.

Ehrlich received the Tiedemann Prize of the Senckenberg Naturforschende Gesellschaft at Frankfurt/Main in 1887, the Prize of Honor at the XVth International Congress of Medicine at Lisbon in 1906, the Liebig Medal of the German Chemical Society in 1911, and in 1914 the Cameron Prize of Edinburgh. In 1908 he shared, with Ilya Ilyich Mechnikov, the Nobel Prize in recognition for his work on immunity.

The Prussian government elected him privy medical counsel in 1897, to a higher rank of the counsel in 1907, and in 1911 he reached the highest rank possible, real privy counsel with the title of excellency. He died on August 20, 1915, from a stroke.

Eijkman, Christiaan (1858–1930) Dutch *Physician* Christiaan Eijkman was born on August 11, 1858, at Nijkerk in Gelderland (The Netherlands) to Christiaan Eijkman, the headmaster of a local school, and Johanna Alida Pool. He received his education at his father's school in Zaandam. In 1875 he entered the Military Medical School of the University of Amsterdam and received training as a medical officer for the Netherlands Indies Army. From 1879 to 1881 he wrote his thesis "On Polarization of the Nerves," which gained him his doctor's degree, with honors, on July 13, 1883. On a trip to the Indies he caught malaria and returned to Europe in 1885.

Eijkman was director of the Geneeskundig Laboratorium (medical laboratory) in Batavia from 1888 to 1896, and during that time he made a number of important researches in nutritional science. In 1893 he discovered that the cause of beriberi was a deficiency of vitamins and not, as thought by the scientific community, of bacterial origin. He discovered vitamin B, and this discovery led to the whole concept of vitamins. For this discovery he was awarded the Nobel Prize in physiology or medicine for 1929.

He wrote two textbooks for his students at the Java Medical School, one on physiology and the other on organic chemistry.

In 1898 he became a professor of hygiene and forensic medicine at Utrecht, but he also engaged in problems of water supply, housing, school hygiene, and physical education. As a member of the Gezondheidsraad (health council) and the Gezondheids Commissie (health commission), he participated in the struggle against alcoholism and tuberculosis. He was also the founder of the Vereeniging tot Bestrijding van de Tuberculose (Society for the struggle against tuberculosis). Eijkman died in Utrecht on November 5, 1930.

Eijkman's syndrome, a complex of nervous symptoms in animals deprived of vitamin B_1, is named for him.

Einthoven, Willem (1860–1927) Dutch *Physiologist* Willem Einthoven was born on May 21, 1860, in Semarang on the island of Java, Indonesia, to Jacob Einthoven, an army medical officer in the Indies, and Louise M. M. C. de Vogel, daughter of the then-director of finance in the Indies.

Upon the death of his father, Einthoven and his family moved to Holland and settled in Utrecht, where he attended school. In 1878 he entered the University of Utrecht as a medical student. In 1885, after receiving his medical doctorate, he was appointed successor to A. Heynsius, professor of physiology at the University of Leiden, where he stayed until his death.

He conducted a great deal of research on the heart. To measure the electric currents created by the heart, he invented a string galvanometer (called the Einthoven galvanometer) and was able to measure the changes of electrical potential caused by contractions of the heart muscle and to record them by creating the electrocardiograph (EKG), a word he coined. The EKG provides a graphic record of the action of the heart. This work earned him the Nobel Prize in physiology or medicine for 1924. He published many scientific papers in journals of the time. He died on September 29, 1927.

electrochemical gradient The relative concentration of charged ions across a membrane. Ions move across the membrane due to the concentration difference on the two sides of the membrane as well as the difference in electrical charge across the membrane.

electrode potential Electrode potential of an electrode is defined as the electromotive force (emf) of a cell in which the electrode on the left is a standard hydrogen electrode and the electrode on the right is the electrode in question.

See also REDOX POTENTIAL.

electrogenic pump Any large, integral membrane protein (pump) that mediates the movement of a substance (ions or molecules) across the plasma membrane against its energy gradient (active transport). The pump, which can be ATP-dependent or Na^+-dependent, moves net electrical charges across the membrane.

electromagnetic spectrum The entire spectrum of radiation arranged according to frequency and wavelength that includes visible light, radio waves, microwaves, infrared, ultraviolet light, X rays, and gamma rays. Wavelengths range from less than a nanometer, i.e., X and gamma rays (1 nanometer is about the length of 10 atoms in a row), to more than a kilometer, i.e., radio waves. Wavelength is directly related to the amount of energy the waves carry. The shorter the radiation's wavelength, the higher its energy. Frequencies of the electromagnetic spectrum range from high (gamma rays) to low (AM radio). All electromagnetic radiation travels through space at the speed of light, or 186,000 miles (300,000 km) per second.

electron A negatively charged subatomic particle of an atom or ion.

electron acceptor A substance that receives electrons in an oxidation-reduction reaction.

electronegativity Each kind of atom has a certain attraction for the electrons involved in a chemical bond. This attraction can be listed numerically on a scale of electronegativity. Since the element fluorine has the greatest attraction for electrons in bond-forming, it has the highest value on the scale. Metals usually have a low electronegativity, while nonmetals usually have high electronegativity. When atoms react with one another, the atom with the higher electronegativity value will always pull the electrons away from the atom that has the lower electronegativity value.

electron magnetic resonance (EMR) spectroscopy *See* ELECTRON PARAMAGNETIC RESONANCE SPECTROSCOPY.

electron microscope (EM) A very large tubular microscope that focuses a highly energetic electron beam instead of light through a specimen, resulting in a resolving power thousands of times greater than that of a regular light microscope. A transmission EM (TEM) is used to study the internal structure of thin sections of cells, while a scanning EM (SEM) is used to study the ultrastructure of surfaces. The transmission electron microscope, the first type of electron microscope, was developed in 1931 by Max Knoll and Ernst Ruska in Germany and was patterned exactly on the light transmission microscope except that it used a focused beam of electrons instead of light to see through the specimen. The first scanning electron microscope was built in 1942, but it was not available commercially until 1965.

electron-nuclear double resonance (ENDOR) A magnetic resonance spectroscopic technique for the determination of HYPERFINE interactions between electrons and nuclear spins. There are two principal techniques. In continuous-wave ENDOR, the intensity of an ELECTRON PARAMAGNETIC RESONANCE signal, partially saturated with microwave power, is measured as radio frequency is applied. In pulsed ENDOR the radio frequency is applied as pulses and the EPR signal is detected as a spin-echo. In each case an enhancement of the EPR signal is observed when the radio frequency is in resonance with the coupled nuclei.

electron paramagnetic resonance (EPR) spectroscopy The form of spectroscopy concerned with microwave-induced transitions between magnetic energy levels of electrons having a net spin and orbital angular momentum. The spectrum is normally obtained by magnetic-field scanning. Also known as electron spin-resonance (ESR) spectroscopy or electron magnetic resonance (EMR) spectroscopy. The frequency (ν) of the oscillating magnetic field to induce transitions between the magnetic energy levels of electrons is measured in gigahertz (GHz) or megahertz (MHz). The following band designations are used: L (1.1 GHz), S (3.0 GHz), X (9.5 GHz), K (22.0 GHz), and Q (35.0 GHz). The static magnetic field at which the EPR spectrometer operates is measured by the magnetic flux density (B), and its recommended unit is the tesla (T). In the absence of nuclear hyperfine interactions, B and ν are related by: $h\nu = g\mu_B B$, where h is the Planck constant, μ_B is the Bohr magneton, and the dimensionless scalar g is called the g-factor. When the paramagnetic species exhibits an ANISOTROPY, the spatial dependency of the g-factor is represented by a 3×3 matrix. The interaction energy between the electron spin and a magnetic nucleus is characterized by the hyperfine coupling constant A. When the paramagnetic species has anisotropy, the hyperfine coupling is expressed by a 3×3 matrix called a hyperfine-coupling matrix. Hyperfine interaction usually results in splitting of lines in an EPR spectrum. The nuclear species giving rise to the hyperfine interaction should be explicitly stated, e.g., "the hyperfine splitting due to ^{65}Cu." When additional hyperfine splittings due to other nuclear species are resolved ("superhyperfine"), the nomenclature should include the designation of the nucleus and the isotope number.

electron spin-echo (ESE) spectroscopy A pulsed technique in ELECTRON PARAMAGNETIC RESONANCE, in some ways analogous to pulsed techniques in NMR (NUCLEAR MAGNETIC RESONANCE SPECTROSCOPY). ESE can be used for measurements of electron spin relaxation times, as they are influenced by neighboring paramagnets or molecular motion. It can also be used to measure anisotropic nuclear hyperfine couplings. The effect is known as electron spin-echo envelope modulation (ESEEM). The intensity of the electron spin-echo resulting from the application of two or more microwave pulses is measured as a function of the temporal spacing between the pulses. The echo intensity is modulated as a result of interactions with the nuclear spins. The frequency-domain spectrum corresponds to hyperfine transition frequencies.

electron spin-echo envelope modulation (ESEEM) *See* ELECTRON SPIN-ECHO SPECTROSCOPY.

electron spin-resonance (ESR) spectroscopy *See* ELECTRON PARAMAGNETIC RESONANCE SPECTROSCOPY.

electron-transfer protein A protein, often containing a metal ion, that oxidizes and reduces other molecules by means of electron transfer.

electron-transport chain   A chain of electron acceptors embedded in the inner membrane of the mitochondrion. These acceptors separate hydrogen protons from their electrons. When electrons enter the transport chain, the electrons lose their energy, and some of it is used to pump protons across the inner membrane of the mitochondria, creating an electrochemical gradient across the inner membrane that provides the energy needed for ATP (adenosine triphosphate) synthesis. The function of this chain is to permit the controlled release of free energy to drive the synthesis of ATP.

See also ATP.

element   A substance consisting of atoms that have the same number of protons in their nuclei. Elements are defined by the number of protons they possess.

elephantiasis (lymphedema filariasis)   A visibly grotesque enlargement and hardening of the skin and subcutaneous tissues, usually in the leg or region of the testis, caused by obstruction of the lymphatic system when the lymph node is infested by the nematode worm, *Wuchereria bancrofti*.

elimination   The process achieving the reduction of the concentration of a XENOBIOTIC compound, including its reduction via METABOLISM.

embryo   The resulting organism that grows from a fertilized egg following rapid development and eventually becomes an offspring (in humans, a baby). In plants, it is the undeveloped plant contained within a seed.

embryo sac   A large cell that develops in the ovule of flowering plants (angiosperms). It contains the egg cell, the female gametophyte, where pollination occurs, and when fertilized it becomes an embryo and eventually a seed. It is formed from the growth and division of the megaspore into a multicellular structure with eight haploid nuclei.

emigration   The process of an individual or group leaving a population.

EMR (electron magnetic resonance)   *See* ELECTRON PARAMAGNETIC RESONANCE SPECTROSCOPY.

emulsion   Droplets of a liquid substance dispersed in another immiscible liquid. Milk in salad dressing is an emulsion.

enantiomer   One of a pair of molecular entities that are mirror images of each other and nonsuperimposable.

endangered species   The classification provided to an animal or plant in danger of extinction within the foreseeable future throughout all or a significant portion of its range.

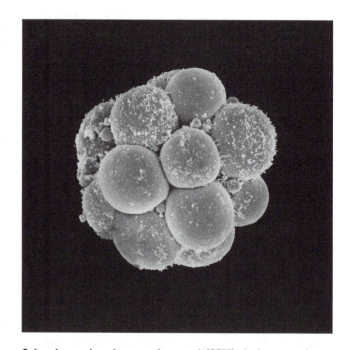

Colored scanning electron micrograph (SEM) of a human embryo at the 16-cell stage, four days after fertilization. Known as a morula, this is a cluster of large rounded cells called blastomeres. The surface of each cell is covered in microvilli. The smaller spherical structures seen will degenerate. This embryo is at the early stage of transformation into a human composed of millions of cells. Here it is in the process of dividing to form a hollow ball of cells (the blastocyst). At this 16-cell stage, the morula has not yet implanted in the uterus (womb). Magnification: ×620 at 6 × 7 cm size; ×2,300 at 8 × 10 in. size; × 960 at 4 × 5 in. size. *(Courtesy © Dr. Yorgos Nikas/Photo Researchers, Inc.)*

The Karner Blue—
New York's Endangered Butterfly

by Robert Dirig

The Karner blue (*Lycaeides melissa samuelis*) is a small, beautifully colored blue butterfly that died out in southern Canada in the early 1990s and was listed as an endangered species by the U.S. government in 1992.

The wings of male Karner blues are deep purplish-blue with a narrow black rim above, but females have wider dark gray borders around blue central areas on all four wings, and a row of bright orange spots on the upper hindwing edges. In both sexes, the wings have elegant white fringes, and are pale gray beneath with arcs of black, white-rimmed dots and orange and satiny blue spots along the hindwing margin. The wings expand about an inch, so the butterfly is quite small.

The Karner blue was first discovered in Canada near London, Ontario, by lepidopterist William Saunders in 1861, and in the United States at Center (now Karner), New York, by Joseph Albert Lintner in 1869. It was originally confused with Scudder's blue (*Lycaeides idas scudderi*), a very similar and closely related but more northern butterfly. Vladimir Nabokov, the world-famous author best known for his controversial novel *Lolita,* also studied butterflies and described the Karner blue as new to science in 1943.

The Karner blue is restricted to a special kind of dry, sandy habitat where wild blue lupine (*Lupinus perennis*), its one caterpillar food plant, grows. Such areas of extensive sand, commonly called "sand plains," are of postglacial origin and occur along major rivers or around large lakes in northeastern and north-central North America.

The type locality, or the place from which the Karner blue was first scientifically described, is the Karner Pine Bush, a large inland pine barrens between Albany and Schenectady, New York. This habitat is unusual in being formed on undulating sand dunes that are stabilized by a low plant cover and widely scattered pitch pines (*Pinus rigida*). Resplendent clumps of wild blue lupine bloom in open areas between shrubby oaks (*Quercus ilicifolia* and *Q. prinoides*), blueberries (*Vaccinium angustifolium* and *V. pallidum*), prairie grasses (*Schizachyrium scoparium*), and other low herbaceous plants that carpet the dunes. Such landscapes are a type of savannah, maintained by a natural fire cycle that keeps them open. Their parklike vegetation is often very beautiful and quite different from the dense forests that surround the sand plains.

The Karner blue occurs very locally throughout its range, with small clusters of populations living many miles apart. The butterfly was historically known from sites in Maine, New Hampshire, Massachusetts, New York, New Jersey, Pennsylvania, Ohio, Indiana, Illinois, Wisconsin, Iowa, Minnesota, Michigan, and Ontario. Many locality records are very old, and the butterfly has since been extirpated in areas such as Brooklyn, in New York City, and the suburbs of Chicago. More recently, Karner blues have died out in New England and from New York City to Illinois along the southern edge of their range, except for northern Indiana. The butterfly is presently known to persist naturally only in Indiana, Michigan, Minnesota, Wisconsin, and the upper Hudson River Valley in New York, with reintroductions attempted in Ohio and planned for several other areas where it formerly lived.

The Karner blue's annual life cycle proceeds like clockwork: the first hatch of Karner blue adults flies in late May and early June, when the lupines bloom. After mating, females lay tiny greenish-white, turban-shaped eggs on lupine plants at the same season. Within a week, minuscule caterpillars hatch from the eggs and begin to feed on lupine leaflets, leaving translucent holes as their unique feeding sign. They are often attended by ants, which feed on a sweet fluid the caterpillars produce in glands on their abdomens, incidentally reducing predation and parasitism. When fully grown three weeks later, the caterpillars are half an inch long, with black heads, velvety green bodies, a dark stripe along their backs, and light stripes along their sides. They crawl off the plant to find a sheltered place in the litter for their chrysalis, which is smooth, bright green, and held to the substrate by a white silken thread around the middle and by microscopic hooks embedded in a silk pad at the tail end. Over the next week the developing butterfly's wings slowly change from green to white to orange and finally to purplish-blue inside the transparent chrysalis skin. The second brood of adults hatches from mid-July to early August, and after mating, females again lay eggs on or near the lupine plants, which by now have largely withered. These summer eggs do not hatch until the following April, when new lupine leaves are pushing through the sand. The tiny spring caterpillars grow, pupate, and produce a new brood of butterflies in late May, finishing the cycle of two full broods per year. These dates are for New York State. The timing of this annual calendar may shift a week or two later at the northern and western edges of the butterfly's range, but the sequence remains the same.

Because this butterfly naturally occurred along or near waterways where major human settlements have grown, the Karner blue has been frequently subjected to urbaniza-

(continues)

The Karner Blue—
New York's Endangered Butterfly
(continued)

tion stresses ever since Europeans colonized North America. Human disturbance and degradation of its habitat are primarily responsible for this butterfly's endangered status throughout its range. A more subtle effect has been disruption of the natural fire cycles that keep its habitats open and sunny. Without fire, the barrens grow up into forests, shading out open areas the butterflies need for mating, feeding, and egg laying, and reducing the lupine plants that are necessary for its caterpillars to survive. Global warming may also be taking its toll: relatively mild winters in the Northeast since the early 1970s have reduced or eliminated the annual snowpack that shelters the overwintering eggs from December to March, forcing them to hatch too early, dessicate on sunbaked spring sand, or lie exposed to predators and parasitoids during this defenseless life stage.

Intensive scientific studies of the Karner blue have been conducted since 1973, when people first realized that it was declining in New York. Private conservation efforts began soon after, starting with the Pine Bush Historic Preservation Project (headed by Don Rittner), the Karner Blue Project (conducted by Robert Dirig and John F. Cryan), and the Xerces Society (led by Robert Michael Pyle and Jo Brewer). Spider Barbour of the rock group Chrysalis composed the song "Shepherd's Purse" in the 1970s to highlight the butterfly's plight. After the Karner blue was classified as threatened or endangered in various states, governmental funding became available, and many ecological studies were conducted by professional scientists. Today the Karner blue is extremely well known biologically (for example, we know the elemental composition of its eggshell), and the butterfly has also been the subject of several studies of preservation strategies. Habitat conservation efforts have continued at Karner (Albany Pine Bush), where approximately 2,750 acres of the type locality have been preserved to date. Managing this large preserve is challenging, as its location between urban centers discourages the fires that are needed to maintain the open vegetation Karner blues require. Additional preserves have been set aside for the butterfly and its habitat in many places throughout its range.

The Xerces Society was founded in 1971 to focus on insect conservation, and initially emphasized imperiled North American butterflies. This group's scope has broadened over the past three decades to include all terrestrial and marine invertebrates, and has had an important, if subtle, impact on North American conservation efforts. It is named for the Xerces blue (*Glaucopsyche xerces*), a California butterfly similar to the Karner blue that became extinct in the 1940s. This vanished insect lived on coastal sand dunes, where its caterpillars fed on a small legume in San Francisco, before its habitat was ruined. Other butterflies that the Xerces Society has championed include the Atala hairstreak (*Eumaeus atala*) and Schaus' swallowtail (*Papilio aristodemus ponceanus*) in southern Florida; and Smith's blue (*Euphilotes enoptes smithi*), El Segundo blue (*Euphilotes battoides allyni*), mission blue (*Icaricia icarioides missionensis*), and the San Bruno elfin (*Callophrys mossii bayensis*) in California. This organization has also expended much effort in trying to help protect the migratory monarch's (*Danaus plexippus*) spectacular overwintering sites in Mexico.

Among these, the Karner blue is an enduring example of ongoing human commitment to preserve an endangered insect in an increasingly crowded world.

—**Robert Dirig** is assistant curator/curator of lichens at Bailey Hortorium Herbarium, Cornell University, in Ithaca, New York.

Endangered Species Act of 1973 (amended)   Federal legislation in the United States intended to provide a means whereby the ecosystems upon which endangered and threatened species depend may be conserved, and to provide programs for the conservation of those species in the hope of preventing extinction of native plants and animals.

endemic species   A species native and confined to a certain region; a species having comparatively restricted distribution.

endergonic reaction   A chemical reaction that consumes energy rather than releases energy. Endergonic reactions are not spontaneous because they do not release energy.

Enders, John Franklin (1897–1985) American *Virologist*   John Franklin Enders was born on February 10, 1897, at West Hartford, Connecticut, to John Ostrom Enders, a banker in Hartford, and Harriet Goulden Enders (née Whitmore). He was educated at the Noah Webster School at Hartford and St. Paul's School in

Concord, New Hampshire. In 1915 he went to Yale University, left to become an air force pilot in 1918, and returned to get his B.A. in 1920. He received a Ph.D. at Harvard in 1930 for a thesis that presented evidence that bacterial anaphylaxis and hypersensitivity of the tuberculin type are distinct phenomena, and he stayed at Harvard until 1946 as a teacher.

In 1938 Enders began the study of some of the mammalian viruses and undertook, in 1941, in collaboration with others, a study of the virus of mumps. This work provided serological tests for the diagnosis of this disease and a skin test for susceptibility to it. It also demonstrated the immunizing effect of inactivated mumps virus and the possibility of attenuating the virulence of this virus by passing it through chick embryos. It showed that mumps often occurs in a form that is not apparent but that nevertheless confers a resistance that is as effective as that conferred by the visible disease.

In 1946 Enders established a laboratory for research in infectious diseases at the Children's Medical Center at Boston. The understanding of viruses at the time was scant, and development of an antipolio vaccine depended on gaining the ability to grow sufficient quantities of the polio virus under laboratory control. The stumbling block was that poliovirus cultures could be kept alive for a useful length of time only in nerve tissue, and that was hard to obtain and maintain.

Enders, along with T. H. WELLER and F. C. ROBBINS, found that viruses could be grown on tissues treated with penicillin to retard bacterial growth, and they were also successful in growing mumps and polio viruses as well. The ability to grow and study polio led to the development of a vaccine later by Salk and Sabin. The research opened the way to other vaccines against highly contagious childhood diseases such as measles, German measles (rubella), and mumps. Enders, Robbins, and Weller shared the 1954 Nobel Prize in medicine for this pioneering work.

Enders was a member of many organizations. Considered one of the most important contributors of the 20th century, Enders also provided insight to links between viruses and cancer, and the pattern and process of tumor growth. He died on September 8, 1985.

endocrine gland   A ductless organ that produces and secretes hormones into the bloodstream.

See also GLANDS.

endocrine system   A collection of glands that work interdependently and produce hormones that regulate the body's growth, metabolism, and sexual development and function. The endocrine system consists of: two adrenal glands, located on the top of each kidney; the pancreas, found in the abdominal cavity behind the stomach; the parathyroid and thyroid, located at the base of the neck; the pituitary, located at the base of the brain; and the ovaries and testes, the female and male sex glands.

Each of the endocrine glands produces hormones that are targeted to a particular area of the body and are released into the bloodstream and serve to regulate the activity of various organs, tissues, and body functions.

endocytosis   A process by which liquids or solid particles are taken up by a cell through invagination of the plasma membrane. The plasma membrane creates a "well" in which the substances settle, become surrounded, and are then pinched off into a vesicle that can be transported through the cell.

See also PHAGOCYTOSIS; PINOCYTOSIS.

endoderm   One of three primary germ layers in embryonic development (along with mesoderm and ectoderm). The endoderm is the inner layer of cells and gives rise to organs and tissues associated with digestion and respiration.

endodermis   A parenchyma tissue that regulates the transport of materials into the vascular bundles of most roots, stems, and leaves. It surrounds the vascular cylinder; is especially prominent in roots; and has suberized Casparian strips, a band of suberin (waxy substance) within the anticlinal walls. It is the innermost layer of the cortex in plant roots.

endogenous   Originating internally. In the description of metal ion COORDINATION in metalloproteins, endogenous refers to internal, or protein-derived, LIGANDs.

endomembrane system   The collection or network of membranous organelles, such as the endoplasmic reticulum and the Golgi apparatus, that are inside a eukaryotic cell; divides the cytoplasm into compartments with various functions, with the compartments related via direct physical contact or by the use of membranous vesicles.

endometrium The lining (mucous membrane) of the uterus and cervix. The endometrium becomes thicker as the menstrual cycle advances in preparation for a fertilized egg. If fertilization does not occur, the endometrium is shed with the menstrual flow. It consists of the stratum functionale, a thick part of endometrium that is lost during menstruation, and the stratum basal, a layer retained during menstruation that serves as a stem source for regeneration of the upper stratum functionale.

A cancerous growth in the endometrium is called endometrial cancer, and an overgrowth in the endometrium, called endometrial hyperplasia, can cause abnormal menstrual bleeding and become precancerous.

endoparasite Any parasitic organism that lives and feeds from inside its host.
See also ECTOPARASITE.

endoplasmic reticulum (**ER**) An extensive convoluted membranous network in the cytoplasm of eukaryotic cells containing two types. The first is a rough endoplasmic reticulum, so called because it appears to be rough due to its surface being covered with ribosomes, that functions to help cells process proteins in sacs called cisternae. The second type of ER is smooth ER and helps cells to process fats. ER generally contains enzymes to break down both proteins and fats.
See also CELL.

ENDOR *See* ELECTRON-NUCLEAR DOUBLE RESONANCE.

endorphin (**endogenous morphine or opiod**) A class of endogenous (made in the body) hormones produced in the brain and anterior pituitary that are chemically similar to opiate drugs (such as morphine) and are released to cope with acute stress and to deal with pain.

endoskeleton An internal skeleton.
See also SKELETON.

endosperm A nutrient, food-storage tissue, formed from double fertilization (sperm cell fuses to two polar nuclei) in the seeds of angiosperms, which nourishes the developing embryo.

endospore A thick-coated, environmentally resistant protective seedlike cell produced within a bacterial cell that is exposed to harsh conditions. In mycology, it is the term for spores formed on the inside of a sporangium; a spore produced within a spherule.

endosymbiotic theory A theory on the evolution of eukaryotic cells. Originally mitochondria and chloroplasts were free-living self-replicating cells that developed a symbiotic relationship with prokaryote cells and eventually lost their independence.

endothelium The simple thin layer of endothelial cells that lines blood and lymph vessels. It plays a number of roles, including acting as a selective barrier for molecules and cells between the blood and surrounding tissues, and secreting and modifying several veinous signaling molecules. The endothelium also helps to make up the blood-brain barrier between the central nervous system and the rest of the body; summons and captures white blood cells (leukocytes) to the site of infections; regulates coagulation of the blood at trauma sites; controls contraction and relaxation of veins; and regulates the growth of the veinous muscular cells, among others.

It is also the term used for the innermost layer of the eye's cornea, one cell layer thick (5–10 microns or 0.005–0.01 millimeters), that provides hydration balance to maintain the cornea's transparency.

endothelium-derived relaxing factor (EDRF) The factor originally described as EDRF is NO·, produced by a specific P-450-type of ENZYME from arginine upon response of a cell to a biological signal (molecule). Different types of cells respond differently to the presence of NO·.

See also CYTOCHROME P-450.

endotherm A warm-blooded animal, one is which the internal temperature does not fluctuate with temperature of environment, but is maintained by a constant internal temperature regulated by metabolic processes. Examples include birds and mammals.

endothermic The state of being warm-blooded or producing heat internally. In chemistry, it is a reaction where heat enters into a system, with the energy absorbed by a reactant.

endotoxin A large toxic molecule consisting of polysaccharide, lipid A, and other components found in the outer cell wall of specific gram-negative bacteria. Also called pyrogen or lipopolysaccharide.

energy Classically defined as the capacity for doing work, energy can occur in many forms such as heat (thermal), light, movement (mechanical), electrical, chemical, sound, or radiation. The first law of thermodynamics is often called the Law of Conservation of Energy and states that energy cannot be created or destroyed but only transformed from one form into another.

enhancer A regulatory element of a gene. A site on DNA that increases transcription of a region even if it is distant from the transcribed region. One gene can have many enhancers.

entatic state A state of an atom or group that, due to its binding in a protein, has its geometric or electronic condition adapted for function. Derived from the Greek *entasis,* meaning tension.

enterobactin A SIDEROPHORE found in enteric bacteria such as *Escherichia coli;* sometimes called enterochelin.

enterochelin *See* ENTEROBACTIN.

entomology The scientific study of the world of insects; a branch of zoology.

entomophilous Refers to a flower pollinated by insects.

entropy The amount of energy in a closed system that is not available for doing work; disorder and randomness in a system. The higher the entropy, the less energy available for work. The Second Law of Thermodynamics states that the entropy of the universe will always increase.

environment The total living and nonliving conditions of an organism's internal and external surroundings that affect an organism's complete life span.

environmental grain Describes an organism's own perception of its environment and how it will react to it; a scale based on the use of space in relation to the size of an organism. Grains can be coarse (large patches) or fine (small patches).

enzootic Affecting animals living in a specific area or limited region. Slime-blotch disease caused by *Brooklynella hostilis* and its associates caused a Caribbean-wide mass fish mortality in 1980 and similar ones in south Florida and Bermuda. In 1990, enzootic pneumonia, caused by *Mycoplasma hyopneumoniae* was evident in 80 percent of Iowa farms, and on these farms, 32 percent of the pigs were actively infected.

enzyme A macromolecule that functions as a BIO-CATALYST by increasing the reaction rate, frequently containing or requiring one or more metal ions. In general, an enzyme catalyzes only one reaction type (reaction specificity) and operates on only a narrow range of SUBSTRATES (substrate specificity). Substrate molecules are attacked at the same site (regiospecificity), and only one, or preferentially one of the ENANTIOMERs of chiral substrate or of RACEMIC mixtures, is attacked (enantiospecificity).

See also CHIRALITY; COENZYME.

enzyme induction   The process whereby an (inducible) ENZYME is synthesized in response to a specific inducer molecule. The inducer molecule (often a substrate that needs the catalytic activity of the inducible enzyme for its METABOLISM) combines with a repressor and thereby prevents the blocking of an operator by the repressor leading to the translation of the gene for the enzyme. An inducible enzyme is one whose synthesis does not occur unless a specific chemical (inducer) is present, which is often the substrate of that enzyme.

enzyme repression   The mode by which the synthesis of an ENZYME is prevented by repressor molecules.

In many cases, the end product of a synthesis chain (e.g., an amino acid) acts as a feedback corepressor by combining with an intracellular aporepressor protein, so that this complex is able to block the function of an operator. As a result, the whole operation is prevented from being transcribed into mRNA, and the expression of all enzymes necessary for the synthesis of the end-product enzyme is abolished.

Eocene   Part of the Tertiary period during the Cenozoic era, lasting from about 54.8 to 33.7 million years ago. Most of the orders of truly warm-blooded mammals were present by the early Eocene.

See also GEOLOGICAL TIME.

eosinophil   One of the five different types of white blood cell (WBC) belonging to the subgroup of WBCs called polymorphonuclear leukocytes. Characterized by large red (i.e., eosinophilic) cytoplasmic granules.

Eosinophil function is incompletely understood. They are prominent at sites of allergic reactions and with parasitic larvae infections (helminths). Eosinophil secretory products inactivate many of the chemical mediators of inflammation and destroy cancer cells. This phenomenon is most obvious with mast-cell-derived mediators. Mast cells produce a chemotactic factor for eosinophils.

Eosinophils are produced in the bone marrow, then migrate to tissues throughout the body. When a foreign substance enters the body, lymphocytes and neutrophils release certain substances to attract eosinophils, which release toxic substances to kill the invader.

See also EOSINOPHILIA.

eosinophil chemotactic factor of anaphylaxis (ECF-A)   A substrate released from mast cells and basophils during anaphylaxis, which attracts eosinophils. A tetrapeptide mediator of immediate hypersensitivity.

eosinophilia (blood eosinophilia)   An abnormally high number of eosinophils in the blood. Not a disease in itself but usually a response to a disease. An elevated number of eosinophils usually indicates a response to abnormal cells, parasites, or allergens.

See also EOSINOPHIL.

ephemeral   Transitory, lasting for a brief time.

epidermis   Both plants and animals have epidermis, the "skin." Epidermis serves as a protective layer against invasion of foreign substances both chemical and animal (parasites).

In plants, it protects against desiccation; participates in gas exchange and secretion of metabolic compounds; absorbs water; and is the site of receptors for light and mechanical stimuli.

In mammals, the epidermis is a superficial layer of the skin and is subdivided into five layers or strata—the stratum corneum, the stratum basale, the stratum spinosum, the stratum granulosum, and the stratum lucidum—each with their own functions.

epigenesis   The complete and progressive development and differentiation that starts from the beginning of a fertilized egg or spore through each stage of change, guided by genetics and environment until the final adult stage is completed. A process where a genotype becomes expressed and transformed into a final phenotype.

epiglottis A leaflike cartilaginous flap that closes and covers the glottis (middle part of the larynx) to prevent food and other objects from entering the trachea and lungs while ingesting.

epilepsy A neurological or brain condition in which a person has a tendency to have repeated seizures. Clusters of nerve cells, or neurons, in the brain sometimes signal abnormally. It affects more than 2 million Americans, with over 180,000 new cases each year.

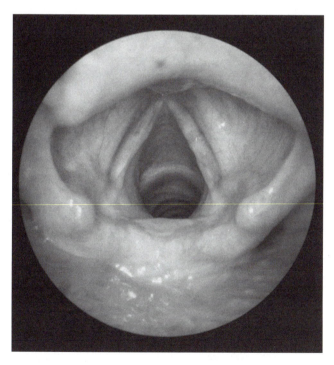

Endoscope view of a healthy larynx (voice box) showing resting vocal cords (v-shaped, center left and right). The vocal cords are responsible for the production of sound. Normally the epiglottis, a leaflike cartilaginous flap, closes and covers the glottis (middle part of the larynx) to prevent food and other objects from entering the trachea and lungs while ingesting. (Courtesy © CNRI/Photo Researchers, Inc.)

epinephrine Another name for adrenaline. A hormone and neurotransmitter secreted by the adrenal gland (adrenal medulla) to react to stress, exercise, low blood glucose. It is also a major component of the fight-or-flight reaction, the reaction that happens in the body when, faced with a sudden unexpected threat or stress situation, both epinephrine and norepinephrine are released.

The drug of choice for the treatment of anaphylaxis. Individuals who are allergic to insect stings and certain foods should always carry a self-injecting syringe of epinephrine.

Epinephrine increases the speed and force of heartbeats and, therefore, the work that can be done by the heart. It dilates the airways to improve breathing and narrows blood vessels in the skin and intestine so that an increased flow of blood reaches the muscles and allows them to cope with the demands of exercise. Usually treatment with this hormone stops an anaphylactic reaction. Epinephrine has been produced synthetically as a drug since 1900.

epiparasite Any organism that extracts nutrients from its host plant by means of intermediates.

epiphyte Any nonparasitic plant, fungus, or microorganism that grows on the surface of another plant for support but provides its own nourishment. Epiphytes can form "mats" that contain a surface of canopy plants with suspended soil and other material. A facultative epiphyte is one that commonly grows epiphytically and terrestrially, but will usually exhibit a preference for one or the other habit in a particular habitat.

episome A plasmid (circular piece of DNA) that can attach to and integrate its DNA in a cell and at other times exist freely and still replicate itself alone, e.g., certain bacterial viruses.

epitasis Interaction between nonallelic genes, with one gene altering the expression of the other gene.

epithelial tissues Closely packed layers of epithelial cells, a membranelike tissue that covers the body and lines body cavities, such as the gastrointestinal tract and the lining of the lung. Epidermal growth factor (EGF) is a hormone that causes epithelial tissues, such as skin and the cells lining the gastrointestinal tract and lungs, to grow and heal.

epitope (**antigenic determinant**) These are particular chemical groups on a molecule that are antigenic, i.e., that elicit a specific immune response.

epizootic A rapid spread of a disease throughout an area affecting an animal group, e.g., rabies (disease affecting raccoons, fox) or epizootic catarrhal enteritis (disease affecting mink). When it occurs in humans, it is called an epidemic.

epoch A period or date of time, shorter than and part of an era, that is used in geological time tables to locate historical events. Usually refers to an event (mountain building, appearance of a species, etc.). Also called a series.

See also ERA; GEOLOGICAL TIME.

EPR *See* ELECTRON PARAMAGNETIC RESONANCE SPECTROSCOPY.

equator The area around the earth with a latitude of 0° that divides the Northern and Southern Hemispheres. It lies equidistant from the poles.

equilibrium constant *See* ACIDITY CONSTANT; STABILITY CONSTANT.

era A period or date of time used in geological time tables to locate historical events. Usually refers to longer periods of time and marks a new or distinctive period.

See also EPOCH; GEOLOGICAL TIME.

Erlanger, Joseph (1874–1965) American *Neuroscientist* Joseph Erlanger was born on January 5, 1874,

in San Francisco, California, to Herman and Sarah Erlanger. He received a B.S. in chemistry at the University of California and later attended Johns Hopkins University to study medicine, receiving an M.D. degree in 1899. He was appointed assistant in the department of physiology at the medical school, after spending a year of hospital training at Johns Hopkins Hospital, until 1906, moving up successively as instructor, associate, and associate professor. He was then appointed the first professor of physiology in the newly established Medical School of the University of Wisconsin. In 1910 he was appointed professor of physiology in the reorganized Medical School of the Washington University, St. Louis, retiring in 1946 as chairman of the school.

In 1922, in collaboration with his student Herbert Gasser, Erlanger adapted the cathode-ray oscillograph for the study of nerve-action potentials. They amplified the electrical responses of single nerve fibers and analyzed them by the use of the oscilloscope. The characteristic wave pattern of an impulse generated in a stimulated nerve fiber could be observed on the screen and the components of the nerve's response studied.

Erlanger and Gasser were given the Nobel Prize in medicine or physiology in 1944 for this work. Erlanger later worked on the metabolism of dogs with shortened intestines, on traumatic shock, and on the mechanism of the production of sound in arteries.

With Gasser he wrote *Electrical Signs of Nervous Activity* (1937). He died on December 5, 1965, in St. Louis.

erythrocyte A concave red blood cell that functions totally within the cardiovascular system. It does not have a nucleus or cytoplasmic organelles and produces little enzyme activity. It contains the red pigment hemoglobin, an oxygen-binding protein, and the cell functions as an efficient vessel for the exchange of respiratory gas. Originates from bone marrow in adult humans.

See also BLOOD.

Escherichia coli A gram-negative, rodlike bacterium that forms acid and gas in the presence of carbohydrates and is commonly found in human intestines and in many other animals. It can be pathogenic and is implicated in a number of food-borne illnesses, with an estimated 10,000 to 20,000 cases of infection occurring in the United States each year. There are hundreds of strains of this one species.

ESE   *See* ELECTRON SPIN-ECHO SPECTROSCOPY.

ESEEM (Electron spin-echo envelope modulation) *See* ELECTRON SPIN-ECHO SPECTROSCOPY.

esophagus   The muscular tube of the digestive tract between the throat (pharynx) and stomach.

ESR (electron spin resonance)   *See* ELECTRON PARA-MAGNETIC RESONANCE SPECTROSCOPY.

essential amino acids   Amino acids that cannot be synthesized in the human body and must be provided from another source (food). These amino acids are histidine, isoleucine, leucine, lysine, methionine, phenylalanine, threonine, tryptophan, and valine.
See also AMINO ACID.

estivation (aestivation)   A state of stagnation or dormancy with slow metabolism (no eating, moving, or growing) during periods of hot temperature and little water supply; a physiological condition for survival.

estrogens   Primary female sex hormones. Estrogens cause growth and development of female sex organs and support the maintenance of sexual characteristics, including growth of underarm and pubic hair and shaping of body contours and skeleton; increase secretions from the cervix and growth of the endometrium (inner lining) of the uterus; and reduce concentrations of bad cholesterol (LDL cholesterol) while increasing good cholesterol (HDL). Estrogen is produced in the ovary by the developing follicle and by the corpus luteum.
See also HORMONE.

estrous cycle (heat cycle)   The period from one ovulation to the next in female mammals; a period of sexual receptiveness preceding ovulation; in humans it occurs every 21 to 23 days. It is characterized by rising and falling levels of estrogens and progesterone in the bloodstream.

estrus   The "heat" cycle in reproduction, the time when the female is sexually receptive.

ethology   The study of natural animal behavior.

ethylene (C_2H_4)   A reactive chemical made from natural gas or crude-oil components (occurs naturally in both petroleum and natural gas) that acts as a plant hormone, the only gaseous hormone. It is used for accelerating fruit ripening (bananas); maturing citrus fruit color; increasing the growth rate of seedlings, vegetables, and fruit trees; leaf abscission; and aging.

etiology (aetiology)   The scientific study or theory of the causes of a certain disease.

euchromatin   Within a nucleus of eukaryotes there are two types of a mixture of nucleic acid and protein called chromatin that make up a chromosome: euchromatin and heterochromatin. During interphase, the genetically active euchromatin is uncoiled and is available for transcription, while heterochromatin is denser and usually not transcribed.

eudismic ratio   The POTENCY of the EUTOMER relative to that of the DISTOMER.

eukaryotes   Organisms whose cells have their GENETIC material packed in a membrane-surrounded, structurally discrete nucleus and who have well-developed cell organelles.

eumetazoa   A subkingdom of the animalia kingdom that includes all animals with the exception of sponges; animals with cells that form tissues and organs, a mouth, and digestive tract. Two branches exist, the radiata and the bilateria. The radiata have radial symmetry, i.e., all longitudinal planes are equal around a central body axis, while the bilateria are animals that have bilateral symmetry, i.e., they have a definite front and rear, and left and right body surfaces.

Eurasia Europe plus Asia considered as one continent. Used in political, economic, and geographical terms.

eusocial A social system of insects, belonging to the order Isoptera (termites) and the order Hymenoptera (ants, bees, and wasps), in which the individuals cooperate in caring for the young after one female produces offspring. There is a reproductive division of labor, and previous generations aid in rearing. This trait also occurs in two species of mammals (mole rats).

eutherian mammals (placental mammals) The female has a placenta that is connected to an embryo within the uterus that supplies it with nutrients and oxygen and acts as an excretory system. Humans are eutherian mammals.

eutomer The enantiomer of a chiral compound that is the more potent for a particular action.
See also DISTOMER.

eutrophication The accelerated loading or dumping of nutrients in a lake by natural or human-induced causes. Natural eutrophication changes the character of a lake very gradually, sometimes taking centuries, but humanmade or cultural eutrophication speeds up the aging of a lake, changing its qualities quickly, often in a matter of years.

eutrophic lake Any lake that has an excessive supply of nutrients, usually nitrates and phosphates. Eutrophic lakes are usually not deep, contain abundant algae or rooted plants, and contain limited oxygen in the bottom layer of water.
See also MESOTROPHIC LAKE; OLIGOTROPHIC LAKE.

evaporative cooling Temperature reduction when water absorbs latent heat from the surrounding air as it evaporates. Similarly, "cooling" of the skin from the evaporation of sweat is evaporative cooling and is a process for the body to lose excess heat.

evolution The long process of change that occurs in populations of organisms. It began with the first life forms on Earth and created the diversity of life forms that exist today and that will exist in the future.

evolutionary species concept A species comprises the totality of individuals that share a common evolutionary history. A species is a lineage evolving separately from others.

evolve Change slowly.

Ewens–Bassett number *See* OXIDATION NUMBER.

EXAFS *See* EXTENDED X-RAY ABSORPTION FINE STRUCTURE.

exaptation The adoption of an attribute that had one function in an ancestral form but now has a new and different form, e.g., swim bladders becoming lungs, or three jaw bones of mammal ancestors becoming the middle bones of the ear. Formerly called preadaptation.

excitatory postsynaptic potential (EPSP) Electrical change in the membrane of a postsynaptic neuron caused by binding of an excitatory neurotransmitter from a presynaptic cell to a postsynaptic receptor. Promotes firing of an action potential in the postsynaptic cell.

excretion   The process of separating and removing waste products of metabolism from the body through the discharge of urine, feces, or expired air.

exegetic reaction   A spontaneous reaction in which energy flows out of the system; a decrease in free energy. A reaction that liberates heat.

exobiology   The study of the origin of life other than on planet Earth.

exocytosis   The process in which a cell discharges large substances to the outside using secretory vesicles, storage organelles, that are then fused with the plasma membrane where they open for export.

exogenous   Originating externally. In the context of metalloprotein LIGANDs, exogenous describes ligands added from an external source, such as CO or O_2.

exon   A section of DNA that carries the coding SEQUENCE for a protein or part of it. Exons are separated by intervening, noncoding sequences (called INTRONs). In EUKARYOTES, most GENEs consist of a number of exons.

exoskeleton   The hard external skeleton made from chitin and connective tissue that attaches it to the underlying parts of a body of animals such as arthropods (insects, spiders, crabs, lobsters). Serves as protection, antidessicant, and sensory interface with the environment.
   See also SKELETON.

exothermic   A reaction that produces heat and absorbs heat from the surroundings.

exotoxin   A toxic substance produced by bacteria and then released outside its cell into its environment.

exponential population growth   Rapid population growth; populations increase at a constant proportion from one generation to the next. For example, the human population is doubling every 40 years. The rate of increase is not limited by environmental factors, only biotic or intrinsic factors. If birth rates exceed death rates, population size will increase exponentially; likewise if death rates exceed birth, population size will decrease exponentially. Also known as J-shaped population growth.

expression   The cellular production of the protein encoded by a particular GENE. The process includes TRANSCRIPTION of DNA, processing of the resulting mRNA product, and its TRANSLATION into an active protein. A recombinant gene inserted into a host cell by means of a vector is said to be expressed if the synthesis of the encoded polypeptide can be demonstrated. For the expression of metalloproteins, usually other gene products will be required.

extended X-ray absorption fine structure (EXAFS)   EXAFS effects arise because of electron scattering by atoms surrounding a particular atom of interest as that special atom absorbs X rays and emits electrons. The atom of interest absorbs photons at a characteristic wavelength, and the emitted electrons, undergoing constructive or destructive interference as they are scattered by the surrounding atoms, modulate the absorption spectrum. The modulation frequency corresponds directly to the distance of the surrounding atoms, while the amplitude is related to the type and number of atoms. EXAFS studies are a probe of the local structure. EXAFS can be applied to systems that have local structure, but not necessarily long-range structure, such as noncrystalline materials. In particular, bond lengths and local symmetry (COORDINATION numbers) can be derived. The X-ray absorption spectrum can also show detailed structure below the absorption edge. This X-ray absorption near-edge structure (XANES) arises from excitation of core electrons to high-level vacant orbitals.

extinct species   A species no longer in existence.
   See also ENDANGERED SPECIES.

extirpated species A species no longer surviving in regions that were once part of its range.

extracellular matrix (ECM) Material produced by animal cells and secreted into the surrounding area, serving as a glue to hold cells together in tissues. It is composed of proteoglycans, polysaccharides, and proteins. Plays a role in cell shape, growth, migration, and differentiation.

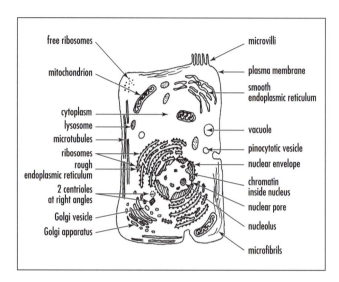

free ribosomes
mitochondrion
cytoplasm
lysosome
microtubules
ribosomes
rough
endoplasmic reticulum
2 centrioles
at right angles
Golgi vesicle
Golgi apparatus

microvilli
plasma membrane
smooth
endoplasmic reticulum
vacuole
pinocytotic vesicle
nuclear envelope
chromatin
inside nucleus
nuclear pore
nucleolus
microfibrils

Material produced by animal cells and secreted into the surrounding area serves as a glue to hold cells together in tissues.

extraembryonic membranes The YOLK SAC, AMNION, CHORION, AND ALLANTOIS, four membranes that support and nourish the developing embryo in reptiles, birds, and mammals. The allantois performs gas exchange and is a repository for the embryo's nitrogenous waste. It is involved in the development of the urinary bladder. The chorion is the outermost layer and contributes to the formation of the placenta. The amnion, the innermost layer, forms a fluid-filled sac around the embryo to protect it from jarring. The yolk sac surrounds the yolk and is the site of blood-cell formation and germ-cell formation, which are the predecessors of male and female gametes.

extrinsic asthma Asthma triggered by external agents such as pollen or chemicals. Most cases of extrinsic asthma have an allergic origin and are caused by an IgE-mediated response to an inhaled allergen. This is the type of asthma commonly diagnosed in early life. Many patients with extrinsic asthma respond to immunotherapy.

F-430 A tetrapyrrole structure containing nickel, a component of the ENZYME methyl-coenzyme M reductase, which is involved in the formation of methane in methanogenic bacteria. The highly reduced macrocyclic structure, related to PORPHYRINs and CORRINs, is termed a CORPHIN.

See also METHANOGENS; OXIDOREDUCTASE.

facilitated diffusion A process by which carrier proteins, also called permeases or transporters or ion channels, in the cell membrane transport substances such as glucose, sodium, and chloride ions into or out of cells down a concentration (electrochemical) gradient; does not require the use of metabolic energy.

See also ACTIVE TRANSPORT; VOLTAGE-GATED CHANNELS.

facultative anaerobe A facultative anaerobe is a microorganism that makes ATP by aerobic respiration if oxygen is present but, if absent, switches to fermentation under anaerobic conditions.

facultative organism Any organism that changes a metabolic pathway to another when needed.

facultative saprophyte Any organism that is usually parasitic but can also live as a SAPROPHYTE.

facultative symbiont Any organism that chooses a symbiotic relationship with a host only if the relationship presents itself but is not physiologically required to do so for survival.

Fahrenheit, Daniel Gabriel (1686–1736) German *Instrument Maker, Physicist* Daniel Gabriel Fahrenheit, a German instrument maker and physicist, was

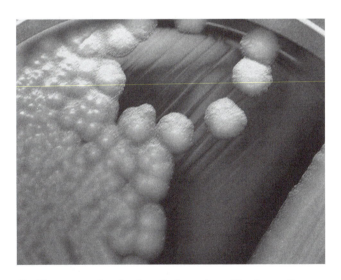

Bacillus cereus showing hemolysis on sheep-blood agar. *B. cereus* is a gram-positive beta hemolytic bacteria that can live in an environment with or without the presence of oxygen (i.e., a facultative anaerobe). *(Centers for Disease Control/Courtesy of Larry Stauffer, Oregon State Public Health Laboratory)*

born in Danzig, Germany (now Gdansk, Poland), in 1686, the oldest of five children. Fahrenheit's major contributions lay in the creation of the first accurate thermometers in 1709 and a temperature scale in 1724 that bears his name today.

When he was 15 years of age, his parents died of mushroom poisoning. The city council placed the four younger Fahrenheit orphans in foster homes and apprenticed Daniel to a merchant who taught him bookkeeping. He was sent to Amsterdam around 1714, where he learned of the Florentine thermometer, invented in Italy 60 years prior in 1654 by the grand duke of Tuscany, Ferdinand II (1610–70), a member of the powerful Medici family. For some unknown reason, it sparked his curiosity and he decided to make thermometers for a living. He abandoned his bookkeeping apprenticeship, whereby Dutch authorities issued warrants for his arrest. While on the run, he spent several years traveling around Europe and meeting scientists, such as Danish astronomer Olaus Roemer. Eventually he returned to Amsterdam in 1717 and remained in the Netherlands for the rest of his life.

What seems so simple today—having a fixed scale and fixed points on a thermometer—was not obvious in Fahrenheit's time, when several makers used different types of scales and liquids for measuring. In 1694 Carlo Renaldini, a member of the Academia del Cimento and professor of philosophy at the University of Pisa, was the first to suggest taking the boiling and freezing points of water as the fixed points. The academy was founded by Prince Leopoldo de Medici and the Grand Duke Ferdinand II in 1657 with the purpose of examining the natural philosophy of Aristotle. The academy was active sporadically over 10 years and concluded its work in 1667 with the publication of the *Saggi di Naturali Esperienze.*

Unfortunately, Florentine thermometers, or any thermometers of the time, were not very accurate; no two thermometers gave the same temperature, since there was no universal acceptance of liquid type or agreement on what to use for a scale. Makers of Florentine thermometers marked the low end of the scale as the coldest day in Florence that year and the high end of the scale as the hottest day. Because temperature fluctuations naturally occur over the years, no two thermometers gave the same temperature. For several years Fahrenheit experimented with this problem, finally devising an accurate alcohol thermometer in

1709 and the first mercury or "quicksilver" thermometer in 1714.

Fahrenheit's first thermometers, from about 1709 to 1715, contained a column of alcohol that directly expanded and contracted, based on a design made by Danish astronomer Olaus Romer in 1708, which Fahrenheit personally reviewed. Romer used alcohol (actually wine) as the liquid, but his thermometer had two fixed reference points. He selected 60 degrees for the temperature of boiling water and 7.5 degrees for melting ice.

Fahrenheit eventually devised a temperature scale for his alcohol thermometers with three points calibrated at 32 degrees for freezing water, 96 degrees for body temperature (based on the thermometer being in a healthy man's mouth or under the armpit), and zero degrees fixed at the freezing point of ice and salt, believed at the time to be the coldest possible temperature. The scale was etched in 12 major points (with zero, four, and 12 as three points) and eight gradations between the major points, giving him a total of 96 points for his scale for body temperature on his thermometer.

Because his thermometers showed such consistency in their measurements, mathematician Christian Wolf at Halle, Prussia, devoted a whole paper in an edition of *Acta Eruditorum,* one of the most important international journals of the time, on two of Fahrenheit's thermometers that were given to him in 1714. From 1682 until it ceased publication in 1731, the Latin *Acta Eruditorum,* published monthly in Leipzig and supported by the duke of Saxony, was one of the most important international journals. The periodical was founded by Otto Mencke, professor of morals and practical philosophy, and mathematician Gottfried Wilhelm Leibnitz. Written in Latin, the journal covered science and social science and was primarily a vehicle for reviewing books. In 1724 Fahrenheit published a paper, "Experimenta circa gradum caloris liquorum nonnullorum ebullientium instituta (Experiments done on the degree of heat of a few boiling liquids), in the Royal Society's publication *Philosophical Transactions* and was admitted to the Royal Society the same year.

Fahrenheit decided to substitute mercury for the alcohol because its rate of expansion was more constant than that of alcohol and could be used over a wider range of temperatures. Fahrenheit, like ISAAC NEWTON before him, realized that it was more accurate to base the thermometer on a substance that changed consistently based on temperature instead of simply on the hottest or coldest day of the year, like the Florentine models. Mercury also had a much wider temperature range than alcohol. The choice of mercury as a benchmark was contrary to the common thought at the time, promoted by Halley as late as 1693, who believed that mercury could not be used for thermometers because of its low coefficient of expansion.

Fahrenheit later adjusted his temperature scale to ignore body temperature as a fixed point, bringing the scale to just the freezing and boiling of water. After his death, scientists recalibrated his thermometer so that the boiling point of water was the highest point, changing it to 212 degrees, as Fahrenheit had earlier indicated in a publication on the boiling points of various liquids. The freezing point became 32 degrees, and body temperature became 98.6 degrees. This is the scale that is presently used in thermometers in the United States and some English-speaking countries, although most scientists use the Celsius scale.

By 1779 there were some 19 different scales being used for thermometers, but it was Fahrenheit, along with astronomer ANDERS CELSIUS and Jean Christin—whose scales were presented in 1742 and 1743—who helped finally set the standards for an accurate thermometer that are still used today. Besides making thermometers, Fahrenheit also was the first to show that the boiling point of liquids varies at different atmospheric pressures, and he suggested this as a principle for the construction of barometers. Among his other contributions were a pumping device for draining the Dutch polders and a hygrometer for measuring atmospheric humidity.

Fahrenheit died on September 16, 1736, in The Hague at the age of 50 years. There is virtually no one in the English-speaking countries today who does not have a thermometer with his initial (F) on it.

See also CELSIUS SCALE.

family The taxonomic category between order and tribe, but if no tribe exists, then it is the category between order and genus. Also a social unit related by marriage, descent, or kinship.

farsightedness A condition in eyesight where distant objects can be seen better than objects that are closer. It is the inability of images to focus properly on the retina

of the eye. The eye is too short or the cornea is too flat, so that the images focus beyond the retina and cause close objects to appear blurry. Also called hyperopia or presbyopia, when the lens of the eye begins to lose elasticity (normal aging process).

fat (general) Any substance made up of lipids or fatty acids that supply calories to the body and can be found in solid or liquid form (e.g., margarine, vegetable oil); three fatty acids linked to a glycerol molecule form fat.

fat (triacylglycerol) Triacylglycerols are storage lipids, comprising three fatty acids attached to a glycerol molecule, found mostly stored in adipose (fat) cells and tissues. They are highly concentrated regions of metabolic energy. Because there are abundant reduced CH groups available in fats for oxidation-required energy production, they are excellent storage containers of energy. Fats can be found in plants, animals, and animal plasma lipoproteins for lipid transport. Formerly known as triglyceride.

fatty acid Fatty acids are the components of two lipid types mostly found in cells in the form of large lipids or small amounts in free form: storage fats and structural phospholipids. They consist of long hydrocarbon chains of varying length (from four to 24 carbon atoms), containing a terminal carbonyl group at one end and may be saturated (has only a single carbon-to-carbon bond) or unsaturated (one or more double or triple carbon-to-carbon bonds). The number and location of double bonds also vary for the different fatty acids. More than 70 different kinds have been found in cells. Saturated fatty acids have higher levels of blood cholesterol, since they have a regulating effect on its synthesis, but unsaturated ones do not have that effect and thus they are more often promoted nutritionally. Some fatty acids are palmitic acid, palmitoleic acid, alpha-linolenic acid, eleostearic acid, linoleic acid, oleic acid, and elaidic acid. Three fatty acids linked to a glycerol molecule form fat.

fauna All wild birds and all wild animals (both aquatic and terrestrial); includes wild mammals, reptiles, amphibians, and aquatic and nonaquatic invertebrate animals, and all such wild animals' eggs, larvae, pupae, or other immature stage and young.

feedback inhibition (end-product inhibition)   A way for the end product of a cell's biosynthetic pathway to stop the activity of the first enzymes in that pathway, thereby controlling the enzymatic activity; it stops the synthesis of the product.

female   Sex classification by gender. The individual in a sexually reproducing species that produces eggs. Female mammals, for example, nourish their young with milk. In humans, females have two X chromosomes.

FeMo-cofactor   An inorganic CLUSTER that is found in the FeMo protein of the molybdenum-NITROGENASE and is essential for the catalytic reduction of N_2 to ammonia. This cluster contains Fe, Mo, and S in a 7:1:9 ratio. The structure of the COFACTOR within the FeMo protein can be described in terms of two cuboidal SUBUNITs, Fe_4S_3 and $MoFe_3S_3$ bridged by three S^{2-} ions and "anchored" to the protein by a histidine bound via an imidazole group to the Mo atom and by a cysteine bound via a deprotonated SH group to an Fe atom of the Fe_4S_3 subunit. The Mo atom at the periphery of the molecule is six-coordinate and, in addition to the three sulfido LIGANDs and the histidine imidazole, is also bound to two oxygen atoms from an (R)-homocitrate molecule.
See also COORDINATION.

Fenton reaction $Fe^{2+} + H_2O_2$ Fe^{3+} + OH· + OH^-. This is the iron-salt-dependent decomposition of dihydrogen peroxide, generating the highly reactive hydroxyl radical, possibly via an oxoiron(IV) intermediate. Addition of a reducing agent such as ascorbate leads to formation of an acyclic compound, which increases the damage to biological molecules.
See also HABER-WEISS REACTION.

fermentation   The anaerobic decomposition of complex organic substances by microorganisms such as bacteria, molds, or yeast, called ferments, on a fermentation substrate that produce simpler substances or some other desired effect, such as the yielding of ethanol and carbon dioxide from yeast for commercial purposes, the production of ATP and energy production, and the development of antibiotics and enzymes. Fermentation is used by microflora of the large intestine to break down indigestible carbohydrates.

Large fermentors are used to culture microorganisms for the production of some commercially valuable products such as bread, beer, wine, and other beverages.

ferredoxin   A protein containing more than one iron and ACID-LABILE SULFIDE that displays electron-transfer activity but not classical ENZYME function.
See also HIPIP.

ferriheme   An iron(III) PORPHYRIN COORDINATION complex.

ferritin   An iron storage protein consisting of a shell of 24 protein SUBUNITs encapsulating up to 4,500 iron atoms in the form of a hydrated iron(III) oxide.

ferrochelatase   An ENZYME that catalyzes the insertion of iron into PROTOPORPHYRIN IX to form HEME. The mammalian enzyme contains an IRON-SULFUR CLUSTER.

ferroheme   An iron(II) PORPHYRIN COORDINATION complex.

ferromagnetic   If there is coupling between the individual magnetic dipole moments of a PARAMAGNETIC sample, spontaneous ordering of the moments will occur at low temperatures. If this ordering results in an electronic ground state in which the moments are aligned in the same direction (parallel), the substance is said to be "ferromagnetic." If the ordering results in an electronic ground state in which the moments are aligned in opposite directions, the substance is said to be "antiferromagnetic."

fertilization The combining of two gametes from different sexes to form a zygote, e.g., the penetration of sperm into the egg and the resulting combining of genetic material from both that develops into an embryo. The process involves karyogamy, the fusion of nuclei of both gametes, and plasmogamy, the fusion of cytoplasm. Each gamete contains a haploid set of chromosomes, with the resulting nucleus containing a diploid set of chromosomes. Fertilization can also be self-induced by the fusion of male and female gametes from the same euploid (nucleus of a cell contains exact multiples of the haploid number of chromosomes) organism; cross fertilized by the fusion of male and female gametes from different euploid individuals; or double fertilized, in which two separate sperm cells unite with two cells in the embryo sac to form the zygote and endosperm, such as in angiosperms.

In agriculture, fertilization means the application of nutrients, a fertilizer, to soil in order to promote growth and development of domestic or crop plants.

[2Fe-2S] Designation of a two-iron, two-labile-sulfur CLUSTER in a protein, comprising two sulfido-bridged iron atoms. The oxidation levels of the clusters are indicated by adding the charges on the iron and sulfide atoms, i.e., $[2Fe-2S]^{2+}$; $[2Fe-2S]^{+}$. The alternative designation, which conforms to inorganic chemical convention, is to include the charges on the LIGANDs; this is more appropriate where the ligands are other than the usual cysteine sulfurs, such as in the RIESKE IRON–SULFUR PROTEINS.

See also FERREDOXIN.

[4Fe-4S] Designation of a four-iron, four-labile-sulfur CLUSTER in a protein. (*See* [2FE-2S].) Possible oxidation levels of the clusters are $[4Fe-4S]^{3+}$; $[4Fe-4S]^{2+}$; $[4Fe-4S]^{+}$.

See also FERREDOXIN; HIPIP.

fetus An unborn offspring in the postembryonic stage where the major features of the organism can be seen.

F factor A bacterial plasmid, which is a piece of DNA that is able to replicate independently of the chromosome, that allows a prokaryote (cell with no nucleus) to join together with and pass DNA. An episome that can replicate by itself or in integrated form and move from one bacterium to another while conjugating. A circular piece of DNA that can replicate independently of the bacterial chromosome or integrate and replicate as part of the chromosome.

fiber A long-walled plant cell that is often dead at maturity, is lignified, and reinforces the xylem of angiosperms, giving elasticity, flexibility, tensile

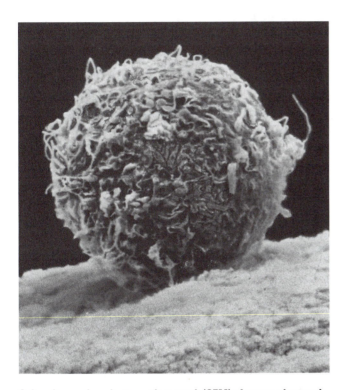

Colored scanning electron micrograph (SEM) of sperm clustered around a human egg (ovum) during fertilization. The rounded egg (at center) is seen on human tissue. Sperm attached to its surface appear as fine hairlike structures; each sperm has a rounded head and a long tail. They are penetrating the thick spongy surface of the zona pellucida of the egg, a surface layer that attracts sperm to the egg and enables the sperm to attach. The human female usually produces a single large egg, and only one of the millions of male sperm may penetrate the egg's wall to fuse with the egg nucleus. Once fertilized, the egg begins its process of growth by cell division. *(Courtesy © K.H. Kjeldsen/Photo Researchers, Inc.)*

strength, and mechanical support to plant structure. Also part of sclerenchyma tissue, which is thickened cell walls of lignin, composed of both sclereids, short cells, and the longer fibers, and lacking a living protoplast when mature.

In human nutrition, fiber is a carbohydrate that resists the action of digestive enzymes and passes through the human digestive system virtually unchanged, without being broken down into nutrients. There are insoluble fibers, found in wholegrain products and vegetables, that help the digestive system by moving stools through the digestive tract by keeping them soft. Soluble fiber slows the digestive process and is water-soluble. Found in beans, fruits, and oat products, it is thought to help lower blood fats and blood glucose (sugar).

Fiber is also a slender, elongated natural or synthetic filament capable of being spun into yarn, e.g., cotton.

Fibiger, Johannes Andreas Grib (1867–1928) Danish *Pathologist* Johannes Andreas Grib Fibiger was born in Silkeborg, Denmark, on April 23, 1867, to C. E. A. Fibiger, a local medical practitioner, and Elfride Muller, a writer.

Fibiger studied under bacteriologists Robert KOCH and Emil von Behring, and from 1891 to 1894 he was assistant to Professor C. J. Salomonsen at the department of bacteriology at the University of Copenhagen. He received his doctorate from the University of Copenhagen in 1895 based on research into the bacteriology of diphtheria.

He was appointed prosector at the university's Institute of Pathological Anatomy (1897–1900), principal of the Laboratory of Clinical Bacteriology of the Army (1890–1905), and in 1905 became the director of the central laboratory of the army and consultant physician to the Army Medical Service.

Fibiger's early research dealt with diphtheria and tuberculosis, and he developed laboratory methods for growing the causing bacteria as well as a serum to protect against the disease. Fibiger achieved the first controlled induction of cancer in laboratory animals, after research in studying tumors in the stomachs of animals, by feeding mice and rats with cockroaches infected with a worm. His work led others to pursue the research on chemical carcinogens and led to the development of modern cancer research.

Fibiger was a founding member and joint editor of the *Acta Pathologica et Microbiologica Scandinavica,* and coeditor of *Ziegler's Beiträge zur pathologischen Anatomie und zur allgemeinen Pathologie.* He received the 1927 Nobel Prize in physiology or medicine for his work on cancer, specifically for his "discovery of the Spiroptera carcinoma." Fibiger died on January 30, 1928, in Copenhagen.

fibril   A small or microscopic thread of cellulose that is part of the cellulose matrix of plant cell walls. The contractile unit of a muscle cell or a bundle of filaments in a striated muscle cell; the thin fibrous structure of a nerve; a long fine hair or fiber; many fibrils bundle together to form a fiber. Makes up the smallest unit of paper fibers. Also a linear feature in the H alpha chromosphere of the Sun, found near strong sunspots and plages or in filament channels. Fibrils parallel strong magnetic fields.

fibrin   An insoluble stringy protein derived from fibrinogen that facilitates blood clotting by forming threads and creating the mesh around the clot. A blood clot is also called a fibrin clot. Coagulation begins usually with an injury to some part of the body. The body forms a clot from a mixture of the blood protein fibrin and platelets. After the bleeding stops, a blood protein dissolves the clot by breaking down the fibrin into tiny fragments.

fibroblast (**fibrocyte**)   A flat, elongated, branched, irregular and motile cell type found in vertebrate connective tissue that produces extracellular collagen and elastin fibers; spindlelike with long cytoplasmic extensions at each end and with oval, vesicular nuclei; most abundant cell type found in the skin. Fibroblasts differentiate into chondroblasts that secrete cartilage matrix, collagenoblasts that proliferate at chronic inflammation sites, and osteoblasts that secrete bone matrix. They form the fibrous tissues in the body, tendons, and aponeuroses, the shiny, broad sheets of connective tissue that bind muscle fibers together to form muscles, as well as supporting and binding tissues.

fight-or-flight reaction The reaction in the body when faced with a sudden and unexpected threat or stress. The reaction is immediate to either run or stay and fight. In humans, a sudden release of the hormones epinephrine and norepinephrine increases blood flow to the muscles and increases blood pressure. The resulting increase in muscle strength and mental ability prepares the body for either reaction that is chosen. In other animals such as the wood thrush (*Hylocichla mustelina*), flight is preferred over fight.

filial generation (**offspring generation**) The successive generations of progeny in a controlled series of crosses, beginning with two specific parents (the P generation), and intercrossing the progeny of each new generation. F1 is the first offspring or filial generation between any two parents, the first generation of descent; F2 is the second (grandchildren); and so on.

filter feeding The filtering of suspended food particles from a water current by using gill rakers or similar organs.

fingerprinting In genetics, the identification of multiple specific alleles on a person's DNA to produce a unique identifier for that person; used in forensics. There are six steps to DNA fingerprinting. First the DNA must be isolated and removed from the cells of the animal or plant. Then special enzymes, called restriction enzymes, are used to cut the DNA at specific places, and the DNA are sorted by size. The DNA pieces are then transferred to a nylon sheet, which is then probed. The fingerprint is generated by adding tagged probes to the nylon sheet, and each probe sticks in only one or two specific places, wherever the sequences match. The final DNA fingerprint is created by using several different probes, with the resulting end product looking like a grocery store bar code. DNA fingerprinting is increasingly being used in criminal cases, and people have been freed from prison based on DNA fingerprinting.

See also GALTON, SIR FRANCIS.

Finsen, Niels Ryberg (1860–1904) *Danish Physician* Niels Ryberg Finsen was born on December 15, 1860, in the capital city Thorshavn in the Faroe Islands (Denmark) to Johanne Fröman and Hannes Steingrim Finsen, an Icelandic family that could trace its ancestry back to the 10th century and occupied many of the highest positions in the administration of the Faroe Islands. He received his early education in schools at Thorshavn and then at Herlufsholm in Denmark.

In 1882 Finsen went to Copenhagen to study medicine. After taking his final examination in 1890, he became prosector of anatomy at the University of Copenhagen until 1893. He continued with private tutoring of medical students to make a moderate income.

By 1883 he was diagnosed with Pick's disease, characterized by progressive thickening of the connective tissue of certain membranes in the liver, the heart, and the spleen, with long-term impairment of the functions of these organs. He also developed symptoms of heart trouble and ascites, and became more and more of an invalid until finally during his last years he was confined to a wheelchair. It did not prevent him from making contributions to medicine.

He was instrumental in discovering the effects of light—and in particular ultraviolet light (then called red light)—as phototherapy against diseases such as lupus vulgaris in 1893. In 1895 he made a great breakthrough that established his international reputation by introducing the revolutionary carbon-arc treatment (Finsen's therapy) of lupus. In 1896 he founded the Finsen Medical Light Institute (now the Finsen Institute) in Copenhagen.

He received the Nobel Prize in physiology or medicine on December 10, 1903, for his work in treating diseases with light.

Among the many publications by Finsen, two are especially noteworthy: *Om Lysets Indvirkninger paa Huden* (On the effects of light on the skin) appeared in 1893, and the classical treatise *Om Anvendelse i Medicinen af koncentrerede kemiske Lysstraaler* (The use of concentrated chemical light rays in medicine) was published in 1896. The results of much of his research are contained in the communications published by his institute. Finsen tried to combat his illness in various ways, including keeping a diet poor in salt during his last years. This led to his last publication, a thorough study of *En Ophobning af Salt i Organismen* (An accumulation of salt in the organism) in 1904.

In 1899 he became Knight of the Order of Dannebrog, and a few years later the Silver Cross was added. He was a member or honorary member of numerous societies in Scandinavia, Iceland, Russia, and Germany. He received a Danish gold medal for merit, and in 1904 the Cameron Prize was given to him from the University of Edinburgh.

In 1892 Finsen married Ingeborg Balslev, the daughter of Bishop Balslev at Ribe. They had four children. Finsen died on September 24, 1904.

firefly   Commonly called a lightning bug, they are neither flies nor bugs. They belong to the order Coleoptera, family Lampyridae, which are beetles. These small flying beetles produce their own light, from a chemical called luciferase, from structures in their abdomen. Females of some species, which are wingless, and many larvae also produce light and are called glowworms. Fireflies can be seen in early summer (late May), appearing at dusk. Males and females attract each other with a flashing green light in their abdomens. The wingless females flash from the ground and the males look for them. There are more than 2,000 species of firefly in temperate and tropical environments worldwide.

first law of thermodynamics   Simply put, energy can neither be created nor destroyed, only transformed or transferred from one molecule to another; in effect, the total amount of energy in the universe is constant. Also known as the Law of Conservation of Energy. Thermodynamics is the study of the conversion of energy between heat and other forms, e.g., mechanical.

See also SECOND LAW OF THERMODYNAMICS.

fission (binary fission)   Asexual reproduction or division of a single-celled individual, such as a prokaryote, into two new single-celled individuals of equal size and genetic composition, without mitosis occurring. From the Latin *fissilis*, meaning "easily split."

fixation   The complete prevalence of one gene form (allele), resulting in the total exclusion of the other. Genes that confer a reproductive advantage generally go to fixation.

fixed action pattern (FAP)   A series of innate behavior patterns (a fixed action) in response to a specific stimulus (called a sign stimulus or innate releaser) that continues until the response is completed. FAPs are genetic and not individually learned. For example, a group of spined larvae of the buck moth (*Hemileuca maia*) will all instantly raise their bodies and thrash back and forth when a predator (bird) approaches.

flaccid   Limp, soft condition, e.g., walled cells are flaccid in isotonic surroundings; low turgid pressure; opposite of turgid.

flagellum   A long whiplike structure that is used to propel certain kinds of prokaryote and eukaryote cells. The cells can have an individual flagellum or a few flagella per cell. In prokaryote organisms the flagellum is composed of a protein called flagellin. In the eukaryote organism, it is longer than a CILIUM but has the same construction of nine outer double microtubules and two inner single microtubules.

flanking region   The DNA sequences extending on either side of a specific gene or locus; a region preceding or following the transcribed region. The 3' flanking region (downstream flanking region) is found immediately distal (distant) to the part of a gene that specifies the mRNA and where a variety of regulatory sequences are located. The 3' flanking region often contains sequences that affect the formation of the 3' end of the message and may contain enhancers or other sites to which proteins may bind. The 5' flanking region flanks the position that corresponds to the 5' end of the mRNA and is that part of DNA that precedes the transcription-start site for a particular gene. The 5' flanking region contains the promoter (transcription control region) and other enhancers or protein binding sites.

flatworms   Organisms that comprise the phylum Platyhelminthes. These are normally hermaphroditic organisms that have flat bodies and are bilaterally symmetrical, with defined head and tail, centralized nervous system, and eyespots (light-sensitive cells). They include flukes (trematodes), tapeworms (Cestoda), and free-living flatworms (Turbellaria), and it is estimated that more than 20,000 species exist. Millions of humans are host to these parasites.

flavin A PROSTHETIC GROUP found in flavoproteins and involved in biological oxidation and reduction. Forms the basis of natural yellow pigments like riboflavin.

flea A major group of bloodsucking insects that feed on animals, belonging to the order Siphonaptera. There are about 2,000 known species existing on all continents. Some species are vectors for diseases. They are wingless, flattened-body types with legs with long claws. They can jump from 14 to 16 inches.

While they tend to be associated with pets such as cats and dogs (*Ctenocephalides canis* [dog flea] and *Ctenocephalides felis* [cat flea]), they do include humans as hosts.

Fleming, Sir Alexander (1881–1955) British *Bacteriologist* Sir Alexander Fleming was born on a farm at Lochfield near Darvel in Ayrshire, Scotland, on August 6, 1881. He attended Louden Moor School, Darvel School, and Kilmarnock Academy before moving to London, where he attended the Polytechnic Institute. He spent four years in a shipping office before entering St. Mary's Medical School, London University, where he received an M.B., B.S., with gold medal in 1908, and became a lecturer at St. Mary's until 1914, when he served during World War I, returning to St. Mary's in 1918. He was elected professor of the school in 1928 and emeritus professor of bacteriology, University of London, in 1948.

Fleming was interested in the natural bacterial action of the blood and in antiseptics, and he worked on antibacterial substances that would not be toxic to animal tissues. In 1921 he discovered an important bacteriolytic substance that he named lysozyme. In 1928 he made his most important discovery while working on an influenza virus. He noticed that mold had developed accidentally on a staphylococcus culture plate and that the mold had created a bacteria-free circle around itself. Further experiments found that a mold culture prevented growth of staphylococci, even when diluted 800 times. He named the active substance penicillin.

Sir Alexander wrote numerous papers on bacteriology, immunology, and chemotherapy, including original descriptions of lysozyme and penicillin. He was the recipient of numerous awards and honors in scientific societies worldwide. Fleming shared the Nobel Prize in physiology or medicine in 1945 with Ernst Boris CHAIN and Howard Walter FLOREY, who both (from 1939) carried Fleming's basic discovery forward in the isolation, purification, testing, and quantity production of penicillin. Fleming died on March 11, 1955, and is buried in St. Paul's Cathedral.

flicker fusion rate (critical flicker frequency) The rate beyond which the human eye can no longer recognize discontinuous changes in brightness as a flicker, i.e., the rate is the frequency at which the "flicker" of an image cannot be distinguished as an individual event. The flicker fusion rate (FFR) is 31.25 Hz, or 60 frames per second (bright light) and 24 frames per second (dim light) in humans. When a frame rate is above this number, the eye sees the signal as a consistent image (as on television). A fly has an FFR of 300 frames per second.

flora The term for all plants in a given location or, collectively, on the planet.

Florey, Sir Howard Walter (1898–1968) Australian *Pathologist* Sir Howard Walter Florey was born on September 24, 1898, in Adelaide, South Australia, to Joseph and Bertha Mary Florey. His early education was at St. Peter's Collegiate School, Adelaide, and then Adelaide University, where he graduated M.B., B.S., in 1921. He was awarded a Rhodes Scholarship to Magdalen College, Oxford, leading to the degrees of B.Sc. and M.A. in 1924. He then attended Cambridge as a John Lucas Walker student.

In 1925, he visited the United States on a Rockefeller traveling fellowship for a year, returning in 1926 to a fellowship at Gonville and Caius College, Cambridge, receiving a Ph.D. in 1927. At this time he also held the Freedom Research Fellowship at the London Hospital. In 1927, he was appointed Huddersfield Lecturer in Special Pathology at Cambridge. In 1931 he succeeded to the Joseph Hunter Chair of Pathology at the University of Sheffield.

In 1935 he became professor of pathology and a fellow of Lincoln College, Oxford. He was made an honorary fellow of Gonville and Caius College, Cambridge, in 1946, and an honorary fellow of Magdalen College, Oxford, in 1952. From 1945 to 1957 he was involved in the planning of the John Curtin School of Medical Research in the new Australian National University. In 1962 he was made provost of Queen's College, Oxford.

During World War II he was appointed honorary consultant in pathology to the army, and in 1944 he became Nuffield visiting professor to Australia and New Zealand.

His collaboration with Ernst Boris CHAIN, which began in 1938, led to the systematic investigation of the properties of naturally occurring antibacterial substances. Lysozyme, an antibacterial substance found in saliva and human tears, discovered by Sir Alexander FLEMING, was their original interest, but they moved to substances now known as antibiotics. The work on penicillin was a result.

In 1939 Florey and Chain headed a team of British scientists, financed by a grant from the Rockefeller Foundation, whose efforts led to the successful small-scale manufacture of penicillin. They showed that penicillin could protect against infection but that the concentration of penicillin in the human body—and the length of time of treatment—were important factors for successful treatment. In 1940 a report was issued describing how penicillin had been found to be a chemotherapeutic agent capable of killing sensitive germs in the living body. An effort was made to create sufficient quantities for use in World War II to treat war wounds, and it is estimated to have saved thousands of lives. In 1945 Florey was awarded a Nobel Prize in medicine with Alexander Fleming and Ernst Chain.

Florey was a contributor to and editor of *Antibiotics* (1949). He was also coauthor of a book of lectures on general pathology and has had many papers published on physiology and pathology.

In 1944 he was created a knight bachelor. When a life peerage was conferred on him in 1965, he chose to be styled Lord Florey of Adelaide and Marston. He was provost of Queen's College, Oxford, from 1962 until he died on February 21, 1968.

flower   The reproductive part of a plant. Can be both male and female, producing both pollen and ovule. Flowers are the most commonly used part in identifying a plant and are often showy and colorful.

fluid feeder   An animal that lives by sucking nutrient-rich fluids from another living organism. The two main ways to fluid-feed are piercing and sucking, and cutting and licking. Examples of insects that pierce and suck are platyhelminths, nematodes, annelids, and arthropods, which all have distinct mouth parts that bore into their prey and then suck out the prey's body fluids with a pharynx. Secreted enzymes help aid in the digestion of the fluids. Piercing by insects typically involves the use of a proboscis formed by the maxillae and composed of two canals. The first canal carries in the prey's blood, and the other delivers saliva and anticoagulants.

The cutting-and-licking technique is used by black flies and vampire bats, who cut the prey's body with teeth or sharp mouthparts and then lick the fluids while injecting anticoagulants to prevent clotting.

fluid mosaic model   The model proposes that a plasma membrane surrounds all cells and is composed of about half lipids, mostly phospholipids and cholesterol, and half proteins, with the proteins and phospholipids floating around the membrane in constant motion unless they bind to something. By being fluid, the lipid molecules can move to open up as a channel whereby substances can enter or leave. The protein molecules in the membrane act as carrier, channel, or active transport mechanisms for larger molecules that must enter or leave the cell.

fluke   An organism belonging to the phylum Platyhelminthes, a flatworm of the class Trematoda. Flukes are flat, unsegmented, and parasitic. Two orders exist, the Mongenea (monogenetic flukes) and Digenea (digenetic flukes). Humans become hosts for *Schistosoma mansoni* (human blood fluke) and *Fasciola hepatica* (sheep liver fluke).

Also a single lobe of a whale's tail.

folate coenzymes   A group of heterocyclic compounds that are based on the 4-(2-amino-3,4-dihydro-

4-oxopteridin-6-ylmethylamino) benzoic acid (pteroic acid) and conjugated with one or more L-glutamate units. Folate derivatives are important in DNA synthesis and erythrocyte formation. Folate deficiency leads to ANEMIA.

folivore   An animal whose primary source of food is foliage. For example, the larvae of the buck moth (*Hemileuca maia* [Drury]) eats only the leaves of oak, favoring scrub, live, blackjack, and post oaks; the Karner blue butterfly (*Lycaeides melissa samuelis*) larvae feed only on the leaves of wild blue lupine (*Lupinus perennis*).
See also HERBIVORE.

follicle   Any enclosing cluster or jacket of cells, or a small sac or pore, that protects and nourishes within it a cell or structure. A fluid-filled follicle in the ovary harbors the developing egg cell. When the follicle ruptures (ovulation), an egg is released. A hair follicle envelops the root of hair.

Food and Drug Administration (FDA)   A U.S. federal agency responsible for regulating the development, use, and safety of drugs, medical devices, food, cosmetics, and related products.

The U.S. Food and Drug Administration is a scientific, regulatory, and public health agency that oversees items accounting for 25 cents of every dollar spent by consumers. Its jurisdiction encompasses most food products (other than meat and poultry); human and animal drugs; therapeutic agents of biological origin; medical devices; radiation-emitting products for consumer, medical, and occupational use; cosmetics; and animal feed. The agency grew from a single chemist in the U.S. Department of Agriculture in 1862 to a staff of approximately 9,100 employees and a budget of $1.294 billion in 2001, comprising chemists, pharmacologists, physicians, microbiologists, veterinarians, pharmacists, lawyers, and many others. About one-third of the agency's employees are stationed outside of the Washington, D.C., area, staffing over 150 field offices and laboratories, including five regional offices and 20 district offices. Agency scientists evaluate applications for new human drugs and biologics, complex medical devices, food and color additives, infant formulas, and animal drugs. Also, the FDA monitors the manufacture, import, transport, storage, and sale of about $1 trillion worth of products annually at a cost to taxpayers of about $3 per person. Investigators and inspectors visit more than 16,000 facilities a year and arrange with state governments to help increase the number of facilities checked.

food chain   The energy path in a community by way of food from those who produce it to those that feed on them. For example, plants are eaten by herbivores that are eaten by carnivores. Food chains that are interconnected are called food webs.

forensics   The use of social and physical sciences to combat crime, e.g., the science of using DNA for identification. It has been used to identify victims; establish paternity in child-support cases; and prove the presence of a suspect at a crime scene. Forensic science can be used for issues from burglary to environmental protection.

formation constant   *See* STABILITY CONSTANT.

formula   An exact representation of the structure of a molecule, ion, or compound showing the proportion of atoms that compose the material, e.g., H_2O.

Forssmann, Werner Theodor Otto (1904–1979) German *Surgeon* Werner Theodor Otto Forssmann was born in Berlin on August 29, 1904, to Julius Forssmann and Emmy Hindenberg. He was educated at the Askanische Gymnasium (secondary grammar school) in Berlin. In 1922 he went to the University of Berlin to study medicine, passing his state examination in 1929. For his clinical training he attended the University Medical Clinic and in 1929 went to the August Victoria Home at Eberswalde near Berlin.

He developed the first technique for the catheterization of the heart by inserting a cannula into his own antecubital vein, through which he passed a catheter for 65 cm. He then walked into the X-ray department to have a photograph taken of the catheter lying in his right auricle. He abandoned cardiology after being ridiculed for this act. André F. Cournand and Dickinson W. Richards perfected this procedure. He was appointed chief of the surgical clinic of the city hospital at Dresden-Friedrichstadt and at the Robert Koch Hospital, Berlin. During World War II he became a prisoner of war until his release in 1945, when he went into practice with his wife. Beginning in 1950 he practiced as a urological specialist at Bad Kreuznach. In 1958 he was chief of the surgical division of the Evangelical Hospital at Düsseldorf until 1970.

In 1956 he was awarded, together with André COURNAND and Dickinson W. RICHARDS, the Nobel Prize in physiology or medicine for their work in development of cardiac catheterization. He was also appointed honorary professor of surgery and urology at the Johannes Gutenberg University, Mainz. He was awarded many honors and belonged to a number of scientific organizations during his career. He died on June 1, 1979, in Schopfheim, in the Black Forest in West Germany.

fossil Preserved remains or imprints of once-living plants or animals or their tracks, or burrows, or products (e.g., dung).

founder effect When a small population migrates from a larger population, becomes isolated, and forms a new population, the genetic constitution of the new population is that of a few of the pioneers, not the main population source; the genetic drift observed in a population founded by a small nonrepresentative sample of a larger population; it is the difference between the gene pool of a population as a whole and that of a newly isolated population of the same species.
See also GENETIC DRIFT.

fragile X syndrome It is the most common form of genetically inherited mental retardation. Named for its association with a malformed X chromosome tip, the frequency of the syndrome is greater in males than in females, occurring in approximately 1 in 1,000 male births and 1 in 2,500 female births. In 1991 the causative gene FMR-1 (fragile X mental retardation) was discovered. Fragile X is the most common inherit-

ed cause of learning disability and affects boys and girls of all ethnic groups.

fragmentation A mechanism of asexual reproduction in which the parent plant or animal separates into parts that re-form whole organisms.

frameshift mutation A mutation via an addition of a pair or pairs of nucleotides that changes the codon reading frame of mRNA by inserting or deleting nucleotides.

fraternal In offspring, twins that are not identical. Identical twins occur when both fetuses come from the division of a single fertilized egg and have separate placentas. Fraternal twins can be either same or opposite sex.

free energy Energy readily available for producing change in a system.

free radical A molecule that contains at least one unpaired electron; highly reactive chemical that usually exists only for a short time. Formed in the body during oxidation, a normal by-product of metabolism, they can bind with electrons from other molecules and can cause cellular damage by disrupting normal cellular processes, but can be kept in check by antioxidants such as certain enzymes or vitamins (C and E).

freshwater The Earth is mostly water, which covers 74 percent of its surface. Freshwater accounts for only 3 percent of the total water. Freshwater is water that contains less than 1,000 milligrams per liter (mg/L) of dissolved solids. The United States Geological Survey (USGS) states that, generally, more than 500 mg/L of dissolved solids is undesirable for drinking and many industrial uses.

Fritts, Harold Clark (1928–) *American Botanist, Dendrochronologist* Harold Fritts was born on

December 17, 1928, in Rochester, New York, to Edwin C. Fritts, a physicist at Eastman Kodak Company, and Ava Washburn Fritts. As a young boy he was interested in natural history and weather and even had a subscription to daily weather maps. Along with his maps, he constructed a weather vane that read out wind directions in his room. Fritts attended Oberlin College in Oberlin, Ohio, from 1948 to 1951 and received a B.A. in botany. From 1951 to 1956 he attended Ohio State University in Columbus and received an M.S. in botany in 1953 and a Ph.D. in 1956.

Fritts made major contributions in understanding how trees respond to daily climatic factors and how they record that information in ring structure. He developed a method to statistically record a tree's response to changes in climate. Using that information he developed a method to reconstruct climate from past tree rings and to reconstruct spatial arrays of past climate from spatial arrays of tree-ring data. He also developed a biophysical model of tree-ring structure response to daily weather conditions. This work has laid the groundwork for much current dendroclimatic reconstruction work.

Fritts authored nearly 60 pioneering scientific papers on dendrochronology, including the bible of the field, *Tree Rings and Climate* in 1976, one of the most cited books on the subject. In 1965 he was elected a fellow in the American Association for the Advancement of Science. He received a John Simon Guggenheim fellowship in botany in 1968. In 1982 he was given the Award for Outstanding Achievement in Bioclimatology from the American Meteorological Society, and in 1990 he received the Award of Appreciation from the dendrochronological community, in Lund, Sweden.

Fritts pioneered the understanding of the biological relationships and reconstruction of past climate from tree-ring chronologies. He currently is engaged in some scientific writing and is finishing work on the tree ring model.

See also DENDROCHRONOLOGY.

frond The leaf of a fern or palm. Consists of the stipe (petiole or stalk of the fruiting body) and blade, the expanded portion of the frond. Also used to describe the main part of a kelp plant.

frugivore An organism that generally eats fruits, e.g., the fruit bat.
See also HERBIVORE.

fruit A mature or ripened ovary or cluster of ovaries in a flower.

fruiting body The organ in which meiosis occurs and sexual spores are produced in fungi and mycobacteria. They are distinct in size, shape, and coloration for each species.

functional group Organic compounds are thought of as consisting of a relatively unreactive backbone, for example a chain of sp^3 (three p orbitals with the s orbital) hybridized carbon atoms, and one of several functional groups. The functional group is an atom, or a group of atoms, that has similar chemical properties whenever it occurs in different compounds. It defines the characteristic physical and chemical properties of families of organic compounds.

Fungi A kingdom of heterotrophic, single-celled, multinucleated, or multicellular organisms that include yeasts, molds, and mushrooms; organisms that lack chlorophyll, cannot photosynthesize, and get their nutrients directly from other organisms by being parasites or from dead organic matter, acting as saprophytes. Molds, yeasts, mildews, rusts, smuts, and mushrooms are all fungi. Fungi have a true nucleus enclosed in a membrane and chitin in the cell wall. There are about 8,000 fungi known to attack plants. Some fungi are pathogenic to humans and other animals. Some molds, in particular, release toxic chemicals called mycotoxins that can result in poisoning or death.
See also ALGAE.

fur (ferric uptake regulator) The iron uptake regulating protein present in PROKARYOTEs, which binds simultaneously Fe and DNA, thereby preventing the biosynthesis of ENZYMEs for the production of SCAVENGER chelates (SIDEROPHORES).
See also CHELATION.

G

gall (hypertrophies) An abnormal swelling, growth, or tumor found on certain meristematic (growing) plant tissues caused by another organism such as parasites, insects, bacteria, fungi, viruses, injuries, or chemicals. There are hundreds of types of galls, and insects and mites are the most common organisms that cause them. The plant interacts with the attacking organism and provides raw materials to construct the gall via abnormal tissue growth. Many galls provide food and protective housing for various species of insects, and the resulting larvae that hatch are used as food by birds. Galls have been used for dyes, tannin for leather processing, medicines, and even food. There are over 1,400 species of insects that produce galls, and these insects collectively are called cecidozoa. Gallic acid was first isolated from oak leaf galls by the Swedish chemist Karl Scheele in 1786.

Galton, Sir Francis (1822–1911) British *Anthropologist, Explorer* Francis Galton has the distinction of being the half cousin of another prominent scientist of the 19th century, Charles Darwin. Galton is known as the founder of biometry and eugenics.

He was born in 1822, the youngest of seven children, into a wealthy Quaker family in Sparkbrook, near Birmingham, to Samuel Tertius Galton, a banker, and Frances Anne Violetta Darwin, the half sister of the physician and poet Erasmus Darwin, father of Charles Darwin, who would later influence greatly the mind of Francis.

He was homeschooled by his invalid sister Adele until he was five and was reading at an early age, appearing to have close to instant recall. He later attended King Edward's School in Birmingham between 1836 and 1838, and then became an assistant to the major surgeon in the general hospital of that city at age 16. He continued his medical education by attending King's College in London, and by 1840 he was attending Trinity College in Cambridge, although his attention was moving from medicine to mathematics. He never finished his studies due to a nervous breakdown and the stress from taking care of a terminally ill father.

By 1865 Galton had become keenly interested in genetics and heredity and was influenced by his cousin Charles Darwin's *On the Origin of Species* in 1859. In Galton's *Hereditary Genius* (1869) he presented his evidence that talent is an inherited characteristic. In 1872 he took on religion with *Statistical Inquiries into the Efficacy of Prayer.* Dalton created the study of eugenics, the scientific study of racial improvement, and a term he coined, to increase the betterment of humanity through the improvement of inherited characteristics, or as he defined it: "the study of agencies, under social control, that may improve or impair the racial qualities of future generations, either physically or mentally." His thoughts on improving human society became widely admired.

Galton contributed to other disciplines and authored several books and many papers. In fact, between 1852 and 1910, he published some 450 papers and books in the fields of travel and geography, anthropology, psychology, heredity, anthropometry, statistics, and more. Twenty-three publications alone were on the subject of meteorology.

He became interested in the use of fingerprinting for identification and published *Finger Prints*, the first comprehensive book on the nature of fingerprints and their use in solving crime. He verified the uniqueness and permanence of fingerprints, and suggested the first system for classifying them based on grouping the patterns into arches, loops, and whorls.

As late as 1901, close to 80 years old, he delivered a lecture "On the Possible Improvement of the Human Breed Under Existing Conditions of Law and Sentiment," to the Anthropological Institute, and he even returned to Egypt for one more visit. He published his autobiography, *Memories of My life*, in 1908 and was knighted the following year. Galton received a number of honors in addition to the ones already cited. He was a member of the Athenaeum Club (1855). He received honorary degrees from Oxford (1894) and Cambridge (1895) and was an honorary fellow of Trinity College (1902). He was awarded several medals that included the Huxley Medal of the Anthropological Institute (1901), the Darwin-Wallace Medal of the Linnaean Society (1908), and three medals from the Royal Society: the Royal (1886), Darwin (1902), and Copley (1910) medals.

Galton lived with a grandniece in his later years, and a month short of his 89th birthday, in 1911, his heart gave out during an attack of bronchitis at Grayshott House, Haslemere, in Surrey. He is buried in the family vault at Claverdon, near Warwick, Warwickshire. *Galtonia candicans*, a white bell-flowered member of the lily family from South Africa, and commonly known as the summer hyacinth, was named for Galton in 1888.

gametangium   A reproductive organ that produces gametes (reproductive cells); nuclei that fuse and produce sexual spores in algae, fungi, mosses, and ferns.

gamete   A haploid (half the number of chromosomes) sex cell, either male (sperm) or female (egg), that fuses with another sex cell during the process of fertilization.

gametophyte   For plants with alternation of generations, the gametophytic generation has haploid nuclei and generates gametes during mitosis.

See also SPOROPHYTE.

gamma band   Identical to SORET BAND.

ganglion   A knot or cyst of fibrous material in fluid in joints and tendons; also a cluster of nerve cells located outside the central nervous system. In invertebrates, ganglia and nerve bundles make up the central nervous system.

See also NEURON.

gap junction   A site between two cells that allows small molecules or ions to cross through and connect between the two cytoplasms; allows electrical potentials between the two cells.

See also CELL.

gastrin   A hormone (linear peptide) produced and regulated by the pyloric gland area of the stomach that stimulates the secretion of gastric acids from the stomach walls and duodenum after eating. It is synthesized in G cells in the gastric pits located in the antrum region of the stomach. It occurs in the body in several forms. Gastrin is released after the eating of food containing peptides, certain amino acids, calcium, coffee, wine, beer, and others.

Too much secretion of gastrin, or hypergastrinemia, is a cause of a severe disease known as Zollinger-Ellison syndrome, which affects both humans and dogs. It creates gastric and duodenal ulceration from excessive and unregulated secretion of gastric acid, but it is also commonly brought on by the action of gastrin-secreting tumors (gastrinomas), which develop in the pancreas or duodenum. The hormone also stimulates the proliferation of gastrointestinal cells and adenocarcinomas (cancer of glandular linings) of the gastrointestinal tract.

gastropod   The most successful and largest class of mollusks (phylum Mollusca). There are more than

35,000 living species, and about half that number in the fossil record. Most gastropods travel by using a large flattened muscular foot and are univalve (one-piece shell), although a few have no shell. Gastropods have a defined head with a mouth and one or a pair of sensory tentacles. Examples of gastropods are snails and slugs.

gastrovascular cavity A body cavity in certain lower invertebrates such as cnidarians and flatworms that serves for both digestion and circulation. It has a single opening that serves as both mouth and anus. Since lower invertebrates do not have a circulatory system, it also functions to distribute nutrients to cells that line the cavity.

gastrula An animal embryo in an early stage of development, between blastula and embryonic axes, forming the characteristic three cell layers of endoderm, mesoderm, and ectoderm, and which will give rise to all of the major tissue systems of the adult animal.
See also EMBRYO.

gastrulation The rearrangement of the germ layers by the blastoderm during animal development to the new positions in the embryo that will produce the three primary germ layers of endoderm, mesoderm, and ectoderm.

gated ion channel A specific ion channel that opens and closes to allow the cell to alter its membrane potential. An ion channel is a membrane protein that forms an aqueous pore so that charged ions can cross through the membrane. There are several types of ion channels. For example, a ligand-gated ion channel is where gating is controlled by binding of a chemical signal (the ligand) to a specific binding site on the channel protein. Other ion channels are voltage gated and mechanically gated.

gel electrophoresis The analytical laboratory process to separate molecules according to their size. The sample is put on an end of a slab of polymer gel, a lyophilic colloid that has coagulated to a jelly. An electric field is applied through the gel, which separates the molecules; small molecules pass easily and move toward the other end faster than larger ones. Eventually all sizes get sorted, since molecules with similar electric charges and density will migrate together at the same rate. There are several types of gel composition, and various chemicals can be added to help separation.

gene Structurally, a basic unit of hereditary material; an ordered SEQUENCE of NUCLEOTIDE bases that encodes one polypeptide chain (via mRNA). The gene includes, however, regions preceding and following the coding region (leader and trailer) as well as (in EUKARYOTES) intervening sequences (INTRONs) between individual

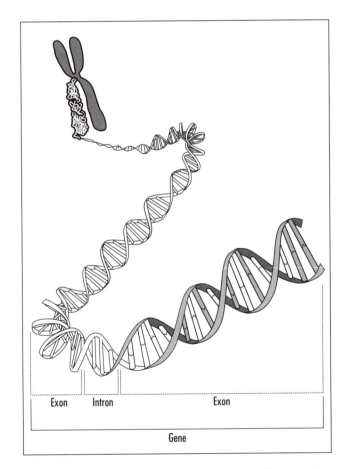

The gene is the functional and physical unit of heredity passed from parent to offspring. Genes are pieces of DNA, and most genes contain the information for making a specific protein. *(Courtesy of Darryl Leja, NHGRI, National Institutes of Health)*

coding segments (EXONs). Functionally, the gene is defined by the *cis-trans* test that determines whether independent MUTATIONs of the same phenotype occur within a single gene or in several genes involved in the same function.

See also CHROMOSOME.

genealogy   The study of one's family and finding and recording the complete history of all ancestors within that family.

gene amplification   The selective increase or production of multiple copies of a specific gene in an organism without a proportional increase in others; specific DNA sequences are replicated disproportionately greater than their representation in the parent molecules. For example, a tumor cell amplifies, or copies, DNA segments naturally as a result of cell signals or sometimes because of environmental events.

gene cloning   A method for making identical copies of a particular DNA; the process of asexually synthesizing multiple copies of a particular DNA sequence, or cells (clones), using a bacteria cell or another organism as a host. The clones are genetically identical to the parent or donor cells. Cloning is used for biomedical research in the form of extracting stem cells in humans with the hope of gaining knowledge on the development of and cure for human diseases. Cloning for producing children and for medical research is controversial, and the ethics of such practices are being debated constantly.

gene expression   A term describing the process of translating information in DNA into an organism's traits. A process by which a gene's code affects the cell in which it is found by synthesizing a protein or RNA product that exerts its effects on the phenotype of the organism. Expressed genes are transcribed into mRNA and translated into protein or transcribed into mRNA but not translated into protein.

gene flow   The exchange of genes between different but usually related populations. Gene flow happens when an individual or group of individuals migrates from one population to another, or vice versa, and interbreeds with its members.

See also GENETIC DRIFT.

gene pool   The total genetic information in all the genes and combinations in a breeding population at a given time.

generation time   The time needed to complete one generation. A generation spans from a given stage in a life cycle to the same stage in the offspring.

gene therapy   A treatment of disease, to correct genetic disorders, by replacing damaged or abnormal genes with new normal ones, or by providing new genetic instructions to help fight disease through the use of recombinant DNA technology. Therapeutic genes are transferred into the patient via a weakened virus, a nonviral vector, or direct delivery of "naked" DNA. Germ line or heritable gene therapy is used for modification of reproductive cells. Somatic cell or noninheritable gene therapy involves those other than reproductive cells.

genetic code   The language of genetics. The instructions in a gene that tell the cell how to make a specific protein. The code defines the series of nucleotides in DNA, read as triplets called codons, that specifies the sequence of amino acids in a protein. The set comprises 64 nucleotide triplets (codons) that specify the 20 amino acids and termination codons (UAA, UAG, UGA).

The code is made up of adenine (A), thymine (T), guanine (G), and cytosine (C), the nucleotide bases of DNA. Each gene's code combines them in various ways to spell out three-letter triplets (codons) that specify which amino acid is needed at each step in making a protein.

See also DEOXYRIBONUCLEIC ACID.

genetic drift   Random changes in allele frequency over time from one generation to another as the genetic makeup of a population drifts at random over time

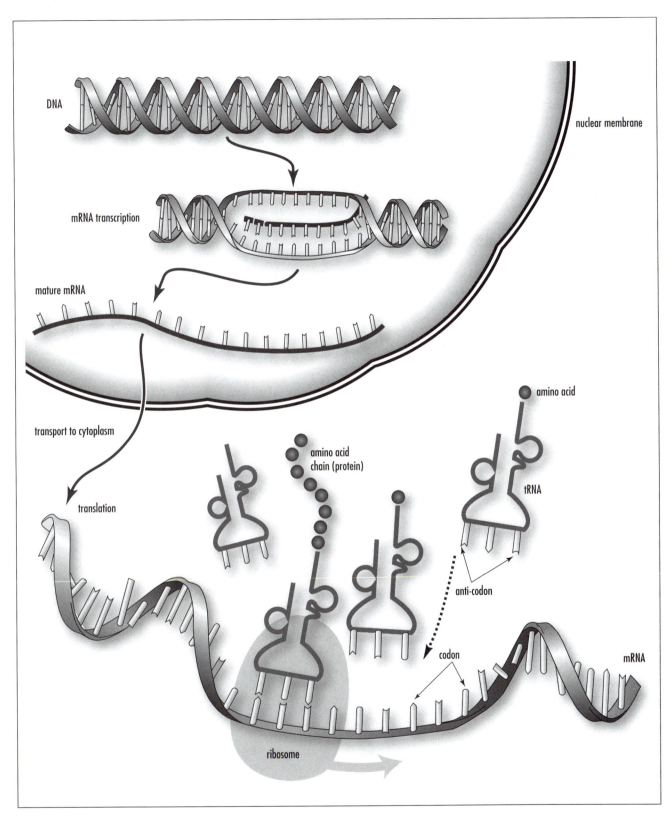

The process by which proteins are made from the instructions encoded in DNA. *(Courtesy of Darryl Leja, NHGRI, National Institutes of Health)*

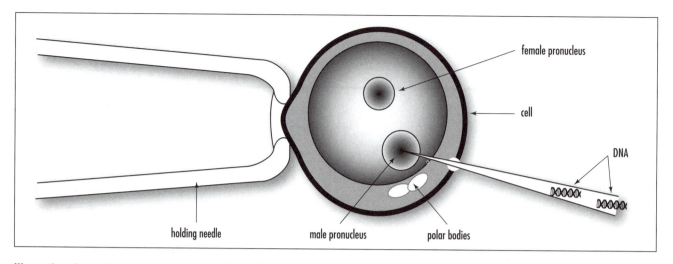

Illustration of gene therapy—an experimentally produced organism in which DNA has been artificially introduced and incorporated into the organism's germ line, usually by injecting the foreign DNA into the nucleus of a fertilized embryo. (Courtesy of Darryl Leja, NHGRI, National Institutes of Health)

instead of being shaped by natural selection in a non-random way. Especially prevalent in small populations, where a particular allele can be eliminated by chance and certain alleles can be favored over time. This can lead to the loss of genetic variability.

genetic engineering   A process that changes the genetic makeup of cells. During the process a gene is isolated, modified, and put back into an individual of the same or different species. The process can be used to introduce or eliminate specific genes. Used in agriculture where plants can be genetically engineered to resist a pest.

genetic map (DNA map)   A chromosome map that shows the order of and distance between genes. Useful for finding inherited diseases by following the inheritance of a DNA marker present in affected individuals. Genetic maps have been used to find the exact chromosomal location of disease genes, including cystic fibrosis, sickle-cell disease, Tay-Sachs disease, fragile X syndrome, and myotonic dystrophy. Genetic maps can be cytogenetic, linkage, or physical. A cytogenetic map produces a visual appearance of a chromosome when stained and examined under a microscope. Important are the visually distinct regions, called light and dark bands, that give each of the chromosomes

a unique appearance. A linkage map is a map of the relative positions of genetic loci on a chromosome, determined on the basis of how often the loci are inherited together. Distance is measured in centimorgans (cM). A physical map of a species shows the specific physical locations of its genes and/or markers on each chromosome.

genetic recombination   The process where offspring have a different genotype from that of the parent due to the recombining of genetic materials, usually caused by crossing over between homologous chromosomes during meiosis, or random orientation of nonhomologous chromosomes pairs, gene conversion, or other means. It occurs during the cell division (meiosis) that occurs during the formation of sperm and egg cells. This shuffling of genetic material increases the potential for genetic diversity.

genetics   The scientific study of heredity and variation.

genome   The complete assemblage of chromosomes and extrachromosomal genes of a cell, organelle, organism, or virus; the complete DNA portion of an organism; the complete set of genes shared by members of any reproductive body such as a population or species.

genomic imprinting   Occurs when DNA receives biochemical marks instructing a cell how and when to express certain genes. Resulting gene expression is usually from one copy of a gene, either from the maternal or paternal gene.

genomic library   A collection of clones made from a set of randomly generated overlapping DNA fragments that represents the entire genome of an organism.

genotype   The complete genetic makeup of an organism, which may not show in physical appearance; the pair of alleles at a particular locus. The percentage of a particular genotype in a population is called the genotype frequency. Also, in taxonomy, the type species of a genus.
   See also PHENOTYPE.

genus   Taxonomic classification of a group of related or similar objects or organisms. A genus has one or more species. Groups of similar genera, plural of genus, make up a family. In the scientific name of an organism, it is the first word followed by a second word to complete the binomial. For example, *Homo* is the genus name for humans, while the entire binomial, *Homo sapiens,* is the species name. Genus and species names are italicized.
   See also TAXON.

geographic range   The total area or range occupied by a species or population.

geological time   The span of time that has passed since the creation of the Earth and its components; a scale used to measure geological events millions of years ago. Measured in chronostratic or relative terms, where subdivisions of the Earth's geology are set in an order based on (a) relative age relationships based on fossil composition and stratigraphic position or (b) chronometric or absolute time where the use of radiometric dating techniques give numerical ages.

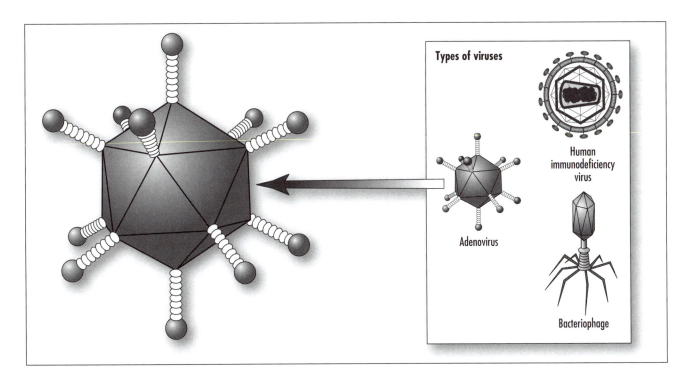

An adenovirus, a grouping of DNA-containing viruses that cause respiratory disease, including one form of the common cold. Adenoviruses can be genetically modified and used in gene therapy to treat cystic fibrosis, cancer, and potentially other diseases. *(Courtesy of Darryl Leja, NHGRI, National Institutes of Health)*

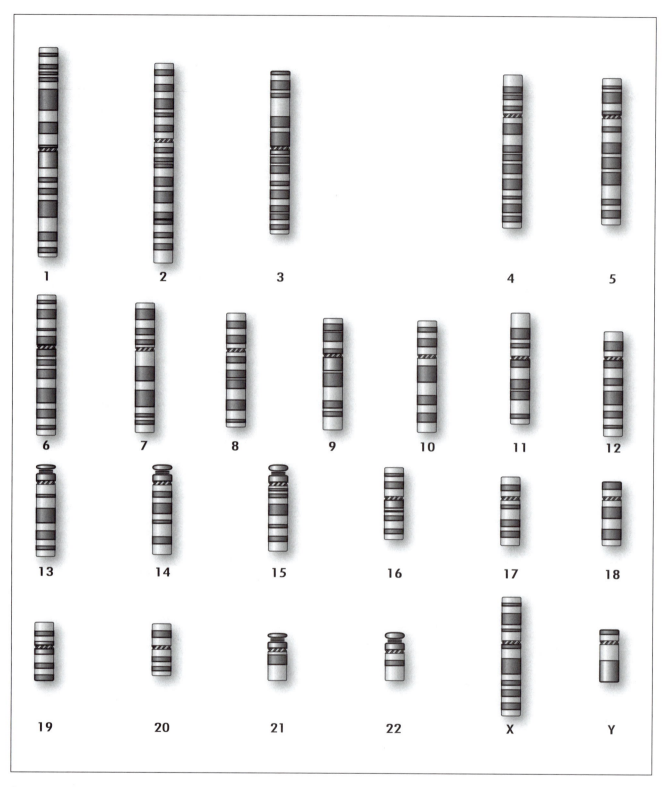

Cytogenetic map showing the visual appearance of a chromosome when stained and examined under a microscope. Particularly important are the visually distinct regions, called light and dark bands, that give each of the chromosomes a unique appearance. This feature allows a person's chromosomes to be studied in a clinical test known as a karyotype, which allows scientists to look for chromosomal alterations. *(Courtesy of Darryl Leja, NHGRI, National Institutes of Health)*

geosynclines A large, down-warped, generally linear basin or trough found in the Earth's crust where large amounts of sediment and volcanic material have accumulated; can be folded into mountains. A miogeocline is a geosyncline in which volcanism is not associated with sedimentation.

geotropism (gravitropism) A plant's response to gravitational effects. A plant's roots grow downward toward the gravitational pull, a characteristic called positive geotropism, while shoots grow upward against gravitational pull, a feature referred to as negative geotropism.

THE GEOLOGICAL TIME SCALE

Era	Period	Epoch	Age (Millions of Years)	First Life Forms	Geology
		Holocene	0.01		
	Quaternary				
		Pleistocene	3	Humans	Ice age
Cenozoic		Pliocene	11	Mastodons	Cascades
	Neogene				
		Miocene	26	Saber-toothed tigers	Alps
	Tertiary	Oligocene	37		
	Paleogene				
		Eocene	54	Whales	
		Paleocene	65	Horses, Alligators	Rockies
	Cretaceous		135		
				Birds	Sierra Nevada
Mesozoic	Jurassic		210	Mammals	Atlantic
				Dinosaurs	
	Triassic		250		
	Permian		280	Reptiles	Appalachians
	Pennsylvanian		310		Ice age
				Trees	
	Carboniferous				
Paleozoic	Mississippian		345	Amphibians	Pangaea
				Insects	
	Devonian		400	Sharks	
	Silurian		435	Land plants	Laursia
	Ordovician		500	Fish	
	Cambrian		570	Sea plants	Gondwana
				Shelled animals	
			700	Invertebrates	
Proterozoic			2500	Metazoans	
			3500	Earliest life	
Archean			4000		Oldest rocks
			4600		Meteorites

A geological time scale is used to measure geological events that occurred millions of years ago.

germination Sprouting of a seed; the first stages in the growth of a seed from a seedling to an adult. After germination the embryonic shoot emerges and grows upward while the embryonic root grows downward. Food for germination is located in the endosperm tissue within the seed and or seed leaves. Sprouting of pollen grains on a stigma and growth of fungus or algal spores are examples of germination.

See also FERTILIZATION.

gestalt Grasping an overall concept without understanding the details; perceiving the whole or patterns over that of the pieces, e.g., the tune of a song. It is based on an object or thing, its context in the environment, and the relationship between it all. *Gestalt* is German for configuration or figure.

g-factor *See* ELECTRON PARAMAGNETIC RESONANCE SPECTROSCOPY.

gibberellins A group of about 50 hormones or growth regulators that primarily stimulate cell division and elongation in plants. Gibberellic acid (GA), the first of this class to be discovered, causes extreme elongation (bolting) of stems. Gibberellins are also involved in flower, fruit, and leaf enhancements, germination, and vernalization (temperature effects).

giemsa stain Stains developed specifically for the phosphate groups of DNA in which the staining of chromosomes produces light and dark bands characteristic for each chromosome, called g-bands. Each homologous chromosome pair has a unique pattern of g-bands, enabling easy recognition of particular chromosomes. Also used for looking for Schuffner's dots, which are small, red-staining granules in red-blood-cell cytoplasm infected with either *Plasmodium vivax* or *P. ovale*.

gigantism Animals that evolve on islands are affected by gigantism or dwarfism, the evolution of body form as either large (e.g., Komodo dragon weighs up to

365 pounds) or small (e.g., Island fox in Channel Islands). Island animal populations tend to acquire different sizes from their mainland counterparts. Gigantism is also the condition of too much growth hormone production in humans where people grow taller than normal.

See also DWARFISM.

gill A respiratory organ in aquatic animals; an outfold of epidermal tissue; the gas-exchange surface of many aquatic animals; a filamentous outgrowth with blood vessels where gas exchanges (oxygen and carbon dioxide) between water and blood. A bony structure supporting the gill filaments is called the gill arch. A flap of bony plates that cover the gills of bony fish is called the gill cover, or operculum.

In fish, gill slits are openings or clefts between the gill arches. Water is taken in by the mouth and then passes through the gill slits and bathes the gills. Gills are also rudimentary grooves in the neck region of embryos of air-breathing vertebrates like humans. A gill is also the part of fungi that contains the basidia, the reproductive cell (meiotangium) that typically produces four spores on the outside.

gizzard Part of an animal gut, e.g., in birds, that is specialized for grinding and mixing food with digestive enzymes. Also called the gastric mill.

glaciation A long period of time characterized by climatic conditions associated with maximum expanse of ice sheets. The process of glaciers spreading over the land. In North America, the most recent glacial event is the Wisconsin glaciation, which began about 80,000 years ago and ended around 10,000 years ago. Most glacial ice today is found in the polar regions, above the Arctic and Antarctic Circles.

gland A group of cells (organ), or a single cell in animals or plants, that is specialized to secrete a specific substance such as a hormone, poison, or other substance. Two types of animal glands are endocrine and exocrine. Endocrine glands place their products directly into the blood stream, while exocrine use a duct or network into the body. Glands important to the human body are the hypothalamus, which secretes hormones to regulate the pituitary gland; the pineal gland, which secretes melatonin, a hormone that deals with daily biological rhythms; the pituitary gland, which secretes hormones that influence other glands and organs that deal with growth and reproduction; the thyroid gland, which regulates metabolism and blood calcium levels; the parathyroid gland, which regulates the use of calcium and phosphorus; the thymus gland, which stimulates the immune system's T cell development; the adrenal gland, which secretes the male hormone, androgens, and aldosterone, which helps maintain the body's salt and potassium balances, and epinephrine (adrenaline) and norepinepherine (noradrenaline); the pancreas gland, which secretes insulin, which controls the use of sugar in the body, and other hormones involved with sugar metabolism; the ovaries, which secrete female hormones such as estrogen, which maintains female traits, and progesterone for pregnancy; the eccrine gland, whose excretory canal emerges directly onto the skin's surface; the exocrine gland, which secretes products that get directly eliminated, at the level of the skin, or through a mucous membrane; and the holocrine gland, whose secretions result from the destruction of the cells forming it.

A plant gland is usually a bump, depression, or appendage on the surface or within that produces a sticky or greasy viscous fluid such as oil or resin. The floral nectary is a gland in the flower that secretes a sugary fluid that pollinators utilize for food, while the extrafloral nectary is a gland on the nonflower part that secretes a sugary fluid that serves the same purpose. A gland-dot is a tiny pore that secretes fluid; glandular hairs bear glands. Oil glands can often be seen on a plant's areole, and irregular (not round) leaf oil glands can be either an island oil gland that is not connected to veinlets or an intersectional oil gland that is connected to veinlets.

An example of alpine conditions and a glacier at full summer retreat. (Glaciers expand and contract with the season.) (Courtesy of Tim McCabe)

glaucoma A disease of the eye characterized by loss of vision due to an increase in the pressure of fluid within the eye (intraocular pressure) that leads to damage to the optic nerve and loss of vision in both eyes. Can eventually lead to blindness. The disease affects about 6 million people worldwide. In the United States, about 3 million people are affected.

glial cell Nonimpulse-conducting cells that make up half the weight of the brain; in the central nervous system they are 10 times as numerous as neurons and act as support cells by forming insulation around the neurons to protect them. Their support functions provide myelin for axons, and they act as housekeepers after cell damage or death by cleaning up. Glial cells also play an important role in the early and continuing development of the brain. A small hormonelike protein, called the glial growth factor, induces the growth of glial cells. A type of brain tumor that forms in the glial tissue is called glioblastoma multiform.

There are several types of glial cells in the central nervous system, including oligodendrocytes, astrocytes, and microglia. It is the oligodendrocytes that produce the fatty protein myelin that insulates the axons by wrapping them in layers of myelin. Star-shaped astrocytes lay down scar tissue on damaged neurons and hold the neurons in place as well as supply potassium and calcium and regulate neurotransmitter levels. Along with microglia cells, astrocytes remove dead cells and other matter from the central nervous system.

Similar functionary types of cells in the peripheral nervous system are the Schwann (provide insulation via myelin) and satellite cells (support cells).

A debilitating disease caused by the demyelinating of neurons is multiple sclerosis, and stem-cell research is being conducted into developing new treatments.

glomerulus   A structure, a tiny ball, between the afferent arterioles and efferent arterioles within the proximal part of the nephron of the kidney; located within the Bowman's capsule. It is composed of a cluster of capillary blood vessels and is involved in the filtration of blood. The glomerulus is a semipermeable structure that allows water and soluble wastes to pass through and then discharges them out of the Bowman's capsule as urine waste at a rate of about 160 liters or 42.7 gallons per 24-hour period. Most of this is reabsorbed back into the blood. The filtered blood then leaves the glomerulus by way of the efferent arteriole to the interlobular vein. Each kidney contains about 1 million glomeruli. Changes in the glomerulus may be responsible for diabetic kidney disease.

Also a nest of nerves in invertebrates found in invertebrate olfactory processing centers; a discrete, globular mesh of densely packed dendrites and axons found in the vertebrate olfactory bulb.

glucagon   A protein hormone released by the pancreas via alpha cells in the islets of Langerhans with the purpose of breaking down glycogen, in the liver, which releases glucose and increases blood levels of glucose. Glucagon works with insulin to maintain normal blood sugar levels.

glucocorticoid   A class of stress-related steroids (hormones) produced by the adrenal glands (cortex) that respond to the stimulation by adrenocorticotropic hormone (ACTH) that comes from the pituitary gland. They are involved in carbohydrate, lipid, and protein metabolism by promoting gluconeogenesis and the formation of glycogen, as well as effects on muscle tone, circulation, blood pressure, and more. They possess anti-inflammatory and immunosuppressive properties. Cortisol (hydrocortisone) is the major natural glucocorticoid.

Synthetically produced ones—mostly derived from cortisol, such as cortinsone, prednisone, prednisolone, methylprednisolone, betamethasone, and dexamethasone—regulate metabolism of lipids, carbohydrates, and protein and work without the use of ACTH.

Glucocorticoids also cause osteoporosis, weight gain, cataracts, heart disease, diabetes, and psychosis.

glucose   A form of six carbon sugar ($C_6H_{12}O_6$) that is the most common energy source and is the usual form in which carbohydrates are assimilated by animals. It is carried through the bloodstream and is made not only

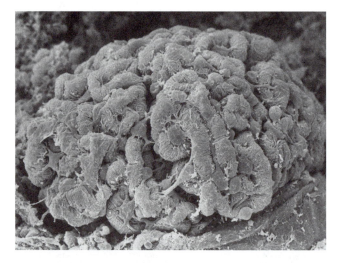

Colored scanning electron micrograph (SEM) of a healthy kidney glomerulus. The glomerulus is a tight ball of capillaries where blood passing through the kidneys is filtered. Fluid passes out of the capillaries into the cavity of the surrounding Bowman's capsule (not seen) and drains into a long tube, from which essential substances and some water are reabsorbed. The remaining unwanted fluid, containing toxins from the blood, drains to the bladder as urine. Magnification unknown. *(Courtesy © Science Photo Library/Photo Researchers, Inc.)*

from carbohydrates but from fats and protein as well. Glucose is known as a dextrorotatory sugar (a chiral molecule that rotates plane-polarized light to the right), which is sweet, colorless, and soluble.

glycocalyx A thick, (7.5–200 nm) extracellular, sticky coating of oligosaccharides linked to plasma membrane glycoprotein and glycolipids; found around the outside of eukaryote cells, and used to adhere to surfaces. Also called the cell coat.

glycogen A large polysaccharide; stored energy found in the muscles and liver. It consists of many monosaccharide glucose molecules linked together and is used as a fuel during exercise, broken down as needed; glycogen is the primary storage form of glucose in animals. Also known as stored sugar or animal starch.

glycolysis The anaerobic pathway or enzymatic conversion (using 11 different enzymes) in the cell's cytoplasm of glucose to simpler compounds. Glucose, a six-carbon sugar, is converted into two molecules of pyruvic acid of three carbons each, with two molecules of NADH and two ATPs as by-products. It is the most universal and basic energy harvesting system; it transforms glucose into lactic acid in muscles and other tissues for energy production when there is not enough oxygen available.

In aerobic respiration, the two pyruvic acids are further used in the KREBS CYCLE.

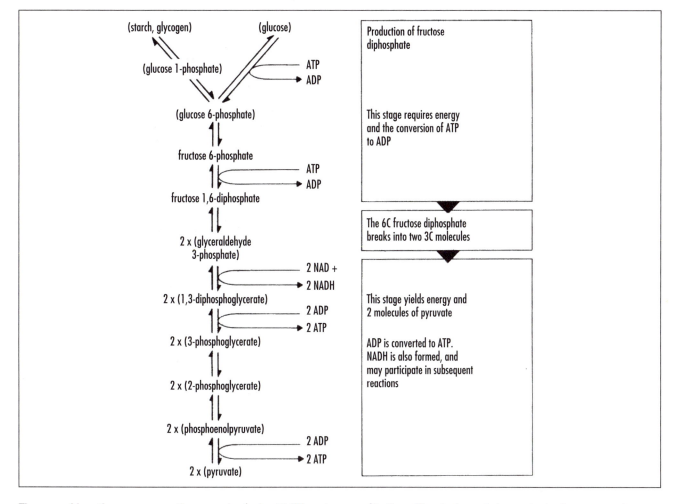

The anaerobic pathway or enzymatic conversion (using 11 different enzymes) in the cell's cytoplasm of glucose to simpler compounds.

glycoprotein (conjugated protein) Glycoproteins are complexes in which carbohydrates are attached covalently to asparagine (N-glycans) or serine/threonine (O-glycans) residues of peptides. A protein coated with a sugar is termed glycosylated and is described or named with the initials "gp" along with its molecular weight, e.g., gp160. Several gps are associated with HIV infection, since they are the outer-coat proteins of HIV: gp41 plays a key role in HIV's infection of CD4+ T cells by facilitating the fusion of the viral and cell membranes. The protein gp120 is one of the proteins that forms the envelope of HIV; it projects from the surface of HIV and binds to the CD4 molecule on helper T cells. GPs are found in mucus and mucins, y-globulins, a1-globulins, a2-globulins, and transferrin, an ion-transporting protein. They act as receptors for molecular signals originating outside the cell. Attachment of oligosaccharides to peptides increases solubility, covers the antigenic domains, and protects the peptide backbone against proteases.

gold drugs Gold COORDINATION compounds used in the treatment of rheumatoid arthritis, examples being auranofin, (tetraacetylthioglucosato-S)(triethylphosphane)gold(I), and myocrisin, disodium thiomalonatogold(I).

Golgi, Camillo (1843–1926) Italian *Medical Teacher* Camillo Golgi was born in Corteno, near Brescia, Italy, on July 7, 1843, the son of a physician. He studied medicine at the University of Pavia, and after graduating in 1865, he continued working in Pavia at the Hospital of St. Matteo. Golgi was influenced by the scientific methods of Giulio Bizzozero, who introduced general pathology in the programs of the medical school at the University of Pavia. In 1872 Golgi accepted the post of chief medical officer at the Hospital for the Chronically Sick at Abbiategrasso and began his investigations into the nervous system.

Golgi returned to the University of Pavia as extraordinary professor of histology, left the university, and returned as the chair for general pathology in 1881, succeeding his teacher Bizzozero. He also married Donna Lina, a niece of Bizzozero.

Golgi developed an interest in the causes of malaria and determined the three forms of the parasite and their associated fevers. He developed a photographic technique to document the most characteristic phases of malaria in 1890.

While Golgi never practiced medicine, he was a famous and popular teacher as director of the Department of General Pathology at St. Matteo Hospital. He also founded and directed the Instituto Sieroterapico-Vaccinogeno of the Province of Pavia. Golgi also became rector of Pavia University and was made a senator of the Kingdom of Italy.

During World War I, he assumed the responsibility for a military hospital in Pavia and created a neuropathological and mechanotherapeutical center for the study and treatment of peripheral nervous lesions and for the rehabilitation of the wounded.

His greatest contribution seems to be his revolutionary method of staining individual nerve and cell structures, known then as the "black reaction" and now called Golgi staining. It allowed a clear visualization of a nerve cell body with all its processes.

Golgi shared the Nobel Prize for 1906 with Santiago Ramón y Cajal for their work on the structure of the nervous system. He retired in 1918 but remained as professor emeritus at the University of Pavia. The Historical Museum at the University of Pavia dedicated a hall to Golgi, where more than 80 certificates of honorary degrees, diplomas, and awards are exhibited.

Golgi's discovery of the black reaction and further research provided a major contribution to the advancement of the knowledge on the structural organization of nerve tissues. He also described the morphological features of glial cells and the relationships between their processes and blood vessels; described two fundamental types of nerve cells today called Golgi type I and Golgi type II; discovered the Golgi tendon organs; explained the cycle of plasmodium (malaria) and the cell organelle, today called the Golgi apparatus.

Golgi died at Pavia on January 21, 1926. In 1994 the Italian Ufficio Principale Filatelico issued a stamp to celebrate his work.

Golgi apparatus or complex Part of a cell, a cup- or disclike organelle in cells, usually near the nucleus and composed of a number of flattened or folded sacs, called cisternae, with vacuoles and vesicles. They act as an assembly line in sorting, modifying, and packaging proteins and lipids produced on the endoplasmic reticulum, located in the cytoplasm of the cell, for various parts of the cell. Named after the Nobel Prize recipient, Italian neurologist and histologist Camillo Golgi (1843–1926). They are the formation site of the carbohydrate side chains of GLYCOPROTEINS and mucopolysaccharides. The vacuoles release these by migrating through the cell membrane. Some of the vesicles send molecules to the cellular membrane, where they are excreted, and some are used for selective excretion.

gonadotropins   A group of hormones that are produced in the pituitary gland and regulate the development and function of the testis and ovary. The group includes the follicle-stimulating hormone (FSH), which promotes male and female gamete formation, and luteinizing hormone (LH), which stimulates the secretion of the male and female testosterone and estrogen. Gonadotropin itself is controlled by the gonadotropin releasing hormone (GnRH), a hormone that controls the production and release of gonadotropins and is secreted by the hypothalamus every 90 minutes or so, which enables the pituitary to secrete LH and FSH.

gonads   The male and female sex organs. In the male, they are glands located inside the scrotum, behind and below the penis, which produce sperm and are the primary source of testosterone. Also called the testes. In the female, they are ovaries, two almond-sized glands located on either side of the uterus. They produce and store the oocytes and the female sex hormones estrogen and progesterone.

Gondwanaland (Gondwana)   The large southern protocontinent, derived from the supercontinent Pangaea, that, when fragmented, formed Africa, South America, Antarctica, Australia, and India during the Paleozoic era more than 200 million years ago as an event of plate tectonics (continental drift).

G protein   A class of heterotrimeric proteins important in signaling pathways in the plasma membrane in mammalian cells. Regulated by the guanine nucleotides GDP (guanosine diphosphate) and GTP (guanosine triphosphate), they participate in cell signal pathways by usually binding a hormone or signal ligand to a seven-pass transmembrane receptor protein; activates intracellular messenger systems when the signaling molecule (typically a hormone) binds to the transmembrane receptor. The proteins are located on the inner surface of the plasma membrane and transmit signals from outside the membrane, via transmembrane receptors, to adenylate cyclase, which then catalyzes the formation of the second messenger, cyclic adenosine monophosphate (AMP), inside the cell.

graded potential   A nerve impulse that is initially proportional to the intensity of the stimulus that produces it, then declines in intensity; membrane potentials that vary in magnitude.

gradualism   An Earth model of evolution that assumes a slow, steady rate of change, with slow steps instead of quick leaps, and with new variation arising by mutation and recombination. A view held by Charles Darwin.

grafting   The process of attaching two different plant parts, usually shoots, to each other to create a single new plant. Used in propagating trees and shrubs.

Gram staining   An important laboratory technique to distinguish between two major bacterial groups, based on stain retention by their cell walls. Bacteria smears are fixed by flaming, then stained with crystal violet followed by iodine solution, and then rinsed with alcohol or acetone, decolorized, and counterstained with safranin. Gram-positive bacteria are stained bright purple or purple-black, while gram-negative bacteria are pink. This staining technique is useful in bacterial taxonomy and identification and in indicating fundamental differences in cell-wall structure. Gram-negative bacteria lack peptidoglycan in the cell wall, while gram-positive bacteria have about 90 percent of their cell wall composed of peptidoglycan.

See also BACTERIA.

granum   A series of disk- or saclike structures called thaylakoid disks—specialized membrane structures located in the inner membrane of chloroplasts—where photosynthesis takes place. They appear as green granules under a microscope and contain the light-reactant chemicals chlorophyll and carotenoid pigments. ATP is generated during photosynthesis by chemiosmosis.

gravid   Used in relation to pregnant insects and meaning heavy with fully developed eggs or denoting an advanced stage of pregnancy.

gravitropism   The ability of an organism or specific cells to respond, e.g., bend, to the gravitational pull; a growth curvature induced by gravity.
See also GEOTROPISM.

Greek letters used
α: *See* HELIX (for ALPHA HELIX) and CYTOCHROME.
β: *See* BETA SHEET, BETA STRAND, BETA TURN, and CYTOCHROME.
γ: *See* SORET BAND (for gamma band).
η: *See* HAPTO and ASYMMETRY PARAMETER.
κ: *See* DONOR ATOM SYMBOL (for kappa convention).
μ: *See* BRIDGING LIGAND (for mu symbol).

greenhouse effect   The warming of an atmosphere by its absorbing and reemitting infrared radiation while allowing shortwave radiation to pass on through.

Certain gaseous components of the atmosphere, called greenhouse gases, transmit the visible portion of solar radiation but absorb specific spectral bands of thermal radiation emitted by the Earth. The theory is that terrain absorbs radiation, heats up, and emits longer wavelength thermal radiation that is prevented from escaping into space by the blanket of carbon dioxide and other greenhouse gases in the atmosphere. As a result, the climate warms. Because atmospheric and oceanic circulations play a central role in the climate of the Earth, improving our knowledge about their interaction is essential.
See also CARBON DIOXIDE.

gross primary productivity (GPP)   The total energy fixed by plants in a community through photosynthesis

(such as repackaging inorganic energy to organic energy) per unit area per unit time; total carbon assimilation by plants; total mass or weight of organic matter created by photosynthesis over a defined time line.

ground meristem   Meristem is embryonic tissue. Ground meristem is one of the primary meristem tissues that are differentiated from the apical meristem. The primary meristem tissues include three different tissues: protoderm, ground meristem, and procambium. The protoderm differentiates into the epidermis; the procambium differentiates into the vascular tissue; and the ground meristem differentiates into three regions: the cortex, which is several layers of parenchyma cells under the epidermis of the stem and root; pith ray, the parenchyma cells between the vascular bundles in the primary dicot stem; and pith, which are parenchyma cells in the center of the primary dicot stem. Parenchyma are isodiametric cells, i.e., approximately equal in length, width, and height, that are thin walled and not extremely specialized. Ground meristem gives rise to cells of the ground tissue system. The ground tissue of a leaf, located between the upper and lower epidermis and specialized for photosynthesis, is called the mesophyll.

ground tissue system   The plant tissue system that forms most of the photosynthetic tissue in leaves, composed mostly of parenchyma cells and some collenchyma (elongate and thick walled in strands) and sclerenchyma (thick, rigid, secondary walls with lignin to provide support) cells. It is found between the epidermis and surrounding the vascular tissue system. Acts also as supporting tissue and in water and food storage.

growth factor   A complex family of organic chemicals, especially polypeptides, that bind to cell surface receptors and act to control new cell division, growth, and maintenance by the bone marrow.

Synthetic growth factors are being used to stimulate normal white blood cell production following cancer treatments and bone marrow transplants.

Examples of growth factors are insulin (including insulinlike growth factor (IGF), GF1, II, all of which are polypeptides similar to insulin); somatomedins, polypeptides made by the liver and fibroblasts that, when released into the blood (stimulated by somatotropin), help cell division and growth by incorporating sulfates into collagen, RNA, and DNA synthesis; HGH (human growth hormone), also called somatotropin, a proteinlike hormone from the pituitary gland that stimulates the liver to produce somatomedins that stimulate growth of bone and muscle; platelet-derived growth factor (PDGF), a glycoprotein that stimulates cell proliferation and chemotaxis in cartilage, bone, and other cell types; fibroblast growth factor, which promotes the proliferation of cells of mesodermal, neuroectodermal, epithelial, or endothelial origin; epidermal growth factor (EGF), important for cell development as it binds to receptors on cell surface to create a growth signal; and granulocyte colony-stimulating factor (G-CSF), a growth factor that promotes production of granulocytes, a type of white blood cell.

guanylate cyclase An ENZYME catalyzing the conversion of guanosine 5'-triphosphate to cyclic guanosine 3',5'-monophosphate, which is involved in cellular REGULATION processes. One member of this class is a HEME-containing enzyme involved in processes regulated by nitrogen monoxide.

guard cell Specialized epidermal cells; two crescent-shaped cells on either side of the pore of a stoma in the stem or leaf epidermis. By changing shape, i.e., by opening and closing via changes in turgor, they regulate gas exchange and water loss by covering or uncovering the pore, which lets oxygen out and carbon dioxide in.

Gullstrand, Allvar (1862–1930) Swedish *Ophthalmologist* Allvar Gullstrand was born on June 5, 1862, in Landskrona and was the eldest son of Dr. Pehr Alfred Gullstrand, principal municipal medical officer, and his wife Sofia Mathilda Korsell. He was educated at schools in Landskrona and Jönköping, and he attended Uppsala University, but left in 1885 to spend a year at Vienna. He then continued his medical studies at Stockholm and presented his doctorate thesis in 1890. He was appointed lecturer in ophthalmology in

1891 and was appointed the first professor of ophthalmology at Uppsala University in 1894. He stayed in that position until 1914, when he was given a personal professorship in physical and physiological optics at Uppsala University, becoming an emeritus professor in 1927.

Gullstrand contributed a great deal to the knowledge of clinical and surgical ophthalmology and of the structure and function of the cornea of the eye, as well as research on astigmatism, and was self-taught in this area. He laid out his ideas in his doctoral thesis in 1890, *Bidrag till astigmatismens teori* (Contribution to the theory of astigmatism) and further refined them in *Allgemeine Theorie der monochromatischen Aberrationen und ihre nächsten Ergebnisse für die Ophthalmologie* (General theory of monochromatic aberrations and their immediate significance for ophthalmology), 1900; *Die reelle optische Abbildung* (The true optical image), 1906; and *Die optische Abbildung in heterogenen Medien und die Dioptrik der Kristallinse des Menschen* (The optical image in heterogeneous media and the dioptrics of the human crystalline lens), 1908. Further important works included *Tatsachen und Fiktionen in der Lehre von der optischen Abbildung* (Facts and fictions in the theory of the optical image), 1907; and *Einführung in die Methoden der Dioptrik der Augen des Menschen* (Introduction to the methods of the dioptrics of the human eyes), 1911.

In 1911 he invented the slit lamp and the reflex-free ophthalmoscope to help study the eye. He introduced a surgical technique for the treatment of symblepharon (a fibrous tract that connects the bulba conjunctiva to conjunctiva on the eyelid) and redefined the theory of accommodation. In 1911 he received the Nobel Prize for his research on dioptrics of the eye, although at first he declined it. Gullstrand died in Stockholm on July 28, 1930. Gullstrand is seen as one of the founders of modern ophthalmology.

guttation A way for plants to expel water in excess of transpiration—principally through hydathodes, a special epidermal structure at the terminations of veins around the margins of the leaves—and especially under conditions of relatively high humidity. The expelled water appears on the ends of leaves mostly on moist cool nights and resembles dew.

gymnosperm A group of flowerless plants that includes pines, conifers, redwoods, firs, yews, and cycads. *Gymnosperm* means "naked seed." Their ovules and the seeds, which develop in them, are born unprotected on the surface of megasporophylls, and are often arranged on cones instead of being enclosed in ovaries as in the flowering plants. They release pollen directly into the air, which finds its way to female ovules and fertilizes them. The earliest examples of gymnosperms appeared during the Devonian period some 408 million years ago and were seed ferns. Gymnosperms have reached as high as 350 feet to as small as 4 inches in height. Worldwide there are 20 plant families with at least 50 percent of their species threatened. Of these, eight are gymnosperm families (including cycads and conifers). This may be due to the fact that many of their species are widely exploited for both timber and horticultural purposes, and, because they are an ancient group, they may not adapt easily to the rapidly changing environment around them. There are only 720 species compared with 250,000 species of angiosperms.

See also ANGIOSPERM.

gynandromorphism A female that develops partly or completely male characteristics. A lateral gynandromorph on one side has the external characters of the male, and the other side has those of the female. Common in bees and silkworms.

See also HERMAPHRODITE.

Haber-Weiss reaction   The Haber-Weiss cycle consists of the following two reactions:

$$H_2O_2 + OH\cdot \rightarrow H_2O + O_2^- + H^+ \text{ and}$$

$$H_2O_2 + O_2^- \rightarrow O_2 + OH^- + OH\cdot$$

The second reaction achieved notoriety as a possible source of hydroxyl radicals. However, it has a negligible rate constant. It is believed that iron(III) complexes can catalyze this reaction: first Fe(III) is reduced by superoxide, followed by oxidation by dihydrogen peroxide.

See also FENTON REACTION.

habituation   Reaction to the repeated presentation of the same stimulus that has no special significance and that causes reduced attention to the stimulus. Short-term habituation lasts for only minutes, while longer-term habituation could last for weeks.

Haeckelian recapitulation (embryonic recapitulation)
A clade goes through, in its development, an ontogenetic stage that is present in the adults of its sister group. Named for Ernst Haeckel (1834–1919), who promoted the idea that the development of the human embryo in the womb is a rerun (recapitulation) of the steps in humanity's rise from a primitive creature. Haeckel's views about evolution have been refuted by most scientists.

half-life   For a given reaction the half-life, $t^{1/2}$, of a reactant is the time required for its concentration to reach a value that is the arithmetic mean of its initial and final (equilibrium) value. For a reactant that is entirely consumed, it is the time taken for the reactant concentration to fall to one-half of its initial value. For a first-order reaction, the half-life of the reactant can be called the half-life of the reaction. In nuclear chemistry, (radioactive) half-life is defined, for a simple radioactive decay process, as the time required for the activity to decrease to half its value by that process.

See also BIOLOGICAL HALF-LIFE; GEOLOGICAL TIME.

haloperoxidase   A PEROXIDASE that catalyzes the oxidative transformation of halides to XO⁻ (X being Cl, Br, or I) or to organic halogen compounds. Most are HEME proteins, but some bromoperoxidases from algae are vanadium-containing ENZYMEs.

halophyte   A plant that has adapted to grow in salt-rich soils or salt-rich air, e.g., glassworts (*Salicornia virginica*). Obligatory halophytes need salt, whereas facultative halophytes can live in freshwater conditions as well.

hansch analysis   The investigation of the quantitative relationship between the biological activity of a series of compounds and their physicochemical substituent or global parameters representing hydrophobic, electronic, steric, and other effects using multiple regression correlation methodology.

haploid cell A cell containing one set of chromosomes. Sperm and egg cells are haploid.

haplozoans Extinct echinoderms of the Cambrian era that include two classes, Cycloidea and Cyamoidea. Cyamoids have bilateral symmetry, and cycloids are dome-shaped with radial symmetry. They are quite uncharacteristic of echinoderms, although they have a skeletal structure that relates them. Otherwise, they have uncertain affinities to all other echinoderms and may provide a link between echinoderms and fish.

hapten A molecule (usually a small organic molecule) that can be bound to an ANTIGENic determinant/epitope. Usually they are too small to give a response of their own. They become antigenic if they are coupled to a suitable macromolecule, such as a protein.

hapto The hapto symbol, η (Greek eta), with numerical superscript, provides a topological description for the bonding of hydrocarbons or other π-electron systems to metals by indicating the connectivity between the LIGAND and the CENTRAL ATOM. The symbol is prefixed to the ligand name or to that portion of the ligand name most appropriate. The superscript numerical index on the right indicates the number of COORDINATING atoms in the ligand that bind to the metal. Examples:

$$[PtCl_2(C_2H_4)(NH_3)] \text{ amminedichloro}$$
$$(\eta^2\text{-ethene})\text{platinum}$$

$$[Fe(\eta^5\text{-}C_5H_5)_2] \text{ bis } (\eta^5\text{-cyclopentadienyl})\text{iron}$$
$$(\text{ferrocene})$$

See also COORDINATION.

hard acid A LEWIS ACID with an acceptor center of low polarizability. It preferentially associates with HARD BASEs rather than with soft bases, in a qualitative sense (sometimes called HSAB rule). Conversely, a soft acid possesses an acceptor center of high polarizability and exhibits the reverse preference for a partner for COORDINATION.

hard base A LEWIS BASE with a donor center of low polarizability; the converse applies to soft bases.
See also HARD ACID.

hard drug A nonmetabolizable compound, characterized either by high lipid solubility and accumulation in adipose tissues and organelles, or by high water solubility. In the lay press, the term *hard drug* refers to a powerful DRUG of abuse such as cocaine or heroin.

Hardy-Weinberg theorem A mathematical theorem that describes a population that is not evolving under certain conditions. First proposed in 1908 by G. H. Hardy, an English mathematician, and B. W. Weinberg, a German physician, although both were working independently of each other. The theory describes the frequencies of various genotypes in a population and the ability to predict how the frequencies will change in future generations based on certain assumptions not being violated, such as the absence of mutation, the absence of migration or selection, random mating, and an infinite population size. In other words, in the absence of any of these factors, the allele frequencies and the frequencies of the different genotypes will remain the same from one generation to the next, and genetic recombination due to sexual reproduction will not result in any changes in allele frequencies or genotype frequencies.

harelip (cleft lip) A congenital (birth) deformity caused by a failure of developing tissues to fuse; a cleft in the upper lip, found on one or both sides of the midline. Often associated with a cleft palate.

haustorium A specialized structure of a parasitic plant or entophyte, an organism that lives at least part of its life cycle within a host plant, that penetrates the living host and absorbs nutrients. In parasitic fungi, it is a hyphal tip that penetrates the host but not the cell membrane.

Haversian system (osteon) A structural composition of compact bone that contains a Haversian canal, a longitudinally arranged vascular channel in the center of the osteon containing blood vessels and nerves; Haversian lamella, the circumferentially arranged layers of osteoid and osteocytes surrounding the canal; and the canaliculi, fine channels through the bone that provide a link between osteocytes.

heart A muscular organ composed of cardiac muscle that pumps blood throughout the body.

heat Kinetic energy in the process of being transferred from one object to another due to a temperature difference. It moves in one of three ways: radiation, conduction, or convection.

heath hen Along with the passenger pigeon, the heath hen (*Tympanuchus cupido cupido*), a relative of the prairie chicken, is an example of extinction by human disturbance. Last observed on March 11, 1932, at James Green's farm in West Tisbury, Martha's Vineyard, Massachusetts, it became the victim of habitat conversion to crops and pastureland, as well as being hunted as game and for sport. It was a ground-dwelling/nesting meadows seedeater ranging from Maine to Chesapeake Bay.

See also EXTINCT SPECIES.

heat-shock proteins (HSPs) A family of closely related proteins, widely distributed in virtually all organisms (plants, animals, microorganisms, and humans). Even though they are found in widely different sources, they show structural similarity. HSP expression increases in response to physiological stresses such as rise in temperature, pH changes, and oxygen deprivation. Many of these stresses can disrupt the three-dimensional structure, or folding, of a cell's proteins, and so HSPs bind to those damaged proteins, helping them refold back into their proper shapes.

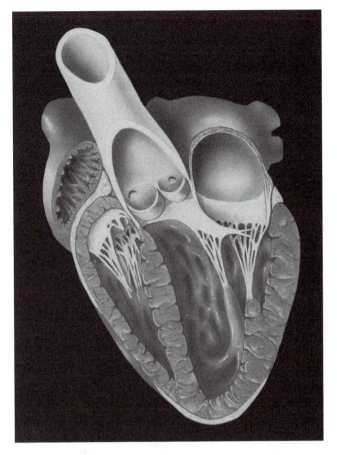

Illustration of a section through a whole human heart showing the internal organization of chambers, valves, connecting arteries, and veins. The heart can be regarded as being two pumps: the right side (left on image) serving the lungs and the left side pushing blood around the general circulation. Oxygenated blood entering the heart from the lungs passes through the left atrium into the left ventricle (bottom right) before being distributed via the aorta (top, center). Deoxygenated blood enters the right atrium through the venae cavae (top left) prior to being pumped to the lungs from the right ventricle. *(Courtesy © David Gifford/ Photo Researchers, Inc.)*

They also help newly synthesized polypeptides fold and prevent premature interactions with other proteins. HSPs, also called chaperones, aid in the transport of proteins throughout the cell's various compartments and aid in the destruction of peptides specific to tumors or pathogens.

hectare A unit of land measurement. One hectare (ha) is equal to 2.5 acres.

Insects and Man—An Exotic Dilemma

By Timothy L. McCabe, Ph.D.

Insects have long been recognized for their benefits as pollinators, honey producers, predators, and parasitoids, as well as for pure aesthetics, such as admiring the grace and beauty of a butterfly. They have also been recognized as vectors of disease, agricultural pests, defoliators of forests, bloodsuckers, and as having other characteristics unwanted or not admired by humans.

Insects have only recently come to be appreciated by the scientific community and public for what they can tell us about the health of the environment. Their sheer diversity, estimated at 30 million species, and numbers (the world's ant biomass equals the total human biomass) are evidence of their significance. In recent times, insects have been used for biocontrol, i.e., using one species to control the population of another. Klamath weed (*Hypericum perforatum,* also known as St. John's-wort) control by a leaf beetle is a famous example. The Klamath weed leaf beetle, *Chrysolina quadrigemina,* intentionally introduced in 1946, feeds on the foliage of this weed in both the adult and larval stages. The Klamath weed leaf beetle has dramatically reduced the abundance and density of this weed in areas where fall temperatures are mild. This biological control agent is responsible for contributing to the impressive decline in St. John's-wort on the Pacific coast.

Introductions have led to mistakes that are equally dramatic. Another leaf beetle, the Colorado potato beetle (*Leptinotarsa decemlineata*), was introduced from the United States to Europe in the late 1920s, where it has become a serious potato pest. A parasitic fly (*Compsilura*) was intentionally introduced to the United States in the 1920s to control a newly established and very serious forest threat, the gypsy moth (*Lymantria dispar*). It took a long time for the fly to become established, and now it is an integral part of our fauna. However, this species of fly has five broods per year, while the gypsy moth only has one. The fly survives for the remainder of the year by attacking hundreds of species of native moths and butterflies, unrelated to the gypsy moth, and has suppressed their numbers to unprecedented low levels. The population of this fly has been estimated at 16,000 individuals per acre in Connecticut. Clearly this "cure" for the gypsy moth was worse than the "disease," having totally unexpected side effects.

The honeybee is widely acclaimed for its ability to pollinate and for its honey production. However, this introduced bee, originally an Old World species, has displaced and probably caused the extinction of several bees native to the New World. The tracheal bee mite, a recent accidental introduction, possibly from the Philippines, has had a catastrophic effect on the honeybee and honey production. The secondary impact of tracheal mites on native bees is an ongoing concern. The "killer" bee, an introduced strain of the honeybee, is another matter entirely.

Not all introductions need be derived from another continent. Geographic barriers, such as the Rocky Mountains and the Great Plains, prevent the natural spread east and west on the North American continent. The western conifer seed bug (*Leptoglossus occidentalis*) was accidentally introduced from the West to a tree nursery in Iowa in 1956. From Iowa, the bug spread throughout the northeastern United States, reaching New York in 1992, where it has become a nuisance due to its proclivity to overwinter in houses.

Insects are continually on the move. Even native populations exhibit periodic instability and can rapidly expand (or contract) a range that had been stable for decades, even centuries. The plain ringlet butterfly (*Coenonympha tullia*) has been moving south from Canada at a pace of nearly 40 miles per year with no end in sight (except eventually an ocean).

Even exotic plants have a profound impact on native plants and consequently on the associated insects. Glaciers "erased" the flora and fauna of the northern states thousands of years ago. Species with poor dispersal mechanisms have still not reinvaded since the glacial retreat 15,000 years ago. Glaciers wiped out earthworms, but introduced earthworms have filled the void. One exotic earthworm species is so adept at removing organics that it actually mineralizes the soil in the process of feeding. Garlic mustard, also an introduction, loves mineralized soils and follows the earthworms. The Virginia white butterfly (*Pieris virginiensis*), a species threatened by competition with the introduced cabbage white butterfly (*Pieris rapae*), has a caterpillar that feeds on a native mustard, but the adult butterfly demonstrates an ovipositional preference for the introduced garlic mustard. Its caterpillar cannot survive on the garlic mustard, and as a consequence of this ovipositional preference, this species is disappearing in the wake of the encroaching garlic mustard. As an added insult, the diminished quality of the forest litter beneath the garlic mustard has led to a dramatic (more than 90 percent) decline in ground beetles, native denizens of the forest floor. This impact makes its way up the food chain as we witness declines in red-backed salamanders that feed on insects in the forest litter.

Ladybugs have been intentionally introduced with wild abandon because of their reputation as voracious aphid eaters. Nearly half of North America's ladybugs are not natives, but introductions. Our native nine-spotted ladybug

(*Coccinella novemnotata*), the "official" New York State insect, was eradicated from New York by the seven-spotted ladybug (*Coccinella septempunctata*), an intentionally introduced European species. No nine-spotted ladybug has been collected in New York State since the 1940s, yet it had been a dominant species before the introduction of this competitor. Another exotic introduction, the Asian ladybug (*Harmonia axyridis*), made its way to the Northeast in the 1990s and has partially displaced the previously introduced seven-spotted ladybug.

It is now somewhat of a rarity to see a native ladybug in New York State. The level at which the various ladybugs compete is not that well understood. The nine-spotted and the seven-spotted are very close relatives, so the fact that they compete directly should not have surprised scientists. What is more surprising is that the distantly related Asian ladybug displaced the nine-spotted and many others, despite the lack of any close relationship. Parasites and diseases of introduced insects are some of the baggage that comes with these introductions. The introductions have evolved a resistance or at least a tolerance to these coevolved agents to which our native species have never been exposed. It is smallpox at an insect level.

All the aforementioned ladybugs eat aphids, but they can be particular about which species and in what habitats. The diversity of native ladybug species was achieved through natural selection. Introductions upset the balance. We can never take back the dozens of introduced ladybugs.

If all life (not just ladybugs) from all of the earth's continents were mixed together tomorrow, it would result in a die-off of diversity that would parallel that of the Devonian extinction that occurred millions of years ago. This is what we are doing, albeit slowly, with accidental and purposeful introductions. Once an undesirable exotic has gained a foothold, we are caught up in a cycle of introducing predators, parasitoids, and diseases in an attempt to suppress or eradicate the original introduction. As shown by these examples, it often backfires.

—**Timothy L. McCabe, Ph.D.,** is curator of entomology at New York State Museum in Albany, New York.

helium   An inert gas; an element with atomic number 2. Helium is produced in stars and is the second most abundant element in the universe. Its atom contains two protons, two neutrons, and two electrons.

See also ELEMENT.

helix   A particular rigid left- or right-handed arrangement of a polymeric chain, characterized by the number of strands, the number (n) of units per turn, and its pitch (p), the distance the helix rises along its axis per full turn. Examples of single-stranded helices are the protein helices: α-helix: $n = 3.6$, $p = 540$ picometer; 3_{10}-helix: $n = 3.0$, $p = 600$ picometer; π-helix: $n = 4.4$, $p = 520$ picometer.

See also DOUBLE HELIX.

helminth   A worm or wormlike organism. Three major helminths exist that affect humans: the nematodes (roundworm), trematodes (flukes), and cestodes (tapeworms).

See also PLATYHELMINTHES.

helper T cell   A type of T cell needed to turn on antibody production by activating cytotoxic T cells and causing other immune responses in the body. They aid in helping B cells to make antibodies against thymus-dependent antigens. TH1 and TH2 helper T cells secrete materials (interleukins and gamma interferon) that help cell-mediated immune response.

heme   A near-planar COORDINATION complex obtained from iron and the dianionic form of PORPHYRIN. Derivatives are known with substitutes at various positions on the ring named a, b, c, d, etc. Heme b, derived from PROTOPORPHYRIN IX, is the most frequently occurring heme.

hemerythrin   A dioxygen-carrying protein from marine invertebrates, containing an oxo-bridged dinuclear iron center.

See also NUCLEARITY.

hemichordates   Consisting of only a few hundred species, the hemichordates include the acorn worms (e.g., *Saccoglossus*), which burrow in sand and mud, and pterobranchs (e.g., *Rhabdopleura*), tiny colonial animals. The hemichordates have a tripartite (three-fold) division of the body, pharyngeal gill slit, a form of dorsal nerve cord, and a similar structure to a notochord. Graptolites, which comprise the class Graptolithina, are common fossils in Ordovician and Silurian rocks and are now considered hemichordates.

hemochromatosis   A genetic condition of massive iron overload leading to cirrhosis and/or other tissue damage attributable to iron.

hemocyanin   A dioxygen-carrying protein (from invertebrates, e.g., arthropods and mollusks), containing dinuclear TYPE 3 COPPER sites.
   See also NUCLEARITY.

hemoglobin   A dioxygen-carrying heme protein of red blood cells, generally consisting of two alpha and two beta SUBUNITs, each containing one molecule of PROTOPORPHYRIN IX.
   See also BLOOD.

hemolymph   The body circulatory fluid found in invertebrates, functionally equivalent to the blood and lymph of the vertebrate circulatory system.

hemophilia   An inherited clotting problem that occurs, with few exceptions, in males. It delays coagulation of the blood, making hemorrhage difficult to control. Hemophilia A, called classical or standard hemophilia, is the most common form of the disorder and is due to a deficiency of a factor called factor VIII (FVIII). Hemophilia B, also called Christmas disease, is due to a deficiency of factor IX (FIX). A person with hemophilia does not bleed any faster than a normal person, but the bleeding continues for a much longer time.
   See also BLOOD.

hemorrhoids (piles)   Abnormally enlarged or dilated veins around the anal opening.

Hench, Philip Showalter (1896–1965) American *Physician* Philip Showalter Hench was born in Pittsburgh, Pennsylvania, on February 28, 1896, to Jacob Bixler Hench and Clara Showalter. After attending local schools he attended Lafayette College, Easton, Pennsylvania, and received a B.A. in 1916. The following year he enlisted in the U.S. Army Medical Corps but was transferred to the reserve corps to finish his medical training. In 1920 he received his doctorate in medicine from the University of Pittsburgh.

   He became the head of the Mayo Clinic Department of Rheumatic Disease in 1926, and by 1947 was professor of medicine. During World War II he became a consultant to the army surgeon general. During this period with the Mayo Clinic he isolated several steroids from the adrenal gland cortex and, working with Edward KENDALL, successfully conducted trials using cortisone on arthritic patients. Hench also treated patients with ACTH, a hormone produced by the pituitary gland that stimulates the

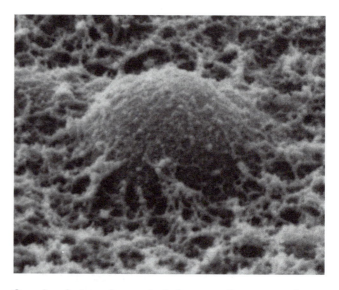

Scanning electron micrograph of a human erythrocyte, or red blood cell, tangled in fibrin. Fibrin is the insoluble form of fibrinogen, a coagulation factor found in solution in blood plasma. Fibrinogen is converted to the insoluble protein fibrin when acted upon by the enzyme thromboplastin. Fibrin forms a fibrous meshwork, the basis of a blood clot, which is the essential mechanism for the arrest of bleeding. Red blood cells contain the pigment hemoglobin, the principal function of which is to transport oxygen around the body. Magnification: ×6900 (at 10 × 8 in. size). *(Courtesy © Dr. Tony Brain/Photo Researchers, Inc.)*

adrenal gland. He shared with Edward C. Kendall and Tadeus REICHSTEIN the 1950 Nobel Prize in physiology or medicine for this pioneering work in the treatment of rheumatoid arthritis with cortisone and ACTH.

He authored many papers in the field of rheumatology, received numerous awards, and belonged to several scientific organizations throughout his life. He died on March 30, 1965, in Ocho Rios, Jamaica.

hepatic portal vessel A system of veins that delivers blood from glands and organs of the gastrointestinal tract (GIT) to the liver. A portal vein enters the liver at the porta hepatis and distributes venules and sinusoids, capillarylike vessels where blood becomes purified, before it goes into the inferior vena cava. The hepatic artery carries oxygen-rich blood to the liver from the heart and mixes with the portal vein in the sinusoids. Thus two filtering systems, the capillaries of the GIT and sinusoids of the liver, perform their tasks on the blood.

herb Part of a plant that is used for medicinal, food, or aromatic properties.

herbivore An animal that only eats plants to obtain its necessary nutrients for survival.

See also CARNIVORE.

hermaphrodite An individual with both male and female sexual reproductive organs and that functions as both male and female, producing both egg and sperm. The individual can be a simultaneous hermaphrodite, having both types of organs at the same time, or a sequential or successive hermaphrodite that has one type early in life and the other type later. If the female part forms first it is called protogynous hermaphroditism; it is called protandrous hermaphroditism if the male forms first.

Some examples of hermaphrodites are most flukes, tapeworms, gastrotriches, earthworms, and even some humans.

In plants, it is when male and female organs occur in the same flower of a single individual.

hernia A protrusion of a tissue or organ or a part of one through the wall of the abdominal cavity or other area. A hiatal hernia has part of the stomach protruding through the diaphragm and up into the chest and affects about 15 percent of the human population.

Herodotus A Greek who lived ca. 400 B.C.E. and observed fossil seashells in the rocks of mountains. He interpreted these remains as once-living marine organisms and concluded that these areas must have been submerged in the past.

Hess, Walter Rudolf (1881–1973) Swiss *Physiologist* Walter Rudolf Hess was born in Frauenfeld, in Aargau Canton, Switzerland, on March 17, 1881, to a teacher of physics. He received a doctor of medicine in 1906 from the University of Zurich.

Originally he began his career as an ophthalmologist for six years (1906–12), but dropped his practice and turned to the study of physiology at the University of Bonn. In 1917 he was nominated director of the Physiological Institute at Zurich, with corresponding teaching responsibilities, and later director of the Physiological Institute (1917–51) at the University of Zürich.

He spent most of his life investigating the responses of behavior, respiration, and blood pressure by stimulating the diencephalon of cats using his own techniques. He was awarded the Nobel Prize in 1949 for his work related to the diencephalon.

Among Hess's books is *The Biology of Mind* (1964). He died on August 12, 1973.

heterochromatin Most cell nuclei contain varying amounts of functional (active) and nonfunctional (inactive) DNA. Functional DNA is called euchromatin, while nonfunctional or inactive is called heterochromatin. The latter is DNA that is so tightly packaged it cannot transcript. Two forms exist: constitutive heterochromatin, where portions of the chromosome are always inactive, and facultative heterochromatin, where portions of the chromosome are active in some cells at one time but are inactive now (such as the Y chromosome and Barr bodies). A gene is closed or inaccessible and not expressed if it is heterochromatin.

heterochrony An evolutionary change in developmental timing in the relative time or rate of appearance or development of a character. The morphological outcomes of changes in rates and timing of development are paedomorphosis (less growth) or peramorphosis (more growth). Peramorphosis is the extended or exaggerated shape of the adult descendant relative to the adult ancestor; its later ontogenetic stages retain characteristics from earlier stages of an ancestor. Paedomorphosis is where the adult descendant retains a more juvenile looking shape or looks more like the juvenile form of its adult ancestor; its development goes further than the ancestor and produces exaggerated adult traits. Paedomorphosis can happen as a result of beginning late (postdisplacement), ending early (progensis), or slowing in the growth rate (neoteny). Likewise, peramorphosis can result from starting early (predisplacement), ending late (hypermorphosis), or having a greater growth rate (acceleration).

heterocyst A large, thick-walled, specialized cell working in anoxic (oxygen absent) conditions that engages in nitrogen fixation from the air on some filamentous cyanobacteria; an autotrophic organism.

heteroecious A parasite that starts its life cycle on one organism and then affects a second host species to complete the cycle, e.g., peach-potato aphid (*Myzus persicae*).

heterogamy Producing gametes of two different types from unlike individuals, e.g., egg and sperm. The tendency for unlike types to mate with unlike types.

heterolysis (heterolytic cleavage or heterolytic fission) The cleavage of a bond so that both bonding electrons remain with one of the two fragments between which the bond is broken.

heteromorphic Having different forms at different periods of the life cycle, as in stages of insect metamorphosis and the life cycle of modern plants, where the sporophyte and gametophyte generations have different morphology.

Heteroptera A suborder known as true bugs. They have very distinctive front wings, called hemelytra. The basal half is leathery and the apical half is membranous. They have elongate, piercing-sucking mouthparts. Worldwide in distribution, there are more than 50,000 species. Two families are ectoparasites. The Cimicidae (bed bugs) live on birds and mammals including humans, and the Polyctenidae (bat bugs) live on bats.

heteroreceptor A RECEPTOR regulating the synthesis and/or the release of mediators other than its own ligand.
See also AUTORECEPTOR.

heterosexual Having an affection for members of the opposite sex, i.e., male attracted to female.
See also HOMOSEXUAL.

heterosis Vigorous, productive hybrids that result from a directed cross between two pure-breeding plant lines.

heterosporous Producing two types of spores differing in size and sex. Plant sporophytes that produce two kinds of spores that develop into either male or female gametophytes.

heterotrophic organisms Organisms that are not able to synthesize cell components from carbon dioxide as a sole carbon source. Heterotrophic organisms use preformed oxidizable organic SUBSTRATEs such as glucose as carbon and energy sources, while energy is gained through chemical processes (chemoheterotrophy) or through light sources (photoheterotrophy).

heterozygote A diploid organism or cell that has inherited different alleles, at a particular locus, from each parent (i.e., Aa individual); a form of polymorphism.

heterozygote advantage (overdominance) The evolutionary mechanism that ensures that eukaryotic heterozygote individuals (*Aa*) leave more offspring than homozygote (*AA* or *aa*) individuals, thereby preserving genetic variation; condition in which heterozygotes have higher fitness than homozygotes.

heterozygous   Two different alleles of a particular gene present within the same cell; a diploid individual having different alleles of one or more genes producing gametes of different genotypes.

Heymans, Corneille Jean-François (1892–1968) Belgian *Physiologist*   Corneille Jean-François Heymans was born in Ghent, Belgium, on March 28, 1892, to J. F. Heymans, a former professor of pharmacology and rector of the University of Ghent, and who founded the J. F. Heymans Institute of Pharmacology and Therapeutics at the same university.

Corneille received his secondary education at the St. Lievens College (Ghent), St. Jozefs College (Turnhout), and St. Barbara College (Ghent). He pursued his medical education at the University of Ghent and received a doctor's degree in 1920. After graduation he worked at various colleges until 1922, when he became lecturer in pharmacodynamics at the University of Ghent. In 1930 he succeeded his father as professor of pharmacology and was appointed head of the department of pharmacology, pharmacodynamics, and toxicology; at the same time he became director of the J. F. Heymans Institute, retiring in 1963.

His research was directed toward the physiology and pharmacology of respiration, blood circulation, metabolism, and pharmacological problems. He discovered chemoreceptors in the cardio-aortic and carotid sinus areas, and made contributions to knowledge of arterial blood pressure and hypertension. He was awarded the Nobel Prize in physiology or medicine in 1938 for his work on the regulatory effect of the cardio-aortic and the carotid sinus areas in the regulation of respiration.

He wrote more than 800 scientific papers, was active in a number of professional organizations, and was publisher and editor-in-chief of the *Archives Internationales de Pharmacodynamie et de Thérapie*, founded in 1895 by his father and Professor E. Gley in Paris. He died on July 18, 1968, in Knokke.

hibernation   A physiological state of dormancy, a sleeplike condition, that lowers body temperature, slows the heart and breathing, and reduces the need for food for extended periods of time, usually during periods of cold. Examples of hibernators are bears, bats, snakes, frogs, squirrels, turtles, and some birds.

high-spin   *See* LOW-SPIN.

Hill, Archibald Vivian (1886–1977) British *Physiologist*   Archibald Vivian Hill was born in Bristol on September 26, 1886. After an early education at Blundell's School, Tiverton, he entered Trinity College, Cambridge, with scholarships. He studied mathematics but was urged to go into physiology by one of his teachers, Walter Morley Fletcher.

In 1909 he began study on the nature of muscular contraction and the dependence of heat production on the length of muscle fiber. From 1911 to 1914, until the start of World War I, he continued his work on the physiology of muscular contraction at Cambridge as well as other studies on nerve impulse, hemoglobin, and calorimetry.

In 1926 he was appointed the Royal Society's Foulerton research professor and was in charge of the biophysics laboratory at University College until 1952.

His work on muscle function, especially the observation and measurement of thermal changes associated with muscle function, was later extended to similar studies on the mechanism of the passage of nerve impulses. He coined the term *oxygen debt* to describe the process of recovery after exercise.

He discovered and measured heat production associated with nerve impulses and analyzed physical and chemical changes associated with nerve excitation, among other studies. In 1922 he won the Nobel Prize in physiology or medicine (with Otto MEYERHOF) for work on chemical and mechanical events in muscle contraction such as the production of heat in muscles. This research helped establish the origin of muscular force in the breakdown of carbohydrates while forming lactic acid in the muscle.

His important works include *Muscular Activity* (1926), *Muscular Movement in Man* (1927), *Living Machinery* (1927), *The Ethical Dilemma of Science and Other Writings* (1960), and *Traits and Trials in Physiology* (1965).

He was a member of several scientific societies and was elected a fellow of the Royal Society in 1918, serving as secretary for the period 1935–45, and foreign secretary in 1946. Hill died on June 3, 1977.

hill topping   A behavior exhibited by butterflies where males and females congregate at a high point in the landscape, increasing each individual butterfly's chance of finding a mate.

hilum   The area where blood vessels, nerves, and ducts enter an organ.

HiPIP   Formerly used abbreviation for high-potential IRON–SULFUR PROTEIN, now classed as a FERREDOXIN. An ELECTRON TRANSFER PROTEIN from photosynthetic and other bacteria, containing a [4FE-4S] CLUSTER that undergoes oxidation-reduction between the $[4Fe\text{-}4S]^{2+}$ and $[4Fe\text{-}4S]^{3+}$ states.
See also PHOTOSYNTHESIS.

hirudin   A nonenzymatic chemical secreted from the leech that prevents blood clotting. Today, the genetically engineered lepirudin and desirudin and the synthetic bivalirudin are used as anticoagulants.

histamine   A hormone and chemical transmitter found in plant and animal tissues. In humans it is involved in local immune response that will cause blood vessels to dilate during an inflammatory response; also regulates stomach acid production, dilates capillaries, and decreases blood pressure. It increases permeability of the walls of blood vessels by vasodilation when released from mast cells and causes the common symptoms of allergies such as running nose and watering eyes. It will also shut the airways in order to prevent allergens from entering, making it difficult to breath. Antihistamines are used to counteract this reaction.

histology   The study of the microscopic structure of plant and animal tissue.

histone   A basic unit of chromatin structure; several types of protein characteristically associated with the DNA in chromosomes in the cell nucleus of eukaryotes. They function to coil DNA into nucleosomes, which are a combination of eight histones (a pair each of H2A, H2B, H3, and H4) wrapped by two turns of a DNA molecule. A high number of positively charged amino acids bind to the negatively charged DNA.

HIV   *See* AIDS.

holoblastic cleavage   A complete and equal division of the egg in an early embryo that has little yolk. Characteristic of amphibians, mammals, nonvertebrate chordates, echinoderms, most mollusks, annelids, flatworms, and nematodes.

holocene   The present epoch of geological time starting approximately 10,000 years ago to the present.
See also GEOLOGICAL TIME.

holoenzyme   An ENZYME containing its characteristic PROSTHETIC GROUP(s) and/or metal(s).

holotype   The exact specimen of a new animal or plant representing what is meant by the new name and designated so by publication. The holotype specimen does not have to be the first ever collected, but it is the official one with which all others are compared.

homeobox (HOX genes)   A short stretch of similar or identical 180-base-pair (nucleotide) sequences of DNA within a homeotic gene in most eukaryotic organisms that plays a major role in controlling body development by regulating patterns of differentiation. Homeotic genes create segments in an embryo that become specific organs or tissues. Homeoboxes determine positional cell differentiation and development. Mutations in these genes will cause one body part to convert into a totally different one.
See also HOMEOTIC GENE.

homeosis The replacement of one body part by another caused by mutations or environmental factors initiating developmental anomalies.

homeostasis The ability of an organism to automatically maintain a constant internal condition regardless of the external environment.

homeothermic The process of maintaining a constant body temperature.

homeotic gene Controls the overall body plan of animals by controlling the developmental fate of groups of cells. A homeotic mutation results in the replacement of one type of body part in place of another.
See also HOMEOBOX.

hominoids A collective term used for humans and apes.

homogamy The tendency for similar types to mate with similar types.

homologous chromosomes A pair of chromosomes (homologues); contains the same length, gene position, centromere location, and same characters at corresponding loci, with one homologue coming from the mother and another from the father. Homologous chromosomes line up with each other and then separate during meiosis.
See also CHROMOSOME.

homologous structures Characteristics or parts in different animals that may have served the same general function and are shared with related species and inherited from a common ancestor. In related species, they may have the same evolutionary origin, but their functions may differ. Examples include the front fins of a whale, forelimb of a bat or horse, and human and chimp arm bones.

homologue Used to describe a compound belonging to a series of compounds differing from each other by a repeating unit, such as a methylene group, a peptide residue, etc. Also refers to one member of a chromosome pair.

homology The similarity of characteristics that result from a shared ancestry; the relationship between structures in different organisms that are united by modification of the same structure, gene, or set of genes of a common ancestor.

homolysis (homolytic cleavage or homolytic fission) The cleavage of a bond so that each of the molecular fragments between which the bond is broken retains one of the bonding electrons.

homonomy (serial homology) Organs that are identical or of similar construction within the same organism (e.g., segments in annelid worms).

homoplasy The possession by two or more species of a similar characteristic that has not evolved in those species from a common ancestor. Instead, these characteristics derive from convergent evolution, parallel evolution, or character reversal. An example is the wings of insects and the wings of the flying dinosaurs, the pterosaurs.

Homoptera An order of insects with beaklike piercing-sucking mouthparts that include cicadas, aphids, tree and leaf hoppers, and scale insects. The forewings are either wholly membranous or wholly leathery. The wings rest on the back in the shape of a tent, e.g., cicadas, frog-hoppers, and aphids. They are found worldwide and are plant feeders.

homosexual Having an affection for members of the same sex, i.e., male attracted to male or female attracted to female (lesbianism).
See also HETEROSEXUAL.

homosporous Plants that produce a single type of spore that develops into a bisexual gametophyte and has both male and female sex organs.

homozygous Having two identical forms of a particular gene.

Hooke, Robert (1635–1703) English *Physicist, Astronomer* Considered one of the greatest scientists of the 17th century, and second only to Sir Isaac Newton, Robert Hooke was born in Freshwater, Isle of Wight, on July 18, 1635, the son of John Hooke, a clergyman.

He entered Westminster School in 1648 at the age of 13, and then attended Christ College, Oxford, in 1653, where many of the best English scientists were congregating, such as Robert Boyle, Christopher Wren (astronomer), John Wilkins (founder of the Royal Society), and William Petty (cartographer). He never received a bachelor's degree, was nominated for the M.A. in 1663 by Lord Clarendon, the chancellor of the university, and given an M.D. at Doctors' Commons in 1691, also by patronage.

Hooke's first publication of his own work in 1661 was a small pamphlet on capillary action. Shortly after, in 1662, he was appointed the first curator of experiments at the newly founded Royal Society of London. The society, also known as The Royal Society of London for Improving Natural Knowledge, was founded on November 28, 1660, to discuss the latest developments in science, philosophy, and the arts. The founding fathers, consisting of 12 men, began after a lecture by Christopher Wren at Gresham College. The group included Wren, Boyle, John Wilkins, Sir Robert Moray, and William Brouncker. This position gave Hooke a unique opportunity to familiarize himself with the latest progress in science.

Part of Hooke's job was to demonstrate and lecture on several experiments at the Royal Society at each weekly meeting. This led him to many observations and inventions in a number of fields, including astronomy, physics, and meteorology. He excelled at this job, and in 1663 Hooke was elected a fellow of the society, becoming not just an employee but on equal footing with the other members.

Hooke took advantage of his experience and position. He invented the first reflecting telescope, the spiral spring in watches, an iris diaphragm for telescopes (now used in cameras instead), the universal joint, the first screw-divided quadrant, a compound microscope, an odometer, a wheel-cutting machine, a hearing aid, a new type of glass, and carriage improvements. Despite all of these accomplishments, he remains one of the most neglected scientists, due to his argumentative style and the apparent retribution by his enemies such as Newton.

Hooke became a professor of physics at Gresham College in 1665 and stayed there for his entire life. It was also where the Royal Society met until after his death. Hooke also served as the society's secretary from 1677 to 1683.

The year 1665 is another milestone year for Hooke, since that is when he published his major work *Micrographia*, the first treatment on microscopy, and where he demonstrated his remarkable powers of observation and his skillful microscopic investigation in the fields of botany, chemistry, and meteorology. Within this work, he made many acute observations, illustrated with intricate drawings, and proposed several theories.

Hooke was the first to discover plant cells and he coined the word *cell*, which he attributed to the porous structure of cork, although he failed to realize that cells were the basic units of life. He made detailed observations, some of the first, on insects, sponges, bryozoans, foraminifera, and even birds. He was the first to examine fossils under a microscope and concluded that many fossils represented organisms that no longer existed on Earth.

Because of his controversies—he had competing claims with Christian Huyghens over the invention of the spring regulator and with Newton, first over optics (1672) and, second, over the formulation of the inverse square law of gravitation (1686)—Hooke fell out of favor in the scientific community. He died in London on March 3, 1703, and was buried in Bishopsgate. However, sometime in the 19th century his bones were removed, and no one knows where he is buried today.

hookworms Tiny parasitic nematode worms belonging to the family Ancylostomatidae. They attach themselves to the intestinal walls of humans with hooked mouthparts. Hookworms (*Necator* and *Ancylostoma* spp.) are responsible for ancylostomiasis.

Hopkins, Frederick Gowland (1861–1947) English *Biochemist* Frederick Gowland Hopkins was born on June 20, 1861, in Eastbourne, England, to a bookseller in Bishopsgate Street, London, who died when Frederick was an infant.

In 1871 he attended the City of London School, and at the early age of 17, he published a paper in *The Entomologist* on the bombardier beetle. He went to University College, London, where he became the assistant to Sir Thomas Stevenson, an expert on poisoning. In 1888 he became a medical student at Guy's Hospital, London.

In 1894 he graduated in medicine and taught physiology and toxicology at Guy's Hospital for four years, and in 1898 he moved to Cambridge. He was appointed fellow and tutor at Emmanuel College, Cambridge.

Hopkins established biochemistry as a field in Great Britain. He discovered how to isolate the amino acid tryptophan and identified its structure, discovered enzymes, and isolated glutathione. For his research on discovering growth-stimulating vitamin, which he called "accessory substances," he was awarded the Nobel Prize in 1929 in medicine or physiology. He actually isolated vitamins C, A, and D.

Hopkins was knighted in 1925 and received the Order of Merit in 1935. Hopkins died in 1947 at the age of 86. The Sir Frederick Gowland Hopkins Memorial Lecture of the Biochemical Society, named in his honor, is presented by a lecturer to assess the impact of recent advances in his or her particular field on developments in biochemistry. The award is made every two to three years and the lecturer is presented with a medal and £1,000.

hormone A substance produced by endocrine glands, released in very low concentration into the bloodstream, and which exerts regulatory effects on specific organs or tissues distant from the site of secretion.

See also GLANDS.

Houssay, Bernardo Alberto (1887–1971) Argentine *Physiologist* Bernardo Alberto Houssay was born in Buenos Aires, Argentina, on April 10, 1887, to Dr. Albert and Clara Houssay (née Laffont), who had come to Argentina from France. His father was a barrister. Houssay's early education was at a private school, the Colegio Británico. He then entered the School of Pharmacy of the University of Buenos Aires at the age of 14, graduating in 1904. He had already begun studying medicine and, in 1907, before completing his studies, took up a post in the department of physiology and began research that resulted in his receiving an M.D. in 1911.

In 1910 he was appointed professor of physiology in the university's school of veterinary medicine. In 1919 he became professor of physiology in the medical school at Buenos Aires University and also organized the Institute of Physiology at the medical school, making it a center with an international reputation. He remained professor and director of the institute until 1943, when the government then in power deprived him of his post, the result of his voicing the opinion that there should be effective democracy in the country. In 1955 a new government reinstated him in the university.

He demonstrated that a hormone secreted by the pituitary prevented metabolism of sugar and that injections of pituitary extract induced symptoms of diabetes. He was awarded the 1947 Nobel Prize in physiology or medicine for this work on the functions of the pituitary gland.

In 1949 he came to the United States as a special research fellow at the National Institutes of Health. During his lifetime, Houssay authored more than 500 papers and several books and won many scientific prizes and awards. He died on September 21, 1971.

Human Genome Project The Human Genome Project (HGP) is an international research effort to determine the DNA sequence of the entire human genome. Contributors to the HGP include the National Institutes of Health (NIH); the U.S. Department of Energy (DOE); numerous universities throughout the United States; and international partners in the United Kingdom, France, Germany, Japan, and China.

Begun in 1990, the U.S. Human Genome Project is a long-term effort coordinated by the Department of Energy and the National Institutes of Health. The goals of the project are to identify all of the approximately 30,000 genes in human DNA; determine the sequences of the 3 billion chemical base pairs that make up human DNA; store this information in databases; improve tools for data analysis; transfer related technologies to

the private sector; and address the ethical, legal, and social issues (ELSI) that may arise from the project.
See also GENE.

humoral immunity   A form of immune reaction that attacks bacteria and viruses found in body fluids using antibodies synthesized by the B lymphocytes that circulate in blood plasma and lymph. These fluids were once called "humors."

Huxley, T. H.   A 19th-century evolutionist who discovered the unbroken horse lineage from the Eocene to the Holocene (recent) epochs, perhaps the longest and most complete evolutionary sequence in the fossil record.

hyaline   A clear or transparent structure such as a wing of a dragonfly (e.g., *Aeshna canadensis* and *Sympetrum vicinum*), or an amorphous texture due to accumulation of intra- or extracellular material.

hybrid   The offspring produced by genetically distinct different parents. Mating can be within species (intraspecific) as well as between species (interspecific).

hybridization   Producing hybrids from interbreeding two species. In genetics, it is the annealing of two complementary strands of DNA, or an RNA strand to a complementary DNA strand.

hybridoma   The fusion of two different cells to create a hybrid cell that secretes a single specific antibody, e.g., the fusion of a spleen cell and a cancer cell, or a T lymphocyte with a lymphoma cell.

hybrid vigor   Increased vitality or success of a hybrid over its inbred parents.

hybrid zone   A geographical territory where previously isolated or genetically distinct populations make contact and form hybrids. Where two geographical races of a single species overlap, hybrids and intermediates can outnumber the pure forms in the overlap region. What determines whether there will be two species or one is the balance between gene flow and selection against hybrids.

hydration   Addition of water or the elements of water (i.e., H and OH) to a molecular entity. The term is also used in a more restricted sense for the process: A (gas) → A (aqueous solution).
See also AQUATION and SOLVATION.

hydrocarbon   A compound made of only carbon and hydrogen.

hydrocephalus   A condition in which the head becomes enlarged and expanded beyond normal size due to congenital or other causes.

hydrogenase   An ENZYME, dihydrogen acceptor OXIDOREDUCTASE, that catalyzes the formation or oxidation of H_2. Hydrogenases are of various types. One class ([Fe]-hydrogenases) contains only IRON–SULFUR CLUSTERS. The other major class ([NiFe]-hydrogenases) has a nickel-containing center and iron–sulfur clusters; a variation of the latter type ([NiFeSe]-hydrogenases) contains selenocysteine.

hydrogen bond   A form of association between an electronegative atom and a hydrogen atom attached to a second, relatively electronegative atom. It is best considered as an electrostatic interaction, heightened by the small size of hydrogen, that permits proximity of the interacting dipoles or charges. Both electronegative atoms are usually (but not necessarily) from the first row of the periodic table, e.g., N, O, or F. Hydrogen bonds can be intermolecular or intramolecular. With a few exceptions, usually involving fluorine, the associated energies are less than 20–25 kJ mol⁻¹ (5–6 kcal mol⁻¹).

A type of bond formed when the partially positive hydrogen atom of a polar covalent bond in one molecule is attracted to the partially negative atom of a polar covalent bond in another.

hydrogen ion (hydron) A single proton with a charge of +1.
See also HYDRON; ION.

hydrolase An ENZYME of EC class 3, also known as a hydro-LYASE, that catalyzes the HYDROLYSIS of a SUBSTRATE.
See also EC NOMENCLATURE FOR ENZYMES.

hydrolysis SOLVOLYSIS by water.

hydron General name for the ion H+, either in natural abundance or where it is not desired to distinguish between the isotopes, such as proton for ^{1}H+, deuteron for ^{2}H+, and triton for ^{3}H+.

hydrophilic "Water loving." The capacity of a molecular entity or of a substituent to interact with polar solvents, in particular with water, or with other polar groups. Hydrophilic molecules dissolve easily in water, but not in fats or oils.

hydrophilicity The tendency of a molecule to be solvated by water.

hydrophobic A molecule or substance that does not associate, bond, or dissolve in water. Hydrophobic molecules dissolve easily in fats and oils.

hydrophobic interaction The tendency of hydrocarbons (or of lipophilic hydrocarbonlike groups in solutes) to form intermolecular aggregates in an aqueous medium as well as analogous intramolecular interactions. The name arises from the attribution of the phenomenon to the apparent repulsion between water and hydrocarbons. Use of the misleading alternative term *hydrophobic bond* is discouraged.

hydrophobicity The association of nonpolar groups or molecules in an aqueous environment that arises from the tendency of water to exclude nonpolar molecules.
See also LIPOPHILICITY.

hydrostatic skeleton A skeletal system created by the pressure caused by fluid-filled closed areas that support rigidity in an organism or one of its parts. Many invertebrates have hydrostatic skeletons. Earthworms are an example.

hydroxyl group A functional group that has a hydrogen atom joined to an oxygen atom by a polar covalent bond (–OH). When put in solution with water, they form alcohols.

hydroxyl ion (–OH). One atom each of oxygen and hydrogen bonded into an ion that carries a negative charge.
See also ION.

Hymenoptera A large order of insects having two pairs of membranous wings (*hymen* means "membrane") coupled by a row of tiny hooks. Examples include ants, bees, sawflies, and wasps.

hyperactive A state of excessive muscular activity or a condition when a particular portion of the body is excessively active, e.g., a gland that produces too much of its particular hormone. Often referred to in attention deficit hyperactivity disorder (ADHD) in children.

hyperendemic Disease organisms that exist in a host population at very high rates. The human papillomavirus is a large group of viruses that are hyperendemic in humans. They cause common warts, such as plantar and genital warts.

hyperfine See ELECTRON PARAMAGNETIC RESONANCE SPECTROSCOPY.

Science and the Spiritual Factor

By John McConnell

For the most part, the scientific community has avoided questions that cannot be measured, analyzed, and evaluated using their present tools and procedures. An example of this is the scientist who stated that she would consider the possible existence of a soul—as soon as they determined the part of the brain that provided communication with the soul.

Most people believe that they have a soul and consider the issues of life after death, the existence and nature of God, and the meaning of life. Religious beliefs and values enable them to relate to the unknown with benefit to their personal values, conduct, and happiness. However, most scientists by their indifference and skepticism have tended to undermine the value of faith and treat it as superstition.

In these matters, hypotheses that cannot be proved or disproved should be judged, or at least acknowledged of value, by the results in the lives of individuals who practice their faith.

The one belief that science has mathematically proved is the existence and benefit of love. (See Von Foerster, Heinz. *Logical Structure of Environment and Its Internal Representation*. Zeeland, Mich.: Herman Miller Inc., 1963.) Love can thus provide the test of hypotheses about phenomena of mind and spirit that presently defy explanation. So now ultimate questions about reality, which remain profound mysteries that cannot be approached by scientific measurements or methods, can nevertheless be recognized and pursued through articles of faith and practice. If reality is consistent, then the truth and the value of a faith or belief can be judged by the increase and depth of love—or creative altruism—in the lives of those who practice at belief.

Prayer to a personal God and practice of the Sermon on the Mount has inspired personal love and courage and led to great peaceful changes for social freedom and justice. (See *This Freedom Whence* by John Wesley Brady, and *Communism and Christ* by Charles Loury.)

Science must no longer negate the values of religious belief, but rather strengthen and support the importance of faith—of using the personal metaphor that increases the well-being of the individual, that deepens relationships with people and kinship with life on Earth.

The scientific approach can at the same time diminish religious intolerance by calling attention to the nature of metaphor or hypothesis. A hypothesis can be exciting and useful and obtain confidence from its supporters who may totally believe in its validity. Nevertheless, by its nature there must be, and can be, recognition and respect for people with a different hypothesis about reality. Where approval for a different religious or philosophical doctrine may be impossible, there still can be deep approval of the love that is motivated and demonstrated in connection with it.

The scientific community should also give importance to any phenomenon that greatly affects human values and potentials, even though scientific explanation eludes its grasp. There is overwhelming evidence of answered prayer in the lives of many people. The incidence of favorable coincidence in deeply dedicated people who pray with fervor and faith should be studied and compared with other people who practice a purely psychological approach to needs. The nature and extent of coincidences that run contrary to probability theory should be more thoroughly explored. Perhaps there is no satisfactory explanation possible. But this should not cause science to ignore the phenomenon or its causes and effects in the lives of people.

Of course, scientists would make a careful distinction between a phenomenon and its effects, on the other hand, explaining what it is and how it works. Great benefit could come from more attention to phenomenology and the many instances of its effects on the lives of people.

The nature and extent of spiritual healing should be more critically examined. While success seems a random effect that is rare, many proven cases defy medical explanation.

By its very nature, any effort to make Earth a healthy, peaceful planet must be achieved through a great spiritual awakening of a kind that will foster the nurture and care of Earth and a creative happy life for all its people.

—**John McConnell** is the original founder of Earth Day; his new program for saving Earth is called the Earth Trustee Agenda (www.earthsite.org).

hyperparasitoid   A parasitoid that lives on another parasitoid, e.g., members of the families Perilampidae, Signiphoridae, and Elasmidae.

hyperplasia   The enlargement of an organ or tissue due to an increase in the number of cells. An example is benign prostatic hyperplasia, a nonmalignant (noncancerous) enlargement of the prostate gland, common in older men.

hyperpolarization An electrical state where the inside of a cell is made more negative relative to the outside than was the case at its resting potential of about −70 mV.

hypertonic solution A solution whose solute concentration is high enough to cause water to move out of cells via osmosis.

hypertrophy A condition where an organ or tissue enlarges or overgrows due to an increase in the size of its cells, not the number of cells.

hyperventilate The act of excessive breathing, causing a loss of carbon dioxide in the blood. Other symptoms include faintness or fainting and numbness around the mouth, fingertips, or toes.

hypha The threadlike, filamentous, absorptive structures of fungi. When combined to form mats, they are called mycelia and are the main body of fungi.

hypogynous ovary (superior ovary) Ovaries that have the calyx, corolla, and androecium attached below the ovary to the receptacle.

hypoosmotic solution A solution whose osmotic pressure is less than that of another solution.

hypothalamus An area in the posterior part of the brain beneath the thalamus that contains nerve cells and controls many autonomic functions. Controls the pituitary gland and is the control site for feeding, drinking, temperature regulation, emotion, and motivation.

hypothesis The formal declaration of the possible explanation of a set of observations that needs to be tested and proved.

hypotonic solution A solution where solute concentration is low enough to cause water to move into cells via osmosis.

I

ileum   The third and last section of the small intestine.

imaginal disk   A region or cluster of cells in the larvae of an insect that are undifferentiated and are determined to form specific organs or tissues during metamorphosis to the adult stage.

imaging   A medical diagnostic technique by which useful organ images are obtained from the radiation emitted by RADIONUCLIDES that are introduced into organs, or from radiation absorbed by atomic nuclei within the organs. Typical examples are imaging obtained by recording the radiation emitted by a radionuclide such as ^{99m}Tc, and the ^{1}H-NMR imaging obtained by whole-body NUCLEAR MAGNETIC RESONANCE measurements.
   See also BONE IMAGING; BRAIN IMAGING; MAGNETIC RESONANCE IMAGING.

immigration   The moving into a location in which the individual is not a native of the area.
   See also EMIGRATION.

immune response   The process by which the body of an organism recognizes and fights invasion of microorganisms, viruses, and other substances (antigens) that may be harmful to the body; the total time from recognition of the intrusion to attack or tolerance of the antigen.

immunoglobulin (Ig)   Also known as antibodies, these are proteins created by plasma cells and B cells that are designed to control the body's immune response by binding to antigens. There are more than 1,000 possible antibody variations and five major types, and each is specific to a particular antigen. Of the five main types—IgA, IgD, IgE, IgG, and IgM—the most common are IgA, IgG, and IgM.
   See also ANTIBODY.

immunogold   A method for visualizing proteins in electron microscopy within a cell using gold particles attached to an ANTIBODY that binds specifically to that protein.

imperfect fungi   Fungi (*deuteromycetes*) that do not have sexually produced spores as part of their life cycle. They cause skin diseases in humans and include the organisms causing ringworm and athlete's foot.

impotence (erectile dysfunction)   Refers to a man's inability to achieve or maintain an erection suitable enough to complete sexual intercourse.

imprinting A type of behavior learned during a certain critical time in development that promotes the learning of behavior and characteristics of the species and is difficult to reverse, regardless of circumstances. An example is the attachment behavior among birds to the mother during the first few hours after hatching. In genetics, it is when an allele at a particular locus is inactivated or altered depending on whether it was inherited by the mother or father.

inbreeding The production of offspring by closely related parents with a high likelihood of carrying similar deleterious recessive mutations that may be expressed in the phenotypes of the offspring. The resulting populations may suffer a higher than average incidence of recessive genetic disorders.

incest When two people too closely related have intercourse, i.e., a parent and child, or two first cousins. Usually socially prohibited.

incomplete dominance A type of inheritance where the heterozygote that has two different alleles (one dominant, one recessive) of a gene pair has a different appearance (phenotype) compared with the homozygous (identical alleles for a given gene) parents.

incomplete flower A flower lacking one of the four major parts: SEPALs, petals, STAMENs, or CARPELs.

incomplete metamorphosis Part of a life cycle of an insect where the nymph stage, or immature form, resembles the adult after hatching and slowly changes into the adult form through a series of molts; does not have a pupa stage. Examples are grasshoppers (Othoptera), aphids, cicadas, and whiteflies. A form of incomplete metamorphosis called gradual metamorphosis is when there is no pupal stage and the nymphs look like the adults minus the wings. Unlike complete metamorphosis, which has four stages (EGG, LARVA, PUPA, and adult), incomplete metamorphosis only has three (no pupa).
See also METAMORPHOSIS.

IND Investigational new drug.

indeterminate cleavage A form of cleavage found in dueterostomes (e.g., chordates and echinoderms) where each cell produced during early cleavage division has the ability to develop into a complete embryo.

indeterminate egg layer The ability to induce birds to lay more eggs by removing or destroying eggs they have already laid. Also called double clutching.

indeterminate growth When an organism continues to grow throughout its life span (ontogeny); characteristic of plants.

indeterminate inflorescence When the central part of the flower is the last to open. Terminal flowers open last and lower flowers open first.

index fossils A biologic method of time correlation using commonly found fossils that are limited to a specific time span. For example, trilobites, though common in the Paleozoic, are not found before the Cambrian period. Some lineages of fossil organisms evolved rapidly, so that the vertical stratigraphic range of the species was short but the species was widespread. A dating technique to correlate the ages of rocks in difference locations (biostratigraphy).
See also GEOLOGICAL TIME.

indigenous An organism that is native and not introduced in a specific environment with certain boundaries.
See also ENDEMIC SPECIES.

induced fit The change in the shape of an enzyme's active site to accommodate and bind firmly to the substrate that enters the site.

induction The process whereby one set of embryonic cells influences the development of another set of embryonic cells.

industrial melanism A natural selection process that developed in the 19th century, when certain species such as moths developed a coloration adaptation to compete with industrial soot pollution (e.g., the blackening of tree bark). England's peppered moth (*Biston betularia*) is the often-used example.

inert STABLE and unreactive under specified conditions.
See also LABILE.

infanticide The purposeful killing of an infant or baby after birth or shortly after.

infectious The ability to transmit a disease; an infectious disease caused by some microbe or agent that is infectious.

inferior ovary (epigynous ovary) Having the calyx, corolla, and androecium appear to rise from the very top of the ovary.

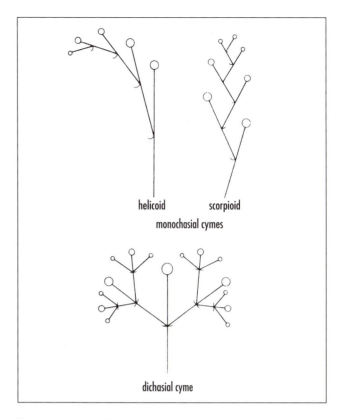

Types of cyrose inflorescence.

inflammatory response The reaction the body takes to invading microorganisms. The response includes: an increase in the blood flow to the infected area to increase the number of leukocytes that can fight the invader; the complementary thinning of the local blood capillary cell walls to allow the increased number of leukocytes to enter along with the leukocytes releasing cytokines, immune-signaling chemicals, to call more leukocytes to the area; and increase in temperature at the infected site.

inflorescence Refers to the various positional and structural arrangements of a flower cluster on a floral axis. The two main categories are racemose (indefinite and not terminating in a flower) and cymose (definite,

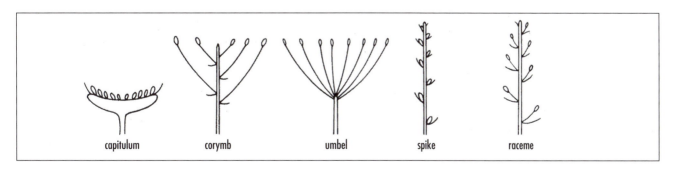

Types of racemose inflorescence.

floral arrangements can be in the form of a panicle [slightly elongated with central axis with branches that branch themselves]. Examples of racemose inflorescence are: raceme (elongated with central axis composed of simple pedicels of about equal length); spike (elongated with central axis with sessile/subsessile flowers); corymb (flat topped with vertical axis with pedicels or branches of unequal length); or umbel (several branches from a common point of the peduncle).

ingestion   The process of obtaining nutrients by heterotrophic organisms by eating other organisms or other organic matter.

inhibition   The decrease in the rate of a reaction brought about by the addition of a substance (INHIBITOR).

inhibitor   A substance that decreases the rate of ENZYME catalysis or other chemical reaction.

inhibitory postsynaptic potential (IPSP)   A small electrical charge of a few millivolts, creating a local hyperpolarization (increase in membrane potential on the negativity of the inside of the neuron) in the membrane of a postsynaptic neuron. Caused when an inhibitory neurotransmitter from a presynaptic cell binds to a postsynaptic receptor, which makes it difficult for a postsynaptic neuron to generate an action potential.

innate behavior   Behavior that is performed without prior learning; considered hardwired in the nervous system. Usually a behavior that is inflexible and built in.

inner cell mass   After a sperm fertilizes an egg, the resulting single cell that is produced is referred to as totipotent, meaning that it has the potential to form an entire organism. Cell division begins immediately, and after a few days the cells begin to specialize, forming a blastocyst, a hollow sphere of cells. The blastocyst has an outer layer of cells, and inside the sphere is the inner cell mass, a cluster of cells that protrude into one end of the cavity and will develop into virtually all of the human tissues.

inositol 1,4,5 triphosphate (IP3)   Acts as a second messenger; increases intracellular calcium and activates calcineurin, an intermediate in the T cell activation pathway; binds to and activates a calcium channel in the endoplasmic reticulum. A second messenger functions as an intermediate between certain nonsteroid hormones and the third messenger, resulting in a rise in cytoplasmic Ca^{2+} concentration.

inquiline   Any organism that lives in the home of another and shares its food or home (e.g., scarab beetle, which lives in ant nests).

Insecta   The class of animals that contain more than 1 million named species and perhaps millions more that are not scientifically named as of yet. Found in almost every habitat worldwide, they are found in every shape and form, small and large. They all share a body composed of a head, thorax, and abdomen; a pair of relatively large compound eyes; usually three ocelli (simple eyes) and pair of antennae located on the head; mouthparts consisting of a labrum, a pair of mandibles and maxillae, a labium, and a tonguelike hypopharynx; two pairs of wings, derived from outgrowths of the body wall, and three pairs of walking legs.

insectivorous   Refers to an animal or plant that eats insects.
See also CARNIVORE.

insertion   A mutation that occurs when one or more nucleotide pairs are added to a gene, causing a frame shift, which is a change in how the information in a gene is translated by the cell. Two other mutations include a deletion mutation, where one or more bases are removed from the DNA sequence of a gene, and substitution mutation, where one base is replaced by another at a single position in the DNA sequence of a gene.
See also DELETION.

instinct 177

This insect (Phasmatodea) from the Bahamas is a good example of cryptic coloration and mimicry. (Courtesy of Tim McCabe)

insertion reaction A chemical reaction or transformation of the general type X–Z + Y → X–Y–Z in which the connecting atom or group Y replaces the bond joining the parts X and Z of the reactant XZ.

insertion sequence A short stretch of mobile bacterial DNA known as a transposon that has the capacity to move between different points within a genome, or change positions; special sequences at their terminal ends allow them to integrate into strands of DNA; usually are inverted repeats (two copies of the same or related sequence of DNA repeated in opposite orientation on the same molecule) at the ends of the insertion sequence; can alter gene expression; involved in transposition.

insight learning Having the ability to perform the appropriate or correct behavior response in a first-time experience on the first try.

instinct An organism's innate, or inborn, intuitive ability to respond to a given stimulus in a fixed way.

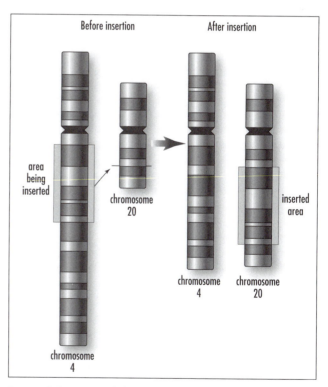

A type of chromosomal abnormality in which a DNA sequence is inserted into a gene, disrupting the normal structure and function of that gene. (Courtesy of Darryl Leja, NHGRI, National Institutes of Health)

insulin A protein hormone produced in the pancreas by beta cells, located in the islets of Langerhans, that stimulates cellular utilization of glucose by body cells, by converting glucose and other carbohydrates to energy, and helps control blood-sugar levels by acting antagonistically with glucagons, the chief source of stored fuel, in the liver. It is released by various signals that are sensitive to the intake and digestion of food. It also acts as an important regulator of protein and lipid metabolism. Insulin is used as a drug to control insulin-dependent diabetes mellitus, a disorder that is caused by the insufficient production of insulin. Without insulin, cells do not absorb glucose. Diabetic individuals can have type I diabetes (juvenile) comprising about 10 percent of the population, or type II diabetes (adult). Presently some 16 million Americans have diabetes, with 1,700 new cases being diagnosed daily. Diabetes has been linked to the development of a variety of diseases including heart disease, stroke, peripheral vascular disease, and neurological disorders.

integral protein A protein of biological membranes that penetrates into or spans the lipid bilayer of a cell membrane. The lipid bilayer is a protective membrane that surrounds the cell and consists of two layers of phospholipids.

integrated pest management (IPM) *See* BIOLOGICAL CONTROL.

inter- A prefix meaning between or among.

interbreed (crossbreed) To breed with another kind or species; hybridize.

intercalation compounds Compounds resulting from inclusion, usually without covalent bonding, of one kind of molecule (the guest molecule) in a matrix of another compound (the host compound), which has a layered structure. The host compound, with a rather rigid structure, may be macromolecular, crystalline, or amorphous.

interferon A chemical messenger of the immune system, composed of a group of cytokine proteins that have antiviral characteristics that are capable of helping the immune response. Three main types of interferon—alpha, beta, and gamma—are produced by virus-infected cells and are released to coat uninfected cells, thus preventing them from becoming infected. Alpha interferon is produced by virus-infected monocytes and lymphocytes, while beta is produced by virus-infected fibroblasts. Gamma is produced by stimulated T and NK cells.

interleukin A group of natural chemical glycoprotein messengers, acting as cytokines, that are secreted by different cells of the immune system to make other cells perform specific cellular functions. Interleukin-1 (IL-1) is released early by monocytes, macrophages, T cells, and other immune cells to fight infection. It stimulates T cell proliferation and protein synthesis and causes fever. Can be cleaved into a peptide involved in cell death (apoptosis). Interleukin-2 (IL-2) is produced by T helper and suppresser lymphocytes. IL-2 increases the expression of natural killer and other cytotoxic cells and stimulates helper T cells to proliferate more rapidly. IL-2 is produced commercially by recombinant DNA technology and used for the treatment of metastatic renal (i.e., kidney) cell cancer. Interleukin-4 (IL-4) is released by lymphocytes (TH-2 subset of T helper lymphocytes) and promotes antibody production by stimulating B cells to proliferate and mature and promotes allergic responses via production of the immunoglobulin IgE. Interleukin-6 (IL-6) affects many different cells in the immune system by inducing differentiation and activation. Interleukin-10 (IL-10), like IL-4, is released by lymphocytes (the TH-2 subset of T helper lymphocytes). IL-10 enhances the humoral response and increases antibody production. Interleukin-12 (IL-12) induces the production of natural killer and other cytotoxic immune cells.

intermediate filament A fibrous protein filament of the cytoskeleton that helps form ropelike bundles in animal cells and is about 10 nm in size, falling in the middle between the size of microtubules and microfilaments; provides tensile strength.

interneuron One of the three types of neuron networks (along with sensory and motor neurons) that allow information flow by way of impulses and action potentials to travel through the nervous system. Found in the central nervous system, they provide the center of a loop that receives upstream input from thousands of motor neurons, then sends the signals to the brain, processes the returns, and then transmits the downstream output to a similar number of sensory neurons for action.

internode The segment on a plant stem between the points where leaves are attached; the region or length of stem between two nodes.

interphase One of the phases of cell division during the process of mitosis. Interphase is the stage between two successive cell divisions and the time when DNA is replicated in the nucleus, followed by mitosis. Interphase itself has several phases: the first gap phase (G1) is the time prior to DNA synthesis where the cell increases in mass; the synthesis or S phase is the time DNA is actually synthesized; and gap or G2 is the phase after DNA syntheses and cell protein synthesis and before the start of prophase.

See also MITOSIS.

interstitial Refers to the space between cells, airways, blood vessels, alveoli, atoms, molecules, and even soil particles.

interstitial cells Cells that exist in the connective tissues between other tissues and structures; cells among the seminiferous tubules, tiny tubes of the testis where sperm cells are produced and that secrete the male sex hormones testosterone and other androgens.

interstitial fluid (**intercellular fluid**) Fluid that fills spaces between cells and provides pathways for the flow of nutrients, gases, and wastes between capillaries and cells.

intertidal zone One of the oceanic zones where the ocean meets land; landform can be submerged or exposed; shallow shore area between low and high tide or water marks.

See also ABYSSAL ZONE; BENTHIC ZONE; PELAGIC ZONE.

intra- A prefix meaning within or inside.

intrinsic activity The maximal stimulatory response induced by a compound in relation to that of a given reference compound.

This term has evolved with common usage. It was introduced by E. J. Ariëns as a proportionality factor between tissue response and RECEPTOR occupancy. The numerical value of intrinsic activity (alpha) could range from unity (for full AGONISTs, i.e., agonist inducing the tissue maximal response) to zero (for ANTAGONISTs),

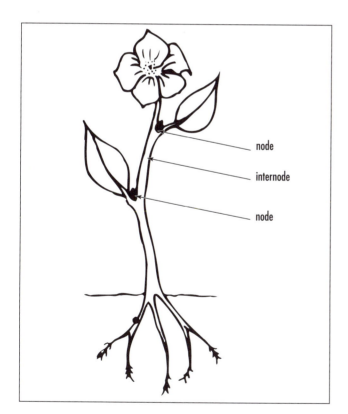

The internode is the segment on a plant stem between the points leaves are attached. It is the region or length of stem between two nodes.

the fractional values within this range denoting PARTIAL AGONISTS. Ariëns's original definition equates the molecular nature of alpha to maximal response only when response is a linear function of receptor occupancy. This function has been verified. Thus, intrinsic activity, which is a DRUG and tissue parameter, cannot be used as a characteristic drug parameter for classification of drugs or drug receptors. For this purpose, a proportionality factor derived by null methods, namely, relative EFFICACY, should be used. Finally, "intrinsic activity" should not be used instead of "intrinsic efficacy." A "partial agonist" should be termed "agonist with intermediate intrinsic efficacy" in a given tissue.

intrinsic rate of increase (rmax)   A mathematical parameter that measures the maximum rate at which a population will grow if resources are unlimited, using birth and death rates in a population as the determinants. It is the difference between the number of births and the number of deaths; the maximum population growth rate.

introgression   The movement of genes from one population into another through hybridization followed by backcrossing.

intron   An intervening section of DNA that occurs almost exclusively within a eukaryotic GENE, but which is not translated to amino acid SEQUENCEs in the gene product. The introns are removed from the premature mRNA through a process called splicing, which leaves the EXONs untouched, to form an active mRNA.

See also EUKARYOTE; MESSENGER RNA; TRANSLATION.

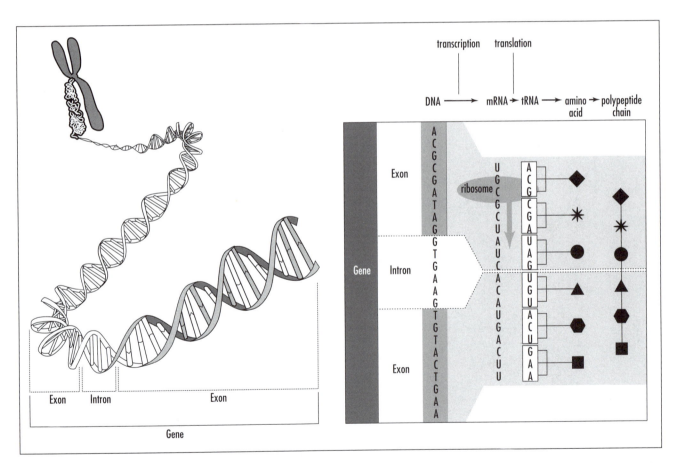

An intron is a noncoding sequence of DNA that is initially copied into RNA but is cut out of the final RNA transcript. *(Courtesy of Darryl Leja, NHGRI, National Institutes of Health)*

inverse agonist (negative antagonist) A DRUG that acts at the same RECEPTOR as that of an AGONIST, yet produces an opposite effect.

See also ANTAGONIST.

inversion A chromosomal aberration or mutation, occurring during meiosis or from mutagens, that involves detaching a chromosome segment, rotating it 180 degrees, and then reinserting it in its original location.

invertebrates Animals that do not have a backbone or notochord. They are cold-blooded and have solid, external skeletons or hydrostatic skeletons. They include small protozoans, sponges, corals, tapeworms, spiders, mollusks, and others. Most animals on the planet (95 percent of all animal species) are invertebrates, and the majority of those, excepting insects, are aquatic.

in vitro fertilization Fertilization outside the body, whereby a male sperm and female egg are combined in a laboratory. The embryo is then artificially transferred into a female's uterus.

ion An atom that acquires a charge by either gaining or losing an electron.

ion channel Enables ions to flow rapidly through membranes in a thermodynamically downhill direction after an electrical or chemical impulse. Their structures usually consist of four to six membrane-spanning DOMAINs. This number determines the size of the pore and thus the size of the ion to be transported.

See also ION PUMP.

ionic bond A chemical bond or link between two atoms due to an attraction between oppositely charged (positive-negative) ions.

ionophore A compound that can carry specific ions through membranes of cells or organelles.

ion pump Enables ions to flow through membranes in a thermodynamically uphill direction by the use of an energy source such as ATP or light. They consist of sugar-containing heteropeptide assemblies that open and close upon the binding and subsequent HYDROLYSIS of ATP, usually transporting more than one ion toward the outside or the inside of the membrane.

See also ADENOSINE TRIPHOSPHATE; ION CHANNEL.

iron-responsive element A specific base SEQUENCE in certain MESSENGER RNAs that code for various proteins of iron METABOLISM, which allows REGULATION at TRANSLATIONal level by the IRON-RESPONSIVE PROTEIN.

iron-responsive protein (IRP) A protein that responds to the level of iron in the cell and regulates the biosynthesis of proteins of iron METABOLISM by binding to the IRON-RESPONSIVE ELEMENT on MESSENGER RNA.

See also REGULATION.

iron-sulfur cluster A unit comprising two or more iron atoms and bridging sulfide LIGANDs in an IRON-SULFUR PROTEIN. The recommended designation of a CLUSTER consists of the iron and sulfide content, in square brackets, e.g., [2FE–2S], [4FE–4S]. The possible oxidation levels are indicated by the net charge excluding the ligands, for example a $[4Fe-4S]_{2+}$ or $[4Fe-4S]_{1+}$ or $[4Fe-4S]_{2+; 1+}$ cluster.

See also BRIDGING LIGAND.

iron-sulfur proteins Proteins in which nonheme iron is coordinated with cysteine sulfur and usually also with inorganic sulfur. Divided into three major categories: RUBREDOXINs; simple iron-sulfur proteins, containing only IRON-SULFUR CLUSTERs; and complex iron-sulfur proteins, containing additional active redox centers such as FLAVIN, molybdenum, or HEME. In most iron-sulfur proteins the clusters function as electron transfer groups, but in others they have other functions such as catalysis of hydratase/dehydratase reactions, maintenance of protein structure, or REGULATION of activity.

See also COORDINATION.

IRP *See* IRON-RESPONSIVE PROTEIN.

irregular Refers to flowers that are bilaterally symmetrical, i.e., that are divisible into equal halves only in one plane.

irruption A rapid and temporary increase in population density, often followed by a mass emigration; common in bird species.

ischemia Local deficiency of blood supply and dioxygen to an organ or tissue owing to constriction of the blood vessels or to an obstruction.

isobacteriochlorin 2,3,7,8-Tetrahydroporphyrin. A reduced PORPHYRIN with two pairs of con-fused saturated carbon atoms (C-2, C-3 and C-7, C-8) in two of the pyrrole rings.
See also BACTERIOCHLORIN.

isoenzymes Multiple forms of ENZYMEs arising from genetically determined differences in PRIMARY STRUCTURE. The term does not apply to those derived by modification of the same primary SEQUENCE.

isogamy Sexual reproduction involving the fusion of gametes that are similar in size or are morphologically indistinguishable, e.g., fungi such as zygomycetes.

isolating mechanism Any environmental, behavioral, mechanical, or physiological barriers or characteristics that will prevent two individuals of different populations from producing viable progeny. Important for the development of new species.

isomerase An ENZYME of EC class 5, which catalyzes the isomerization of a SUBSTRATE.
See also EC NOMENCLATURE FOR ENZYMES.

isomers Compounds that have the same number and type of atoms (same molecular formula) but differ in the way they are combined with each other. They can differ by the bonding sequence, called structural or constitutional isomerism, or the way their atoms are arranged spatially, called stereoisomerism. Other types include conformational, configurational, geometric, optical, enantiomers, and diastereomers.

isomorphic alternation of generations When gametophyte and sporophyte generations are morphologically alike, but differ in the number of chromosomes.

Isoptera Termites, a social order of insects that are soft bodied and comprise about 2,300 species worldwide. They obtain their nutrition from eating wood and digesting cellulose with the help of bacteria and protozoans found in their intestinal area.

isosteres Molecules or ions of similar size containing the same number of atoms and valence electrons, e.g., O^{2-}, F^-, Ne.
See also BIOISOSTERE.

isotonic solutions Solutions having identical osmotic pressures, i.e., a solution where cells do not swell or shrink.

isotope A different form of a single element that has the same number of protons, but has a different number of neutrons in its nucleus. Radioactive isotopes are unstable and break down until they become stable. Carbon 14 is a radioactive isotope of carbon that is used to date fossilized organic matter.

isotropy Lack of ANISOTROPY; the property of molecules and materials of having identical physical properties in all directions.

Ixodes A genus of ticks. *Ixodes scopularis* is the vector for Lyme disease because it can carry a spirochete known as *Borrelia burgdorferi* that is transmitted upon the bite of the tick.

J

joint The contact area between two bones. There are many different types of joint, including hinge, ball and socket, universal, sliding, and slightly movable.

Joly, John An Irish geologist (1857–1933) who, in 1899, tried to determine the Earth's age by calculating how long it would take for the rivers to dump salt into the ocean to reach the present salinity levels. It was based on the assumption that oceans were freshwater when first formed. He estimated 90 million years for the age of the Earth. He was more accurate with dating a geological time period. Working with Sir Ernest Rutherford in Cambridge in 1913, and using the radioactive decay in minerals, he estimated that the beginning of the Devonian period—the time between the Silurian and Carboniferous—was not less than 400 million years ago, an age that is pretty well accepted today. Joly collaborated with Henry Horatio Dixon and was the first to explain how sap rises in plants largely due to evaporation from leaves.

joule (J) A unit of work and energy; 1 J = 0.239 cal; 1 cal = 4.184 J. It is defined as being equal to the work done when the point of application of a force of 1 newton (N) moves in the direction of the force, a distance of 1 meter (m).

junk DNA (noncoding DNA) Genomic DNA that serves no apparent purpose; stretches of DNA that do not code for genes.

Jurassic period The middle period of the Mesozoic era, 213 to 145 million years ago. Age of the dinosaurs. Named after the Jura Mountains between France and Switzerland, where rocks of this age were first studied.

See also GEOLOGICAL TIME.

juvenile A stage in development prior to adult stage.

juvenile hormone A chemical hormone in insects secreted by a pair of endocrine glands, corpora allata, close to the brain. It inhibits metamorphosis and maintains larva or nymph characteristics during development and is responsible for determining the molt type. One of three major insect development hormones. Used as a pesticide, it retards the development of insects.

juxtaglomerular apparatus A group of specialized cells or tissue in a kidney nephron that is located near the point where the afferent arteriole meets the distal tubule. It is composed of the macula densa, a specialized group of cells in the distal tubule, and the juxtaglomerular cells, epithelioid cells in the media of the afferent arterioles just as they enter the glomerulus. The juxtaglomerular apparatus supplies blood to the GLOMERULUS, controls the glomerular filtration rate (volume of plasma filtered through the glomerulus per second), blood pressure, and circulating volume through the release of rennin, which activates angiotensin, a family of peptide hormones that control blood pressure and body fluid levels.

K

kala-azar (Leishmaniasis) A disease found in tropical countries and southern Europe caused by a protozoan parasite, *Leishmania donova,* obtained through the bite of a sand fly. It affects some 1.5 million people each year. Also known as dumdum fever or visceral leishmaniasis.

kame A short hill, ridge, or mound of stratified drift deposited by glacial meltwater.

kappa convention *See* DONOR ATOM SYMBOL.

karyogamy In syngamy, the process of union of two gametes (fertilization), there are two processes. The first step, called plasmogamy, is the fusion of cytoplasm of two cells. The second step is karyogamy, the fusion of the nuclei of two cells. It is the fusion of two compatible haploid nuclei to form one diploid nucleus.

karyotype A method of classifying the paired chromosomes of a cell in relation to number, size, and morphology. A microscopic picture is taken of an individual's chromosome set. The chromosomes are then stained with special color dyes that produce a distinct stripe or banding pattern. The chromosomes are then rearranged according to size from largest to smallest. Used to see gross chromosomal abnormalities with the characteristics of specific diseases.

kelp A group of large brown seaweeds or algae (class Phaeophyceae) that belong to the order Laminariales. They have a heteromorphic alternation of generations, which has two free-living life phases: the first or macroscopic diploid sporophyte generation and a microscopic haploid gametophyte generation. Sporophytes are typically differentiated into a holdfast, stipe, and one or more leaflike blades. They live in cold, nutrient-rich waters throughout the world in shallow open coastal waters. They are photosynthetic, which restricts them to clear shallow water not much deeper than 15–40 meters.

See also ALGAE.

Kendall, Edward Calvin (1886–1972) American *Biochemist* Edward Calvin Kendall was born on March 8, 1886, in South Norwalk, Connecticut. He was educated at Columbia University and obtained a B.S. in 1908, M.S. in chemistry in 1909, and a Ph.D. in chemistry in 1910.

From 1910 until 1911 he was a research chemist for Parke, Davis and Co., in Detroit, Michigan, and conducted research on the thyroid gland, continuing the work from 1911 until 1914 at St. Luke's Hospital, New York.

In 1914 he was appointed head of the biochemistry section in the graduate school of the Mayo Foundation, Rochester, that is part of the University of Minnesota. In 1915 he was appointed director of the division of biochemistry and subsequently professor of physiological

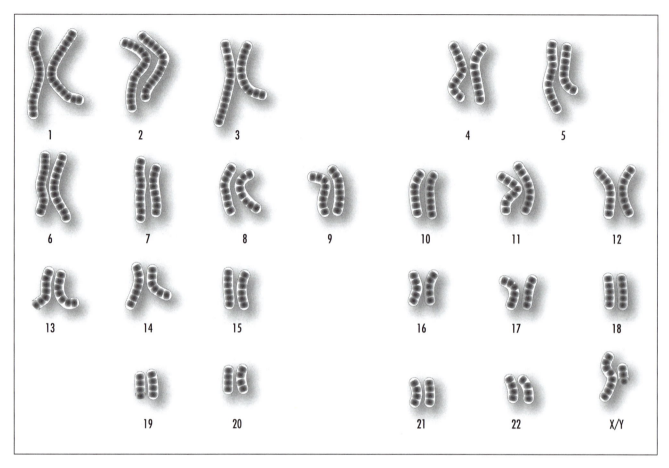

The term *karyotype* refers to the chromosomal complement of an individual, including the number of chromosomes and any abnormalities. The term is also used to refer to a photograph of an individual's chromosomes. *(Courtesy Darryl Leja, NHGRI, National Institutes of Health)*

chemistry. In 1951 he retired from the Mayo Foundation and accepted the position of visiting professor in the department of biochemistry at Princeton University.

In 1914 he isolated thyroxine, the active principle of the thyroid gland, and also discovered the crystallization and chemical nature of glutathione, and conducted work on the oxidation systems in animals.

Kendall isolated and identified a series of compounds from the adrenal gland cortex, and while working at Merk & Co., Inc., he prepared cortisone by partial synthesis. He also investigated the effects of cortisone and of adrenocorticotropic hormone (ACTH) on rheumatoid arthritis with Philip S. HENCH, H. F. Polley, and C. H. Slocumb. Kendall and Hench, along with Tadeus REICHSTEIN, shared the Nobel Prize in physiology or medicine in 1950 for this work. Kendall received many awards and honors. He died on May 4, 1972.

keratin   A tough, insoluble, fibrous protein with high sulfur content that forms the main structure and protective barrier or cytoskeleton of epidermal cells and is the chief constituent of skin, hair, nails, and enamel of the teeth. It is produced by keratinocytes, the most abundant cells in the epidermis (95 percent). Keratin makes up 30 percent of the cellular protein of all living epidermal cells. The high amount of sulfur content is due to the presence of the amino acid cystine.

keystone predator   A dominant species as predator that maintains species richness in a community through selective predation of the best competitors in the community, and as a result maintains populations of less competitive species.

The ocher star (*Pisaster orchraceous*) is a keystone predator because it prevents mussels, its prey, from taking over intertidal areas. Sea otters are a keystone predator in kelp beds, since they eat the urchins that feed on kelp, whose beds maintain a diversity of other organisms.

The removal of a keystone predator from an ecosystem causes a reduction of the species diversity among its former prey.

Keystone species increase or decrease the diversity of a system because they play a dominant role and affect many other organisms, including the death and disappearance of the dependent species.

kidney An organ, paired, in vertebrates that regulates secretion and osmoregulation as part of the urinary system. Filtration takes place at the site of the glomerulus in the Bowman's capsule, while the loops of Henle are responsible for taking waste material to be excreted. Millions of nephrons within the renal cortex and renal medulla filter the blood and regulate the volume and composition of body fluids during the formation of urine. A ureter from each kidney carries urine produced in the kidney to the bladder for elimination.

kilocalorie (kcal) One kilocalorie is equal to 1,000 calories, a unit of heat equal to the amount of heat required to raise the temperature of 1 kilogram of water by 1 degree at one atmosphere pressure; used to measure the energy value in food and labor; usually just called calorie: 1 kilocalorie (kcal) = 1 Calorie (Cal) = 1,000 calories (cal). However, in the International System of Units (ISU), the universal unit of energy is the joule (J). One kilocalorie = 4.184 kilojoules (kJ).

kilogram The basic unit of mass (not of weight or of force). A kilogram is equal to the mass of 1.000028 cubic decimeters of water at the temperature of its maximum density.

kinematics A division of mechanics that studies objects in motion (like an animal's gait) and their changes in position and the effects of motion on distance and time of travel. Differs from the study of mechanics, which also includes the effects of mass and force.

kinesis When an animal finds itself in an unwanted environment, kinesis is its ability to respond to the intensity of this stimulus by undirected movement in a random way (trial and error).

kinetic energy Energy of motion; kinetic energy depends on the object's mass and velocity and can be described mathematically as K.E. = $1/2mv^2$. Moving matter, be it a rolling rock, flowing water, or falling ball, transfers a portion of its kinetic energy to other matter. For example, an inelastic collision is one in which at least a portion of the kinetic energy of the colliding particles is lost. Potential energy, energy stored in a body, can be converted to kinetic energy.

kinetochore A specialized region or structure on the centromere of chromosomes, the region that joins two sister chromatids; links each sister chromatid to the mitotic spindle. The mitotic spindle is the specialized region on the chromatid where kinetochores and sister chromatids attach. When a chromosome replicates after mitosis or meiosis, it produces two side-by-side chromatids, with each eventually becoming a separate chromosome.

See also CENTROMERE.

kinetochore fibers The microtubules that connect kinetochores to spindle polar fibers.

kinetosome (basal body) The structure at the base of a flagellum or cilium that rises from the centriole and consists of a cylinder composed of nine longitudinally oriented, evenly spaced, triplet microtubules surrounding one central pair, called the axoneme. Usually found in pairs. Also called a blepharoplast. The type of kinetosomes—based on number, structure, and position—are used to type ciliates.

kingdom Taxonomic name used to organize, classify, and identify plants and animals. There are five taxonomic kingdoms: Monera, Protista, Plantae, Fungi, and Animalia. Only the domain is higher in ranking. This system of ranking, called the Linnaean system, was developed by the Swedish scientist Carolus Linnaeus (1707–78), who developed a two-name system, binomial nomenclature (genus and species), for identifying and classifying all living things. The system is based on a hierarchical structure in which organisms are sorted by kingdom, phylum, class, order, family, genus, and species. Organisms belonging to the same kingdom do not have to be very similar, but organisms belonging to the same species are very similar and can reproduce and create offspring.

See also TAXON.

kinocilium A long cilium at the apex of a hair cell, along with other microvilla, that is used in the process of hearing. It senses movement of cupula and is important for balance.

kinship The act of organizing individuals into social groups, roles, and categories based on parentage, marriage, or other criteria.

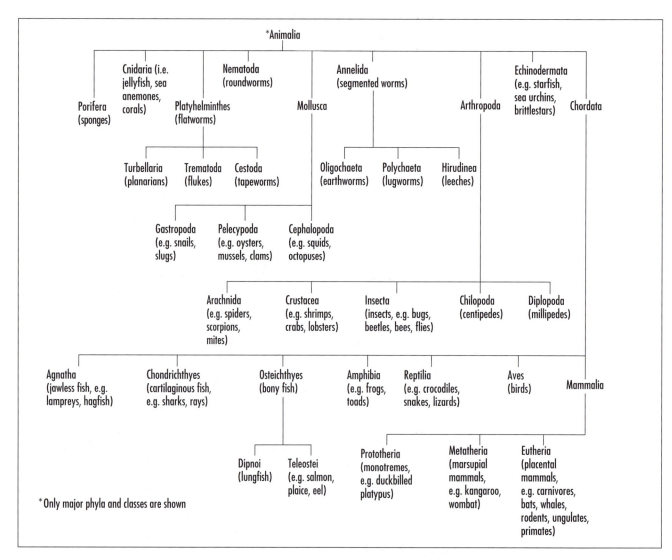

The animal kingdom is one of five taxonomic groupings of living organisms, the others being Monera, Protista, Plantae, and Fungi.

klinotaxis A movement in a specific direction relative to a given stimulus, either directly toward or away from the source.

Koch, Robert (1843–1910) German *Bacteriologist*
Robert Koch was born on December 11, 1843, in Clausthal in the Upper Harz Mountains and is considered one of the fathers of modern medical bacteriology (along with Louis Pasteur). The son of a mining engineer, he taught himself to read by age five. He attended the local high school and developed an interest in biology.

In 1862 Koch attended the University of Göttingen to study medicine under professor of anatomy Jacob Henle. In 1840 Henle had published that living, parasitic organisms caused infectious diseases. In 1866 Koch married Emmy Fraats and they had one daughter. After receiving his M.D. degree, Koch went to Berlin for further study. In 1867, after a period as assistant in the general hospital at Hamburg, he settled into general practice at Langenhagen and, in 1869, at Rackwitz, in the province of Posen. There he passed his district medical officer's examination, but in 1870 he volunteered to serve in the Franco-Prussian War. From 1872 to 1880 he was district medical officer for Wollstein and while there carried out his groundbreaking research.

Studying from his four-room flat, with a microscope and other equipment he purchased, he began research on anthrax, earlier discovered by other scientists to be caused by a bacillus. Koch wanted to prove whether or not anthrax was actually caused by the bacillus, and he used mice and bacilli taken from the spleens of dead farm animals. He found that the mice did die from the bacillus and proved that the blood of anthrax-infected animals transmitted it. He also showed that the bacilli could produce resistance spores when environmental conditions were unfavorable, and that they could reemerge as viable anthrax-causing organisms when conditions changed. Koch's work became internationally known when it was published in 1876. He was the first person to establish a definite causal connection between a particular disease and a particular bacillus. He continued working on methods of studying bacteria and in 1878 published his results that included how to control infections caused by bacteria.

In 1880 he was appointed a member of the Reichs-Gesundheitsamt (Imperial Health Bureau) in Berlin and continued to refine his methods in bacteriological research. He invented new methods of cultivating pure cultures of bacteria on solid media and on agar kept in the special kind of flat dish invented by his colleague Petri, which is still in common use (the Petri dish). He also developed new methods of staining bacteria, making them easily visible for identification. All of this work helped establish the methods to study pathogenic bacteria easily in pure cultures. Koch also laid down the conditions, known as Koch's postulates, that must be satisfied before it can be accepted that particular bacteria cause particular diseases. These postulates were:

- The specific organism should be shown to be present in every case of the disease.
- The specific microorganism should be isolated from the diseased animal and grown in pure culture on artificial laboratory media such as in a Petri dish.
- The freshly isolated microorganism, when inoculated into a healthy laboratory animal, should cause the same disease seen in the original animal.
- The microorganism should be recovered from the experimentally infected animal.

Koch also discovered the tubercle bacillus, the cause of tuberculosis, and developed a method of growing it in pure culture, and in 1882, he published his now classical work on the bacillus. In 1883 he was sent to Egypt as leader of the German Cholera Commission to investigate a cholera outbreak and soon discovered the bacteria that causes cholera and brought back pure cultures of it to Germany. He also studied cholera in India.

Koch formulated rules for the control of epidemics of cholera in 1893 and formed the basis of the methods of control. His work on cholera was rewarded with a prize of 100,000 German marks.

In 1885 Koch was appointed professor of hygiene in the University of Berlin and director of the newly established Institute of Hygiene in the university. Five years later he was appointed brigadier general (*generalarzt*) class I and freeman of the city of Berlin. In 1891 he became an honorary professor of the medical faculty of Berlin and director of the new Institute for Infectious Diseases. In 1893 Koch married Hedwig Freiberg. Koch continued to travel and explore the causes of many other diseases.

Koch was the recipient of many prizes and medals; honorary doctorates from the Universities of Heidelberg and Bologna; honorary citizenships of Berlin, Wollstein, and his native Clausthal; and honorary memberships of learned societies and academies in Berlin, Vienna, Posen, Perugia, Naples, and New York. He was awarded the German Order of the Crown, the Grand Cross of the German Order of the Red Eagle, and Orders from Russia and Turkey.

In 1905 Koch was awarded the Nobel Prize in physiology or medicine for his pioneering work on tuberculosis and continued his experimental work on bacteriology and serology. He died on May 27, 1910, in Baden-Baden.

Though he made great strides in the study of tuberculosis, today, 2 billion people worldwide suffer with latent infection. There are 8 million new cases a year, and up to 2 million annual deaths.

Kocher, Theodor Emil (1841–1917) Swiss *Surgeon* Theodor Kocher was born on August 25, 1841, in Berne, Switzerland, to a Swiss engineer. He received his medical doctorate in Berne in 1865. In 1872 he became ordinary professor of surgery and director of the University Surgical Clinic at Berne, where he remained for the rest of his career.

He discovered a new method for the reduction of dislocations of the shoulder in 1902, which he explained in his *Mobilization of the Duodenum.*

He published numerous works on the thyroid gland, hemostasis, antiseptic treatments, surgical infectious diseases, gunshot wounds, and more. His book *Chirurgische Operationslehre* (Theory on surgical operations) reached six editions and was translated into many languages. His book *Erkrankungen der Schilddrüse* (Diseases of the thyroid gland) discussed the etiology, symptology, and treatment of goiters. His new ideas on the physiology and pathology of the thyroid gland caused great controversy, but after many successful surgeries, his work on goiter treatment became world known and accepted. In 1909 he received the Nobel Prize for this work and donated the prize money to create a research institute. Kocher devised many new surgical techniques, instruments, and appliances that carry his name, such as forceps (Kocher and Ochsner/Kocher forceps, Kocher tweezers), Kocher incision (in gallbladder surgery), the Kocher vein (thyroid), Kocherization (surgery technique), and Kocher's test (for thyroid-tracheomalacia).

The Kocher-Debré-Semélaigne syndrome is named for him and Robert Debré and Georges Semélaigne. There is also Kocher's reflex (contraction of abdominal muscles following moderate compression of the testicle) and Kocher's sign (eyelid movement). Kocher died in Berne on July 27, 1917.

Koch's postulates   Criteria proposed in the 19th century by Nobel Prize winner (1905) Robert Koch to determine whether a microbe is the cause of a particular infection. He laid down the following postulates, which must be satisfied before it can be accepted that particular bacteria cause particular diseases.

- The specific organism should be shown to be present in every case of the disease.
- The specific microorganism should be isolated from the diseased animal and grown in pure culture on artificial laboratory media such as in a Petri dish.
- The freshly isolated microorganism, when inoculated into a healthy laboratory animal, should cause the same disease seen in the original animal.
- The microorganism should be recovered from the experimentally infected animal.

See also KOCH, ROBERT.

Kossel, Ludwig Karl Martin Leonhard Albrecht (1853–1927) German *Chemist, Medical doctor* Ludwig Karl Martin Leonhard Albrecht Kossel was born in Rostock on September 16, 1853, the eldest son of Albrecht Kossel, a merchant and Prussian consul, and his wife Clara Jeppe. He attended the secondary school in Rostock and went to the newly founded University of Strassburg in 1872 to study medicine. He received his doctor of medicine in 1878.

Kossel specialized in chemistry of tissues and cells (physiological chemistry), and by the 1870s he had begun his investigations into the constitution of the cell nucleus. He isolated nucleoproteins from the heads of fish sperm cells in 1879. By the 1890s he had focused on the study of the proteins. In 1910 he received the Nobel Prize in physiology or medicine for his contributions in cell chemistry and work on proteins.

Among his important publications are *Untersuchungen über die Nukleine und ihre Spaltungsproducte* (Investigations into the nucleins and their cleavage products), 1881; *Die Gewebe des menschlichen Körpers und ihre mikroskopische Untersuchung* (The tissues in the human body and their microscopic investigation), 1889–91, in two volumes, with Behrens and Schieerdecker; and *Leitfaden für medizinisch-chemische Kurse* (Textbook for medical-chemical courses), 1888. He was also the author of *Die Probleme der Biochemie* (The problems of biochemistry), 1908; and *Die Beziehungen der Chemie zur Physiologie* (The relationships between chemistry and physiology), 1913.

Kossel had honorary doctorates from the Universities of Cambridge, Dublin, Ghent, Greifswald, St. Andrews, and Edinburgh, and he was a member of various scientific societies, including the Royal Swedish Academy of Sciences and the Royal Society of Sciences of Uppsala. Albrecht Kossel died on July 5, 1927.

Krakatoa   An island volcano along the Indonesian arc, between the islands of Sumatra and Java. Krakatoa erupted in 1883 in one of the largest eruptions in history and was heard as far away as Madagascar (2,200 miles). The volcanic dust veil that created spectacular atmospheric effects, like vivid red sunsets, acted as a solar radiation filter, lowering global temperatures as much as 1.2°C in the year after the eruption. Temperatures did not return to normal until 1888.

Krebs, Sir Hans Adolf (1900–1981) German *Biochemist* Sir Hans Adolf Krebs was born in Hildesheim, Germany, on August 25, 1900, to Georg Krebs, M.D., an ear, nose, and throat surgeon of that city, and his wife Alma (née Davidson).

Krebs was educated at the Gymnasium Andreanum at Hildesheim. Between 1918 and 1923 he studied medicine at the Universities of Göttingen, Freiburg-im-Breisgau, and Berlin. He received an M.D. degree at the University of Hamburg in 1925. In 1926 he was appointed assistant to Professor Otto Warburg at the Kaiser Wilhelm Institute for Biology at Berlin-Dahlem, where he remained until 1930. He was forced to leave Germany in 1933 because of his Jewish background.

In 1934 he was appointed demonstrator of biochemistry at the University of Cambridge, and the following year was appointed lecturer in pharmacology at the University of Sheffield. In 1938 he became the newly founded lecturer-in-charge of the Department of Biochemistry. In 1939 he became an English citizen. By 1945 he was a professor and director of the Medical Research Council's research unit established in the department. In 1954 he was appointed Whitley Professor of Biochemistry in the University of Oxford, and the Medical Research Council's Unit for Research in Cell Metabolism was transferred to Oxford.

At the University of Freiburg in 1932, he discovered a series of chemical reactions (now known as the urea cycle) by which ammonia is converted to urea in mammalian tissue. For his discoveries of chemical reactions in living organisms now known as the citric acid cycle or the Krebs cycle, he was awarded the 1953 Nobel Prize in physiology or medicine. These reactions involve the conversion, in the presence of oxygen, of substances that are formed by the breakdown of sugars, fats, and protein components to carbon dioxide, water, and energy-rich compounds.

Krebs was a member of many scientific societies, winning many awards and citations for his work, and published works including *Energy Transformations in Living Matter* (1957) with British biochemist Hans Kornberg. He was knighted in 1958. He died on November 22, 1981, in Oxford, England.

Krebs cycle   A biochemical cycle in the second stage of cellular respiration involving eight steps that completes the metabolic breakdown of glucose molecules to carbon dioxide. Acetyl coenzyme A (CoA) is combined with oxaloacetate to form citric acid. Citric acid is then converted into a number of other chemicals, and carbon dioxide is released. The process takes place within the mitochondrion. Also called the citric acid cycle or tricarboxylic acid (TCA) cycle, it was conceived and published by scientist Hans Adolf Krebs in 1957.

Krogh, Schack August Steenberg (1874–1949) Danish *Physiologist* Schack August Steenberg Krogh was born in Grenaa, Jutland, Denmark, on November 15, 1874, to Viggo Krogh, a shipbuilder, later brewery master, and Marie Drechmann. He earned his upper secondary school diploma at the Cathedral School of Århus in 1893 and entered the University of Copenhagen in 1893. He began his

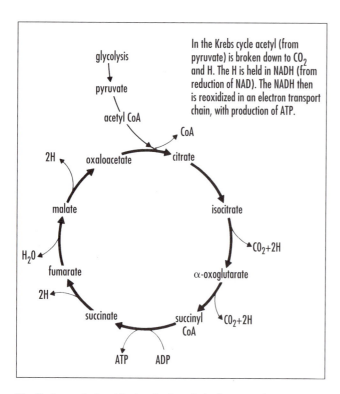

In the Krebs cycle acetyl (from pyruvate) is broken down to CO_2 and H. The H is held in NADH (from reduction of NAD). The NADH then is reoxidized in an electron transport chain, with production of ATP.

The Krebs cycle is a biochemical cycle in the second stage of cellular respiration, involving eight steps, that completes the metabolic breakdown of glucose molecules to carbon dioxide.

studies in medicine but turned to zoology, an early childhood interest.

In 1897 he began work for Christian Bohr at the laboratory of medical physiology and later became his assistant. In 1916 he became chair of the zoophysiology department at the University of Copenhagen, a position he held until he retired in 1945.

In 1904 he constructed a microtonometer, making it possible to determine the gas content of blood with great precision. He established that the movement of gases between the alveoli and the blood occurs through passive diffusion. With exercise physiologist Johannes Lindhard, Krogh began the field of exercise physiology in Scandinavia between 1910 and 1915. He developed an automatically controlled bicycle ergometer (1910–11) that is still in use today for exercise. He also demonstrated that muscles burn fat during exercise, and included in his inventions was a climate chamber that could be carried to measure continuous temperature and humidity over a 24-hour period. For his work on capillaries, discovering capillary-motor-regulating mechanisms, he received the Nobel Prize in physiology or medicine in 1920. Some of his important writings include *The Anatomy and Physiology of Capillaries*

An example of the scrubby condition of trees at the tree line, known as krummholz, in the White Mountains of New Hampshire. *(Courtesy of Tim McCabe)*

(1922, rev. ed. 1959), *Osmotic Regulation in Aquatic Animals* (1939), and *Comparative Physiology of Respiratory Mechanisms* (1941). He died on September 13, 1949, in Copenhagen.

krummholz Trees that grow near the tree line and that are profoundly shaped by wind and ice, often twisted and flaglike or stunted. *Krummholz* literally means "crooked wood."

K-selection Natural selection, or pattern of reproduction, that occurs in populations at or near the carrying capacity (K) of a normally stable environment. Characteristics of these populations are slow reproduction, increased longevity of individuals, delayed maturity of offspring, or reduction in the number of offspring, with increased level of parental care. In effect, the K-selection strategy is to ensure maximum survival of the individual over that of reproductive capacity. A prime example is humans.

See also R-SELECTION.

k-T boundary A transitional stage in geology from the end of the Cretaceous (k) to the beginning of the Tertiary (T), approximately 65 million years ago. Also pertains to a site or area showing evidence of rocks and other materials from both periods. There is a superabundance of iridium and osmium isotopes typical of meteorites in this zone, and this has led some to suggest that the mass extinction of species including dinosaurs was caused by a large Earth collision with asteroids or meteorites.

kuru (trembling disease) A slowly progressive and fatal disease of the brain; a human transmissible spongiform encephalopathy (TSE) caused by a prion. Found in New Guinea and transmitted when people handled and ate the brains of their dead relatives, a cannibalistic ritual in Papua New Guinea.

L

labile The term has loosely been used to describe either a relatively unstable and transient chemical species or a relatively STABLE but reactive species.

See also INERT.

lability Instability; refers to a state that is easily altered or modified. A phenotype's capacity to respond differentially to varying environmental conditions. Also called phenotypic plasticity.

laccase A copper-containing ENZYME, 1,4-benzenediol oxidase, found in higher plants and microorganisms. Laccases are MULTICOPPER OXIDASES of wide specificity that carry out one-electron oxidation of phenolic and related compounds and reduce O_2 to water. The enzymes are polymeric and generally contain one each of TYPE 1, TYPE 2, TYPE 3 COPPER centers per SUBUNIT, where the type 2 and type 3 are close together, forming a trinuclear copper CLUSTER.

See also NUCLEARITY.

lactate Alternate name for lactic acid, a chemical created from sugars when broken down for energy in the absence of oxygen.

lacteal Numerous small lymphatic vessels in the interior of each intestinal villus, small fingerlike projections of the mucosal layer of the small intestine, that picks up chyme, a thick semifluid mass of partially digested food, and passes it through the lymph system via the thoracic duct to the blood system; aids in the absorption of fats.

lactoferrin An iron-binding protein from milk, structurally similar to the TRANSFERRINS.

lagging strand In DNA synthesis, polymerization occurs both in and away from the nearest replication fork. One of two newly created DNA strands is the lagging strand. Found at the replication fork on linear chromosomes, it is synthesized in the direction away from the replication fork. It is the synthesis of a new strand of replication DNA by the creation of short segments of various lengths that are later joined together covalently by the enzyme DNA ligase.

It is made in discontinuous lengths, called Ozark fragments, in the 5' to 3' growing-tip (number of atoms in the sugar residues) direction during DNA polymerization, and these are joined covalently at a later time. Since it is not synthesized continuously, it is called discontinuous replication.

The difference between the lagging and leading strands is due to the orientation of the parent template strands. The leading-strand template is facing a 5' to 3' direction, but the lagging strand itself is oriented in the 3' to 5' direction, so the DNA polymerase responsible for adding nucleotides has to move backwards away from the replication fork; synthesis is therefore not continuous, but in repeated steps.

There is a gap between the final lagging strand segment and the end of the chromosome. As a result, the 5' end of the lagging strand will lose some nucleotide every time a cell replicates its DNA.

See also LEADING STRAND.

Lamarckism The idea, promoted by Jean Baptiste de Lamarck, that acquired traits can be passed from parent to offspring, i.e., that characteristics or traits acquired during a single lifetime can be transmitted directly to offspring.

land bridges Pieces of land once connecting the continents that have since sunk into the sea as part of a general cooling and contraction of the Earth. Believed to have served as migratory passages for animals, plants, and humans.

Landsteiner, Karl (1868–1943) Austrian *Biochemist* Karl Landsteiner was born in Vienna on June 14, 1868, to Leopold Landsteiner, a journalist and newspaper publisher, and Fanny Hess. Landsteiner studied medicine at the University of Vienna, graduating in 1891.

From 1898 until 1908 he held the post of assistant in the university department of pathological anatomy in Vienna. In 1908 he received the appointment as prosector in the Wilhelminaspital in Vienna, and in 1911 he became professor of pathological anatomy at the University of Vienna.

Landsteiner, with a number of collaborators, published many papers on his findings in anatomy and immunology, such as the immunology of syphilis and the Wassermann reaction, and he discovered the immunological factors, which he named haptens. He also laid the foundations of the cause and immunology of poliomyelitis.

His 1901 discovery of the major blood groups and his development of the ABO system of blood typing and 1909 classification of the blood of human beings into the now well-known A, B, AB, and O groups, as well as the M and N groups, made blood transfusion a routine medical practice. For this work, he received the Nobel Prize for physiology or medicine in 1930. In 1936 he wrote *The Specificity of Serological Reactions*, a classic text that helped to establish the science of immunochemistry. In 1940 he discovered the Rh factor, the protein on the surface of red blood cells that determines if the blood type is positive (Rh-positive) or negative (Rh-negative). If the mother has a negative Rh factor (Rh-negative) and the father and fetus are Rh-positive, the mother can become Rh-sensitized and produce antibodies to combat fetal blood cells that cross the placenta into her bloodstream. These antibodies can destroy the fetus's Rh-positive blood cells, putting it at serious risk of anemia.

In 1939 he became emeritus professor at the Rockefeller Institute. On June 24, 1943, he had a heart attack in his laboratory and died two days later.

larva The solitary live, but sexually immature, form of a variety of animal life cycles, e.g., butterflies, flies, wasps, that may, after reaching adulthood, be completely different in morphology, habitat requirements, and food needs. Some larvae have other names, like maggots for flies, grub for beetle larvae, tadpoles for frogs, etc. Insect larvae may molt, i.e., shed layers of skin, several times during their development.

latentiated drug *See* DRUG LATENTIATION.

lateral line system A sensory system composed of a longitudinal row of porelike openings that open into tubes in skin on the sides of fish and larval amphibians that is used to detect water and electrical disturbance in their surroundings. The ampullary organs, or ampullae of Lorenzina in fish, detect weak electrical currents generated by other animals, while the neuromasts—a small group of pairs of oppositely oriented hair sensory cells embedded in a gel-filled cupula on either side of the skin surface or in pit organs—detect the direction of water movement. The lateral canal is the portion of the system located on the head, beginning at the junction with the supra and infraorbital canals behind the mid-eye, and extending back to the rear of the head connecting to the lateral line.

lateral meristem An embryonic tissue, the meristem, on the portion of a plant that gives rise to secondary growth such as the cambium, vascular cambium, and cork cambium. Runs most of the length of stems and roots. Also called the cambium.

Laurasia Laurasia was the northern supercontinent formed after PANGAEA broke up during the JURASSIC PERIOD some 180 million years ago and formed the present continents of North America, Europe, and Asia as well as the land masses of Greenland and Iceland.

lava Igneous rock, magma that exits volcanoes and vents and reaches the exterior or surface of the land or seabed.

Laveran, Charles-Louis-Alphonse (1845–1922) French *Physician, Protozoologist* Charles-Louis-Alphonse Laveran was born in Paris on June 18, 1845, in the house at 19 rue de l'Est, to Dr. Louis Théodore Laveran, an army doctor and professor at the École de Val-de-Grâce, and Guénard de la Tour, the daughter and granddaughter of high-ranking army commanders.

After completing his education in Paris at the Collège Saint Baube and later at the Lycée Louis-le-Grand, he applied to the Public Health School at Strasbourg in 1863 and attended the school for four years. In 1866 he was appointed a resident medical student in the Strasbourg civil hospitals, and the following year he submitted a thesis on the regeneration of nerves. In 1874 he was appointed to the chair of military diseases and epidemics at the École de Val-de-Grâce, previously occupied by his father, and in 1878, when his period of office had ended, he moved to Bône in Algeria until 1883. It was during this period that he carried out his chief researches on the human malarial parasites, first at Bône and later at Constantine.

In 1880 Laveran examined blood samples from malarial patients and discovered amoebalike organisms growing within red blood cells, and he noticed that they divided, formed spores, and invaded unaffected blood cells. He also noted that the spores were released in each affected red cell at the same time and corresponded with a fresh attack of fever in the patient. His researches confirmed that the blood parasites that he had described were in fact the cause of malaria, but his first publications on the malaria parasites were received with skepticism until scientists around the world published confirmative research results. In 1889 the Academy of Sciences awarded him the Bréant Prize for his discovery.

In 1896 he entered the Pasteur Institute as chief of the Honorary Service, and from 1897 until 1907 he carried out many original research projects on endoglobular haematozoa and on sporozoa and trypanosomes. In 1907 he was awarded the Nobel Prize for his work on protozoa in causing diseases. In 1908 he founded the Société de Pathologie Exotique, over which he presided for 12 years.

He was the first to express the view that the malarial parasite must be a parasite of *Culicidae,* and after this view had been proved by the research of Ronald Ross, he played a large part in the enquiry on the relationships between *Anopheles* and malaria in the campaign undertaken against endemic disease in swamps, notably in Corsica and Algeria.

From 1900 on, he studied the trypanosomes and published, either independently or in collaboration with others, a large number of papers on these blood parasites. His research concentrated on the trypanosomes of the rat; the trypanosomes that cause nagana and surra; the trypanosome of horses in Gambia; a trypanosome of cattle in the Transvaal; the trypanosomiases of the Upper Niger; the trypanosomes of birds, chelonians, batrachians, and fishes; and finally the trypanosome that causes the endemic disease of equatorial Africa known as sleeping sickness. For 27 years he worked on pathogenic protozoa and the field he opened up by his discovery of the malarial parasites. He died on May 18, 1922.

law of equal segregation Gregor Mendel's first law, which says that two copies of a gene separate during meiosis and end up in different gametes (sperm and ova) and are passed on to offspring.

law of independent assortment Gregor Mendel's second law, which says that after gametes form, the separation of alleles for one gene is independent of the separation of alleles for other genes; genes located on nonhomologous chromosomes are independent from one another. This is not true for genes that are linked, where genes are located close to one another on the same chromosome.

lead discovery The process of identifying active new chemical entities that, by subsequent modification, can be transformed into a clinically useful DRUG.

lead generation The term applied to strategies developed to identify compounds that possess a desired but nonoptimized biological activity.

leading strand One of two new DNA strands that is being replicated continuously, unlike the lagging strand. The strand is made in the 5' to 3' direction, placing an –OH group at the 3' end for continuous polymerization (adding nucleotides) at the 3' growing tip by DNA polymerase that moves forward when the template strands unwind at the replication fork. It is the DNA polymerase that plays the pivotal role in the process of life, since it is responsible for duplicating the genetic information.
See also LAGGING STRAND.

lead optimization The synthetic modification of a biologically active compound to fulfill all stereoelectronic, physicochemical, pharmacokinetic, and toxicologic requirements for clinical usefulness.

leaf primordia Young leaves at the tip of a shoot formed by the shoot apical meristem.

leaf veins A netlike network called reticulate venation in vascular tissue in dicots. The veins are in parallel venation to each other in monocots.

learned behavior Behavior that is not fixed but acquired by trial and error or by observing others.

leghemoglobin A monomeric HEMOGLOBIN synthesized in the root nodules of leguminous plants that are host to nitrogen-fixing bacteria. Has a high affinity for dioxygen and serves as an oxygen supply for the bacteria.
See also NITROGEN FIXATION.

legume A pod-bearing plant that is a member of the Leguminosae (Fabaceae) or pea or bean family. These plants form symbiotic relationships with certain nitrogen-fixing bacteria (rizobia) in their root nodules to acquire nitrogen for growth. Examples include beans, peas, lentils, alfalfa, clover, and wildflowers like wild blue lupine (*Lupinus perennis*).

lek A place where males display in groups and females choose for the purpose of fertilization; a special kind of polygynous mating system.

lemur Primates (prosimians) of the superfamily Lemuroidea that live in trees; found only on the island of Madagascar and the adjacent tiny Comoro Islands. Related to monkeys, they have large eyes, foxlike faces, and long furry tails (not prehensile). There are only about 30 to 50 living species, 17 of which are endangered. They are small, ranging from about an ounce up to about the size of a house cat.

lenticle A pore where gases are exchanged in the stems of woody plants.

Leonardo da Vinci (1452–1519) Italian *Scientist, Artist* Leonardo da Vinci, one of the greatest minds of all time, was born on April 15, 1452, near the town of Anchiano near Vinci. He was an illegitimate child of a notary, Piero da Vinci, and a peasant woman named Caterina. In his teenage years he became an apprentice in one of the best art studios in Italy in 1469, that of

Andrea Verrocchio, a leading Renaissance master of that time. During this time da Vinci drew *La valle dell'Arno* (The Arno Valley) in 1473 and painted an angel in Verrocchio's *Baptism of Christ* (1475). In 1478 da Vinci became an independent master. Da Vinci is famous for his works of art such as the Mona Lisa, but he is also as famous for his visionary drawings of instruments and machines of the future. He was an artist, scientist, engineer, and architect. He was also one of the first to take detailed observations and to experiment in a scientific manner.

In his later notebook the *Codex Leicester,* one finds the largest assemblage of da Vinci's studies relating to astronomy, meteorology, paleontology, geography, and geology. It reveals that his profound scientific observations far outweigh anyone else of his time and documents his passion for research and invention. His interest in light and shadow led him to notice how the earth, moon, and planets all reflect sunlight, for example.

The central topic of the *Codex Leicester* is the "Body of the Earth" and in particular its transformations and movement of water. This study includes a discussion on the light of the moon, the color of the atmosphere, canals and flood control, the effect of the moon on the tides, and modern theories of the formation of continents.

He was well acquainted with knowledge about his local rocks and fossils (Cenozoic mollusks) and was uncannily prescient in his interpretation of how fossils were found in mountains, theorizing that they had once been living organisms in seas before mountains were raised. Da Vinci contributed a great deal of knowledge to human anatomy with detailed drawings and notes in his anatomical notebooks after having dissected some 19 human cadavers.

Unfortunately, many of his scientific projects and treatises were never completed, since he recorded his technical notes and sketches in numerous notebooks and used mirror script (his writing had to be read in a mirror to be deciphered). It was centuries before the genius of da Vinci was recognized. He died at the age of 67 on May 2, 1519, at Cloux, near Amboise, France.

Lepidoptera The taxonomic order that includes butterflies, skippers, and moths. Butterflies have antennae, compound eyes, six pair of legs, a hard exoskeleton, and a body that is divided into a head, thorax, and abdomen. The butterfly's outer body is covered by tiny sensory hairs and has wings covered by scales.

lesion A visible region, such as a wound or fissure, where there is abnormal tissue change in a body part; a structural change due to wound, injury, or disease.

lethal mutation A mutant form of a gene that will result in the inviability (death) of the organism if expressed in its phenotype. A conditional-lethal mutation is lethal under one condition but not under another.

leukemia Cancer of the developing blood cells in the bone marrow. The rampant overproduction of white blood cells (leukocytes). A fatal cancer that is diagnosed yearly in 29,000 adults and 2,000 children in the United States. There are several types of leukemia.

See also CANCER.

leukocyte (**white blood cell**) A type of blood cell that contains a nucleus but has no pigment. White blood cells (WBC) are found in the blood and are involved in defending the body against infective organisms and foreign substances. They are produced in the bone marrow along with other types of blood cells. There are five types of white blood cell within two main groups, and each has its own characteristics. The polymorphonuclear or granulocyte group comprises the neutrophils, eosinophils, and basophils. The mononuclear group comprises the monocytes and lymphocytes.

White blood cells are the main attackers of foreign substances as part of the immune system.

Neutrophils move out of the blood vessels into the infected tissue and engulf the foreign substances (phagocytosis). Eosinophils migrate to body tissues and release toxic substances to kill foreign substances. Basophils, also called granular leukocytes, digest foreign objects from granules containing toxic chemicals. Monocytes, which contain chemicals and enzymes, ingest dead cells through phagocytosis and develop into macrophages (large white blood cells) as they migrate into various tissues. Lymphocytes, which inhabit the blood, produce antibodies and cells that go after foreign substances. Lymphocytes subtypes are B cells, T cells, NK cells, and null cells.

See also BLOOD.

Silk Degrees: A Tale of Moths and People, Part One

By James G. (Spider) Barbour

Silk moths are found around the world, in all hemispheres, mostly in the tropics and temperate zones. Silk moths, a group containing not only many of the world's most useful insects, but also many of the largest and most spectacularly beautiful, have had a long and positive impact on people's lives. The silk moths comprise two families, the Bombycine silk moths (family Bombycidae) and the Saturniids or giant silk moths (family Saturniidae).

Collectively, they inhabit nearly every type of ecosystem within their tremendous geographic range, from forests to prairies to mountains to deserts. Most silk moths are medium to large species, with a good number of enormous ones and very few small types. Like all lepidoptera (butterflies and moths), any silk moth has a four-stage life cycle, or complete metamorphosis, in which it begins life as an egg; hatches into a hungry, fast-growing caterpillar; enters a resting stage as a pupa; and transforms at last into a winged, reproductive adult. Most adult silk moths do not eat at all, having built up an ample supply of fat reserves in the caterpillar stage. Silk moth caterpillars are eating machines, growing from tiny hatchlings smaller than rice grains into fat sausagelike creatures, the largest the size of fingers. Most feed on leaves or some other plant material, tending to move from branch to branch, eating a leaf here and a leaf there, in order to avoid consuming their cover and camouflage.

After feeding, some species burrow into the ground or simply crawl under a sheltering object, molting one last time to form the sleeping pupa, a limbless ovoid or cylindrical thing, capable of nothing more than wiggling and twisting. To further protect this vulnerable life stage, many caterpillars provide a shield for the pupa, a cocoon made of silk, which is issued from large glands within the caterpillar's body and spun out through organs near its mouth called spinnerettes. Depending on the species and its habitat, a cocoon may be spun inside a rolled leaf, attached to a twig, spun among grass blades, placed under overhanging rocks, or deposited in the grooves of a tree trunk's rough bark. In temperate climates, the pupae typically remain dormant within their cocoons, the moths developing in the warmth of spring and emerging in summer. Some genera, such as the buck moths (*Hemileuca*), overwinter as eggs, deposited by the autumn-emerging females in thick rings on twigs or stems of the host plants.

The material of cocoons has been a part of human culture for thousands of years, probably having been discovered and used before the beginning of recorded history for purposes we can only guess at. The technology for the manufacture of silk fabric and the farming of silkworms (sericulture) dates back at least 4,000 years to ancient China, where it was a closely guarded secret until the third century B.C.E., when the Japanese acquired it. From there, silk production and the making of silk garments spread widely.

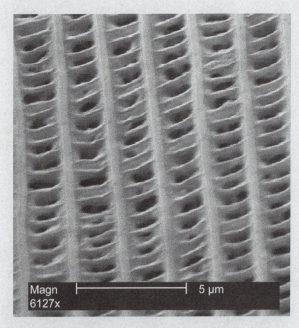

A highly magnified single-scale strutlike pattern from a butterfly's wing magnified ×6,127. Note the struts and perforations of the individual scale. This helps to promote heightened aerodynamic lift during the insect's flight while also reducing the weight of the wing mechanism. *(Courtesy of Janice Carr, Centers for Disease Control and Prevention)*

The first requirement of silk-making is the gathering of moth cocoons for processing into long threads for weaving the fabric. The earliest source of cocoons was almost certainly local wild moth populations. Silk-makers must have soon learned which moth species produced the best and most easily processed silk. The next step may have been some form of silk-moth domestication. In its simplest form, this might have meant securing eggs from fertile females and raising the caterpillars in protective custody on appropriate food plants, either living plants or fresh-cut foliage, in much the way amateur silk-moth enthusiasts raise them today. In time, however, more sophisticated, assembly-line-style rearing techniques were developed in which caterpillars were reared en masse indoors in containers.

Only one species of silk moth, the mulberry silkworm moth (*Bombyx mori*), a member of the family Bombycidae, has been completely domesticated. This species appears to have been selectively bred for rearing the caterpillars on flat screen trays stacked in tiers, an extremely space-efficient means of production. Over time, silk producers developed caterpillars lacking the velcrolike ends of the abdominal legs (called "prolegs," as they are not true legs), so that they did not crawl off the breeding trays, but stayed there feeding as long as they were supplied with leaves. A singular advantage to this species is that the cocoon, after the gummy sericin is boiled away, unravels

Banded-purple butterflies gathering at a mud puddle in the Adirondacks in order to siphon up sodium ions that are needed for metabolism. *(Courtesy of Tim McCabe)*

in one long, unbroken thread of fibrin, an extremely strong and supple protein. The cocoons of saturniid silk moths cannot be so easily untangled, nor is the resulting silk of such refined quality. Saturniid sericulture utilizes cocoons of several Asian species of the genus *Antheraea,* which also contains the familiar North American polyphemus moth (*Antheraea polyphemus*). In China, the coarse silk from *Antheraea* and a few other genera has long been used for everyday garments, including sturdy work clothes.

In North America, early colonial settlements in the Carolinas and Georgia produced silk for local use using the mulberry silkworm. Later, in the middle 19th century, commercial silk production was promoted in Europe and North America as a viable industry, but startups in the United States, mostly along the east and west coasts, proved too costly against foreign competition, bringing American sericulture to an early end. Its legacy remains, however, in the form of one lovely moth and its host plant, and in the form of an unlovely moth that haunts the Northeast to this day. A major silk promoter, the Frenchman Léopold Trouvelot, imported both moths in the 1860s.

The good moth is the ailanthus silk moth (*Samia cynthia*), which feeds in the wild only on the tree imported as food for its caterpillars, the tree-of-heaven (*Ailanthus altissima*), also known as "stink-tree" for the musty smell of the male flowers. The moth and the tree grow in many cities, mostly along the east coast, with scattered occurrences as distant as Pittsburgh and St. Louis.

Trouvelot's other introduction is the notorious gypsy moth (*Lymantra dispar*). Having escaped from Trouvelot's unsuccessful crossbreeding experiments with the mulberry silk moth (*Bombyx mori*), the gypsy moth invaded the local New England forests and spread, becoming perhaps the most devastating forest insect pest in the Northeast, with periodic outbreaks defoliating millions of acres of trees. Most trees recover, and the gypsy moth populations plummet after one to three years, so there is no lasting damage. The ailanthus silk moth is well behaved by comparison, never spreading from cities into the countryside, nor attacking trees other than its natural host. Though tree-of-heaven is widespread in disturbed portions of rural areas (along highways and power-line cuts, for example), the moth is never found in these wilder places, only in sooty railroad yards, unkempt used car lots and junk yards, and similar sites. Probably the caterpillars are so vulnerable to our native predators and parasites that inevitably all are killed

(continues)

Silk Degrees: A Tale of Moths and People, Part One
(continued)

except in the "enemy-free space" (Tuskes et al., 1996) of urban environments.

As food for human consumption, giant silk moths hold an important place in some cultures, and they might in Western culture were it not for an ubiquitous and intense aversion to eating insects among Europeans and Euro-Americans. Native American cultures of the Southwest included in their regular diet the caterpillars and pupae of the pinion pine moths (genus *Coloradia*). Throughout southern Africa, larvae and pupae of the native silk moths are eaten, fully integrated into local cuisines as appetizers, main dishes, and ingredients in soups and stews. Caterpillars of some species, such as the widespread and often abundant "Mopane worm" (*Gonimbrasia belina*), are frequently sold at markets. In China, surplus pupae from silk farms were commonly eaten in soups and rice-based dishes.

Today, as much silk-moth rearing occurs for enjoyment as for silk production. Breeders sell and swap cocoons and eggs, and share rearing tips and experiences. The world's silk-moth lovers have become more closely bonded by the Internet, with Web sites and forums uniting moth buffs from many nations. "The World's Largest Saturniidae Site," maintained by Bill Oehlke, a silk-moth fanatic who picked up the habit from his father, Don Oehlke, can be visited at http://www.silkmoths.bizland.com/indexos.htm.

—James G. (Spider) Barbour,
a field biologist specializing in entomology and ecology, has 25 years of field experience in the Hudson Valley, especially in the Catskills, Hudson Highlands, and Westchester County. He conducts botanical, rare species, wetland, plant community, and insect surveys. He and his wife, Anita, are the authors of *Wild Flora of the Northeast*.

Sassafras and Its Lepidopteran Cohorts, or Bigger and Better Caterpillars through Chemistry

by Timothy L. McCabe, Ph.D.

Joyfull News of the Newe Founde World was published in 1577, the first European pharmacopoeia to catalog materials from the Americas. In this work by Spanish physician Nicolás Monardes, we are told of the use of sassafras root bark for fever, liver discomfort, headache, bronchial congestion, stomach ailments, kidney stones, gout, toothache, arthritis, constipation, and infertility. Sassafras root bark became a principal export to Europe, second only to tobacco.

Sassafras is a member of the family Lauraceae, which comprises approximately 40 genera and more than 2,000 species. Besides sassafras, other familiar members include the cinnamon trees of the Orient, the West Indian avocado pear, and spicebush. Early accounts of New World explorers cite enormous populations of the sassafras tree along the entire Eastern Seaboard. Our native sassafras occurs from Maine to Michigan, and south to Florida and Texas.

Sassafras oil possesses a sweet, woody flavor and was widely employed to flavor toothpaste and soft drinks, particularly root beer, prior to 1960. It was discovered to be a carcinogen and banned from products after 1960. A hallucinogenic drug known as Ecstasy is derived from safrole, a secondary plant substance obtained from the roots of sassafras. Today, sassafras tea, produced by steeping the young roots, remains a fixture in Appalachian folk culture as a "spring tonic" and "blood thinner."

There are an estimated 100,000–400,000 different secondary compounds in plants. Secondary plant substances are those products not known to have any function in plant growth or metabolism. They were once thought to represent waste or by-products of plant metabolism. Current thinking associates these substances with chemical defenses against herbivores. Ironically, these same plant chemicals sometimes provide the herbivore with protection: monarch caterpillars store cardenolides, vertebrate heart poisons, obtained from their milkweed host.

Plants can even manufacture insect hormones. The "paper factor" story is a famous example. K. Slma came from Czechoslovakia to spend a year in C. M. Williams's laboratory at Harvard. He brought along his favorite laboratory insect, *Pyrrhocoris apterus,* a native European bug. Growing in Williams's laboratory, the bugs went through an extra stage to form giant, sexually immature, sixth-instar nymphs. These ultimately died without becoming sexually mature. All aspects of their rearing were studiously examined, and the culprit was determined to be the paper toweling used to line the cages. The toweling was derived from the balsam fir, a pulp tree indigenous to North America. This was the first indication of the existence of juvenile hormones naturally occurring in plants. Molting hormone, alpha ecdysone, has been found in ferns and various other plants. Apparently alpha ecdysone may not be at high enough concentrations to disrupt insect metabolism in every plant where it occurs; this has been demonstrated in ferns. Perhaps it represents an intermediate condition in the continual struggle for survival between plant and herbivore.

Secondary plant substances can also serve as cues for the insect to feed. Sinigrin, a mustard oil glycoside, is found in the turnip aphid's natural foodplants. When sinigrin is sprayed on plants not acceptable to the aphids, feeding commences and the aphid often does as well or better than on its normal host. A parasitoid of the cabbage aphid locates its prey by the smell of sinigrin. Highly aromatic camphor compounds, safrole (oil of sassafras) and syringin (a glycoside), are all secondary plant substances characteristic of sassafras. Syringin, a glycoside common in lilac (genus *Syringa*), should not be confused with the previously mentioned sinigrin. R. D. Gibbs in his *Chemotaxonomly of Flowering Plants,* published in 1974, states: "Syringin is relatively rare in the Plant Kingdom, and positive results from the Syringin Test are therefore of considerable interest."

The geographic distribution of the Saturniidae, as well as that of other members of the superfamily, combined with high endemism and disjunct distributions, are all indicative of an ancient lineage. Saturniidae and Apatelodidae, being ancient lineages, coevolved with the Lauraceae (among the most primitive flowering plants) and secondarily adapted to modern plants. *Apatelodes torrefacta* is still on sassafras, but its largely sympatric relative, *Olceclostera angelica,* is on ash and lilac, a peculiar association repeated by other sassafras lepidopterans, notably the promethea moth and the tiger swallowtail. The latter two also use cherry as does *A. torrefacta.* Ash, lilac, and cherry possess syringin, a glycoside that has been found in virtually all Oleaceae (includes lilac and ash), cherry, most Lauraceae (including spicebush and sassafras), as well as Magnoliaceae (magnolia). Privet also has syringin and is widely used by moth breeders as a substitute plant for rearing exotic saturniids.

Few lepidopterans feed on sassafras. Significantly, those that do are among our largest species (giant silk moths) or are large members of their respective families (e.g., *Epimecis hortaria* in the Geometridae). By some mechanism, possibly secondary plant substances, sassafras may be driving lepidoptera toward "bigness." Swallowtails reared on sassafras produced heavier pupae than those reared on other, even preferred, host plants. Despite their long association, smaller members of the Saturniidae, e.g., *Anisota,* do not utilize sassafras. W. T. M. Forbes in *The Lepidoptera of New York and Neighboring States* (1923) cites only four records for all microlepidopteran families on sassafras. This is in contrast to the list for the Saturniidae: *Automeris io, Callosamia promethea, C. angulifera, Hyalophora cecropia, Antheraea polyphemus, Eacles imperialis,* and *Citheronia regalis.*

Strikingly few lepidopterans have secondarily adapted to sassafras. The Noctuidae, a diverse family (over 3,000 species in the United States) of recent lineage, has only one host record for sassafras, *Hypsoropha hormos.* This species may have made the jump from persimmon (Ebenaceae), upon which it and its congeners normally feed, to sassafras by seeking kaempferol, a phenolic acid held in common by both plant families. Persimmon is a favorite of the luna moth. The luna, as a large member of the Saturniidae, is unusual in not using sassafras. Another secondary specialist is the spicebush swallowtail that feeds on both spicebush and sassafras, which have syringin in common. Syringin is rare in plants and makes it difficult to fathom the pathway that led the spicebush swallowtail to sassafras. Perhaps safrole is the common link in this case. Sassafras and pipevine both possess safrole. Pipevine (*Aristolochia*) serves as a host for the pipevine swallowtail, while sassafras serves as a host for the spicebush swallowtail. The pipevine and the spicebush swallowtails even mimic one another.

For a tree, sassafras does very well on poor soils. It possesses a very deep taproot. One should seek saplings less than a foot tall for transplant or else plan to do a lot of digging. It is a favorite deer browse and quickly becomes eliminated in the understory where deer are dominant. The variously lobed leaves are unusual and nearly unique for the Northeast. Sassafras leaves on the forest floor will attract adult lepidopterans that probe the decomposing tissues. Such feeding behavior has also been observed with fallen cherry leaves.

—**Timothy L. McCabe, Ph.D.,** is curator of entomology, New York State Museum in Albany, New York.

Lewis acid   A molecular entity that is an electron-pair acceptor and therefore is able to react with a LEWIS BASE to form a LEWIS ADDUCT by sharing the electron pair furnished by the Lewis base.

Lewis adduct   The adduct formed between a LEWIS ACID and a LEWIS BASE. An adduct is formed by the union of two molecules held together by a coordinate covalent bond.

See also COORDINATION.

Lewis base   A molecular entity able to provide a pair of electrons and thus capable of COORDINATION to a LEWIS ACID, thereby producing a LEWIS ADDUCT.

lice   Small (<$\frac{1}{6}$ in. long) wingless insects (Arthropoda: Insecta) that have three pairs of legs, and a body composed of a head, thorax, and abdomen. They have flattened, elongate bodies and somewhat oval heads and a three-stage life cycle (egg, nymph, and adult). The nymphs molt three times before reaching maturity. There are two suborders of lice, Mallophagorida (chewing lice, broad heads with biting mouthparts) and Anoplurorida (sucking lice, narrow heads with sucking mouthparts), and they are found worldwide.

Human lice come in three forms: head lice (*Pediculus humanus capitis*), body lice (*Pediculus humanus humanus*), and pubic lice (*Pediculus pubis*). They feed exclusively on humans, and their survival depends on temperature and humidity conditions associated with human bodies. They desiccate rapidly if they are away from humans for more than 24–36 hours.

lichen   Lichens are two organisms in one: Having no roots, stems, or leaves, they actually are symbiotic associations of a fungus and a photosynthesizing partner that results in a stable thallus of specific structure. The fungus is almost always an ascomycete, very rarely a basidiomycete (about 20 basidiolichens are known). Lichens are classified as fungi, and their scientific names formally refer to the fungal partner (mycobiont). The photosynthesizing partner (photobiont) can be a green alga, such as *Trebouxia* or *Trentepohlia,* or a cyanobacterium ("blue-green

alga"), such as *Nostoc* or *Scytonema*. The symbiotic unit has a definite, reproducible form that is different from the nonlichenized fungus or alga. A few lichens have two photobionts, and the fungus looks different where each occurs within the thallus; i.e., it may be flat where the green alga grows but have dark bumps where the cyanobacterium occurs, as in *Peltigera leucophlebia*. The lichen symbiosis thus involves closely merged members from usually two but sometimes three biotic kingdoms.

Worldwide in distribution, lichens can be found on bare soil or sand, grow over mosses and low plants, or become attached to rocks, trees, or almost any other substrate. They are good measures of air quality, since they have great sensitivity to sulfur dioxide (SO_2), an indicator of local pollution and acid rain. In addition, their tolerance and accumulation of metals, including heavy metals and radioactive isotopes, are useful indicators of industrial pollution.

Lichens can be classified into several types based on appearance and structure: Fruticose (shrublike) lichens, such as British soldiers (*Cladonia cristatella*) and common reindeer lichens (*Cladina rangiferina*), include species that look like tiny trees or columns but also include old man's beard (genus Usnea), which can dangle from tree branches. Foliose lichens, such as the puffed shield lichen (*Hypogymnia physodes*) are flat and leaflike. Crustose species, such as the map lichen (*Rhizocarpon geographicum*), form a "crust" on rocks, wood, or soils, and are very abundant. Squamulose lichens, somewhat similar to crusts, are composed of tiny overlapping or shingle-like flakes; examples are the primary (basal) thalli of *Cladonia*, and the oyster lichen (*Hypocenomyce scalaris*), which grows on acidic wood. Umbilicate lichens have leaflike thalli that attach to rocks at a central point; smooth rock tripe (*Umbilicaria mammulata*) is an example. Jelly lichens, such as the blue jellyskin (*Leptogium cyanescens*), have a flabby thallus and a cyanobacterial photobiont, and usually occur in damp habitats.

Many lichens have a layered internal morphology, with the bulk of the thallus formed of fungal filaments and the photobiont often restricted to a distinct layer. Fungal hyphae are densely arranged in the upper cortex to exclude other organisms and regulate light intensity to the green algal cells. These are often located just below the cortex. Underlying the algae is the medulla, another layer of much less compact fungal filaments.

Some lichens are held on their substrate by medullary fungal hyphae, but many foliose lichens have a lower cortex and special structures called rhizines to hold them in place on tree bark or rock.

There are only about 100 algal partners, but there are as many fungal partners as there are lichen species, which number about 14,000. The fungus receives sugars and other nutrients from the photobiont and provides a stable and secure structure in which it lives, provides water, and shields the photobiont from the sun's desiccating rays. Lichens reproduce vegetatively in three basic ways: fragmentation of the thallus, and by two kinds of tiny vegetative propagules that contain both partners—soredia, the more common, and isidia, which are slightly larger and heavier. Isidia and soredia are dispersed by wind, water, and animal agents. Lichens are slow growers and are particularly good pioneer species in disturbed habitats.

Lichens have been used as food, especially by caribou in winter; as medicine and dyes; as dwarf "trees" in model railroad setups and architectural plans; and in holiday wreaths and arrangements. And many harsh landscapes are much enriched by their subtle beauty.

See also ALGAE.

life table A tabular and numerical representation of mortality of births for each age group and includes various groups of information such as probabilities of death, survival, and life expectancies.

ligament One component of the musculoskeletal system that also consists of tendons, bones, and muscles. Ligaments are collagen-rich connective structures, a band of fibrous tissue that attaches and binds bone to bone and to cartilage and that supports organs in place.

ligand A molecule, ion, hormone, or compound that binds to a specific receptor site that binds to another molecule.

The atoms or groups of atoms bound to the CENTRAL ATOM. The root of the word is sometimes converted into the verb *to ligate*, meaning to coordinate as a ligand, and the derived participles *ligating* and *ligated*. This use should not be confused with its use to describe the action of LIGASES (a class of ENZYMES).

The names for anionic ligands, whether inorganic or organic, end in -o. In general, if the anion name ends in -ide, or -ate, the final -e is replaced by -o, giving -ido, and -ato, respectively. Neutral and cationic ligand names are used without modification. Ligands bonded by a single carbon atom to metals are regarded as radical substituents, their names being derived from the parent hydrocarbon, from which one hydrogen atom has been removed. In general, the final letter -e of the name is replaced by -yl.

In biochemistry the term *ligand* has been used more widely: if it is possible or convenient to regard part of a polyatomic molecular entity as central, then the atoms or groups or molecules bound to that part may be called ligands.

See also COORDINATION.

ligand field Ligand field theory is a modified CRYSTAL FIELD theory that assigns certain parameters as variables rather than taking them as equal to the values found for free ions, thereby taking into account the potential covalent character of the metal-LIGAND bond.

ligand gated ion channel receptor Ion channels are specialized pores in the cell membrane that help control and transfer electrical impulses (action potentials) in the cell. They regulate the flow of sodium, potassium, and calcium ions into and out of the cell. The ligand gated ion channel receptor is a signal receptor protein in a cell membrane that can act as a channel for the passage of a specific ion across the membrane. When activated by a signal molecule, it allows or blocks the passage of the ion. This results in a change in ion concentration that often affects cell functioning.

ligase An ENZYME of EC class 6, also known as a synthetase, that catalyzes the formation of a bond between two SUBSTRATE molecules coupled with the HYDROLYSIS of a diphosphate bond of a NUCLEOSIDE triphosphate or similar cosubstrate.

See also EC NOMENCLATURE FOR ENZYMES.

ligating See LIGAND.

light microscope   A common laboratory instrument that uses optics to bend visible light to magnify images of specimens placed on an attached platform, or other viewing area.

light reactions   A major component of photosynthesis in which a group of chemical reactions occur in the thylakoid membranes of chloroplasts that harvest energy from the sun to produce energy-packed chemical bonds of ATP and NADPH and that give off oxygen as a by-product.

lignin   A complex amorphous polymer in the secondary cell wall (middle lamella) of dead woody plant cells that cements or naturally binds cell walls to help make them rigid. Highly resistant to decomposition by chemical or enzymatic action, it is the major source material for coal. It also acts as support for cellulose fibers. Cells that contain lignin are fibers, sclerids, vessels, and tracheids.

limbic system   The limbic system underlies the corpus callosum, an area that provides communication and links to the two cerebral hemispheres, and is a collective term referring to several brain parts, an interconnected neural network or collection of bodies including the hippocampus, cingulate cortex, and the amygdala. The limbic structures are important in the regulation of visceral motor activity and emotional expression. The amygdala deals with emotion, while the cingulate cortex plays a role in emotional behavior, and the hippocampus deals with spatial and memory functions.

linkage   Refers to the tendency for two genes that reside next to each other on a chromosome to remain together, "linked," during reproduction. The closer they are to each other, the lower the chance of separation during DNA replication, and therefore the greater the chance that they will be inherited with each other. Linked genes are an exception of Mendel's laws of inheritance, specifically the law of independent assortment, which states that pairs of genes segregate independently of each other when germ cells are formed.

linkage map   A map that shows the relative positions of genetic loci on a chromosome, determined by how often the loci are inherited together. Linked genes are usually inherited together and are located close to each other on the chromosome. The distance is measured in centimorgans (cM). Genes that are located on different chromosomes follow Mendel's principle of independent assortment, while genes that are on the same chromosome do not sort independently. The closer the two genes are to each other on a chromosome, the greater the chance that they will remain together during meiosis instead of crossing over. Researchers can construct a genetic map showing the relative positions of the genes on the chromosomes by comparing appearance of a trait to appearance of marker phenotypes. The distance between closely spaced genes can be expressed in "map units" reflecting relative frequency of recombination.

linked genes   Genes that are located close to each other on the same chromosome and do not show independent assortment. This results in parental allele combinations occurring greater than recombinant arrangements. A set of closely linked genes that are inherited together are called a haplotype. Some diseases are a result of abnormalities in certain gene combinations. X-linked diseases such as hemophilia, color blindness, and some muscular dystrophies are caused by genes located on the X chromosome. Sex-linked diseases are caused by a mutant gene on part of the X chromosome and affects men, since they do not have a Y chromosome of the XY pair that has a compensating normal gene.

lipid   A large group of hydrophobic (water insoluble) molecules that are the building blocks of cell membranes and liposomes (lipid vesicles) and contain fatty acids; the principal components of fats, oils, waxes, triglycerides, and cholesterol. They are insoluble in water but soluble in solvents such as alcohol and ether. The phospholipid bilayer of the plasma membrane is a double layer of phospholipid molecules arranged so that the hydrophobic "tails" lie between the hydrophilic "heads." Also known as fat, they easily store in the body and are an important source of fuel for the body.
See also FAT.

Lipmann, Fritz Albert (1899–1986) German *Biochemist* Fritz Albert Lipmann was born on June 12, 1899, in Koenigsberg, Germany, to Leopold Lipmann and his wife Gertrud Lachmanski.

Between 1917 and 1922 he was educated at the Universities of Koenigsberg, Berlin, and Munich, where he studied medicine and received an M.D. degree in 1924 at Berlin. In 1926 he was an assistant in Otto MEYERHOF's laboratory at the Kaiser Wilhelm Institute, Berlin, and received a Ph.D. in 1927. He then went with Meyerhof to Heidelberg to conduct research on the biochemical reactions occurring in muscle.

In 1930 Lipmann went back to the Kaiser Wilhelm Institute in Berlin and then to a new institute in Copenhagen in 1932. Between 1931 and 1932 he served as a Rockefeller fellow at the Rockefeller Institute in New York and identified serine phosphate as the constituent of phosphoproteins that contains the phosphate.

He went to Copenhagen in 1932 as a research associate in the Biological Institute of the Carlsberg Foundation. In 1939 he came to America and became research associate in the department of biochemistry at Cornell Medical School, New York, and in 1941 joined the research staff of the Massachusetts General Hospital in Boston, first as a research associate in the department of surgery, then heading his own group in the Biochemical Research Laboratory of the hospital. In 1944 he became an American citizen. In 1949 he became professor of biological chemistry at Harvard Medical School, Boston. In 1957 he was appointed a member and professor of the Rockefeller Institute, New York.

In 1947 he isolated and named coenzyme A (or CoA). He later determined the molecular structure (1953) of this factor that is now known to be bound to acetic acid as the end product of sugar and fat breakdown in the absence of oxygen. It is one of the most important substances involved in cellular metabolism, since it helps in converting amino acids, steroids, fatty acids, and hemoglobins into energy. For his discovery of this coenzyme he was awarded the 1953 Nobel Prize in physiology or medicine. He died on July 24, 1986, in Poughkeepsie, New York.

lipophilicity Represents the AFFINITY of a molecule or a moiety (portion of a molecular structure) for a lipophilic (fat soluble) environment. It is commonly measured by its distribution behavior in a biphasic system, either liquid-liquid (e.g., partition coefficient in octan-1-ol/water) or solid-liquid (retention on reversed-phase high-performance liquid chromatography (RP-HPLC) or thin-layer chromatography (TLC) system).

See also HYDROPHOBICITY.

lipoprotein Since lipids are hydrophobic (water insoluble), certain lipids like cholesterol and triglycerides are coated or bonded with a protein so they can be carried in the blood. Since it is not possible to determine the exact lipoprotein content in blood due to the variety of lipoproteins, the medical profession talks about low-density lipoproteins (LDLs) and high-density lipoproteins (HDLs) that transport fats and cholesterol through the blood.

lipoxygenase A nonheme iron ENZYME that catalyzes the INSERTION of O_2 into polyunsaturated fatty acids to form hydroperoxy derivatives.

lithosphere The solid inorganic uppermost portion or mantle of the Earth that includes the surface land and that of the ocean basin and is about 60 miles thick.

littoral zone The shallow shoreward region of a freshwater body, just beyond the breaker zone, and where light penetrates to the bottom sediments, giving rise to a zone that is colonized by rooted plants called helophytes; a region of a lake or pond where the water is less than 6 meters deep; in oceanography, the line extending from the high water line to about 200 meters; also called the intertidal zone where submersion of tides is a normal event. The near-surface open water surrounded by the littoral zone is the limnetic zone, which gets ample light and is dominated by plankton. The littoral system is divided into a eulittoral (lower, middle, and upper) and a sublittoral (or subtidal, or supratidal) zone, the zone exposed to air only at its upper limit by the lowest spring tides. They are separated at a depth of about 50 meters. The term is also frequently used interchangeably with *intertidal zone*.

liverwort A green photosynthetic bryophyte belonging to the family Hepaticae (division Hepatophyta). A small, simple plant that lives in moist, shady areas with wide, flat leaves that lie close to the ground. Liverworts reproduce with spores and have one of two forms: thalloid liverworts (Marchantiidae) and leafy liverworts (Jungermanniidae). Liverworts have no roots, but have thin (one-cell thick) rootlike structures known as rhizoids that serve for attachment and water absorption. Liverworts comprise two separate generations: the gametophyte generation and the sporophyte generation. There are around 6,000 species of liverworts worldwide, with some 4,000 species belonging to the Jungermanniales.

lizards Lizards are vertebrates that belong to the class Reptilia, which contains some 27 families. Lizards and snakes are scaled reptiles. Lizards were first found in the Jurassic period about 213 million years ago. There are about 3,500 species of lizards worldwide living in warm tropics and subtropical climates, as well as in temperate regions, and they can be as large as the 10-foot Kimodo dragon. They have long bodies, typically with four legs and a tail, have movable eyelids, and external ear openings. They can lose a tail and regenerate a new one. Lizards may be the most successful reptiles living today.

The fringe-toed lizard, which lives in the Palen Dunes of California, is an example of Reptilia as well as an endangered species and an insectivore. *(Courtesy of Tim McCabe)*

locus The specific position or location on a chromosome that is occupied by a gene. The plural form is loci.

Loewi, Otto (1873–1961) Austrian *Physician, Pharmacologist* Otto Loewi was born on June 3, 1873, in Frankfurt-am-Main, Germany, to Jacob Loewi, a merchant, and Anna Willstätter. He attended the humanistic gymnasium (grammar school) locally from 1881 to 1890 and entered the Universities of Munich and Strassburg as a medical student in 1891. In 1896 he received a doctor's degree at Strassburg University.

After spending a few months working in the biochemical institute of Franz Hofmeister in Strassburg, he became an assistant to Carl von Noorden, clinician at the city hospital in Frankfurt during 1897–98. In 1898 he became an assistant of Professor Hans Horst Meyer, a renowned pharmacologist at the University of Marburg-an-der-Lahn and a professor of pharmacology in Vienna. In 1905 Loewi became associate professor at Meyer's laboratory, and in 1909 he was appointed to the chair of pharmacology in Graz. In 1940 he moved to the United States and became research professor at the School of Medicine of New York University, New York City, where he remained until his death.

His neurological researches during the period 1921–26 provided the first proof that chemicals were involved in the transmission of impulses from one nerve cell to another and from neuron to the responsive organ. It was for his discovery of the chemical transmission of nerve impulses that he received the Nobel Prize in physiology or medicine in 1936, jointly with Sir Henry DALE. Loewi spent his years investigating the physiology and pharmacology of metabolism, the kidneys, the heart, and the nervous system. He became an American citizen in 1946 and died on December 25, 1961.

logistic growth A model developed by Belgian mathematician Pierre Verhulst (1838) that states that the growth rate is dependent on population density and restricted by carrying capacity. Growth is represented by an S-curve, and the growth rate declines as the population increases. The pattern of growth is a slow start, an explosive middle growth period, and then a flattening of the curve as growth slows.

long-day plant A plant affected by photoperiodism. A plant that needs more light than dark for flowering. Long-day plants are spinach, lettuce, and wheat. Others such as calceolaria, philodendron, and tuberous begonias need 14 to 18 hours of light for flowering. Long-day plants form flowers during day lengths of more than 12 hours and are both indoor and outdoor plant types.

See also PHOTOPERIODISM.

longevity The act of living for a long period of time.

loop of Henle One of the six structural and functional parts of the kidney's nephron, along with Bowman's capsule, glomerulus, proximal convoluted tubule, distal convoluted tubule, and collecting tubule. The long hairpin turn, or loop of Henle, extends through the medulla from the end of the proximal convoluted tubule to the start of the distal convoluted tubule. It has a descending limb called the proxima straight tubule that reabsorbs water, and an ascending limb that reabsorbs NaCl, ending with the distal straight tubule. It is the major site of water and salt reabsorption. Some parts of it are permeable to water and impermeable to materials such as salt or ammonia in the urine, and vice versa. Also called the *ansa nephroni*.

low-spin In any COORDINATION entity with a particular d^n ($1 < n < 9$) configuration and a particular geometry, if the n electrons are distributed so that they occupy the lowest possible energy levels, the entity is a low-spin complex. If some of the higher-energy d orbitals are occupied before all the lower-energy ones are completely filled, then the entity is a high-spin complex.

luminescent The act of emitting light without causing heat (called cold light). Bioluminescence is the act of producing light by biological organisms, e.g., firefly; luminescent bacteria (*Photobacterium phosphoreum*); and fox fire in the form of *Clitocybeilludens* (*Omphalotus olerius*), *Panellus stypticus*, and *Armillaria mellea*. The lanthanides are a special group of elements (elements cerium [Ce, atomic no. 58] through lutetium [Lu, atomic no. 71] within the periodic table that have trivalent cations that emit light. When these elements are absorbed into materials, the materials can become luminescent after being excited by an electrical current (electroluminescence) or by absorbing light (photoluminescence). As the ions relax to their ground state, they release light.

lumper A taxonomist who prefers to classify organisms into relatively small groups, emphasizing similarities. Opposite of a splitter.

See also SPLITTER.

lung The basic respiratory organ of air-breathing vertebrates. The basic function of the respiratory system is for lungs to supply oxygen to tissues and remove excess carbon dioxide from the blood. This is accomplished by inspiration, the movement of air into the lungs, and expiration, movement out of the lungs. Exchange of gases occurs at the internal surface of the lungs by diffusing oxygen from the lungs into the

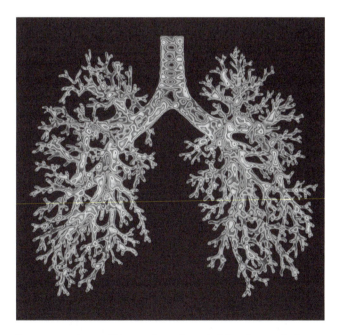

Computer-enhanced image of a resin cast of the airways in the lungs. The trachea (windpipe, top center) divides into two bronchi, which divide further into small bronchioles. The bronchioles terminate in alveoli (not seen), grapelike clusters of air sacs surrounded by blood vessels. Here the blood takes up oxygen and releases carbon dioxide to be exhaled. *(Courtesy © Alfred Pasieka/Photo Researchers, Inc.)*

blood, while carbon dioxide (CO_2) diffuses from the blood into the lungs. Gas is transported in the blood, and the circulatory system distributes oxygen throughout the body while collecting CO_2 for return to the lungs. Ventilation, or airflow to the lungs, and perfusion, blood flow through lung capillaries, are the main factors that determine the overall pulmonary function. The ventilation/ perfusion ratio needs to be fairly constant in all parts of the lung for the most effective transfer of respiratory gases.

Lungs are saclike structures of varying complexity, depending on the organism. They are connected to the outside by a series of tubes and a small opening. In humans, the lungs can be found in the thoracic cavity and consist of the internal airways; the alveoli, tiny, thin-walled, inflatable sacs where oxygen and carbon dioxide are exchanged; the pulmonary circulatory vessels; and elastic connective tissues. The exchange of gases in the lungs takes place by breathing, chest movements that inhale air and exhale or force gases such as carbon dioxide out.

lyase An ENZYME of EC class 4 that catalyzes the separation of a bond in a SUBSTRATE molecule.
See also EC NOMENCLATURE FOR ENZYMES.

lymph A clear, colorless, watery fluid that contains white blood cells and antibodies. Lymph bathes the tissues, passes through lymph-node filters and returns to the blood stream. The lymphatic system is a system of vessels and lymph nodes that returns fluid and protein to the blood, and it is separate from but parallel to the circulatory system. Lymph flows through lymph nodes and is transformed into lymphocytes, a special kind of white blood cell that helps create the immune system and comprises up to 50 percent of all white blood cells.

lymphocyte A white blood cell that identifies foreign objects such as bacteria and viruses in the body and produces antibodies and cells that target them. There are different types of lymphocytes, the most important of which are: B cells that produce specific antibodies to destroy foreign objects; T cells that attack cells that are infected by viruses, cancer cells, and other foreign tissue; NK (natural killer) cells that kill cancer cells and cells

infected by viruses through phagocytosis; and null cells, also NK-type cells that attack certain targets. These cells occupy less than 1 percent in circulating blood as they travel from the blood to the lymph and lymph nodes and back into the circulating blood. Other types of lymphocytes include acidophilic or eosinophilic, agranular, nongranular or lymphoid, basophilic, granular, heterophilic, neutrophilic, polymorphonuclear, and activated.

lysis Biological term for destruction or killing, as in bacteriolysis or the rupturing of a cell membrane with resulting loss of cytoplasm.

lysogenic cycle A stage in viral development when a virus inserts its genome in a host genome and lies dormant until outside factors initiate the new genetic material to be transcribed. It then goes into a lytic cycle, a massive replication of new viruses, and lyses (destroys the cell membrane) the host cell.

lysosome A small organelle found in eukaryotic cells that is surrounded by a membrane and contains digestive hydrolytic enzymes and chemicals that are strongly acidic. Lysosomes fuse with the vascular membrane, grab the food, and digest it, breaking it down into usable parts. Lysosomes also work with leukocytes by digesting leukocytes that have engulfed foreign objects or cleaned up damaged cells.

lysozyme A basic enzyme or antibacterial product that can be found in tears, saliva, perspiration, the nasal cavity, egg whites, animal fluids, and some plant tissues. It attacks the cells walls of gram-positive bacteria by cleaving the muramic acid [beta(1-4)-N-acetylglucosamine] linkage in the cell walls of this class of bacteria.

lytic cycle The viral replication that releases new phages (particles) through the lysis of the host cell.
See also LYSOGENIC CYCLE.

lytic virus A virus that causes lysis of the host cell.

M

Macleod, John James Richard (1876–1935) Scottish *Physiologist* John James Richard Macleod was born on September 6, 1876, in Cluny, near Dunkeld, Perthshire, Scotland, to the Rev. Robert Macleod. Macleod went to the grammar school at Aberdeen and later entered the Marischal College of the University of Aberdeen to study medicine. In 1898 he worked for a year at the Institute for Physiology at the University of Leipzig, and the following year he was appointed demonstrator of physiology at the London Hospital Medical School.

In 1903 he was appointed professor of physiology at the Western Reserve University at Cleveland, Ohio, in the United States. In 1918 he became professor of physiology at the University of Toronto, Canada, and served as director of the physiological laboratory and as an associate dean of the faculty of medicine. In 1928 he was appointed Regius professor of physiology at the University of Aberdeen, a position he held until his death.

For his work on the discovery of insulin with Frederick BANTING, he was awarded the Nobel Prize in physiology or medicine for 1923.

Macleod conducted research in carbohydrate metabolism, focusing especially on diabetes, and published some 37 papers on carbohydrate metabolism and 12 papers on experimentally produced glycosuria (sugar in the urine).

He wrote 11 books and monographs, including *Recent Advances in Physiology* (with Sir Leonard Hill) (1905); *Physiology and Biochemistry of Modern Medicine; Diabetes: Its Pathological Physiology* (1925); *Carbohydrate Metabolism and Insulin* (1926); and the Vanuxem lectures, published in 1928 as the *Fuel of Life*. He died on March 16, 1935.

macroevolution Evolution that deals with large-scale and complex changes such as the rise of species, mass extinctions, and evolutionary trends.

macromolecule A large molecule of high molecular mass composed of more than 100 repeated monomers, single chemical units of lower relative mass; a polymer. DNA, proteins, and polysaccharides are examples of macromolecules in living systems; a large complex molecule formed from many simpler molecules.

macrophages Blood cells that are able to ingest a wide variety of particulate materials. They are a type of PHAGOCYTE.

magnetic circular dichroism (MCD) A measurement of CIRCULAR DICHROISM of a material that is induced by a magnetic field applied parallel to the direction of the measuring light beam. Materials that are achiral still exhibit MCD (the Faraday effect), since the magnetic field leads to the lifting of the degeneracy of electronic orbital and spin states and to the mixing of electronic states. MCD is frequently used in combination with absorption and CD studies to affect electronic assignments. The three contributions to the MCD spectrum are the A-term, due to Zeeman splitting of the ground and/or excited degenerate states; the B-term, due to field-induced mixing of states; and the C-term, due to a change in the population of molecules over the Zeeman sublevels of a paramagnetic ground state. The C-term is observed only for molecules with ground-state paramagnetism, and becomes intense at low temperatures; its variation with field and temperature can be analyzed to provide magnetic parameters of the ground state, such as spin, g-factor, and zero-field splitting. Variable-temperature MCD is particularly effective in identifying and assigning electronic transitions originating from paramagnetic CHROMOPHORES.

magnetic resonance imaging (MRI) The visualization of the distribution of nuclear spins (usually water) in a body by using a magnetic field gradient (NMR IMAGING). A similar technique, but less widely used, is to visualize the distribution of paramagnetic centers (EPR imaging).

magnetic susceptibility For paramagnetic materials, the magnetic susceptibility can be measured experimentally and used to give information on the molecular magnetic dipole moment, and hence on the electronic structure of the molecules in the material. The paramagnetic contribution to the molar magnetic susceptibility of a material, χ, is related to the molecular magnetic dipole moment m by the Curie relation: $\chi =$ constant m^2/T.

magnetotactic Ability to orient in a magnetic field.

major histocompatibility complex (MHC) A large cluster of genes on chromosome 6 in humans, encoding cell-surface proteins that play several roles in the immune system. Several classes of protein such as MHC class I and II proteins are encoded in this region. In humans, these are known as human leukocyte antigens (HLA). Class I protein molecules are designated

HLA A, B, or C. Class II molecules are designated DP, DQ, or DR.

- MHC class I molecule. A molecule encoded to genes of the MHC that participates in antigen presentation to cytotoxic T (CD8+) cells.
- MHC class II molecule. A molecule encoded by genes of the MHC that participates in antigen presentation to helper T (CD4+) cells.

The ability of T lymphocytes to respond only when they "see" the appropriate antigen in association with "self" MHC class I or class II proteins on the antigen-presenting cells is called MHC restriction.

malaria A tropical disease caused by a protozoa of the genus *Plasmodium* (*Plasmodium falciparum*) and transmitted to humans by the bite of mosquitoes of the genus *Anopheles*.

malignant Term used to designate a cancerous condition.
See also CANCER.

Mallophaga One of the insect orders, known as chewing lice, with three families, Menoponidae, Philopteridae, and Trichodectidae, and made up of the chewing lice, characterized by flattened, wingless bodies, chewing mouthparts, and gradual metamorphosis. Commonly found on dogs and cats.

Malpighian tubule An excretory organ in arthropods (insects) that correspond functionally to the kidneys of vertebrates. The tubules maintain internal salt and water balance and remove wastes such as urea; uric acid; urates of sodium, calcium, and ammonia; leucin; and various salts of calcium and potassium (oxalates, carbonates, and phosphates). Most of the nitrogen excreted is in the form of uric acid. Opens into the posterior section of the alimentary canal (gut). In some insects, such as *Mermeleon formicarius*, *Chrysopa perla*, and *Euplectrus bicolor,* the Malpighian tubules produce a substance just before the time of pupation that is spun out of the anus in the form of silk threads, with which the cocoon is woven.

Mammalia   A class of warm-blooded animals that have three characteristics not shared by other animals: body hair; the production of milk for nourishment of their infants by mammary glands (teats), which are modified sweat glands; and the three middle ear bones (malleus, incus, and stapes). Placental mammals have a vascular-connected placenta formed between the embryo and mother (e.g., humans, bears, whales). Aplacental mammals do not (e.g., marsupials and monotremes). Most mammals also have differentiated teeth. Found on land and sea, there are around 5,000 species placed in 26 orders.

mange   A partial or complete lack or removal of hair resulting from various disorders or conditions. Also a form of dermatitis caused by species of mites (for example, *Sarcoptes scabiei* causes sarcoptic mange or scabies). Also called alopecia.

mantle   A membranous or muscular outer form of tissue that surrounds the visceral mass in a mollusk and secretes the shell and periostracum, the outermost layer of shell that provides protection.

marine   Refers to sediments or environments in seas or ocean waters.

marsupial   An aplacental mammal whose young are born undeveloped and complete their embryonic development not inside the body of the mother but, rather, inside a maternal pouch called a marupium, located on the outside of the body, in which the young attach to the mother's nipples. Kangaroos, wombats, bandicoots, opossums, and Koala bears are examples. Found only in Australia, Tasmania, New Guinea, and a few nearby islands, except for the opossum, which is the only North American marsupial.

Mastigophora   A phylum composed of the most primitive type of protozoans. Usually parasitic, they have many flagella for movement, and some can form pseudopodia, used for food engulfing or movement, called flagellates. They live inside host organisms to obtain nutrients and cause diseases such as trichomoniasis, giardiasis, trypanosomiasis, and leishmaniasis.

matrilineal   Societies in which descent is traced through mothers rather than through fathers. Property is often passed from mothers to daughters, and the custom of matrilocal residence may be encouraged. Many Native American nations are matrilineal.

matrix   Often described as a scaffolding, support, or cell growth director (intercellular); it is a complex network of nonliving fibrous material of the connective tissues that acts as structural support. Examples include the skin, cartilage, bone, tendon, and muscle.

matter   Any substance that has inertia and occupies physical space; can exist as solid, liquid, gas, plasma, foam, or Bose–Einstein condensate.

mean   A statistical method used to indicate a point on the scale of measures where the population is centered. The mean is the average of the scores in a population.

mechanoreceptor   A specialized sensory receptor that responds to mechanical stimuli, e.g., tension, pressure, or displacement. Examples include the inner-ear hair cells, carotid sinus receptors, and muscle spindles.

Mechnikov, Ilya Ilyich (1845–1916) Russian *Zoologist*   Ilya Ilyich Mechnikov was born on May 16, 1845, in a village near Kharkoff in Russia to an officer of the Imperial Guard, who was a landowner in the Ukraine steppes.

Mechnikov went to school at Kharkoff and was interested in natural history. He attended the University of Kharkoff to study natural sciences. After graduating at Kharkoff, he went to study marine fauna at Heligoland, and then to the University of Giessen, the University of Göttingen, and the Munich Academy. In 1865, while he was at Giessen, he discovered intracellular digestion in one of the flatworms. At Naples he prepared a thesis for his doctorate on the embryonic development of the cuttle-fish *Sepiola* and the crustacean *Nelalia*.

In 1882, in a private laboratory he set up, he discovered the phenomenon of phagocytosis. In 1888 he went to Paris, where Louis Pasteur gave him a laboratory and an appointment in the Pasteur Institute, where he remained for the rest of his life.

Apart from his work on phagocytosis, Mechnikov published many papers on the embryology of invertebrates. These included work on the embryology of insects, published in 1866, and, in 1886, his studies of the embryology of medusae. At the Pasteur Institute in Paris, Mechnikov was engaged in work associated with the establishment of his theory of cellular immunity. He published several papers and two volumes on the comparative pathology of inflammation (1892), and in 1901 he published a treatise entitled *L'Immunité dans les Maladies Infectieuses* (Immunity in infectious diseases). In 1908 he was awarded, together with Paul EHRLICH, the Nobel Prize in physiology or medicine.

He later proved that syphilis can be transmitted to monkeys and took up the study of the flora (lactic-acid-producing bacteria) of the human intestine and developed a theory that senility is due to poisoning of the body by the products of certain of these bacteria.

Mechnikov received many distinctions, among which were the honorary D.Sc. of the University of Cambridge, the Copley Medal of the Royal Society of which he was a foreign member, the honorary memberships of the Academy of Medicine in Paris and the Academies of Sciences and of Medicine in St. Petersburg. In addition, he was a corresponding member of several other societies and a foreign member of the Swedish Medical Society. He died on July 16, 1916.

mediator modulator (**immune modulator; messenger**) An object or substance by which something is mediated, such as:

- A structure of the nervous system that transmits impulses eliciting a specific response.
- A chemical substance (transmitter substance) that induces activity in an excitable tissue, such as nerve or muscle (e.g., hormones).

- A substance released from cells as the result of an antigen-antibody interaction or by the action of antigen with a sensitized lymphocyte (e.g., cytokine).

Concerning mediators of immediate hypersensitivity, the most important include histamine, leukotriene (e.g., SRS-A), ECF-A, PAF, and serotonin. There also exist three classes of lipid mediators that are synthesized by activated mast cells through reactions initiated by the actions of phospholipase A2. These are prostaglandins, leukotrienes, and platelet-activating factors (PAF).

medicinal chemistry A chemistry-based discipline, also involving aspects of biological, medical, and pharmaceutical sciences. It is concerned with the invention, discovery, design, identification, and preparation of biologically active compounds; the study of their METABOLISM; the interpretation of their mode of action at the molecular level; and the construction of STRUCTURE-ACTIVITY RELATIONSHIPS.

medulla oblongata One part of the brain stem, along with the midbrain, pons, and reticular formation; connects the brain to the spinal cord through the foramen magnum. Nerve tissue that deals with vital functions in respiration, circulation (heart rate, blood flow), and vasomotor; controls reflexes such as coughing, sneezing, swallowing, vomiting, and gagging.

medusa One of two basic life cycles of Cnidaria (i.e., hydra, jellyfish, anemones, and corals), the other being polyps. The medusa form is a free-swimming, floating, umbrellalike, flat, mouth-down version. Polyps' forms are sessile. Body parts on both forms are similar. Most medusa have separate sexes. The eggs are fertilized after they are shed into the water and usually grow into polyp form that, in a form of asexual reproduction, may break into multiple medusas.

megapascal (**MPa**) A unit of pressure. 1 MPa = 1,000,000 Pa (pascal). 1 megapascal (MPa) = 10 bars. A value of 1 bar is approximately equal to one atmosphere of pressure.

meiosis The reductive division of diploid cells in ovaries and testes that produce gametes (sperm and ova). Two divisions with several stages take place that result in the production of four daughter cells, each of which contain half (haploid) of the original number of chromosomes.

See also MITOSIS.

melanism A medical condition usually seen in the skin, characterized by abnormal deposits of melanin.

See also INDUSTRIAL MELANISM.

membrane potential The difference in electrical charge (voltage difference) across the cell membrane due

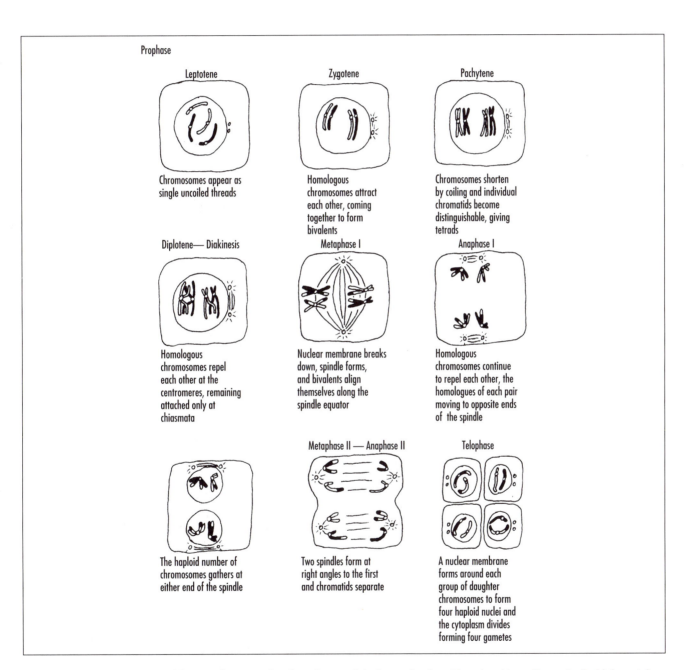

During meiosis, two divisions with several stages take place that result in the production of four daughter cells, each of which contain half (haploid) of the original number of chromosomes.

to a slight excess of positive ions on one side and of negative ions on the other; potential inside a membrane minus the potential outside. A typical membrane potential is –60 mV, where the inside is negative relative to the surrounding fluid, and resting membrane potentials are typically found between –40 and –100 mV.

memory cell Cells (lymphocytes) that have been exposed to specific antigens and remain in the body after an immune response to attack those same antigens if reexposed to them in the future. Memory cells are subsets of T and B cells.

Mendel, Gregor Johann Mendel (1822–84) was an Austrian botanist and monk who was the first to lay the groundwork for the foundation of the science of genetics, using his now famous experiments with breeding peas at his monastery. His groundbreaking research paper "Experiments in Plant Hybridization," was read at a meeting on February 8, 1865. He concluded that genes were not blends of parental traits, but instead were separate physical entities passed individually in specific proportions from one generation to the next.

Menkes' disease A sex-linked inherited disorder, causing defective gastrointestinal absorption of copper and resulting in copper deficiency early in infancy.

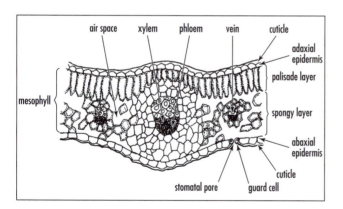

The ground tissue of a leaf, located between the upper and lower epidermis, specialized for photosynthesis is called the mesophyll.

menstrual cycle The cyclical growth and destruction of the female endometrium each month; a stage of the female reproductive cycle. As an egg matures and is released monthly, hormones such as estrogen stimulate the thickening of the endometrium. Progesterone stops the growth of the endometrium and prepares the body for pregnancy. If it does not occur, the endometrium becomes a bloody discharge through the cervix into the vagina, thus ending the menstrual period.

See also ESTROUS CYCLE.

meristem A group of plant cells that can divide indefinitely and can provide new cells for the plant as long as it lives.

meroblastic cleavage (incomplete cleavage) A type of cleavage where only part of the fertilized egg, the blastodisc, goes through division, usually leaving a large concentration of yolk in the egg; observed in avian development.

mesentery A membrane fold (peritoneum) suspending, attaching, and anchoring various organs to the body cavity, such as the small intestine and spleen; provides blood, lymphatic vessels, and nerve supply to and from the organs. The term *mesentery* is used generically describing peritoneal extensions not only from the intestine but from all abdominal and pelvic organs.

mesoderm The middle (mesos) of three germ layers (endo- and ectoderms are the other two) of the early embryo during gastrulation, early mammalian embryonic development, that gives rise to blood, cartilage, circulatory system, connective tissue, gonads, kidneys, and muscle. Three regions of the mesoderm are responsible for specific parts: the epimere or dorsal mesoderm forms somites, discrete clumps of mesoderm, which forms the connective tissue layer of the skin, most of the bony skeleton, and most of the striated musculature (each somite is further split into dermotome, myotome, and sclerotome segments); the mesomere or intermediate mesoderm, which differentiate into the kidney and urogenital structures; and the hypomere, which differentiate into limbs, peritoneum, gonads, heart, blood vessels, and mesenteries.

mesophyll Plant tissue that forms the inner or middle cells of the leaf and lies between the upper and lower epidermis. It contains two types of chlorenchyma cells (cells that contain chlorophyll): the long and vertically arranged palisade cells, the upper layer, which are on top of the round and loosely packed spongy cells in the lower layer, where most of the gas exchange occurs and where photosystem II is most active. It is in both the palisade and spongy cells that photosynthesis takes place by way of chloroplasts.

mesotrophic lake Any lake with a moderate nutrient supply.

Mesozoic era A geological time that extends from the end of the Paleozoic era (230 million years ago) to the beginning of the Cenozoic era (about 65 million years ago). Subdivided into the Triassic, Jurassic, and Cretaceous periods, which includes the age of the dinosaurs.

See also GEOLOGICAL TIME.

messenger RNA (mRNA) An RNA molecule that transfers the coding information for protein synthesis from the chromosomes to the ribosomes. Fragments of ribonucleic acid serve as templates for protein synthesis by carrying genetic information from a strand of DNA to ribosomes for translation into a protein. The information from a particular gene or group of genes is transferred from a strand of DNA by constructing a complementary strand of RNA through transcription. Transfer RNA (tRNA), composed of three nucleotide segments attached to specific amino acids, correctly match with a template strand of mRNA, lining up the correct order of amino acids and bonding them, via translation in the ribosome with rRNA (ribosomal RNA), to form a protein.

See also DEOXYRIBONUCLEIC ACID; RIBONUCLEIC ACID.

met- A qualifying prefix indicating the oxidized form of the parent protein, e.g., methemoglobin.

metabolism The entire physical and chemical processes involved in the maintenance and reproduction of life in which nutrients are broken down to generate energy and to give simpler molecules (CATABOLISM), which by themselves may be used to form more complex molecules (ANABOLISM).

In the case of heterotrophic organisms, the energy evolving from catabolic processes is made available for use by the organism.

In medicinal chemistry, the term *metabolism* refers to the biotransformation of xenobiotics and particularly DRUGS.

See also BIOTRANSFORMATION; XENOBIOTIC.

metabolite Any intermediate or product resulting from METABOLISM.

metalloenzyme An ENZYME that, in the active state, contains one or more metal ions that are essential for its biological function.

metallo-immunoassay A technique in which ANTIGEN-ANTIBODY recognition is used, with attachment of a metal ion or metal complex to the antibody. The specific absorption or (radioactive) emission of the metal is then used as a probe for the location of the recognition sites.

See also IMAGING; RADIONUCLIDE.

metallothionein A small, cysteine-rich protein that binds heavy metal ions, such as zinc, cadmium, and copper in the form of CLUSTERS.

metamorphosis The change and reorganization of the tissues and body shape during the development of an animal from larva to adult. Metamorphosis can be complete or incomplete, i.e., lacking a pupal stage. Simple or gradual metamorphosis is an incomplete

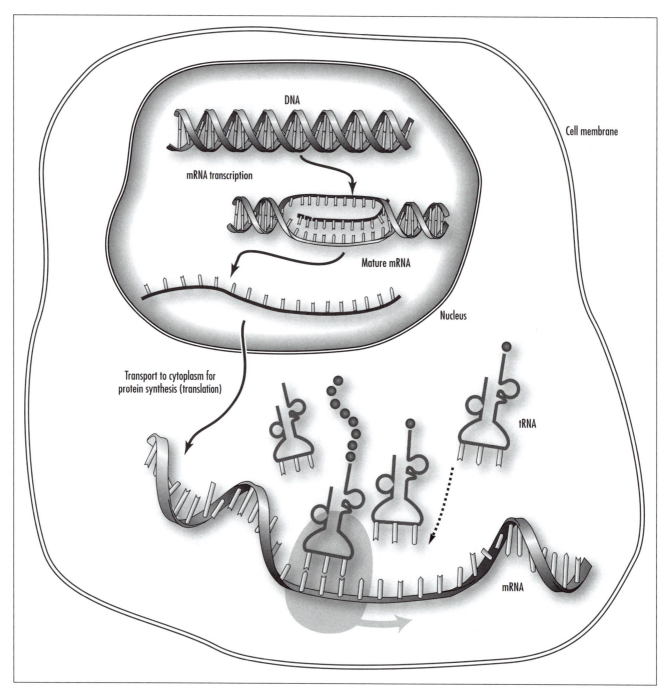

Template for protein synthesis. Each set of three bases, called codons, specifies a certain protein in the sequence of amino acids that comprise the protein. The sequence of a strand of mRNA is based on the sequence of a complementary strand of DNA. *(Courtesy of Darryl Leja, NHGRI, National Institutes of Health)*

metamorphosis with no pupa and with young immature forms looking similar to the adult minus wings.

See also INCOMPLETE METAMORPHOSIS.

metanephridium Excretory organ or tubule surrounded by capillaries in invertebrates (e.g., annelid worms) in which nephrostomes, the internal ciliated funnel-shaped opening, collects body fluids and nitrogenous compounds and discharges at the other end, the nephridiopore. Also reabsorbs fatty acids, water, and amino acids back into circulation.

metaphase A development stage in mitosis or meiosis. Characterized by chromosomes aligning along the equatorial plane of the cell.

See also MEIOSIS; MITOSIS.

metapopulation Groups of local species populations where each group occupies separate habitat patches that often are connected by corridors allowing migration between them.

metastable *See* STABLE.

metastasis The spread of cancer cells from one part of the body to another.

See also CANCER.

Metazoa The kingdom that includes all multicellular organisms. It includes vertebrates and invertebrates.

methane mono-oxygenase A METALLOENZYME that converts methane and dioxygen to methanol using NADH as COSUBSTRATE. Two types are known, one containing a dinuclear oxo-bridged iron center; the other is a copper protein.

See also NUCLEARITY.

methanogens Strictly ANAEROBIC ARCHAEA that are able to use a variety of SUBSTRATES (e.g., dihydrogen,

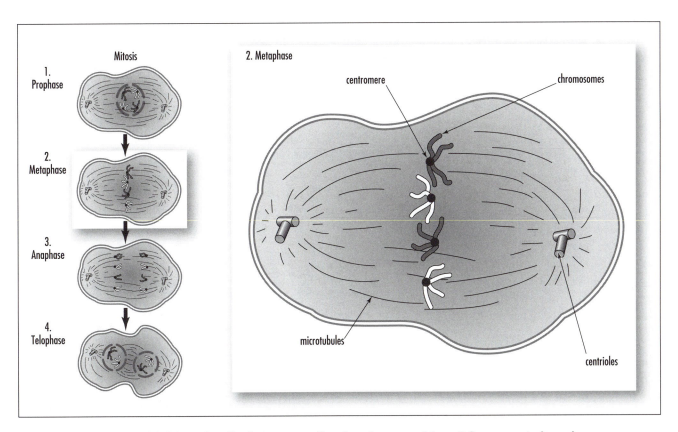

The phase of mitosis, or cell division, when the chromosomes align along the center of the cell. Because metaphase chromosomes are highly condensed, scientists use these chromosomes for gene mapping and identifying chromosomal aberrations. *(Courtesy of Darryl Leja, NHGRI, National Institutes of Health)*

formate, methanol, methylamine, carbon monoxide, or acetate) as electron donors for the reduction of carbon dioxide to methane.

me-too drug   A compound that is structurally very similar to already known DRUGS, with only minor pharmacological differences.

Meyerhof, Otto Fritz (1884–1951) German *Physiologist, Chemist*   Otto Fritz Meyerhof was born on April 12, 1884, in Hannover to Felix Meyerhof, a merchant, and Bettina May. He went to the Wilhelms Gymnasium (classical secondary school) in Berlin, leaving at age 14 only to have kidney problems two years later, which kept him confined for a long period. He eventually studied medicine at Freiburg, Berlin, Strassburg, and Heidelberg and graduated in 1909. From 1912 he worked at the University of Kiel, becoming a professor in 1918.

Meyerhof conducted experiments on the energy changes in cellular respiration. For his discovery of the fixed relationship between the consumption of oxygen and the metabolism of lactic acid in the muscle, he was awarded, together with the English physiologist A.V. HILL, the Nobel Prize in physiology or medicine in 1922. In 1925 Meyerhof successfully extracted the enzymes that convert glycogen to lactic acid from the muscle. He introduced the term *glycolysis* to describe the anaerobic degradation of glycogen to lactic acid, and showed the cyclic nature of energy transformations in living cells. This metabolic pathway of glycolysis—conversion of glucose to lactic acid—is now known as the Embden-Meyerhof pathway, after Meyerhof and Gustav George Embden.

During World War II, he went to the United States and became research professor of physiological chemistry, a position created for him by the University of Pennsylvania and the Rockefeller Foundation. He died from a heart attack on October 6, 1951.

Michaelis-Menten kinetics   The dependence of the initial rate of conversion of a SUBSTRATE (S) of the product (P) by an ENZYME or other catalyst (E). The simplest mechanism:

$$E + S \underset{k_{-1}}{\overset{k_1}{\rightleftharpoons}} ES \underset{k_{-2}}{\overset{k_2}{\rightleftharpoons}} E + P$$

yields, under initial STEADY STATE conditions, and [P] = 0, the Michaelis-Menten equation,

$$v = \frac{V[S]}{K_m + [S]}$$

where v is the rate of conversion (Ms^{-1}), $V = k_2[E]$ is the maximum rate at [S] = ∞ for a particular enzyme/catalyst concentration, k_2 is the turnover number (s^{-1}), and $K_m = (k_{-1} + k_2)/k_1$ is the Michaelis constant under the conditions used. In the case of an impure enzyme or catalyst, [E] is given as gl^{-1} instead of M. This equation leads to a hyperbolic dependence of v upon [S], which is frequently observed in practice even when [S] is not in great excess over [E].

microevolution   The smallest scale of evolution; changes within a species; a change in allele or genotype frequencies over time.

See also MACROEVOLUTION.

microfilament (**actin filament**)   A minute solid helical rod—about 7 nm in diameter, composed of the protein actin found in most eukaryotic cell cytoplasm—that makes up part of the cytoskeleton and often is found in association with microtubules. Plays a role in cell mobility, cytokinesis, and, with myosin, part of the contractile mechanism of skeletal muscle. One of three protein filaments of the cytoskeleton, along with microtubules and intermediate filaments. The cytoskeleton provides structural support for the cell and movement of organelles, chromosomes, as well as the cell itself.

micronutrient   A compound essential for cellular growth, being present in concentrations in minute amounts in the growth medium.

See also TRACE ELEMENTS.

microtubule   A lengthy hollow cylindrical structure composed of the protein tubulin. One of three protein filaments of the cytoskeleton, along with microfilaments and intermediate filaments, it is also found in cilia, flagella, and centrioles. The cytoskeleton provides structural support for the cell and movement of organelles, chromosomes, as well as the cell itself. Microtubules also form the spindle fibers of mitosis.

microvillus Very small hairlike or fingerlike projection from the surface of some types of epithelial cells, particular those in the small intestine, where they serve to increase surface area.

middle lamella The gel-like pectin layer between adjacent plant cell walls that binds or cements the plant cells together; in woody tissues, lignin replaces pectin.

mimicry The ability of an individual to look or share similar traits of an individual of a different species to protect it from predation. It can be in the form of protective coloration, imitation of characteristics, or deception. Two forms of mimicry are common. BATESIAN MIMICRY is where the mimic, which is palatable to a predator, resembles an unpalatable species; and MÜLLERIAN MIMICRY is where two or more unpalatable species have a reduced predation rate due to their similarity, so that predators avoid them to a greater degree than they would individually. The viceroy butterfly, which mimics the monarch, is an example of Batesian mimicry, while the queen butterfly (*Danaus gilippus*), which is a poisonous butterfly, mimics the poisonous monarch and is an example of Müllerian mimicry.

mineral A naturally occurring homogeneous solid, inorganically formed, with a definite chemical composition, usually crystalline in form, and an ordered atomic arrangement, e.g., quartz.

mineralocorticoid Any of the group of C21 (21 carbon) corticosteroids, principally aldosterone, that are predominantly involved in the regulation of electrolyte and water balance through their effect on ion transport

Caterpillar of *Papilio troilus* from Albany pine bush. It is a mimic of South American tree snakes that migratory birds have learned to avoid. Illustrates Batesian mimicry and protective coloration. *(Courtesy of Tim McCabe)*

in epithelial cells of the renal tubules. This results in retention of sodium and loss of potassium. Some also possess varying degrees of glucocorticoid activity. Their secretion is regulated principally by plasma volume, serum potassium concentration, and angiotensin II and, to a lesser extent, by anterior pituitary ACTH (adreno-corticotropic hormone).

minimum dynamic area   The smallest area of habitat necessary to sustain a viable population.

minimum viable population (MVP)   The smallest isolated population having the best chance of surviving for x years, regardless of natural catastrophes or future demographic, environmental, and genetic variables.

Minot, George Richards (1885–1950) American *Pathologist* George Richards Minot was born on December 2, 1885, in Boston, Massachusetts, to James Jackson Minot, a physician, and Elizabeth Whitney. He attended Harvard University and received a B.A. degree in 1908, an M.D. in 1912, and an honorary degree of Sc.D. in 1928. In 1915 he was appointed assistant in medicine at the Harvard Medical School and the Massachusetts General Hospital. From 1928 to 1948 he was professor of medicine at Harvard and director of the Thorndike Memorial Laboratory, Boston City Hospital.

Earlier research revealed that anemia in dogs, induced by excessive bleeding, is reversed by a diet of raw liver. In 1926 Minot and William MURPHY found that ingestion of a half pound of raw liver a day dramatically reversed pernicious anemia in human beings. He received (with George WHIPPLE and William Murphy) the Nobel Prize for physiology or medicine in 1934 for the introduction of a raw-liver diet in the treatment of pernicious anemia, which up to that time was almost always a fatal disease. Today, 10,000 lives a year are saved in the United States alone because of this discovery.

With chemist Edwin Cohn, they prepared liver extracts that, when taken orally, constituted the primary treatment for pernicious anemia until 1948, when vitamin B was discovered. Minot died on February 25, 1950, in Brookline, Massachusetts.

Miocene   A geological age that extends from the end of the Oligocene epoch (22.5 million years ago) to the beginning of the Pliocene epoch (5 million years ago).
See also GEOLOGICAL TIME.

missense mutation   One of four types of point mutations. A point mutation is when a triplet of three nucleotides (codon) has the base sequence permanently changed. A missense mutation is when a change in the base sequence converts a codon from one amino acid to a codon for a different amino acid. The other three point mutations are nonsense (codon for a specific amino acid is converted to a chain-terminating codon), silent (converts a codon for an amino acid to another codon that specifies the same amino acid), and frameshift (nucleotide is deleted or added to the coding portion of a gene) mutations.

Mississippian age   The first of the two geologic ages of the Carboniferous period, extending from about 345 to 310 million years ago. The Pennsylvanian is the second (310 to 280 million years ago).
See also GEOLOGICAL TIME.

mites   Mites and ticks belong to the order Acari and are the most diverse and abundant of all arachnids. Very small in size, usually less than a millimeter in length, they are ubiquitous, found in almost every part of the world, and account for 30,000 species in at least 50 families. Many are parasitic and cause disease.

mitochondria   Cytoplasmic organelles of most eukaryotic cells, they are surrounded by a double membrane and produce ADENOSINE 5'-TRIPHOSPHATE as useful energy for the cell by oxidative PHOSPHORYLATION. The proteins for the ATP-generating electron transport of the respiration chain are located in the inner mitochondrial membrane. Mitochondria contain many ENZYMES of the citric acid cycle and for fatty acid ß-oxidation. They also contain DNA, which encodes some of their proteins, the remainder being encoded by nuclear DNA.
See also CYTOPLASM; EUKARYOTES.

mitochondrial matrix Each mitochondrion is surrounded by a double membrane. The aqueous matrix is bounded within the inner membrane and contains ribosomes and oxidative enzymes.

mitosis The cell division process in eukaryotic cells that replicates chromosomes so that two daughter cells get equally distributed genetic material from a parent cell, making them identical to each other and the parent. It is a four-step process that includes prophase (prometaphase), metaphase, anaphase, and telophase. Interphase is the time in the cell cycle when DNA is replicated in the nucleus.

See also MEIOSIS.

mixed valency This is one of several names, such as "mixed oxidation state" or "nonintegral oxidation state," used to describe COORDINATION compounds and CLUSTERs in which a metal is present in more than

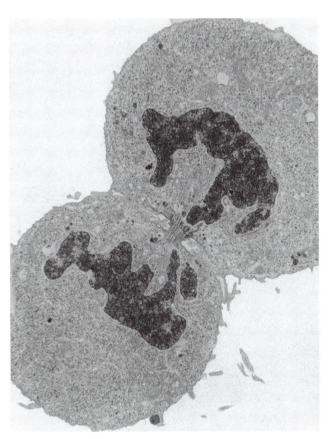

Transmission electron micrograph of cytokinesis (division of a cell's cytoplasm) after mitosis of a human embryonic kidney cell. This is a tissue culture cell. In this late phase of cell division (telophase) the nucleus has divided into two (dark areas). Each daughter nucleus contains genetic material identical to that of the mother cell. Spindle microtubules can be seen in the cytoplasmic bridge between the two cells. The spindle is involved in separating chromosomes during division of the nucleus. At telophase, the cell membrane is drawn in to form a cleavage furrow, which will break to leave two daughter cells. Magnification: ×8500 at 8 × 10-inch size. *(Courtesy © Dr. Gopal Murti/Photo Researchers, Inc.)*

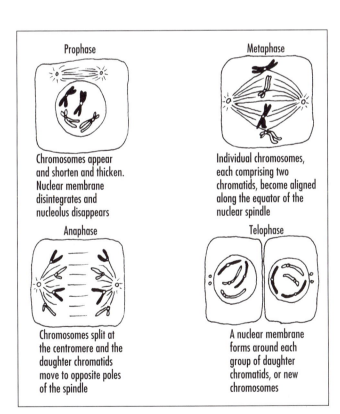

Prophase

Chromosomes appear and shorten and thicken. Nuclear membrane disintegrates and nucleolus disappears

Metaphase

Individual chromosomes, each comprising two chromatids, become aligned along the equator of the nuclear spindle

Anaphase

Chromosomes split at the centromere and the daughter chromatids move to opposite poles of the spindle

Telophase

A nuclear membrane forms around each group of daughter chromatids, or new chromosomes

Mitosis is a four-step process that includes prophase (prometaphase), metaphase, anaphase, and telophase.

one level of oxidation. The importance in biology is due to the often-complete delocalization of the valence electrons over the cluster, allowing efficient electron-transfer processes.

See also OXIDATION NUMBER.

mobbing The display of a flock of small birds attacking a larger predatory bird to keep it away from nests, a defensive posturing.

model A synthetic COORDINATION entity that closely approaches the properties of a metal ion in a protein and yields useful information concerning biological structure and function. Given the fact that the term is also loosely used to describe various types of molecular structures, constructed, for example, in the computer, the term BIOMIMETIC is more appropriate.

modern synthesis The neo-Darwinism theory of evolution. A modern theory about how evolution works at the level of genes, phenotypes, and populations.

molarity The number of moles of solute dissolved in 1 liter of solution.

mold Molds are naturally occurring clusters of microscopic fungi that reproduce by releasing airborne spores. Certain individuals with a mold allergy will develop asthma and nasal symptoms if they breathe in these spores. Many people are allergic to mold. Mold spores are carried in the air and can be present all year long. Mold is most prevalent indoors, in damp locations and in bathrooms, washrooms, fabrics, rugs, stuffed animals, books, wallpaper, and other "organic" materials. Outdoors, mold lives in the soil, on compost, and on damp vegetation.

See also FUNGI.

mole (mol) An amount of substance that contains as many items (such as ions, molecules, etc.) as the number of atoms in exactly 12 grams of carbon (C). The number of molecules contained is equal to 6.022×10^{23} (602,200,000,000,000,000,000,000), known as Avogadro's number. Thus a mole is anything that has Avogadro's number of items in it.

molecular formula The formula of a compound in which the subscripts give the actual number of each element in the formula.

molecular graphics The visualization and manipulation of three-dimensional representations of molecules on a graphical display device.

molecular modeling A technique for the investigation of molecular structures and properties using computational chemistry and graphical visualization techniques in order to provide a plausible three-dimensional representation under a given set of circumstances.

molecule The smallest unit in a chemical element or compound that contains the chemical properties of the element or compound. They are made of atoms held together by chemical bonds that form when they share or exchange electrons. They can vary in complexity from a simple sharing of two atoms, such as oxygen, O_2, to a more complex substance such as nitroglycerin, $C_3H_5(NO_3)_3$.

molt The process of periodically losing old skin or exoskeleton and replacing it with a new one. Reptiles shed their skin, birds shed feathers, and butterflies molt several times during their development.

molybdenum cofactor (Moco) The molybdenum complex of the MOLYBDOPTERIN PROSTHETIC GROUP (LIGAND). In the molybdenum COFACTOR, the minimal COORDINATION of the Mo atom is thought to be provided by the chelating dithiolenato group of the molybdopterin and either two oxo or one oxo and one sulfide ligands.

molybdopterin The PROSTHETIC GROUP associated with the Mo atom of the MOLYBDENUM COFACTOR found in all molybdenum-containing ENZYMEs except NITROGENASE. Many of the enzymes catalyze two-electron redox reactions that involve the net exchange of an oxygen atom between a SUBSTRATE and water. The molybdopterin prosthetic group contains a pterin ring bound to a dithiolene functional group on the 6-alkyl side chain. In bacterial enzymes a NUCLEOTIDE is attached to the phosphate group.

Monera The large prokaryotic kingdom that includes archaebacteria, eubacteria, and cyanobacteria, members of which were the first forms of life over 3.5 billion years ago. All bacteria belong to the kingdom Monera.

See also BACTERIA.

Moniz, António Caetano de Abreu Freire Egas
(1874–1955) Portuguese *Neuroscientist* António
Caetano de Abreu Freire Egas Moniz was born in
Avanca, Portugal, on November 29, 1874, to Fernando
de Pina Rezende Abreu and Maria do Rosario de
Almeida e Sousa. He received his early education
from his uncle before joining the Faculty of Medicine
at Coimbra University. He continued his education
in Bordeaux and Paris, received his doctor's degree in
medicine in 1899, and became professor at Coimbra in
1902. In 1911 he became the new chair in neurology at
Lisbon and stayed there until his death.

Moniz also participated in politics in 1903, serving
as a deputy in the Portuguese parliament until 1917,
when he became Portuguese ambassador to Spain. In
1917 he was appointed minister for foreign affairs and
was president of the Portuguese delegation at the Paris
Peace Conference in 1918.

Aside from politics, Moniz discovered cerebral
angiography and prefrontal leucotomy (lobotomy). He
was awarded a Nobel Prize in medicine in 1949 for
developing the first psychosurgery procedure, frontal leucotomy, and a surgical interruption of nerve fibers that
connect the frontal or prefrontal areas of the cerebral
cortex of the brain. He published a number of books,
including *Physiological and Pathological Aspects of Sex
Life,* 1901; *Neurology in War,* 1917; *A Year of Politics,*
1920; *Diagnostics of Cerebral Tumours and Application
of Arterial Encephalography,* 1931; *Tentative Methods in
the Treatment of Certain Psychoses,* 1936; *Prefrontal
Leucotomy: Surgical Treatment of Certain Psychoses,*
1937; *Clinical Cerebral Angiography,* 1938; *Cerebral
Arteriography and Phlebography,* 1940; *On the Side of
Medicine,* 1940; *How I Came to Perform Leucotomy,*
1948; and even *History of Playing-Cards,* 1942.

Moniz received numerous honors during his career
and is considered the father of modern psychosurgery.
He died on December 13, 1955, in Lisbon.

monoamine Small organic molecule containing both
a carboxyl group and an amino group bonded to the
same carbon atom, e.g., histamine, serotonine,
epinephrine, and norepinephrine.

monoclonal Literally, coming from a single clone.
A clone is the progeny of a single cell. In immunology, monoclonal generally describes a preparation
of antibody that is monogenous, or cells of a single
specificity.

monoclonal antibodies Laboratory-produced antibodies, which can be programmed to react against a
specific antigen in order to suppress the immune
response.

monocot Monocotyledonae, a subclass of the
angiosperms, that has as characteristics an embryo
containing one cotyledon, parallel leaf veins, pollen
with single furrow or pore, stem vascular bundles scattered, secondary growth absent, adventitious roots,
and flower parts in multiples of three. Monocots comprise some one-quarter of all flowering plants in the
world, about 65,000 species divided among 9 families:
Gramineae, Liliaceae, Iridaceae, Orchidaceae, Palmae,
Pandanaceae, Agavaceae, Bromeliaceae, Musaceae.
Includes lilies, orchids, palms, and grasses. Economically important monocots such as corn, rice, wheat,
barley, sugar cane, pineapples, dates, and bananas
make up much of the food supply.

Monocots reproduce sexually. The flowering plants
contain female carpels and male stamens. Monocots
are primarily tropical, with the exception of the lilies,
asparagus, and gumflower.

See also ANGIOSPERM.

monoculture The practice of raising only one species
on a large land area. Makes planting and harvesting
easier with a large machine like a combine.

monocyte A white blood cell that can ingest dead or
damaged cells (through phagocytosis) and provide
immunological defenses against many infectious organisms. Monocytes migrate into tissues and develop into
macrophages.

monoecious Plants having separate male and
female flowers on the same plant, e.g., cucumbers,
American beech, black walnut, and corn. Opposite of
DIOECIOUS.

monogamy The practice of having one mate at a time, lifelong pair bonding.

monogyny In social insects, having one queen per nest, as with ants.

monohybrid cross A cross that involves two parents that differ in only one trait.

monomer A basic building block or small organic molecule that makes up a polymer when combined with identical or similar monomers through polymerization. Polymers are important substances in organisms, e.g., proteins are polymers.

mono-oxygenase An ENZYME that catalyzes the INSERTION of one atom of oxygen, derived from O_2, into an aromatic or aliphatic compound. The reaction is coupled to the oxidation of a COSUBSTRATE such as NAD(P)H or 2-oxoglutarate.

monophyletic Refers to a group of organisms that includes the common ancestor and all descendants of this common ancestor. The group of organisms is also called a clade.

monosaccharide A simple sugar such as fructose or glucose that cannot be decomposed by hydrolysis; colorless crystalline substances with a sweet taste that have the same general formula, $C_nH_{2n}O_n$. They are classified by size according to the number of carbon atoms in the chain, such as dioses, two carbon-ring backbone; trioses, three carbon-ring backbone; heptose with seven carbon-ring backbone, etc. Further classified as aldoses (when carbonyl group is an aldehyde) or ketoses (contains a carbonyl [keto] group in its straight-chain form).

monotreme An egg-laying mammal that feeds its young with milk once hatched. Only two forms exist in the world, and both live in Australia: the short-beaked echidna (the spiny anteater) and the duck-billed platypus.

montane A biological zone of altitude found in mountains, above the tree line, and usually based on at least 500 m (1,600 ft.) elevation.

Morgan, Thomas Hunt (1866–1945) American *Zoologist* Thomas Hunt Morgan was born on September 25, 1866, in Lexington, Kentucky, to Charlton Hunt Morgan. He received a B.S. at the University of Kentucky in 1886 and completed postgraduate work at Johns Hopkins University, where he studied morphology with W. K. Brooks and physiology with H. Newell Martin.

In 1890 he obtained a Ph.D. degree at Johns Hopkins University for his work on the evolutionary relationships of pycnogonids (sea spiders).

In 1891 he became associate professor of biology at Bryn Mawr College for women, Pennsylvania, where he stayed until 1904, then becoming professor of experimental zoology at Columbia University, New York. He remained there until 1928, when he was appointed professor of biology and director of the G. Kerckhoff Laboratories at the California Institute of Technology, at Pasadena. Here he remained until 1945. During his later years he had his private laboratory at Corona del Mar, California.

Morgan's early work on genetic linkage with *Drosophila melanogaster* (the fruit fly) was put forward as a theory of the linear arrangement of the genes in the chromosomes. He described the phenomena of linkage and crossing over, which he explained in *Mechanism of Mendelian Heredity* (1915). He was the first to show that variation derives from numerous small mutations. His work also established the use of drosophila as a model for genetic research gene-mapping efforts. For his discoveries concerning the role played by the chromosome in heredity, he was awarded the Nobel Prize in 1933. Morgan is considered a cofounder, along with William Bateson, of modern genetics.

He also wrote several important genetics books: *Heredity and Sex* (1913), *The Physical Basis of Heredity* (1919), *Embryology and Genetics* (1924), *Evolution and Genetics* (1925), *The Theory of the Gene* (1926), *Experimental Embryology* (1927), and *The Scientific Basis of Evolution* (1935). He died on December 4, 1945.

The Morgan is now the unit of measurement of distances along all chromosomes in flies, mice, and humans. The University of Kentucky honored Morgan by naming its school of biological science after him.

morphogen A diffusible protein molecule present in embryonic tissues that, through a concentration gradient, can influence the development process of a cell; different morphogen concentrations specify different cell fates.

morphogenesis The development of body shape and organization of an embryo from fertilized egg to adult.

morphological species concept A way to classify organisms in the same species if they appear identical by anatomical criteria. Recognizing a species based initially on appearance; the individuals of one species look different from the individuals of another.

morphometrics A branch of mathematics that focuses on the study of the metrical and statistical properties of shapes and the changes of geometric objects both organic or inorganic. Biologically relevant when dealing with species that have morphs that appear radically different.

morphotype Reference to a particular morphological appearance of an organism or group.

mosaic development A pattern of development in which body parts are determined from an early stage of cell division; the blastomere fate is established at a very early stage in development. Cells develop more or less independently and are largely unaffected by each other or the environment. When and what the cell becomes is under tight genetic control.

mosaic evolution When different anatomical, physiological, and behavioral features evolve at different rates and at different times. Human evolution and language are examples.

mosquitoes An organism belonging to the dipteran suborder, Nematocera, the more primitive group of flies that also includes groups such as crane flies, midges, gnats, and black flies. This is a large, abundant, and well-known family (Culicidae) whose members are pests to humans and who, with their long proboscises for feeding on the blood of mammals, are vectors for disease. *Aedes aegypti* (L.) is a vector for yellow fever and dengue fever, *Anopheles punctipennis* (Say) is the vector for malaria, and *Culex pipiens* (L.) is the vector for filariasis and encephalitis.

Mössbauer effect Resonance absorption of gamma radiation by specific nuclei arranged in a crystal lattice in such a way that the recoil momentum is shared by many atoms. It is the basis of a form of spectroscopy used for studying COORDINATED metal ions. The principal application in bioinorganic chemistry is ^{57}Fe. The parameters derived from the Mössbauer spectrum (isomer shift, quadrupole splitting, and the HYPERFINE coupling) provide information about the oxidation, spin, and COORDINATION state of the iron.

moth Any of numerous insects of the order Lepidoptera. The nocturnal counterpart to the butterfly. Whereas moths rest with wings in various positions, butterflies rest with the wings folded over the back.

motif A pattern of amino acids in a protein SEQUENCE that has a specific function, e.g., metal binding.
See also CONSENSUS SEQUENCE.

motor neuron A neuron that sends messages from the central nervous system (brain or spinal cord) to smooth muscles, cardiac muscle, or skeletal muscles. There are upper-motor neurons that lie entirely within the central nervous system and cause movement because they terminate on a lower motor neuron; the lower motor neurons cell body lies in the central nervous system, and their axons leave the central nervous system through a foramen (hole in a bone for nerves and blood vessels) and terminate on an effector (e.g., muscle).

Silk Degrees: A Tale of Moths and People, Part Two

By James G. (Spider) Barbour

Recreational domestication of our native moths dates back to horses and buggies. Several devoted American and British silk-moth breeders have committed their knowledge and experience to print. The first popular book devoted to rearing moths, *Caterpillars and Their Moths,* was published in 1902. The authors, two New Englanders named Ida Eliot and Caroline Soule, gave detailed accounts of raising 43 species of moths, most of them native to the eastern United States. Among their successes was a cross-pairing with viable but infertile hybrid offspring between species of two different genera, the spicebush silk moth (*Callosamia promethea*) and the ailanthus silk moth (*Samia cynthia*). Intergeneric hybrids are very rare. Even more remarkable, as mentioned in part one (page 200) in connection with the French silk promoter Léopold Trouvelot, the ailanthus silk moth is a native of China. Though others have tried, to this day this particular feat of old-style genetic engineering has never been repeated.

A similar book appeared 10 years later by the popular Indiana novelist Gene Stratton Porter. *Moths of the Limberlost* was a nonfiction companion to her best-known work of fiction, *A Girl of the Limberlost*. Illustrated with the author's hand-tinted color photographs, this inspired moth tome launched many a young reader into the throes of moth cultivation, so that by the 1940s, standard techniques of silk-moth culture had been developed and disseminated.

The Amateur Entomologists' Society, a British organization, published a compendium of moth-breeding methods in its sixth volume in 1942, edited by Beowulf Cooper. The material in this issue of the AES journal was revised and expanded by W. J. B. Crotch and published as Volume 12, *A Silk Moth Rearer's Handbook,* in 1956.

The first American authors to present these techniques, along with species accounts, were Michael Collins and Robert Weast in *Wild Silk Moths of the United States* in 1961. Paul Villiard, a New York moth breeder, published *Moths and How to Rear Them* in 1969. Therein he lamented the policy of the U.S. Department of Agriculture to restrict importation of nonnative moths, and gave fascinating accounts of his successes and failures with exotic species. (By mail from less restricted European breeders, he managed to import living material of many Asian, African, and tropical American moths.) Attempting to rear the fabulous Madagascar moon moth (*Argema mittrei*) from a dozen eggs he obtained from a British breeder, Villiard offered the caterpillars every plant he could find. Almost all were rejected, and the caterpillars died, except for two that thrived on poison ivy.

In 1996 Cornell University Press published the definitive book (to date) on our native Saturniidae. Authored by Paul Tuskes, James Tuttle, and Michael Collins, *The Wild Silk Moths of North America* presents everything you never imagined there was to know about these spectacular insects, including rearing tips, collecting techniques, ecological data, host plants, range maps, emergence times in different parts of a species's geographic range, and much more.

As odd an obsession as it may seem to some, there are a number of perfectly understandable reasons for breeding moths. Those assembling collections of pinned insects can obtain perfect, unblemished specimens freshly hatched from the cocoon. Researchers can perform a variety of controlled experiments by raising large numbers of caterpillars, for example, comparing growth rates on different caterpillar host plants, testing the effects of artificial photoperiods on caterpillar growth and the length of time spent in the pupal stage, or recording levels of predation on caterpillars placed in the open on different plants or in different environmental settings.

There is even some profit in rearing large numbers of caterpillars—selling the cocoons to collectors or the pupae to research labs for a variety of projects. A recent example is the splicing of a gene from cecropia onto the genome of a potato to impart resistance to fungus. Mostly, though, silk moths are simply fascinating animals in their life habits, their amazing transformations, their large size, and their otherworldly beauty.

Raising moths is not an entirely easy business, though. Obtaining fertile eggs is simple enough, once one has collected a few cocoons. Chances are, in a batch of six or more cocoons, at least one will produce a female. Fresh males and females may mate overnight in an amply sized screen cage. Lacking males, a female may be tethered on a string tied around the thorax between the wings, tied in an open-ended screen cylinder for protection from predators, and left in a tree or shrub overnight. Very likely in the morning there will be a male mating with her, the female having emitted her seductive perfume (a chemical attractant called a pheromone) during the night. Unlike butterflies, most of which cannot be induced to lay eggs on anything but fresh leaves of a specific plant species, silk moths will lay eggs on anything. Most breeders place a fertilized female in a paper shopping bag (plastic will not do), and the moth will lay eggs on the sides of the bag. The bag can be cut into pieces with eggs firmly attached, and placed on appropriate host plants in places judged to be relatively free of predators and parasites.

That's the easy part. Nursing the caterpillars to the cocoon-spinning stage is the hard part. A saturniid caterpillar is a nutrient-rich morsel, full of easily digested calories, an energy bar for a bird, a mouse, or a wasp; or a milkshake for a stink bug, which inserts its proboscis through the caterpillar's skin, secretes a digestive enzyme, and sucks out the resultant soup. For a parasitic wasp or fly, a caterpillar is a living pantry within which its grubs, after hatching from eggs inserted by the female into its victim, feed on the abundant food storage cells, avoiding the internal organs and muscles until the final stage of their lives, when they reduce the hapless caterpillar to a limp bag or dry husk before their own transformation to winged adults. These same natural enemies that reduce a fertile female's 200–400 eggs to two surviving adult moths (on average) are capable of launching assaults on captive caterpillars as well. Containers may protect caterpillars from predators, but not from diseases, accidents, and adverse conditions. Screen-topped containers allow air to circulate, but this causes leaves placed in the containers to dry out, so the caterpillars don't get enough water. Closed containers have sufficient air for several days at least, but this lack of ventilation produces condensation, creating a breeding ground for fungal and bacterial diseases. Open-weave cloth bags ("sleeves" in moth breeder parlance), placed over tree limbs and tied tight around the branch at the open end, guarantee fresh food and fresh air but are subject to attack by birds and wasps, which can peck or chew holes in the sleeves and consume the helpless caterpillars in short order.

Beginners can expect to raise at least several species of silk moth everywhere except in an extremely cold, polar climate. Caterpillars of many species feed on a much wider range of plants than those they feed upon in their native habitats; even some exotic species can be reared on plant genera common almost everywhere, such as oak, apple, willow, and privet. Eggs and cocoons are available in season, for a price, from dealers and amateur breeders. Start with a web search for "moths + livestock" or "lepidoptera + livestock." First, however, you should learn which species will do well on plants available in your area and in your local climate. Before your livestock arrives, be ready with all the rearing gear you will need to care for your captives.

With some time and effort, trial and error, and poking around, one should eventually be able to find native silk moths in wild or weedy areas near home. To the sharp-eyed, a winter walk may yield cocoons hanging from tree twigs, spun among the shoots at the bases of maple and poplar trees, or hidden in tangles of shrubbery. In summer and fall, one can locate caterpillars by looking for missing leaves (nibbled down to the stem) on trees and shrubs, or checking bare ground or pavement under tree branches for caterpillar droppings ("frass"). Once one has female moths, either from discovered cocoons or caterpillars or from purchased stock, one can attract local wild males by tethering the females overnight. It is advisable to purchase a standard work on moth culture before embarking on such an adventure.

A patchwork of open and forested land is probably the best place to look for caterpillars and cocoons, the more varied the better. Wetlands will have different species than mature forests, farm hedgerows, weedy lots, or wind-swept ridges. Silk moths are known to tolerate a fairly high degree of disturbance and tend to prefer "edge habitats." Roadsides, railroad rights-of-way, utility corridors, and even unkempt, weedy corners of industrial yards are tried and true places for collecting wild silk moths in all developmental stages. Even cities have scattered, unmanicured corners where moths eke out a living. The ailanthus silk moth is one saturniid species that country-dwelling silk moth enthusiasts must visit a city to find. In the same urban setting reside several of the native silk moths, particularly cecropia, polyphemus, and promethea. During the early 20th century, many authors reported these species as abundant in eastern and midwestern cities, the caterpillars and cocoons regularly discovered by schoolchildren and curious nature lovers on shade plantings. Toward the end of the century, especially in the 1990s, observers of silk moths reported significant declines and apparent disappearances of common species from urban and suburban areas in many areas of the eastern states. The causes for these declines are uncertain, but the following factors have been suggested: increased lighting at night (which may interfere with reproductive behavior), the release of nonnative parasites to control gypsy moth and other pests, and general increased use and accumulation of insecticides and other toxic chemicals.

Though certain species of silk moths may be common in settled areas, few ever become sufficiently populous to be considered pests. Exceptions include the range caterpillar (*Hemileuca olivae*) in the Plains states, where the caterpillars, arrayed with stinging spines, occasionally are so numerous as to seriously injure the mouths and tongues of grazing cattle. The cecropia moth has been reported as an

(continues)

Silk Degrees: A Tale of Moths and People, Part Two
(continued)

occasional pest in nurseries, where it would take only one or two of the huge caterpillars to defoliate a small sapling.

Whether the long-standing association between moths and humans will continue into the next several centuries is as uncertain as the future of the human species. People continue to alter the Earth in ways detrimental to other species and to ourselves. Very likely the salvation of wild silk moths lies with that segment of the human population that cherishes the moths' existence and keeps the knowledge of their intriguing, mysterious ways.

—**James G. (Spider) Barbour**, a field biologist specializing in entomology and ecology, has 25 years of field experience in the Hudson Valley, especially in the Catskills, Hudson Highlands, and Westchester County. He conducts botanical, rare species, wetland, plant community, and insect surveys. He and his wife, Anita, are the authors of *Wild Flora of the Northeast*.

motor unit   A single motor neuron and the entire complement of muscle cells that receive synaptic connections from it.

MPF (M-phase promoting factor)   A protein complex containing cyclin and a cyclin-dependent protein kinase that triggers a cell to enter mitosis (M phase). Cyclin B

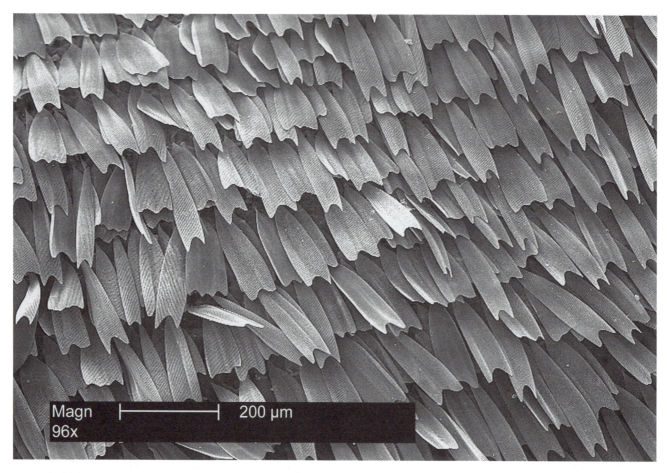

Scalar patterning of a butterfly's wing magnified ×96. Note the shinglelike pattern of the wing, which promotes heightened aerodynamic lift during the insect's flight. *(Courtesy of Janice Carr, Centers for Disease Control and Prevention)*

increases prior to mitosis, combines with the cyclin-dependent protein kinase, and forms an active MPF complex. The MPF phosphorylates lamin, a structural protein of the nuclear envelope that assists in maintaining the nuclear shape, among other substrates, causing the dissolution of the nuclear envelope, which triggers the initial phases of mitosis. Formerly called maturation-promoting factor.

M phase Mitosis in eukaryote cells; when a cell divides into two daughter cells, each with the identical chromosomes as the parent cell.

Muller, Hermann Joseph (1890–1967) American *Geneticist* Hermann Joseph Muller was born in New York City on December 1, 1890. He attended public school in Harlem and later Morris High School in the Bronx. He attended Columbia University in 1907.

He obtained a teaching assistantship in zoology at Columbia (1912–15) and received a Ph.D. in 1916. After three years at the Rice Institute, Houston, Texas, and at Columbia as an instructor, in 1920 Muller became an associate professor (later professor) at the University of Texas, Austin, where he remained until 1932. In 1932 he was awarded a Guggenheim Fellowship and for a year worked at Oscar Vogt's institute in Berlin and then spent three and a half years as senior geneticist at the Institute of Genetics of the Academy of Sciences of the USSR, first in Leningrad and later (1934–37) in Moscow. He moved to the Institute of Animal Genetics, University of Edinburgh (1937–40), and he did both teaching and research at Amherst College as a professor from 1942 to 1945. At Amherst he completed a large-scale experiment showing the relationship of aging to spontaneous mutations. In 1945 he accepted a professorship in the zoology department at Indiana University, Bloomington, Indiana, and retired in 1964.

His method for recognizing spontaneous gene mutation led to his discovery of a technique for artificially inducing mutations using X rays, showing in 1927 that mutations could be induced by radiation. For this discovery, he was awarded the 1946 Nobel Prize in physiology or medicine.

Muller contributed over 300 articles on biological subjects to scientific publications. His principal books are *The Mechanism of Mendelian Heredity* with T. H. Morgan, 1915; *Out of the Night—a Biologist's View of the Future*, 1935, 1936, and 1938; and *Genetics, Medicine and Man* with C. C. Little and L. H. Snyder, 1947. He also wrote articles on the biological effects of atomic radiation. He was awarded numerous honorary degrees and other recognitions. During the 1930s and 1940s, his controversial views on eugenics and unpopular opinions about the hazards of radiation forced him to leave the United States and to work in Russia, but he came back. During the late 1940s his criticisms on nuclear fallout were recognized by his peers.

He was known as the father of radiation genetics and died on April 5, 1967, in Indianapolis.

Müller, Paul Hermann (1899–1965) Swiss *Chemist* Paul Hermann Müller was born in Olten, Solothurn, Switzerland, on January 12, 1899. He attended primary school and the Free Evangelical elementary and secondary schools. He began working in 1916 as a laboratory assistant at Dreyfus and Company, followed by a position as an assistant chemist in the Scientific-Industrial Laboratory of their electrical plant. He attended Basel University and received a Ph.D. in 1925. He became deputy director of scientific research on substances for plant protection in 1946.

Müller began his career with investigations of dyes and tanning agents with the J. R. Geigy Company, Basel (1925–65), and beginning in 1935 concentrated his research to find an "ideal" insecticide, one that had rapid, potent toxicity for the greatest number of insect species, while causing little or no damage to plants and warm-blooded animals. He tested and concluded that dichlorodiphenyltrichloroethane (DDT) was the ideal insecticide.

In 1939 DDT was successfully tested against the Colorado potato beetle by the Swiss government and then by the U.S. Department of Agriculture in 1943.

For this discovery of DDT's potent toxic effects on insects Müller received the Nobel Prize in physiology or medicine. DDT proved to be a two-edged sword. With its chemical derivatives, DDT became the most widely used insecticide for more than 20 years and was a major factor in increased world food production and suppression of insect-borne diseases, but the widespread use of the chemical made it hazardous to wildlife and it was banned in 1970. He died on October 12, 1965, in Basel.

Müllerian mimicry   Refers to a situation where the physical similarity of two or more unpalatable species reduces their predation rate because predators avoid them to a greater degree than they would for the individual species.

See also MIMICRY.

multicopper oxidases   A group of ENZYMEs that oxidize organic SUBSTRATES and reduce dioxygen to water. These contain a combination of copper ions with different spectral features, called TYPE 1 centers, TYPE 2 centers, and TYPE 3 centers, where the type 2 and type 3 sites are clustered together as a triNUCLEAR unit. Well-known examples are LACCASE, ascorbate oxidase, and CERULOPLASMIN.

multienzyme   A protein possessing more than one catalytic function contributed by distinct parts of a polypeptide chain (DOMAINs), or by distinct SUBUNITs, or both.

multigene family   A group of genes that are similar in nucleotide sequence and have evolved from some ancestral gene through duplication.

multiheme   Refers to a protein containing two or more HEME groups.

mummification   An Egyptian process to preserve their dead, where the brain and inner parts are separated from the body and the corpse is processed in a natronizing bath and then covered with protective balms, amulets, finger pieces, or other items, then bandaged to special fabric or sometimes covered with bitumen.

Murphy, William Parry (1892–1987) American *Medical Researcher* William Parry Murphy was born on February 6, 1892, in Stoughton, Wisconsin, to Thomas Francis Murphy and Rose Anna Parry. His father was a congregational minister with various pas-

torates in Wisconsin and Oregon. Parry was educated at the public schools of Wisconsin and Oregon and received a B.A. at the University of Oregon in 1914.

From 1915 to 1917, he taught physics and mathematics at Oregon high schools and spent one year at the University of Oregon Medical School at Portland, where he also acted as a laboratory assistant in the

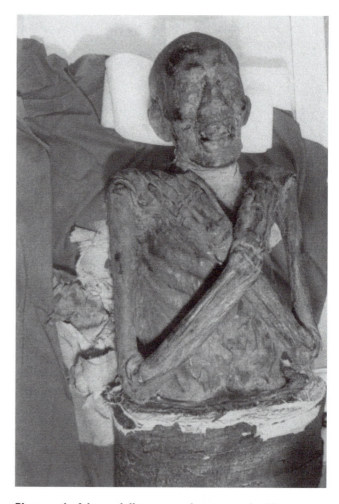

Photograph of the partially uncovered mummy at the Albany Institute of History and Art (AIHA) in Albany, New York. This mummy was obtained partially uncovered in 1909 from the Cairo Museum. His skin is dark from the resin used in the mummification process. The arms are crossed in the typical burial pattern, and there is no defect in the skull because the brain had been removed through the nose. Numerous layers of linen wrapping are visible at the waist. Hieroglyphics on his coffin bear the name of Ankhefenmut and identify him as a priest from the temple of Mut at Karnak in Thebes (according to Ms. Bethyl Mayer, who did background research on the two Albany mummies). He is 5 feet 3 inches tall. *(Courtesy of William Wagle, M.D.)*

Egyptian Mummies:
Brief History and Radiological Studies

By William A. Wagle, M.D.

Egyptians practiced the art of mummification for several thousand years and produced millions of human mummies. Mummification began in its most primitive form in the late fourth millennium B.C.E. (before 3100 B.C.E.) but did not reach its zenith until around 1000 B.C.E. The underlying goal of mummification was to preserve the physical form of the deceased person so that his soul or *ba* would recognize its body and be united with it in the hereafter. Mummification was practiced by various ancient cultures, including the Incas of South America, but the technique was perfected in ancient Egypt, where funeral houses dotted the Nile River from the Mediterranean Sea to the ancient city of Thebes.

The Egyptians believed that after death the body must be intact and have a proper burial for resurrection. Instead of cremation, the bodies were prepared to enter the hereafter as "whole" as possible. During the early dynasties, mummification was a burial form reserved for kings, but by 1000 B.C.E. the practice had become widespread. The pharaohs had the most elaborate burials because it was believed that they became gods after death. High-ranking officials, priests, nobles, and wealthy individuals also underwent mummification.

Mummification was not practiced in prehistoric times, when the dead were simply buried in desert pits. The Egyptians realized that bodies were preserved naturally in the dry, hot sands of the Egyptian desert. The remains of buried individuals were recognizable many years later. According to renowned Egyptologist Bob Brier, "The Egyptians started burying their dead in chambers cut into moist bedrock beneath the sand in order to protect the bodies from robbers or from being uncovered by the shifting desert sands. Without contact with the dehydrating hot sands, they were subject to decay. This probably gave the Egyptians the impetus to invent mummification."

While Egyptians left very little in written documents about the process of mummification, Egyptologists have been able to unlock many of its mysteries with detailed analysis of human mummies. The best ancient source of information on the details of this embalming technique comes from Herodotus, a Greek historian who visited Egypt around 500 B.C.E. He left an account of mummification in a publication called *The History II*.

According to this account, the first step of mummification was the removal of the brain, which the Egyptians felt had no value. This was achieved with a curved metal hook, which was inserted through the nose. The metal could easily fracture the thin ethmoid bone and enter the skull. The curved hook was twisted repeatedly to macerate the brain, which was removed piecemeal through the nose.

According to Herodotus, "The skull is cleared of the rest by rinsing with drugs; next they make a cut along the flank with a sharp Ethiopian stone and take out the whole contents of the abdomen, which they then cleanse, washing it thoroughly with palm wine, and again frequently with an infusion of pounded aromatics. After this they fill the cavity with the purest bruised myrrh, with cassia, and every sort of spice except frankincense, and sew up the opening. Then the body is placed in natrum for 70 days and covered entirely over. After the expiration of this space of time, which must not be exceeded, the body is washed, and wrapped round, from head to foot, with bandages of fine linen cloth, smeared over with gum, which is used generally by the Egyptians in the place of glue, and in this state it is given back to the relations, who enclose it in a wooden case which they have made for this purpose, and shaped into the figure of a man. Then fastening the case, they place it in a sepulchral chamber, upright against the wall. Such is the most costly way of embalming the dead."

Herodotus's account has been confirmed as basically accurate by modern analysis of Egyptian mummies. Mummification evolved considerably during the 2,000 years from its beginning in the Old Kingdom to the time described by Herodotus. All internal organs except the heart, which was considered the seat of emotions and intellect, were removed to prevent decomposition and putrefaction. By tradition, the abdominal incision was made with an Ethiopian stone, which was as sharp as a razor. The lungs, stomach, liver, intestines, and other organs were embalmed with natron, which was the primary drying agent used in mummification. This salt worked by absorbing water from the internal organs, muscles, and blood before microbial action and decomposition could begin. Once the process of enzymatic decomposition of protein begins, there is production of foul-smelling compounds such as hydrogen sulfide, ammonia, and mercaptans. The tissues decompose and rot away.

When the organs were sufficiently dried out, they were placed into ornamental burial jars usually made of limestone. These were called "canopic jars" named in honor of the Greek legend of Canopus, who was buried in Egypt. It is said that he was worshipped in the form of a jar with feet. Each canopic jar contained a different organ. The lid of each jar was carved in the shape of one of the four sons of Horus: Mesti, the human-headed son; Duamutef, the jackal; Hapi, the baboon; and Qebesenef, the hawk.

(continues)

Egyptian Mummies:
Brief History and Radiological Studies
(continued)

During the 21st dynasty (1085 to 945 B.C.E.) embalmers began wrapping the internal organs with long strips of linen and then replacing them back into the body as "visceral packages." This may have been done to dissuade grave robbers from opening or destroying canopic jars in search for jewels and treasures. With this technique there was no need for a canopic jar, and imitation one-piece canopic jars were sometimes placed in the tombs.

The dehydrating agent used in mummification was natron, a naturally occurring salt composed of sodium carbonate, sodium bicarbonate, and sodium chloride (table salt). The two chief sources of natron in ancient Egypt were El Kab, a city in Upper Egypt, and the Wadi Natrun, an area just outside Cairo. In Arabic the word *wadi* means "dry river bed." The chemical symbol for sodium, Na, is derived from the word *natron.*

The body was placed on a slanted embalming table and completely covered with natron. A groove in the dependent end of the table allowed the fluids to drain off. The human body is about 70 percent water, and during mummification there is considerable loss of weight due to loss of water. It was very likely that embalming took place outside in a tent, which allowed better ventilation. Anubis, the jackal-headed god of embalming, is often referred to as being "in his tent."

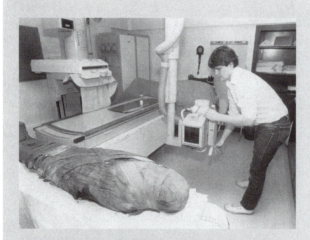

This female mummy at the Albany Institute of History and Art has never been unwrapped and has only been "seen" by X ray and CT. Once believed to be a male priest, the 1988 X-ray and CT studies performed by Dr. Wagle proved it to be the body of a female. She is five feet tall. *(Courtesy of William Wagle, M.D.)*

Once the body was dehydrated, the abdominal and chest cavities were then washed with palm oil and aromatic spices. The abdominal incision site was sewn together, and a metallic amulet, inscribed with the sacred eye of the god Horus, was placed over the incision. The body was then covered with a mixture of cedar oil and spices. A protective coating of dark-colored resin was poured over the body as a sealant to prevent moisture from entering the body. The body was anointed with oil mixed with spices and perfumes. Finally the corpse was wrapped in numerous turns of linen bandages. Laid out end-to-end, the linen would measure hundreds of yards long.

At the time of burial the body was adorned with gold, jewels, and amulets. The amulets were prescribed by the Book of the Dead and were supposed to aid the deceased on his journey to the underworld. These valuable objects, which were placed beneath the wrappings, would become the targets of grave robbers over the ensuing centuries.

The coating resin imparted a dark color to the skin of the mummies. Foreign travelers to Egypt mistakenly assumed that the blackened, solidified resin covering the mummies was bitumen, the mineral formed from pitch. Since the major source of bitumen was a mountain in Persia (Iran), where the substance was called "mummia," these embalmed bodies eventually became known as "mummies."

The process of mummification has been studied for centuries. Visual inspection necessitates that the mummy be unwrapped from its burial cloth. Anatomic dissection of a mummy results in the irrevocable destruction of the "intact body" and violates the ancient Egyptians' desire to remain whole in the hereafter. With the advent of the X ray, mummies could be studied without destroying or even disturbing them. The first published radiograph of mummified remains appeared in 1898, only three years after the discovery of the X ray by Wilhelm Konrad Röntgen in 1895. Other reports followed.

With the invention of computed tomography (CT) in 1972, a much more powerful nondestructive tool became available to scientists and Egyptologists interested in studying human and animal mummies.

CT provides excellent detail of the inside of a mummy. A collaborative project performed between the author and the Albany Institute of History and Art (AIHA) in Albany, New York, in November 1988, involved the plain film and CT analysis of two ancient human Egyptian mummies. Both mummies have been part of AIHA's Egyptian collection since their purchase from the Cairo Museum in 1909. The mummies came from the ancient community of Thebes on the upper Nile, and they lived during the 21st Egyptian dynasty, which was from 1085 to 945 B.C.E.

One of the mummies had been obtained uncovered from the waist up. X rays performed in the Radiology Department at Albany Medical Center showed that this mummy had a male pelvic structure. The other mummy was completely wrapped in the original burial bandage (3,000 years old), and it had always been assumed that this was another male priest. X rays, however, demonstrated a female pelvic structure, proving this to be the mummy of a woman. According to the hieroglyphics on the side of the wooden coffin containing the male mummy, he was a priest named Ankhefenmut. Since the coffin was not sealed at the time it was obtained by the AIHA, absolute identification of this mummy could not be certain. The name and station in life of the female mummy remain a mystery to this day.

X rays of each mummy included the skull, spine, pelvis, and lower extremities. Skeletal structures were amazingly well preserved. The 3,000-year-old bones had an apparently normal density and were not eroded or fractured. CT studies visualized the thoracic and abdominal cavities, cranial vault, and spinal canal. As described in other studies including anatomic dissection and CT, the oval-shaped objects in the thoraco-abdominal cavities contained the dried-out inner organs wrapped in multiple layers of linen, which created a "jelly-roll" appearance on CT. The skull was filled with air and a solidified liquid that corresponds to the resin used in the mummification process. Because the cranial vault and the spinal canal are contiguous, the resin had flowed into both and had solidified.

The most novel finding of this study was the discovery of a well-crafted, two-component great-toe prosthesis (artificial toe) in the female mummy. This prosthesis attaches directly to the first metatarsal bone of the right foot. The distal portion of the prosthesis has a CT density measurement of –600 Hounsfield units, which is that of air. This is probably some low-density air-filled shell that looks like a toe. The more proximal socketlike portion, which holds the toe, measures +1,318 Hounsfield units. This is very high-density material, but it does not appear to be metal on X ray. It may be some form of high-density ceramic, which exactly fits the "toe" and rests upon the first metatarsal bone of the foot. This prosthesis has only been visualized with X ray and CT because the female mummy has never been unwrapped from her burial cloth. The reason for the artificial toe stems from the ancient Egyptian belief that one should enter the hereafter "whole."

A recent report in the British medical journal *The Lancet* describes a wooden prosthesis of the great right toe of a female mummy from the 21st or 22nd dynasty (ca. 1065–740 B.C.E.), which is the same time period as the AIHA mummy.

The amputation site was covered with skin, indicating that it was an intravital (during life) amputation. This prosthesis was composed of three separate components. The main component consisted of a perfectly shaped wooden corpus ($12 \times 3.5 \times 3.5$ cm), which resembled a big toe including the nail. This was attached to two small wooden plates that were fixed to each other by seven leather strings. According to the authors, there were clear marks of use on the sole of the prosthetic toe, indicating that it had been used during life.

In early 1989 a collaborative study was performed between the AIHA, the author, and William Lorensen, Ph.D., of the General Electric Corporate Research & Development in Schenectady, New York. The CT data obtained from the November 1988 mummy studies was used to perform the first-ever computer-assisted "unwrapping" of a mummy. Using a proprietary surface-algorithm program developed by Dr. Lorensen, the mummy's facial linen was removed in a computer program capable of sensing differences in surface shading. (The original study is in the archives of GE Corporate Research & Development.) When multiple images from this study are viewed in ciné mode, there is clearly the visual impression that the linens are actually being removed.

(continues)

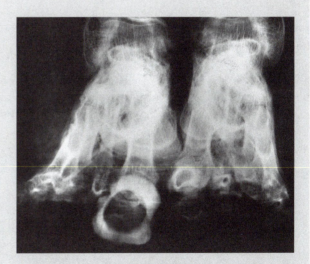

Computed tomography (CT) shows the prosthesis on a mummy's foot. The prosthesis is composed of a high-density U-shaped holder and a low-density "toe." The mummy has never been unwrapped, and only X ray and CT have provided images of one of the most carefully constructed toe prostheses ever seen. This appears in an article written by Dr. Wagle for the *American Journal of Roentgenology* in April 1994. *(Courtesy of William Wagle, M.D.)*

Department of Anatomy. At Harvard Medical School, in Boston, he received an M.D. in 1922.

In 1924 he was appointed assistant in medicine at Harvard, and from 1928 until 1935 he was an instructor in medicine. From 1935 until 1938 he was associate in medicine at Harvard, and from 1948 until 1958 he was a lecturer in medicine, becoming in 1958 a senior associate in medicine and subsequently emeritus lecturer in that subject.

In 1923 he practiced medicine and engaged in research on diabetes mellitus and on diseases of the blood, and he used intramuscular injections of extract of liver for the treatment of pernicious and hypochromic anemia and for granulocytopenia. He was associated with George Richards MINOT and George Hoyt WHIPPLE in work on pernicious anemia and the treatment of it by means of a diet of uncooked liver. For this work he was awarded, with George Richards Minot and George Hoyt

Whipple, the Nobel Prize in physiology or medicine for 1934. Murphy's treatment of the subject was published as *Anemia in Practice: Pernicious Anemia* (1939). In 1958 he was granted emeritus status at both Harvard and the Peter Bent Brigham Hospital. He died at home in Brookline, Massachusetts, on October 9, 1987.

mu (μ) symbol   *See* BRIDGING LIGAND.

mutagen   An agent that causes a permanent heritable change (i.e., a mutation) into the DNA (deoxyribonucleic acid) of an organism.
    See also MUTATION.

mutagenesis   The introduction of permanent heritable changes, i.e., MUTATIONS into the DNA of an

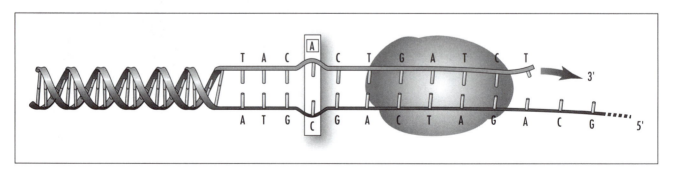

Mutagenesis involves the replacement of one nucleotide in a DNA sequence by another nucleotide or the replacement of one amino acid in a protein by another amino acid. *(Courtesy of Darryl Leja, NHGRI, National Institutes of Health)*

organism. In the case of site-directed mutagenesis, the substitution or modification of a single amino acid at a defined location in a protein is performed by changing one or more BASE PAIRS in the DNA using recombinant DNA technology.

mutation A heritable change in the NUCLEOTIDE SEQUENCE of genomic DNA (or RNA in RNA viruses), or in the number of GENEs or chromosomes in a cell, that can occur spontaneously or be brought about by chemical mutagens or by radiation (induced mutation).

mutualism A state where two different species benefit from their association. Two types exist: symbiotic and nonsymbiotic mutualism. Examples include lichens (algae and fungi), mycorrhizae and rooting plants, yucca plant pollination by the yucca moth, and bees and flowers. In symbiotic mutualism, both individuals interact physically, and their relationship is biologically essential for survival, e.g., the fungus–alga relationship. The more common nonsymbiotic mutualism is when individuals live independent lives but cannot survive without each other, e.g., bees and flowering plants.

mutual prodrug The association in a unique molecule of two, usually synergistic, DRUGs attached to each other, one drug being the carrier for the other and vice versa.

mycelium Threadlike tubes, filaments, or hyphae; the roots of mushrooms; the thalus or vegetative part of a fungus. Bacteria (Actinomycetales) also produce branched mycelium.

mycorrhizae A symbiotic (mutualism) connection between plant roots and the mycelia of some fungi species. The fungus provides water and mineral nutrients to the plant, and the plant provides energy to the fungus. Ectomycorrhizae form between tree species and basidiomycete fungi, and the fungus provides a sheath around the root that it penetrates. Endomycor-

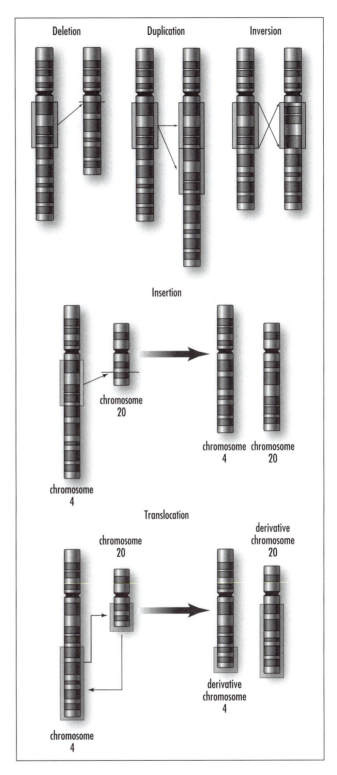

A permanent structural alteration in DNA. In most cases, DNA changes either have no effect or cause harm, but occasionally a mutation can improve an organism's chance of surviving and passing the beneficial change on to its descendants. *(Courtesy of Darryl Leja, NHGRI, National Institutes of Health)*

rhizae form when the fungus hyphae grow between and within the cells of the plant roots. Many species cannot grow without their mycorrhizae connections.

myelin   A lipid that forms a multilayered sheath around some nerve fibers (axons) in the central, autonomic, and peripheral nervous systems.

myelin sheath   Multilayered specialized Schwann cells (nonconducting glial cells) that help provide the efficient movement of signals by coating and insulating sections (internodes) of neurons (axons) in the nervous system.

myiasis   An infection caused by fly maggots. Primary myiasis is when a fly deposits eggs on the host and the maggots feed upon living tissue. Secondary myiasis occurs first when there is a break in the skin that allows maggots the access to tissue.

myocrysin   *See* GOLD DRUGS.

myofibril   A long, cylindrical, contractile filament within muscle or muscle fiber that makes up striated muscle. Each myofibril contains intertwined filaments of muscle proteins, myosin, and actin; called a sacromere. The myofibril bundles have alternate light and dark bands (thick and thin) that contain these protein filaments responsible for the muscle's contractile ability, and these give it the characteristic striated look under a microscope.

myoglobin   A monomeric dioxygen-binding hemeprotein of muscle tissue, structurally similar to a SUBUNIT of HEMOGLOBIN.

myopia (nearsightedness; shortsightedness)   An inherited disorder where a refractive error in the eye means that the shape of the eye does not bend light correctly, resulting in a blurred image; inability to see distant objects clearly because the images are focused in front of the retina.

Transmission electron micrograph of part of a striated muscle fiber (cell) from the human neck. Striated muscle cells are elongated, hence their description as fibers. Each cell has several nuclei (not visible) positioned just below the sarcolemma, the name given to the cell membrane of muscle cells. Running the length of each fiber are contractile protein filaments (actin and myosin myofilaments) arranged in bunches called myofibrils (20–30 myofibrils are visible here). The arrangement of the two types of myofilament causes the prominent transverse banding, which gives striated muscle its name. Magnification: ×8, 100 at 8 × 10-inch size. *(Courtesy © Science Photo Library/Photo Researchers, Inc.)*

myosin   A thick contractile protein found in myofibrils that interacts with actin, another protein, to create contraction in a muscle cell. It is the most abundant protein in muscles fibrils.

myriapods   A group of organisms comprising centipedes and millipedes, which are long, flattened, segmented predators; each segment bears a pair of legs. Centipedes evolved during the Silurian period, and

An Arthropoda and an example of a myriapod from New York's Catskills. Its oviduct is located in the anterior, on the side of the neck. Shown ovipositing in this picture. *(Courtesy of Tim McCabe)*

there are about 2,800 species alive today; there are about 10,000 species of millipedes.

myrmecophile   An organism, such as a beetle, that lives within an ant nest and is either being cared for by the ants or preying upon the ants or their brood for food. *Myrmecophile* is Greek for "ant loving."

N

NAD⁺ Oxidized form of nicotinamide adenine dinucleotide. Note that despite the plus sign in the symbol, the COENZYME is anionic under normal physiological conditions. NAD⁺ is a coenzyme derived from the B vitamin niacin. It is transformed into NADH when it accepts a pair of high-energy electrons for transport in cells and is associated with catabolic and energy-yielding reactions.

NADH Reduced form of nicotinamide adenine dinucleotide (NAD). Called coenzyme I, it is an electron donor essential for a variety of oxidation-reduction reactions.

NADP⁺ Oxidized form of nicotinamide adenine dinucleotide phosphate. Note that despite the plus sign in the symbol, the COENZYME is anionic under normal physiological conditions. An enzyme commonly associated with biosynthetic reactions. NADP is a hydrogen carrier in a wide range of redox reactions.

NADPH Reduced form of nicotinamide adenine dinucleotide phosphate. An energy-rich compound produced by the light-reaction of photosynthesis. It is used to synthesize carbohydrates in the dark-reaction.

naiad (larva, nymph) Used to describe the intermediate stage, between egg and adult, of a dragonfly's development.

natural killer cell (NK cell; NK lymphocyte) A large granular lymphocyte killer cell that attacks and kills tumorous, cancerous, and virus-infected cells through phagocytosis and by using chemicals that destroy the abnormal cells. They are not specifically targeted to any antigen the way other lymphocytes are, and therefore they do not need additional stimulation to attack and kill. Part of the innate immune response.
See also LYMPHOCYTE.

natural selection Survival and reproduction of a genotype that leaves more progenies who possess attributes of fitness, survivability, and adaptation to their environment over that of another genotype. Those better adapted to their environment are likely to increase in frequency over a number of generations over those that are less adapted, and this difference is not due to chance. It is the environment's strong influence on the reproductive success of individuals in a population; one of the major forces of evolution. First promoted by Charles Darwin in the 19th century.
See also EVOLUTION.

NCE *See* NEW CHEMICAL ENTITY.

NDA (new drug application) The process of submitting a new drug for approval. After a new drug application (NDA) is received by the federal agency in charge, it undergoes a technical screening generally referred to as a completeness review and is evaluated to ensure that sufficient data and information have been submitted in each area to justify the filing.

Neanderthal   An early type of human (*Homo sapiens neanderthalensis*) that evolved in Europe about 250,000 years ago and spread to the Middle East. By the last Ice Age, fossil remains reveal a people with faces that had distinct eyebrow ridges, flattened noses, and heavy jaws. Their bodies were short and well built, all features that may have been adaptations to the cold conditions of the last Ice Age. They appear to have survived in parts of Europe until some time after 30,000 years ago and lived for a time side by side with present-day-type humans.
    See also CRO-MAGNON.

Nearctic   Biogeographic region including arctic, temperate, and subtropical North America, reaching south to the northern border of the tropical rain forest in Mexico.

necrosis   Condition describing dead plant or animal tissue due to injury, disease, or treatment.

negative feedback   Describes a situation where an action produces a consequence that affects, or feeds back on, the action. In negative feedback, the consequence stops or reverses the action. In biology, for example, the stopping of the synthesis of an enzyme by the accumulation of the products of the enzyme-mediated reaction is a negative feedback reaction.

nematocyst   A small harpoonlike structure or stinging capsule located in a cnidoblast that coils out; used by coelenterates (e.g., jellyfish) to deliver a toxin to stun or kill its prey.

nematode   Roundworms, simple worms of microscopic size, consisting of an elongate stomach with a reproduction system inside a resistant outer cuticle or outer skin. Most feed on bacteria, fungi, and other soil organisms; however, some are parasitic, obtaining their nutrients from animals (dog heartworm), humans (pinworm), and plants. There are 15,000 known species. They are related to the arthropods and are part of a newly recognized group, the Ecdysozoa, which includes the arthropods and nematodes.

neo-Darwinism   *See* DARWINISM.

neoteny   The retention of immature features in the adult stage.
    See also PAEDOGENESIS.

Neotropic   Geographical region including the West Indies, South America, and Central America north to the northern edge of tropical forests in Mexico.

nephron   A microscopic coiled tubular structure found in each kidney that consists of a GLOMERULUS, a mass of capillaries that filters the blood, and a renal tubule that produces urine for elimination. Each kidney is estimated to contain about 1 million nephrons.

neritic zone   The part of ocean that covers all water to a depth of 600 feet. The entire ocean is called the pelagic zone and is divided into two major zones, the oceanic and neritic. The oceanic covers depths below 600 feet, while the neritic zone covers depths from 0 to 600 feet.

net primary productivity (NPP)   The primary productivity of any community is the total amount of biomass that is produced through photosynthesis per unit area and time by the primary producers—plants. It is usually expressed in units of energy or in units of dry organic matter. The annual primary production around the world is more than 240 billion metric tons of dry plant biomass. Productivity is further divided into gross and net primary productivity. Gross primary productivity (GPP) is the total energy fixed by plants in a community through photosynthesis. However, a portion of the energy is used in plant respiration, so by subtracting this from the gross primary production, the net primary productivity (NPP) is deduced, and this represents the rate of production of biomass available to be consumed by animals and other organisms. The most productive areas on the Earth are those that have higher temperatures, abundant soil nitrogen, and water. These ecosystems include the tropical rain forest, swamps, marshes, estuaries, and deciduous forests.

neural crest The neural crest is formed only in vertebrate embryos during the final stages of neurulation, a process in vertebrates where the ectoderm of the future brain and spinal cord—the neural plate—develops folds (neural folds) and forms the neural tube. The neural crest consists of paired dorsal lateral streaks of cells that migrate away from the neural tube and spread throughout the organism to differentiate into many cell types that contribute to the development of systems such as skin, heart, endocrine, bones of the skull, teeth, and even parts of the peripheral nervous system, to name a few.

neuron The basic data processing unit of the nervous system; a specialized cell that carries information electrically from one part of the body to another by specialized processes or extensions called dendrites and axons. Widely branched dendrites carry nerve impulses toward the cell body, while axons carry them away and speed up transmitting nerve impulses (conduction) from one neuron to another. Each neuron has a nucleus within a cell body.

Neuroptera An order of insects with 18 families worldwide comprising dobsonflies, fishflies, alderflies, spongillaflies, owlflies, snakeflies, ant lions, and lacewings. They are characterized by having usually similar membranous wings with many veins, held in a rooflike position, and chewing or sucking mouthparts. They undergo complete metamorphosis, and they are insectivorous.

neurosecretory cells Modified nerve cells that have endocrine functions. Nerve cells in the hypothalamus that react to other nerve cells and that have characteristics of both neurons and endocrine cells and release hormones (neurohormones) directly into body fluids or storage areas. There are two classes of neurosecretory cells in the hypothalamus. They are located in the median eminence near the adenohypophysis, the anterior glandular lobe of the pituitary, and in the neurohypophysis, the posterior lobe of the gland.

The adenohypophysis arises from epithelial tissue at the roof of the mouth and includes the pars distalis, pars tuberalis, and pars intermedia.

The median eminence is an enlarged area of the infundibulum, the part of the brain between the brainstem and cerebrum. Here nerve cells secrete their releasing and inhibitory hormones that get absorbed into the blood capillaries in the median eminence and carried on in the venous blood down along the infundibulum to the anterior pituitary gland.

The adenohypophysis secretes FSH (follicle-stimulating hormone), LH (luteinizing hormone), ACTH (adrenocorticotropic hormone), TSH (thyroid-stimulating hormone), GH (growth hormone), and prolactin.

The neurohypophysis, or the posterior lobe of the pituitary, is connected to the hypothalamus by the stalklike infundibulum and receives two hormones released by the neurons in the hypothalamus, oxytocin (OT) and vasopressin or ADH (antidiuretic hormone). Oxytocin and ADH are stored in the posterior pituitary. The neurohypophysis is an outgrowth of nervous tissue from the floor of the brain in the region of the hypothalamus. It is made up of the pars nervosa and is connected to the brain by the infundibulum. Oxytocin acts on myoepithelial cells in the mammary glands and uterine muscles, causing both to contract. ADH stimulates the kidneys to reabsorb more water.

The hypothalamus contains mammillary bodies, the median eminence, and infundibulum. It contains nuclei that produce the neurohormones. In the supraoptic region are cell bodies of neurons in the paraventricular nucleus that synthesize oxytocin, while those in the supraoptic nucleus synthesize ADH.

Neurosecretory cells are also found in other animals such as mollusks, nematodes, and platyhelminthes, to name a few.

neurotransmitter   A chemical made of amino acids and peptides that switch on or off nerve impulses across the synapse between neurons. Excitatory neurotransmitters stimulate the target cell, while inhibitory ones inhibit the target cells. Examples of neurotransmitters are acetylcholine, dopamine, noradrenaline, and serotonin.

Acetylcholine is the most abundant neurotransmitter in the body and the primary neurotransmitter between neurons and muscles. It controls the stomach, spleen, bladder, liver, sweat glands, blood vessels, heart, and others. Dopamine is essential to the normal functioning of the central nervous system. Noradrenaline, or norepinephrine, act in the sympathetic nervous system and produce powerful vasoconstriction. Serotonin is associated with the sleep cycle.

neutral variation   Genetic diversity that appears to offer no selective advantage.

neutron   An atomic particle found in the nuclei of atoms that is similar to a proton but has no electric charge.
See also ELECTRON.

new chemical entity   A compound not previously described in the literature.

niche   A habitat providing a particular set of environmental conditions needed for the survival of a given species. Species that occupy different niches may coexist side by side in a stable manner with no competition. However, if two species occupy the same niche, i.e., if they require the same resources, there will be competition, and the weaker of the two will become extirpated. Evolutionary effects can make a species adapt to a specialized niche that has a particular set of abiotic and biotic factors within the habitat.

Nicolle, Charles-Jules-Henry (1866–1936) French *Biologist* Charles-Jules-Henry Nicolle was born in Rouen, France, Seine-Maritime, on September 21, 1866, to Eugène Nicolle, a local doctor. He entered a local medical school, where he studied for three years and then began working in Paris hospitals.

He received his M.D. degree in 1893 and returned to Rouen, becoming director of the bacteriological laboratory in 1896. He left in 1903 to become director of the Pasteur Institute in Tunis, a position he held until his death in 1936.

Early in his career, Nicolle worked on cancer, and at Rouen he investigated the preparation of diphtheria antiserum. In North Africa, under his influence, the Institute at Tunis quickly became a world-famous center for bacteriological research and for the production of vaccines and serums to combat most of the prevalent infectious diseases.

He discovered in 1909 that the body louse transmits typhus fever, and this discovery was vital to prevention of the disease during the two world wars. He also made several contributions on knowledge and prevention of Malta fever, tick fever, scarlet fever, rinderpest, measles, influenza, tuberculosis, and trachoma. In 1928 he was awarded the Nobel Prize in physiology or medicine for his work on typhus.

Nicolle wrote several important books, including *La Nature, conception et morale biologiques* (1934); *Responsabilités de la Médecine* (1936), and *La Destinée humaine* (1937).

He was also a philosopher and writer of stories, such as *Le Pâtissier de Bellone* (1913), *Les Feuilles de la Sagittaire* (1920), *La Narquoise* (1922), *Les Menus Plaisirs de l'Ennui* (1924), *Les deux Larrons* (1929), and *Les Contes de Marmouse et ses hôtes* (1930). He died on February 28, 1936, in Tunis (Tunisia).

nif   A set of about 20 GENEs required for the assembly of the NITROGENASE ENZYME complex.

nitrate reductase   A METALLOENZYME, containing molybdenum that reduces nitrate to nitrite.

nitrite reductase   A METALLOENZYME that reduces nitrite. DISSIMILATORY nitrite reductases contain copper and reduce nitrite to nitrogen monoxide. Assimilatory nitrite reductases contain SIROHEME and IRON-SULFUR CLUSTER and reduce nitrite to ammonia.
See also ASSIMILATION.

nitrogenase An ENZYME complex from bacteria that catalyzes the reduction of dinitrogen to ammonia: $N_2 + 8e^- + 10H^+ \rightarrow 2NH_4 + H_2$ with the simultaneous HYDROLYSIS of at least 16 ATP molecules. The electron donor is reduced FERREDOXIN or flavodoxin. Dihydrogen is always a coproduct of the reaction. Ethyne (acetylene) can also be reduced to ethene (ethylene) and in some cases ethane. All nitrogenases are IRON-SULFUR PROTEINS. Three different types, which differ in the type of COFACTOR present, have been identified: molybdenum-nitrogenase (the most common, which contains the iron-molybdenum cofactor), vanadium-nitrogenase, and iron-only nitrogenase.

See also FEMO-COFACTOR; REDUCTION.

nitrogen fixation The natural process where atmospheric nitrogen, N_2, is converted to compounds that can be easily utilized by plants. All organisms require nitrogen compounds, but few are able to utilize N_2, a relatively inert and unreactive form and, unfortunately, the most readily available. Most organisms require fixed forms such as NH_3, NO_3, NO_2, or organic N. Bacteria perform nitrogen fixation by combining the nitrogen with hydrogen to form nitrates (NH_3) in the soil, which plants can then use. Cyanobacteria (blue-green algae) and bacteria (e.g., *Rhizobium* spp.; *Azotobacter* spp.) associated with legumes (like peas) can fix N_2 by reducing it to ammoniacal (ammonialike) N, mostly in the form of amino acids.

The ASSIMILATION of dinitrogen through microbial reduction to ammonia and conversion into organonitrogen compounds such as amino acids. Only a limited number of microorganisms are able to fix nitrogen.

See also NITROGENASE.

NMR *See* NUCLEAR MAGNETIC RESONANCE SPECTROSCOPY.

nocturnal An animal or plant that is active during the night rather than the day.

node A point on a plant stem where a leaf is attached.

nodes of Ranvier These are gaps or segments between the neuron's myelin sheath wrapped around

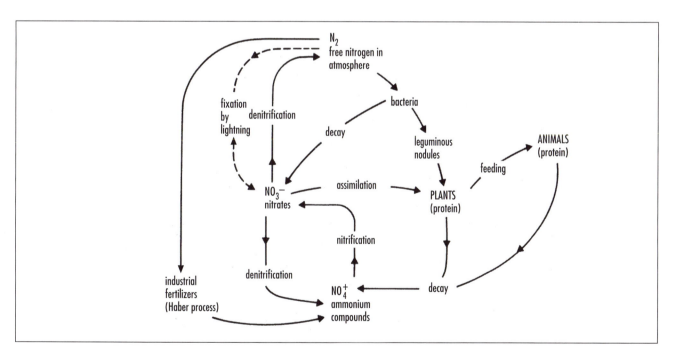

The circulation of nitrogen between organisms and the environment.

an axon. The nodes contain voltage-sensitive sodium channels or gates that generate action potentials. The neural signal jumps from node to node by a process called saltatory conduction, which creates a faster transmission than if the action potential traveled the entire length of the axon.

nomenclature   *See* BINOMIAL.

nonclassical isostere   *See* BIOISOSTERE.

noncompetitive inhibitor   An inhibitor that binds to an enzyme at some location other than the active site and changes the enzyme's shape so that it becomes inactive or less active; binding to a location remote from the active site and changing its conformation so that it no longer binds to the substrate.

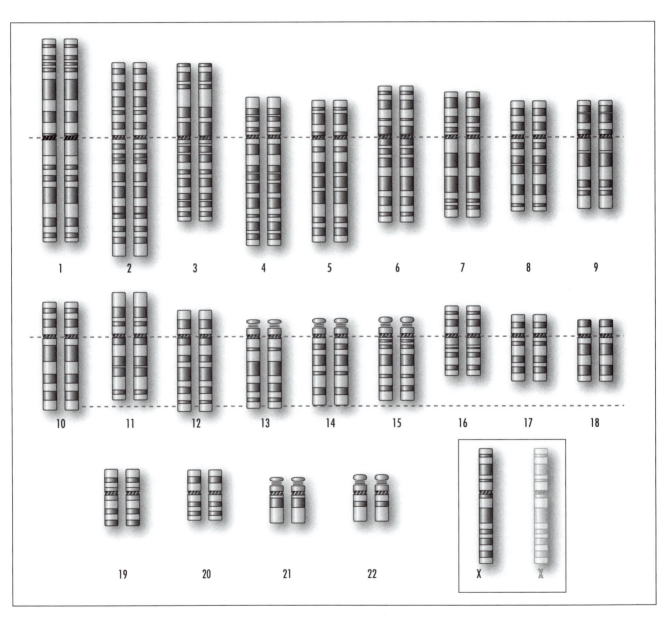

Monosomy is the condition of possessing only one copy of a particular chromosome instead of the normal two copies. *(Courtesy of Darryl Leja, NHGRI, National Institutes of Health)*

noncyclic electron flow The first stage of photosynthesis; begins when light energy enters a cluster of pigment molecules (called the photosystem) located in the thylakoid. The light-induced flow of electrons from water to NADP in oxygen-evolving photosynthesis involving both photosystems I and II. Photosystems are large complexes of proteins and chlorophyll that capture energy from sunlight. Both systems I and II include special forms of chlorophyll A. Photosystem I, or P-700, includes chlorophyll A pigment with a specific absorbance of 700 nm (red light). Photosystem II, or P-680, contains the reaction center responsible for oxygen evolution and contains a special chlorophyll A that absorbs light at 680 nm (red light). If the photochemical reactions in photosystem II are inhibited, photosystem I is inhibited as well.

noncyclic photophosphorylation The formation of ATP by NONCYCLIC ELECTRON FLOW.

nondisjunction The failure of paired chromosomes to separate normally and migrate to opposite poles after cell division, thereby giving rise to cells that have too many or too few chromosomes. This results in monosomy, where one member of a chromosome pair is missing, or trisomy, where there is an extra chromosome in any of the chromosome pairs.

nonpolar covalent bond A covalent bond formed by the equal sharing of electrons between two atoms with the same electronegativity. Electronegativity is the tendency of an atom to attract electrons to itself in a covalent bond.

nonsense mutation A mutation where one of the mRNA sequences (UAA, UAG, UGA) signals the termination of translation, the process whereby the genetic code carried by mRNA directs the synthesis of proteins from amino acids; a codon is changed to a stop codon, prematurely stopping polypeptide chain synthesis. These three nonsense codons are amber (UAG), ocher (UAA), and opal (UGA).

See also MESSENGER RNA.

norm of reaction An expression or variation in pattern among phenotypes produced by a single genotype caused by a variation in environmental conditions. While phenotypes are an expression of genes in an organism that give it certain traits and characters (anatomical structure, physiology, and behavior) that can be measured, sometimes the traits are actually norms of reaction, i.e., traits expressed in different ways in different environments. For example, human body weight or height is different in identical environments or on different diets. Take a set of genes that produce a particular trait together, list all the possible environmental conditions that they can survive in, and the norm of reaction is the total variation in that trait in all of the survivable environmental conditions.

notochord A rodlike cord, a rudimentary skeleton, composed of stiff cartilage in chordates that runs lengthwise under the dorsal (top) surface of the body. It forms the support structure of the body, supporting the nerve cord. It is replaced by the vertebral column later in the development of most chordates; it persists in primitive fishes such as hagfish, coelacanth, and chimaeras. The phylum Chordata comprises animals characterized by bodies that have elongated bilateral symmetry and that, in some part of their development, have a notochord and gill slits (or pouches), and usually have a head, tail, and digestive system with openings at both ends. Chordates include the fish, amphibians, reptiles, birds, and mammals, but not all chordates are vertebrates, e.g., invertebrate tunicates and lancelets.

N-terminal amino acid residue *See* AMINO ACID RESIDUE.

nuclear envelope (nuclear membrane) A two-membrane structure that surrounds the nucleus. The space between the two membranes is referred to as the perinuclear space. The outermost membrane meets the rough endoplasmic reticulum and has ribosomes attached. The inner membrane lies next to a dense filamentous network called the nuclear lamina and surrounds the nucleus, except where there are nuclear pores. Nuclear pores are found where the inner and outer membranes are joined. The space is filled with filamentous material. The pores are involved in regulating the transport of materials between the nucleus and the cytoplasm.

nuclearity   The number of CENTRAL ATOMS joined in a single COORDINATION entity by BRIDGING LIGANDS or metal-metal bonds is indicated by dinuclear, trinuclear, tetranuclear, polynuclear, etc.

nuclear magnetic resonance (NMR) spectroscopy
NMR spectroscopy makes it possible to discriminate nuclei, typically protons, in different chemical environments. The electron distribution gives rise to a chemical shift of the resonance frequency. The chemical shift, δ,

of a nucleus is expressed in parts per million (ppm) by its frequency, ν_n, relative to a standard, ν_{ref}, and defined as $\delta = 10^6 (\nu_n - \nu_{ref})/\nu_o$, where ν_o is the operating frequency of the spectrometer. It is an indication of the chemical state of the group containing the nucleus. More information is derived from the SPIN–SPIN COUPLINGS between nuclei, which give rise to multiple patterns. Greater detail can be derived from two- or three-dimensional techniques. These use pulses of radiation at different nuclear frequencies, after which the response of the spin system is recorded as a free-induction decay (FID). Multidimensional techniques, such as COSY and NOESY, make it possible to deduce the structure of a relatively complex molecule such as a small protein (molecular weight up to 25,000). In proteins containing paramagnetic centers, nuclear HYPERFINE interactions can give rise to relatively large shifts of resonant frequencies, known as contact and pseudo-

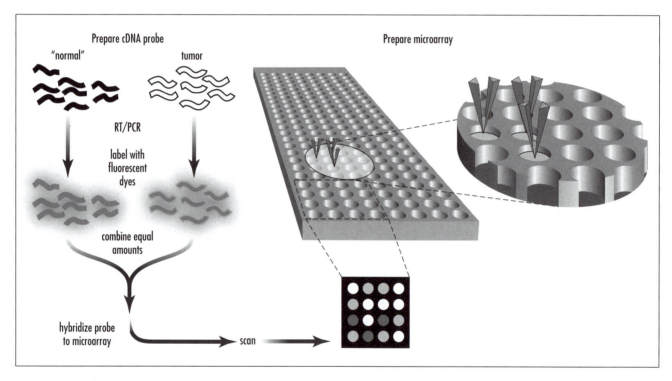

The nucleic acid probe provides a new way of studying how large numbers of genes interact with each other and how a cell's regulatory networks control vast batteries of genes simultaneously. The method uses a robot to precisely apply tiny droplets containing functional DNA to glass slides. Researchers then attach flourescent labels to DNA from the cell they are studying. The labeled probes are allowed to bind to complementary DNA strands on the slides. The slides are put into a scanning microscope that can measure the brightness of each fluorescent dot. The brightness reveals how much of a specific DNA fragment is present, an indicator of how active it is. *(Courtesy of Darryl Leja, NHGRI, National Institutes of Health)*

contact (dipolar) shifts, and considerable increases in the nuclear spin relaxation rates. From this type of measurement, structural information can be obtained about the paramagnetic site.

nuclear pores Openings in the membrane of a cell's nuclear envelope that allow the exchange of materials between the nucleus and the cytoplasm.

nucleation The process by which nuclei are formed; defined as the smallest solid-phase aggregate of atoms, molecules, or ions that is formed during a precipitation and that is capable of spontaneous growth.

nucleic acid probe (DNA probe) A single strand of DNA that is labeled or tagged with a fluorescent or radioactive substance and binds specifically to a complementary DNA sequence. It is used to detect its incorporation through hybridization with another DNA sample. Nuclear acid probes can provide rapid identification of certain species like mycobacterium.

nucleic acids Macromolecules composed of SEQUENCES of NUCLEOTIDES that perform several functions in living cells—e.g., the storage of genetic information and its transfer from one generation to the next (DNA), and the EXPRESSION of this information in protein synthesis (mRNA, tRNA)—and can act as functional components of subcellular units such as RIBOSOMES (rRNA). RNA contains D-ribose, while DNA contains 2-deoxy-D-ribose as the sugar component. Currently, synthetic nucleic acids can be made consisting of hundreds of nucleotides.
See also GENETIC CODE; OLIGONUCLEOTIDE.

nucleobases See NUCLEOSIDES.

nucleoid The irregularly shaped, aggregate mass of DNA that makes up the chromosome in a prokaryotic cell; not bound by a membrane. Found in the nucleoid region of the cell.

nucleolus A somewhat round structure in the nucleus that forms at the nuclear organizer, a specific chromosomal region, consisting of ribosomal RNA (rRNA) and protein. It disappears during nuclear division in late prophase, is completely absent during meta and anaphase, and then reappears during telophase. The nucleus controls the synthesis of proteins in the cytoplasm through messenger RNA (mRNA). Messenger

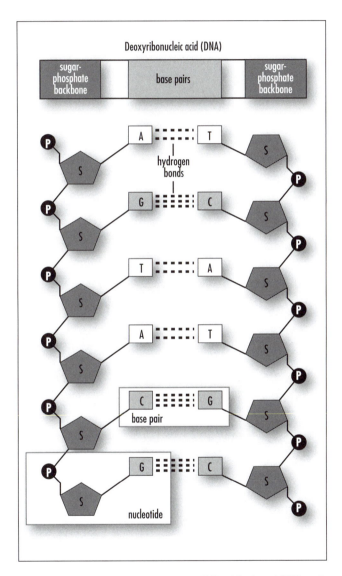

One of the structural components, or building blocks, of DNA and RNA, a nucleotide consists of a base (one of four chemicals: adenine, thymine, guanine, and cytosine) plus a molecule of sugar and one of phosphoric acid. *(Courtesy Darryl Leja, NHGRI, National Institutes of Health)*

RNA is produced in the nucleolus of the cell and travels to the cytoplasm through the pores of the nuclear envelope.

nucleosides Compounds in which a purine or pyrimidine base is ß-*N*-glycosidically bound to C-1 of either 2-deoxy-D-ribose or of D-ribose, but without any phosphate groups. The common nucleosides in biological systems are adenosine, guanosine, cytidine, and uridine (which contain ribose), and deoxyadenosine, deoxyguanosine, deoxycytidine, and thymidine (which contain deoxyribose).

See also NUCLEOTIDES.

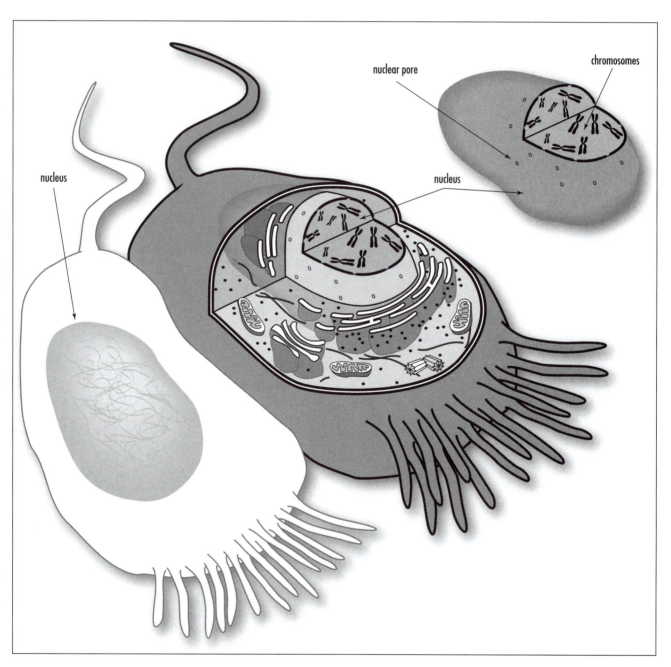

The nucleus is the central cell structure that houses the chromosomes. *(Courtesy Darryl Leja, NHGRI, National Institutes of Health)*

nucleosome A beadlike subunit of chromatin, the material chromosomes are made of, consisting of about 150–200 base pairs of DNA wrapped around a core complex of eight histone proteins (two molecules each of histones H2a, H2b, H3, and H4). Histones are a type of protein associated with DNA in chromosomes in the nucleus, and they function to coil the DNA into nucleosomes.

nucleotides Nucleosides with one or more phosphate groups esterified mainly to the 3- or the 5- position of the sugar moiety. Nucleotides found in cells are adenylic acid, guanylic acid, uridylic acid, cytidylic acid, deoxyadenylic acid, deoxyguanylic acid, deoxycytidylic acid, and thymidylic acid. A nucleotide is a nucleoside in which the primary hydroxy group of either 2-deoxy-D-ribose or of D-ribose is esterified by orthophosphoric acid.

See also ADENOSINE 5'-TRIPHOSPHATE; NAD+; NADP+; NUCLEOSIDES.

nucleus There are four definitions:

1. The round structure within a cell enclosed in a double membrane containing chromosomes in which DNA replication and transcription take place.
2. A collection of nerve cells in the brain.
3. A positively charged core of an atom containing protons and neutrons.
4. The frozen center of a comet's head that contains most of the comet's mass.

null cell A class of lymphocytes characterized by a lack of surface markers specific for either T or B lymphocytes.

numerical taxonomy *See* PHENETICS.

nuptial flight The mating flight of insects (e.g., queen with males).

nut A simple, dry, indehiscent fruit with a bony shell.

nyctinasty A movement made by a plant partly in response to a nondirectional stimulus such as day and night conditions.

nymph A development stage in insects that is immature and undergoes incomplete metamorphosis. Does not have a pupal stage.

obligate aerobe Any organism that must utilize atmospheric oxygen in its metabolic pathways and for cellular respiration, and cannot survive without it. The adjective *obligate* refers to an environmental factor.
See also AEROBE.

obligate anaerobe Any organism where atmospheric oxygen is toxic to its growth; growth can occur only in an anaerobic environment. The adjective *obligate* refers to an environmental factor.
See also ANAEROBE.

oceanic zone The ocean is divided into zones. The whole mass of water is called the pelagic. This is divided into two major subzones: the neritic zone, which covers all water to a depth of 600 feet, and the oceanic zone, which covers all water below 600 feet. The oceanic zone is further divided into subzones. The mesopelagic (semidark waters) covers the depths from 650 feet to 3,200 feet, which is the middle layer between the upper (sunlit 650 feet) epipelagic and the lower (cold and dark) bathypelagic.
See also PELAGIC ZONE.

octahedron *See* COORDINATION.

ODMR *See* OPTICALLY DETECTED MAGNETIC RESONANCE.

OEC *See* OXYGEN-EVOLVING COMPLEX.

olfaction The process of smell. In humans, chemoreceptors are located in a patch of tissue about the size of a postage stamp, called the olfactory epithelium, that is located high in the nasal cavity.

oligochaete Worms of the phylum Annelida, class Oligochaetae. Chiefly terrestrial and freshwater worms with distinct body segmentation and no apparent head. The earthworm is a familiar example.

oligogyny Multiple egg-laying queens that each have their own territory within a colony.

oligonucleotide Macromolecules composed of short SEQUENCES of NUCLEOTIDES that are usually synthetically prepared and used, for example, in site-directed MUTAGENESIS.

oligotrophic lake A condition of a lake that has low concentrations of nutrients and algae resulting in clear blue conditions. Contrast with mesotrophic lakes, which have a moderate nutrient condition, and eutrophic lakes, which have excessive levels of nutrients.

ommatidium   A single unit, or visual section, of a compound eye such as that found in insects. It resembles a single simple eye that descends directly into the eye core. Each ommatidium contains a corneal lens, crystalline cone, and pigment and retinula cells and functions as a single eye sending an image to the brain. The insect's brain must take all the messages from each ommatidia and piece the image together.

omnivore   An animal that eats both plant and animal material. Humans are omnivores.
  See also CARNIVORE.

oncogene   A normal cellular gene that, when inappropriately expressed or mutated, can transform eukaryotic cells into tumor cells. A gene that controls cell growth but also is responsible for directing the uncontrolled growth of tumor or cancer if it is damaged by means such as an inheritance defect, mutation, or environmental exposure such as to carcinogens. There are dozens of oncogenes known, and they function in a variety of ways, but their commonality is the overexpression that interferes with the normal regulation of cell growth.

ontogeny   The life history of one individual and its stages as it evolves from zygote to adult. The opposite of phylogeny, which is the history of a group.

oogamy   A process such as the fusion or union between a small motile or flagellated sperm (gamete) and large nonmotile egg (gamete).

oogenesis   The process and development of producing a female ovum; developing a diploid cell into a haploid egg.

open circulatory system   A type of circulatory system where the internal transport of blood flows through the body cavity and bathes the organs directly and not through a system of vessels. Humans have a closed circulatory system, which is a type of circulatory system where the blood flows through a system of vessels and the heart. Examples include horseshoe crabs, lobsters, and insects.

operant conditioning   A type of associative learning behavior also called trial-and-error learning, or instrumental conditioning. It is a method to modify behavior (an operant) that utilizes contingencies between the response and the presentation of the reinforcer. Based on the 1938 experiments of Burrhus Friederich Skinner (rats pressing lever for food) and published in his book *The Behavior of Organisms*.

operon   A functional unit consisting of a PROMOTER, an operator, and a number of structural GENEs, found mainly in PROKARYOTEs. An example is the operon NIF. The structural genes commonly code for several functionally related ENZYMEs, and although they are transcribed as one (polycistronic) mRNA, each has its separate TRANSLATION initiation site. In the typical operon, the operator region acts as a controlling element in switching on or off the synthesis of mRNA.
  Also refers to a group or sequence of closely linked genes that function as a unit in synthesizing enzymes needed for biosynthesis of a molecule and that is controlled by operator and repressor genes; common in bacteria and phages. An operator gene is the region of the chromosome, next to the operon, where a repressor protein binds to prevent transcription of the operon. The repressor gene protein binds to an operator adjacent to the structural gene, preventing the transcription of the operon.

opsonization   The modification of the surface of a bacterium by coating or deposition of an opsonin (an antibody or complement), which is a group of proteins that lyse organisms. An opsonin coats a bacterium in order to enhance its ability to be eaten (phagocytosis) by macrophages and other leukocytes; an immune response.

optically detected magnetic resonance (ODMR)   A double-resonance technique in which transitions between spin sublevels are detected by optical means. Usually these are sublevels of a triplet, and the transitions are induced by microwaves.

order A taxonomic grouping between class and family. The order consists of groups that are more alike than those in a class.
See also TAXON.

organ A specialized combination of two or more different tissues that performs a particular function. Several organs work together in an organ system to perform a set of coordinated functions. There are ten major organ systems in the human body, and each system is made up of one or more organs. These systems include the circulatory, digestive, endocrine, integumentary, muscular, nervous, reproductive, respiratory, skeletal, and urinary systems.

organelle The "organs" of the cell. Any membrane-bound or nonmembrane-bound structure that is specialized in performing a specific role in the cell. Examples of organelles are chloroplasts, centrosomes, Golgi apparatus, ribosomes, mitochondria, and even the nucleus.

organic chemistry The study of carbon (organic) compounds; used to study the complex nature of living things. Organic compounds are composed mostly of carbon, hydrogen, oxygen, and nitrogen atoms bonded together. One of two main divisions of chemistry. The other is inorganic chemistry. Branches of these two include analytical, biochemical, and physical chemistry.

organism A living entity.

organ of Corti The organ within the cochlea that contains thousands of hairlike receptor cells that respond to different sound frequencies and convert them to nerve impulses through the auditory nerve to the brain.

organogenesis The formation and advanced period of embryonic development of plants and animals when organs are formed from the primary germ layer. Its study is called organogeny.

orgasm An involuntary and rhythmic contraction characterized by strong feelings of pleasure by both sexes during human sexual activity; the highest point of sexual excitement.

orphan drug A DRUG for the treatment of a rare disease for which reasonable recovery of the sponsoring firm's research and development expenditure is not expected within a reasonable time. The term is also used to describe substances intended for such uses.

osmoconformer Not actively changing internal osmolarity (total solute concentration) because an animal is isotonic (body fluids are of equal concentration with respect to osmotic pressure) with the environment.

osmolarity Solute concentration expressed as molarity. Molarity is the moles of solute dissolved in 1 liter of solution. A solute is, in the case of a gas or solid dissolved in liquid, the gas or solid, but in other examples it is the part that has the smaller amount. For example, a beaker of salt water would have the water as the solvent and the salt as the solute. The mole is defined as the number of carbon 12 atoms in 12 grams of carbon 12. It allows scientists to weigh substances and tell how many particles are in that substance.
See also MOLARITY.

osmoregulation A process to control water balance in a cell or organism with respect to the surrounding environment using osmosis. The ability by which organisms maintain a stable solute concentration by maintaining osmotic pressure on each side of a semipermeable membrane.

osmoregulator An organism that must take in or discharge excess water because its body fluids have a different osmolarity than the environment.

osmosis The diffusion or movement of water across a selectively permeable membrane from one aqueous system to another of different concentration. Water moves from areas of high-water/low-solute concentration to areas of low-water/high-solute concentration.

osmotic pressure Pressure that is generated by a solution moving by osmosis into and out of a cell and caused by a concentration gradient.

Osteichthyes A class of fish—"bony fish"; Osteichthyes is the largest and most diverse taxon of all vertebrates. Found as early as the lower Devonian period, there are two subclasses, Actinopterygii (e.g., ray fin: sturgeon, tuna, catfish) and Sarcopterygii (fleshy fin: lungfish, coelacanths). Two of the most successful groups of vertebrates ever known, with some species adapted to breathe air. Their endoskeleton is made of bone.

osteoporosis A decrease in bone mass and bone density with associated increased risk of fracture, especially of wrists, hips, and spines. Of the 10 million Americans estimated to have osteoporosis, 8 million are women and 2 million are men.

ostracoderm A primitive and extinct fish without jaws and encased in an armor of bony plates and dermal scales. It lived from the early Ordovician to the late Devonian (470 to 370 million years ago). Also known as agnathans, fossils have been found in both North America and Europe. Pteraspids, cyathaspids, and amphiaspids are examples.

ostracods Aquatic crustaceans with seven pairs of appendages, each specialized for different tasks, that live inside a calcified carapace made of two valves. Sexually dimorphic, i.e., males and females are different shapes. There are more than 50,000 named species.

outgroup During a phylogenetic analysis, an outgroup is any group that is not included in the study group but is related to the group under study. Used for comparative purposes. An outgroup is chosen because it is related to the taxon under study, but it has an ancestor in a more distant past than the taxon being classified.

ovarian cycle Determines the menstrual cycle in the ovary; a regular cyclic event that comprises two phases and ovulation: the follicular phase (before ovulation), ovulation itself, and the luteal phase (after ovulation but before a new follicular phase). It normally lasts for 11 to 16 days and is regulated by hormones.

ovary The female organ or gonad, located on each side of the uterus, that produces oocytes that develop into mature eggs. The ovaries are connected by the fallopian tubes, also called the oviducts. They also produce female estrogens (estradiol and progesterone) that are responsible for secondary sexual characteristics.

oviduct Another name for fallopian tubes, hollow organs about 12 centimeters long (about 6 inches) that join the ovary to the uterus and that transport the ovum.

oviparous Refers to a process in which a female animal produces eggs that develop and hatch outside the body. In oviparous species, fertilization occurs when the sperm meet the ova as they pass through the oviduct. Insects such as butterflies and moths are oviparous. They lay their eggs on leaves in which the larvae hatch.

ovotransferrin An iron-binding protein from eggs, structurally similar to the TRANSFERRINs.

ovoviviparous Animals that reproduce by eggs that remain in the mother's uterus until they are ready to hatch. The young emerge alive with only a membrane to break away from. There is no umbilical cord attached to a placenta. Some fish and reptiles are ovoviviparous.

ovulation   The process in which a mature ovary follicle opens and releases an egg (secondary oocyte) enclosed in a mucouslike material. In mammals, one egg is released each menstrual cycle.

See also OVARY.

ovule   A protective structure in seed plants where the female gametophyte develops, and where fertilization occurs. The integument, a layer of tissue, surrounds the ovule and it becomes a seed.

ovum   Alternative term for egg. It is the secondary oocyte in mammals. Unfertilized haploid nonmotile egg cell.

oxidase   An ENZYME that catalyzes the oxidation of SUBSTRATEs by O_2.

oxidation   A process where one or more electrons are lost, and the oxidation state of some atom increases. It can occur only in combination with reduction, a process where electrons are gained and the oxidation state of some atom decreases.

oxidation number   The oxidation number of an element in any chemical entity is the number of charges that would remain on a given atom if the pairs of electrons in each bond to that atom were assigned to the more electronegative member of the bond pair. The oxidation number of an element is indicated by a roman numeral placed in parentheses immediately following the name (modified if necessary by an appropriate ending) of the element to which it refers. The oxidation number can be positive, negative, or zero. Zero, not a roman numeral, is represented by the usual cipher, 0. The positive sign is never used. An oxidation number is always positive unless the minus sign is explicitly used. Note that it cannot be nonintegral. Nonintegral numbers may seem appropriate in some cases where a charge is spread over more than one atom, but such a use is not encouraged. In such ambiguous cases, the charge number, which designates ionic charge, can be used. A charge number is a number in parentheses written without a space immediately after the name of an ion, and whose magnitude is the ionic charge. Thus the number may refer to cations or anions, but never to neutral species. The charge is written in Arabic numerals and followed by the sign of the charge.

In a COORDINATION entity, the oxidation number of the CENTRAL ATOM is defined as the charge it would bear if all the LIGANDs were removed along with the electron pairs that were shared with the central atom. Neutral ligands are formally removed in their closed-shell configurations. Where it is not feasible or reasonable to define an oxidation state for each individual member of a group or CLUSTER, it is again recommended that the overall oxidation level of the group be defined by a formal ionic charge, the net charge on the coordination entity.

See also MIXED VALENCY.

oxidative addition   The INSERTION of a metal of a COORDINATION entity into a covalent bond involving formally an overall two-electron loss on one metal or a one-electron loss on each of two metals.

oxidative phosphorylation   An aerobic process of energy harnessing by the production of ATP (energy) in mitochondria by enzymatic phosphorylation of ADP coupled to an electron transport chain (ETC). The ETC is a series of mitochondrial enzymes (protein carrier molecules) in the mitochondrial membranes. As high-energy electrons are shuttled down the chain via NADH and FADH2 (flavin adenine dinucleotide) to oxygen molecules, they produce ATP and water.

oxidizing agent   An atom or ion that causes another to be oxidized, and therefore the agent to become reduced. It is a reactant that accepts electrons from another reactant. Oxygen, chlorine, ozone, and peroxide compounds are examples of oxidizing agents.

oxidoreductase   An ENZYME of EC class 1, which catalyzes an oxidation-reduction reaction.

See also EC NOMENCLATURE FOR ENZYMES.

oxygen One of the most important elements for biological systems and for other processes, such as reacting with other substances to release energy. One tree can produce enough oxygen in one week to meet the demands of a person's daily oxygen need. Oxygen is needed in oxidation-reduction reactions within cells. Cellular respiration is the process that releases energy by breaking down food molecules in the presence of oxygen. Atomic symbol is O; atomic number is 8.

oxygen-evolving complex (OEC) The ENZYME that catalyzes the formation of O_2 in PHOTOSYNTHESIS. Contains a CLUSTER of probably four manganese ions.

ozone (O_3) A form of oxygen containing three atoms instead of the common two; formed by ultraviolet radiation reacting with oxygen. Ozone accounts for the distinctive odor of the air after a thunderstorm or around electrical equipment, first reported as early as 1785; ozone's chemical constitution was established in 1872. The ozone layer in the upper atmosphere blocks harmful ultraviolet radiation that normally causes skin cancer. Ozone is an oxidizer and a disinfectant, and it forms hydrogen peroxide when mixed with water. The Earth's ozone layer protects all life from the sun's harmful radiation, but human activities have damaged this shield. The United States, in cooperation with over 140 other countries, is phasing out the production of ozone-depleting substances in an effort to safeguard the ozone layer.

See also OXYGEN.

pacemaker Another name for the sinoatrial (SA) node.

See also SA NODE.

paedogenesis The ability to reproduce while still in immature or larval stage; acceleration of reproductive ability.

paedomorphosis Retainment of ancestral juvenile characteristics in the adult form. Can occur through neoteny or progenesis (acceleration of gonad development).

paleontology The scientific study of past life forms, both plant and animal, in their geological and paleoenvironmental context. It can be in the form of study of fossils of organisms or their by-products. The field has several subdisciplines, including micropaleontology, paleobotany, palynology, invertebrate and vertebrate paleontology, human paleontology or paleoanthropology, taphonomy, and ichnology. James Hall (1811–98), New York State's first paleontologist, is considered the father of American paleontology.

See also GEOLOGICAL TIME.

Pangaea The name given to the one huge landmass, or supercontinent, that existed during the Permian (280

to 248 million years ago) through the JURASSIC PERIOD (206 to 144 million years ago). Pangaea began to break apart during the Jurassic, forming two more large continents, GONDWANALAND and LAURASIA. Gondwanaland, or Gondwana, formed the southern supercontinent, and its remnants are now the continents South America, Africa, India, Australia, and Antarctica. Laurasia, the northern supercontinent, has as its remnants current North America, Europe, Asia, Greenland, and Iceland.

paraphyletic Refers to a group of individuals (taxon) that includes the most recent common ancestor of all of its members, but does not include all of the descendants of that most recent common ancestor.

parasite Any plant or animal that lives on or in another organism, the host, to obtain its nutrients and eventually harming or killing the host; only the parasite benefits. Examples of parasites are ticks, fleas, trematodes, lice, *Giardia lamblia, Sacculina,* and plants such as rafflesia and dodder (*Cuscuta* sp.) Parasites can be ectoparasites (living on the host, e.g., tick) or endoparasites (living within the host, e.g., tapeworm).

parasitism The symbiotic relationship between two organisms where one species, the parasite, benefits, but the other, the host, is harmed. If the parasite kills the host it endangers its own survival.

parasitoid An organism that lives in or on the body of a single host individual during its development and eventually kills the host. Falls between a predator and a true parasite.

parasympathetic division (craniosacral division) One of two divisions of the autonomic nervous system. The parasympathetic deals with conserving energy, digestion of food, and excretory functions; encourages sedentary functions as opposed to the sympathetic division; increases the nutrient content in the blood, which stimulates growth and storage of energy reserves. Other responses include a decreased heart rate and airway and pupil diameters.

parathyroid glands A set of four glands, two in the left lobe of the thyroid gland and two in the right lobe. They function to control blood calcium levels by secreting parathyroid hormone (PTH). If too much calcium is allowed in the blood, a condition known as hyperparathyroidism occurs; if the blood calcium is too low, a condition known as hypoparathyroidism can exist. The parathyroid hormone also stimulates absorption of food by the intestines and conservation of calcium by the kidneys.

Parazoa The animal subkingdom of sponges, the phylum Porifera. Mostly mouthless marine multicellular sessile animals that feed by drawing in microorganisms through their pores (suspension feeders). Four classes of sponges exist: Calcarea (calcareous sponges), Hexactinellida (glass sponges), and Demospongiae (the largest and most diverse, 90 percent of known species), and Sclerospongiae (mostly fossil records). There are about 5,000 species of sponges.

parenchyma One of the three types of plant tissue; unspecialized, composed of large thin-walled cells forming the greater part of leaves, roots, the pulp of fruit, and the pith of stems; Has an abundance of plastids, chloroplast-containing cells able to perform photosynthesis in leaves and stems (when filled with chloroplasts, called chlorenchyma). The cells of a plant's "ground tissue," composed of parenchyma cells that function in assimilation and storage, are called assimilates. Parenchyma tissues are the mesophyll, an important assimilation tissue; the palisade parenchyma, directly beneath the epidermis of the upper part of a leaf; and the spongy parenchyma, which fills the spaces beneath the palisade parenchyma. Parenchyma are the only cells that can engage in mitosis and are the only type of cell found in apical meristems. Parenchyma found in the air-filled floating leaves of aquatic plants are called aerenchyma.

parthenogenesis A form of reproduction in many lower animals in which the egg develops into a new individual without fertilization. In certain social insects, such as ants and honeybees, the unfertilized eggs develop into male drones.

partial agonist An AGONIST that is unable to induce maximal activation of a RECEPTOR population, regardless of the amount of DRUG applied.
See also INTRINSIC ACTIVITY.

partial pressure Each gas in a mixture of gases exerts a pressure called the partial pressure. It is the pressure exerted by one gas in a mixture of gases.

passive transport (diffusion) A molecule or ion that crosses a biological membrane by moving down a concentration or electrochemical gradient with no expenditure of metabolic energy. Passive transport, in the same direction as a concentration gradient, can occur spontaneously, or proteins can mediate passive transport and provide the pathway for this movement across the lipid bilayer without supplying energy for the action. These proteins are called channels if they mediate ions and permeases if they mediate large molecules. This type of transport always operates from regions of greater concentration to regions of lesser concentration.
See also ACTIVE TRANSPORT.

pattern formation The direction given to cells to form a specific three-dimensional structure in shaping the development of an organism and its parts.

pattern recognition The identification of patterns in large data sets using appropriate mathematical methodologies.

Pavlov, Ivan Petrovich (1849–1936) Russian *Physiologist* Ivan Petrovich Pavlov was born on September 14, 1849, in Ryazan, to Peter Dmitrievich Pavlov, a village priest. He was educated first at the church school in Ryazan, the Ryazan Ecclesiastical High School, and then at the local theological seminary.

Pavlov abandoned religion for science when he was inspired by the progressive ideas of the Russian literary critic D. I. Pisarev, by Ivan M. Sechenov, the father of Russian physiology, and by the works of Charles Darwin. In 1870 he enrolled in the physics and mathematics faculty to take the course in natural science.

Pavlov became interested in physiology. While taking the course in natural science, he and a fellow student wrote a treatise on the physiology of the pancreatic nerves that won wide acclaim, and Pavlov was awarded a gold medal.

In 1875 Pavlov received the degree of candidate of natural sciences. However, his overwhelming interest in physiology forced him to continue his studies and attend the Academy of Medical Surgery. Pavlov won a fellowship at the academy, and with a position as director of the physiological laboratory at the clinic of the famous Russian clinician, S. P. Botkin, he was able to continue his research work. In 1881 Pavlov married Seraphima (Sara) Vasilievna Karchevskaya, a teacher, the daughter of a doctor. They had four sons and a daughter. In 1883 he presented his thesis, entitled "The centrifugal nerves of the heart," and laid down the basic principles on the trophic function of the nervous system. Pavlov showed that there existed a basic pattern in the reflex regulation of the activity of the circulatory organs.

In 1890 Pavlov organized and directed the department of physiology at the Institute of Experimental Medicine and spent the next 45 years making it one of the most important centers of physiological research. Also, in 1890 he was appointed professor of pharmacology at the Military Medical Academy, and five years later he was appointed to the then-vacant chair of physiology, which he held until 1925.

Between the years 1891 and 1900, Pavlov did the bulk of his research on the physiology of digestion at the Institute of Experimental Medicine and demonstrated that the nervous system played the dominant part in regulating the digestive process. Pavlov promoted his research in lectures that he delivered in 1895 and published under the title *Lektsii o rabote glavnykh pishchevaritelnyteh zhelez* (Lectures on the function of the principal digestive glands) in 1897.

His research on the physiology of digestion led to the development of the first experimental model of learning, called classical conditioning. Pavlov's research into the physiology of digestion led to the study of conditioned reflexes. In a now-classic experiment, he trained a hungry dog to salivate at the sound of a bell, which was previously associated with the sight of food. He implanted small stomach pouches in dogs to measure the secretion of gastric juices produced when the dogs began to eat.

In 1901 he was elected a corresponding member of the Russian Academy of Sciences, and in 1904 he was awarded a Nobel Prize for his work in digestion. In 1907 he was elected academician of the Russian Academy of Sciences, and in 1912 he was given an honorary doctorate at Cambridge University. He held many other honorary memberships of various scientific societies abroad. Finally, upon the recommendation of the Medical Academy of Paris, he was awarded the Order of the Legion of Honor (1915).

After the Russian Revolution, a special government decree signed by Lenin on January 24, 1921, noted "the outstanding scientific services of Academician I. P. Pavlov, which are of enormous significance to the working class of the whole world."

The Communist Party and the Soviet government gave Pavlov and his collaborators unlimited scope for scientific research and built him a laboratory. Pavlov summarized his discoveries in his book, *Conditioned Reflexes*. Pavlov died in Leningrad on February 27, 1936.

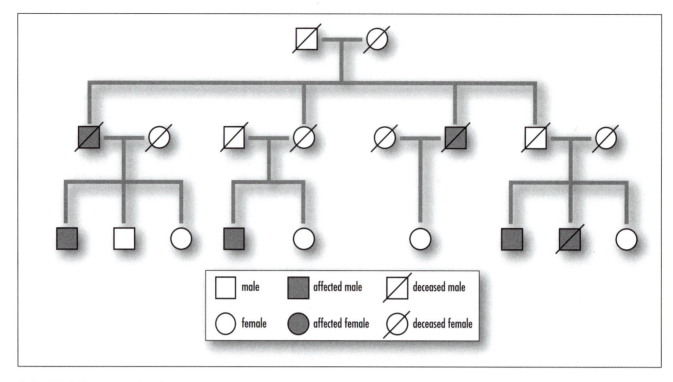

A simplified diagram of a family's genealogy that shows family members' relationships to each other and how a particular trait or disease has been inherited. *(Courtesy Darryl Leja, NHGRI, National Institutes of Health)*

pedigree   The line of descent; a documented list or table of ancestors, as in a genealogical record of an animal or person.

pelagic zone   The ocean is divided into zones. The whole mass of water is called the pelagic. This is divided into two major subzones: the neritic zone, which covers all water to a depth of 600 feet, and the oceanic zone, which covers all water below 600 feet. The oceanic zone is further divided into subzones. The mesopelagic (semidark waters) covers the depths from 650 feet to 3,200 feet, which is the middle layer between the upper (sunlit 650 feet) epipelagic and the lower (cold and dark) bathypelagic.

peptide bond   The bond that links amino acids together. Created by a condensation reaction between the alpha-amino group of one amino acid and the alpha-carboxyl group of another amino acid; a covalent bond.

peptidoglycan   A thick, rigid-layer, cross-linked polysaccharide-peptide complex that is found in the walls of bacteria. It is composed of an overlapping lattice of two sugars, N-acetyl glucosamine (NAG) and N-acetyl muramic acid (NAM), that are cross-linked by amino acid bridges that are found only in the cell walls of bacteria. This elaborate, covalently cross-linked structure provides great strength of the cell wall.

peptidomimetic   A compound containing nonpeptidic structural elements that is capable of mimicking or antagonizing the biological action(s) of a natural parent peptide. A peptidomimetic no longer has classical peptide characteristics, such as enzymatically scissile peptidic bonds.
   See also PEPTOID.

peptoid   A PEPTIDOMIMETIC that results from the oligomeric assembly of N-substituted glycines.

perception The brain's interpretation of a sensory stimulus; conscious mental awareness.

perennial A plant that lives through many seasons, such as a tree.

pericycle Plant cells just inside the endodermis of the root that remain meristematic, allowing new lateral roots to grow (only the pericycle cells nearest the internal xylem poles perform this function). Located between the endodermis and phloem. One of four outer layers of the root (the others being epidermis, cortex, endodermis) that surround the vascular tissue in the middle of a root.

periderm Secondary tissue produced by the cork cambium that, when mature, is made up of dead cells (called cork) that are composed of a waterproof substance called suberin. Acts as a protective shield against water loss, extreme temperatures, and as a pathogen barrier. Usually called bark.

peripheral nervous system Part of the nervous system that consists of (a) sensory neurons composed of stimulus receptors that notify the central nervous system that stimuli have occurred and (b) motor neurons, called effectors, that travel from the central nervous system to make muscles and glands react. The peripheral nervous system is subdivided into the

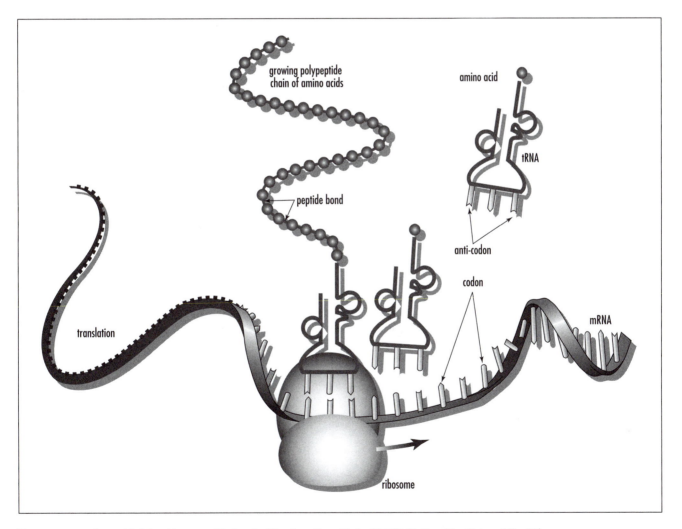

Two or more amino acids joined by a peptide bond. *(Courtesy Darryl Leja, NHGRI, National Institutes of Health)*

sensory-somatic nervous system and the autonomic nervous system.

periplasm   The fluid occupying the space between the inside and outside cellular membranes of bacteria.

peristalsis   Involuntary rhythmic waves or movements of longitudinal and circular muscles in the digestive tract (stomach, intestines, esophagus) in short or long duration.

peroxidase   A heme protein (donor: hydrogen peroxide OXIDOREDUCTASE, EC class 1.11.1) that catalyzes the one-electron oxidation of a SUBSTRATE by dihydrogen peroxide. Substrates for different peroxidases include various organic compounds, CYTOCHROME-c, halides, and Mn^{2+}.
    See also EC NOMENCLATURE FOR ENZYMES.

peroxisome   A single-membrane enzyme-containing organelle present in most eukaryotic cells. The peroxisome is involved in metabolic processes such as the ß-oxidation of long- and very-long-chain fatty acids, bile acid synthesis, cholesterol synthesis, plasmalogen synthesis, amino acid metabolism, and purine metabolism.

petiole   A leaf stalk that attaches to the stem.

Pfeiffer's rule   States that in a series of chiral compounds the EUDISMIC RATIO increases with increasing POTENCY of the EUTOMER.

phage (bacteriophage)   A type of virus that attacks bacteria.
    See also BACTERIA.

phagocyte   A cell that is able to ingest, and often to digest, large particles such as bacteria and dead tissue cells.

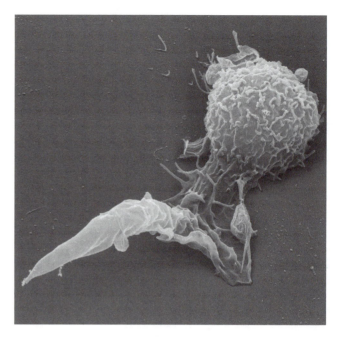

Scanning electron micrograph (SEM) of a macrophage, a white blood cell (upper right), engulfing a protozoan, *Leishmania mexicana* (lower left). This parasitic protozoan causes the disease leishmaniasis. Here, the macrophage has extended part of its body to surround and capture the *Leishmania.* This process is called phagocytosis. Leishmaniasis is transmitted by the bites of infected sandflies, causing a skin ulcer at the site of the bite. The more serious form, kala-azar, can be fatal. Macrophages are part of the immune system that keep the body free of invading organisms. Magnification unknown. *(Courtesy © Science Photo Library/Photo Researchers, Inc.)*

phagocytosis   The act of ingestion and digestion of microorganisms, insoluble particles, damaged or dead host cells, and cell debris by specific types of cells called phagocytes (e.g., macrophages, neutrophils); a form of endocytosis.
    See also ENDOCYTOSIS.

pharmacokinetics   The study of absorption, distribution, METABOLISM, and excretion (ADME) of bioactive compounds in a higher organism.
    See also DRUG DISPOSITION.

pharmacophore (pharmacophoric pattern)   The ensemble of steric and electronic features that is necessary to ensure the optimal supramolecular interactions with a specific biological target structure and to trigger (or to block) its biological response.

A pharmacophore does not represent a real molecule or a real association of functional groups but, rather, is a purely abstract concept that accounts for the common molecular interaction capacities of a group of compounds toward their target structure. The pharmacophore can be considered as the largest common denominator shared by a set of active molecules. This definition discards a misuse often found in the MEDICINAL CHEMISTRY literature, which consists of naming as pharmacophores simple chemical functionalities such as guanidines, sulfonamides, or dihydroimidazoles (formerly imidazolines), or typical structural skeletons such as flavones, phenothiazines, prostaglandins, or steroids.

pharmacophoric descriptors   Used to define a PHARMACOPHORE, including H-bonding and hydrophobic and electrostatic interaction sites, defined by atoms, ring centers, and virtual points.

pharynx   A short section of fibromuscular tube that is the common opening for the digestive and respiratory systems. Located in the throat at the convergence of the nasal passage and oral cavity it opens to the larynx (respiratory system) and the esophagus (digestive system). Composed of three parts: the nasopharynx (behind the nose and above the soft palate; tonsils), the oropharynx (back of the mouth and soft palate, tonsils, and posterior third of the tongue wall of the throat), and the hypopharynx or laryngopharynx (lower part of the throat behind the larynx and above the esophagus). The pharynx receives air from the nasal cavity, while air, food, and water enter from the mouth.

phenetics (numerical taxonomy)   A former school of taxonomy that classified organisms on the basis of overall morphological or genetic similarity that involved observable similarities and differences without considering whether or not the organisms were related. It involved clustering groups into types and represented a nonphylogenetic approach to biological classification.

phenotype   The outward observable features, functions, or behaviors of an organism, based on the coding of the genotype.
See also GENOTYPE.

pheromone   A volatile chemical secreted and sent externally by an organism to send information to members of the same species via olfactory senses that induce a physiological or behavioral response, such as sexual attraction.

phloem   The principal food-conducting living tissue of vascular plants; one of the vascular tissues in plants, the other being the xylem, that make up the vascular bundle. Composed of food-conducting sieve elements: sieve cells and sieve-tube members. Sieve cells are found in gymnosperms, while angiosperms have sieve-tube members.
See also XYLEM.

phosphatase   An ENZYME that catalyzes the hydrolysis of orthophosphoric monoesters. Alkaline phosphatases (EC 3.1.3.1) have an optimum pH above 7 and are zinc-containing proteins. Acid phosphatases (EC 3.1.3.2) have an optimum pH below 7, and some of these contain a dinuclear center of iron, or iron and zinc.
See also EC NOMENCLATURE FOR ENZYMES; NUCLEARITY.

phosphate group   Oxygenated phosphorus ($-PO_4$) that is attached to a carbon chain; important in energy transfer from ATP in cell signal transduction, the biochemical communication from one part of the cell to another; also part of a DNA nucleotide.

phospholipase A (phosphatide acylhydrolase)   Catalyzes the hydrolysis of one of the acyl groups of phosphoglycerides or glycerophosphatidates. Phospholipase

A1 hydrolyzes the acyl group attached to the 1-position, while phospholipase A2 hydrolyzes the acyl group attached to the 2-position.

phospholipases   A class of enzymes that catalyze the hydrolysis of phosphoglycerides or glycerophosphatidates.

phospholipids   The main component of cell walls; an amphiphilic molecule (lipid). A glycerol skeleton is attached to two fatty acids and a phosphate group, and onto the phosphate is attached one of three nitrogen groups, so both phosphate and nitrogen groups make the "polar head" larger and more polar. The phosphate part of the molecule is water soluble, while the fatty acid chains are fat soluble. The phospholipids have a polar hydrophilic head (phosphate) and nonpolar hydrophobic tail (fatty acids). When in water, phospholipids sort into spherical bilayers; the phosphate groups point to the cell exterior and interior, while the fatty acid groups point to the interior of the membrane.

Examples include lecithin, cephalin and sphingomyelin, phosphatidic acid, plasmalogen. Two types of phospholipids exist: glycerophospholipid and sphingosyl phosphatide. A synthetic phospholipid, alkylphosphocholine, has been used in biological and therapeutic areas.

Structurally, phospholipids are similar to triglycerides, except that a phosphate group replaces one of the fatty acids.

phosphorylation   A process involving the transfer of a phosphate group (catalyzed by ENZYMES) from a donor to a suitable acceptor; in general, an ester linkage is formed, for example: ATP + alcohol → ADP + phosphate ester

photic zone   The upper layer within bodies of water, reaching down to about 200 meters, where sunlight penetrates and promotes the production of photosynthesis; the richest and most diverse area of the ocean. A region where photosynthetic floating creatures (phytoplankton) are primary producers as well as a major food source. The littoral zone and much of the sublittoral zone fall within the photic zone.

photoautotroph   An organism that uses sunlight to provide energy and uses carbon dioxide as the chief source of carbon, such as photosynthetic bacteria, cyanobacteria, algae. Green plants are photoautotrophs.

photoheterotroph   Like photoautotrophs, any organism that uses light as a source of energy but must use organic compounds as a source of carbon, for example, green and purple nonsulfur bacteria.

photolysis   A light-induced bond cleavage. The term is often used incorrectly to describe irradiation of a sample.

photon   Name given to a quantum or packet of energy emitted in the form of electromagnetic radiation. A particle of light and gamma and X rays are examples.

photoperiodism   The physiological response to length of day and night in a 24-hour period, such as the flowering or budding in plants.

photophosphorylation   The process of creating adenosine 5-triphosphate (ATP) from ADP and phosphate by using the energy of the sun. Takes place in the thylakoid membrane of chloroplasts.

photorespiration   A process that decreases photosynthesis and occurs when carbon dioxide levels are low inside a plant's leaves; a process that uses oxygen, releases carbon dioxide, and generates no ATP in the process. Occurs on hot days when oxygen concentrations in leaves exceed carbon dioxide levels due to closed stomata.

photosynthesis   A metabolic process in plants and certain bacteria that uses light energy absorbed by CHLOROPHYLL and other photosynthetic pigments for the reduction of CO_2, followed by the formation of organic compounds.

See also METABOLISM; PHOTOSYSTEM.

photosystem A membrane-bound protein complex in plants and photosynthetic bacteria, responsible for light harvesting and primary electron transfer. Comprises light-harvesting pigments such as CHLOROPHYLL; a primary electron-transfer center and a secondary electron carrier. In green-plant photosynthesis, photosystem I transfers electrons from PLASTOCYANIN to a [2FE-2S] FERREDOXIN and contains IRON-SULFUR PROTEINS. Photosystem II transfers electrons from the OXYGEN-EVOLVING COMPLEX to plastoquinone and contains an iron center.
See also PHOTOSYNTHESIS.

phototropism Growth movement by plants that is induced by light. If growth is toward the light source it is called positive phototropism; if it is away from the source it is termed negative phototropism.

pH scale The concentration of hydrogen ions in a solution, based on a scale from 0 to 14. Low pH corresponds to high hydrogen ion concentration, and high pH refers to low hydrogen concentration. A substance added to water that increases the concentration of hydrogen ions (i.e., lowers the pH) is called an acid, while a substance that reduces the concentration of hydrogen ions (i.e., raises the pH) is called a base. Acid in the stomach has a pH of 1, while a liquid drainer has a pH of 14. Pure water is neutral with a pH of 7. Compounds called buffers can be added to a solution that will resist pH changes when an acid or base is added.
See also ACID; BASE.

phylogeny The evolutionary tree that connects a group of organisms.

phylum A taxonomic category between kingdom and class.
See also TAXON.

phytoalexin A toxic substance that acts like an antibiotic that is produced by plants to inhibit or kill the growth of microorganisms, such as certain fungi that would otherwise infect them; e.g., pisatin (produced by peas), phaseollin (produced by beans [*Phaseolus*]), camalexin (produced by *Arabidopsis thaliana*), resveratrol (grapes).

phytochelatin A peptide of higher plants consisting of polymers of 2–11 glutathione (γ, glutamyl, cysteinyl, glycine) groups, which binds heavy metals.

phytochrome (red-light-sensitive system) Photoreceptor proteins that regulate light-dependent growth processes; absorbs red and far-red light in a reversible system; one of the two light-sensing systems involved in photoperiodism and photomorphogenesis. Plant responses regulated by phytochrome include photoperiodic induction of flowering, chloroplast development (minus chlorophyll synthesis), leaf senescence, leaf abscission, seed germination, and flower induction.

picket-fence porphyrin A PORPHYRIN with a protective enclosure for binding oxygen at one side of the ring that is used to mimic the dioxygen-carrying properties of the HEME group.
See also BIOMIMETIC.

pilus A hairlike projection, composed of a protein pilin and found on the surface of certain gram-negative bacteria, that functions to adhere to carbohydrate receptors on host cells. Specialized sex pili conjugate with other bacteria to transfer DNA. Also called fimbriae.

pineal gland (pineal body; epiphysis) A small endocrine gland shaped like a pine cone and located in the middle of the brain. It functions as a regulator of the biological clock. At night, the hormone serotonin converts into melatonin (n-acetyl-5-methoxy-tryptamine [NA-5-MT]) by enzymatic interaction. Melatonin is a neurotransmitter and neurohormone.

The pineal helps to regulate the function of all organs of the endocrine system in the body (pituitary gland, thyroid + parathyroid glands, thymus, pancreas, and ovaries/testes) that secrete their hormones to the blood. The pituitary gland stimulates the secretion of these hormones, but the pineal gland stops it via melatonin. If there is too much hormone, the pineal releases melatonin to counteract; too much stress in the body, serotonin triggers the release of adrenaline. The pineal is also sensitive to light, and it releases melatonin at night and is inhibited during the day.

New studies regarding the effect of electromagnetic frequency (EMF) on the body (from cell phones, high voltage lines, etc.) also relates to the pineal gland, which is sensitive to EMF and seems to suppress the activity of the pineal gland by reducing melatonin production.

The pineal gland is suspected to play a role in a number of problems including cancer, sexual dysfunction, hypertension, and the decline in melatonin has been suggested to be a trigger for the aging process.

pinocytosis An active transport process by which liquids or very small particles are ingested into the cell by endocytosis. The cytoplasmic membrane invaginates (forms "pockets"), fills with liquid or material, and pinches off into a pinocytic vesicle or vacuole that can then be transported.

pith The core of plant stems; location of vascular systems containing parenchyma cells and ground tissue. Also to kill by severing the spinal cord.

pituitary gland A two-lobed, pea-size gland at the base of the brain and attached to the hypothalamus that controls the endocrine system. The major divisions of the gland are the anterior lobe, or adenohypophysis, and the posterior lobe, the neurohypophysis. Each lobe produces hormones. The anterior lobe produces growth hormone, thyroid-stimulating hormone, adrenocorticotropic hormone, prolactin, luteinizing hormone (LH), and follicle-stimulating hormone (FSH), the last two called gonadotropins because they stimulate the gonads. The posterior lobe releases antidiuretic hormone (vasopressin) and oxytocin. These hormones regulate many body functions from growth to birth contractions.

placebo An inert substance or dosage form that is identical in appearance, flavor, and odor to the active substance or dosage form. It is used as a negative control in a BIOASSAY or in a clinical study.

placenta A structure that develops in the uterus during pregnancy that provides a blood supply and nutrients for the fetus and eliminates waste; formed from the uterine lining and embryonic membranes. In humans, it is also referred to as the afterbirth because it is ejected after the baby in a normal vaginal birth. If the placenta is abnormally low in the uterus and covering the uterus, a pregnancy-related condition called placenta praevia occurs and usually necessitates delivery by a cesarean section.

placental mammal Any mammals that bear their young live and are nourished before birth in the mother's uterus through a placenta. There are about 4,000 species, from bats to cats to humans.

placoderm A member of an extinct class of early hinged-jaw fishlike vertebrates covered with a tough outer body armor of bony plates with paired fins; lived during the Silurian and Devonian periods (438 to 360 million years ago). Examples include the antiarchs and dinichthyids.

plankton Plankton includes mostly small-sized plants called phytoplankton (e.g., diatoms) and animals called zooplankton (e.g., radiolarians) that drift and float along with the tides and currents of water bodies. Their name comes from the Greek meaning "drifter" or "wanderer." Phytoplankton produce their own food by photosynthesis and are primary producers and food supply for a host of other organisms. Plankton are also an oxygen producer, generating according to some estimates as much as 80 percent of the Earth's oxygen supply.

planula The free-swimming, flat, ciliated larvae of the coelenterates.

plasma In biology, this term has the following three meanings:

1. Fluid component of blood in which the blood cells and platelets are suspended (blood plasma). Note the distinction between plasma, which describes a part of the blood (the fluid part of blood, outside the blood cells), and serum, which describes a fraction derived from blood by a manipulation (the fluid that separates when blood coagulates).
2. Fluid component of semen produced by the accessory glands, the seminal vesicles, the prostate, and the bulbourethral gland.
3. Cell substance outside the nucleus (CYTOPLASM).

plasma cell An antibody-producing B cell that has reached the end of its differentiation pathway. B cells are white blood cells that develop from B stem cells into plasma cells that produce immunoglobulins (antibodies).

plasma membrane An interface and permeability-limiting membrane composed of lipids and proteins that act as a selective barrier for the cell's interior cytoplasm.

plasmid An extrachromosomal GENETIC element consisting generally of circular double-stranded DNA, which can replicate independently of chromosomal DNA. R plasmids are responsible for the mutual transfer of antibiotic resistance among microbes. Plasmids are used as vectors for CLONING DNA in bacteria or yeast host cells.
See also CLONING VECTOR.

plasmodesma Living bridges between cell walls; small tubes or openings lined with plasma membrane between cell walls that connect each cell to one another and are believed to allow molecules to pass through.

plasmogamy A process of cytoplasm fusion between two cells; the first step in syngamy (sexual reproduction).

plasmolysis A process caused by diffusion when the cell membrane shrinks away from its cell wall, with the resulting vacuole and cytoplasm shrinking due to the presence of a foreign material, like salt, becoming too abundant; water is drawn from the cell into the extracellular area, and the cell becomes flaccid after losing its internal turgor.

plastid A type of plant cytoplasmic organelle that develops from a precursor small and colorless undifferentiated organelle, the proplastid. During cell differentiation, proplastids differentiate into particular plastid types according to the type of cell in which they are located in response to the particular metabolic demands. Plastids are essential components for plant cell function.

Plastids develop into specialized functional types. They divide by binary fission or budding. Several types of plastids exist: amyloplast or leucoplast (starch synthesis and storage), chloroplast (photosynthesis), chromoplast (plant color), etioplast (night plants, can develop chloroplasts), proteoplast or proteinoplast (storage), elaioplast (oil storage).

plastocyanin An ELECTRON TRANSFER PROTEIN, containing a TYPE 1 COPPER site, involved in plant and cyanobacterial PHOTOSYNTHESIS, which transfers electrons to PHOTOSYSTEM I.

platelet (thrombocyte) Disk-shaped, colorless blood cells produced by the bone marrow (from megakaryocytes); contains numerous proinflammatory mediators and functions to stop bleeding and allow damaged areas to clot. A normal platelet count is 150,000–400,000 mm^3 (millimeters cubed).
See also BLOOD.

platelet-activating factor (PAF) A cytokine mediator of immediate hypersensitivity, perhaps even the most important, which produces inflammation.

pleated sheet (beta pleated sheet) One type of several secondary structures, the three-dimensional arrangements (folding, twisting, coiling) of polypeptide chains in a protein. They can be in the form of a helix, random coil, or pleated sheet; linked by hydrogen bonds in the peptide backbone; sheets are formed when the polypeptide chains fold back and forth, or when two parts are parallel to each other and bonded; also called the beta pleated sheet.

pleiotropy   The ability of a single gene to affect many phenotypic traits. Jonathan Hodgkin from the MRC Laboratory of Molecular Biology, in Cambridge, United Kingdom, has characterized several types of pleiotropy:

Artefactual   Adjacent but functionally unrelated genes affected by the same mutation, e.g., claret.

Secondary   Simple primary biochemical disorder leading to complex final phenotype, e.g., phenylketonurea.

Adoptive   One gene product used for quite different chemical purposes in different tissues, e.g., e-crystallin.

Parsimonious   One gene product used for identical chemical purposes in multiple pathways, e.g., gpb-1.

Opportunistic   One gene product playing a secondary role in addition to its main function, e.g., sisB/AS-C.

Combinatorial   One gene product employed in various ways, and with distinct properties, depending on its different protein partners, e.g., unc-86.

Unifying   One gene, or cluster of adjacent genes, encoding multiple chemical activities that support a common biological function, e.g., cha-1 unc-17.

plesiomorphic character   The ancestral character of a homologue; the descendant character is termed the apomorphic character. Two characters in two taxa are homologues if they are the same as the character that is found in the ancestry of the two taxa, or if they have characters that have an ancestor/descendant relationship described as preexisting or novel (plesiomorphic and apomorphic character).

All taxa are mixtures of ancestral and derived characters. Ancestral character states are those similar to the remote ancestor, while derived characters are those that have undergone recent change. Taxa can show either the ancestral (or plesiomorphic) character state or the derived (apomorphic) character state for a particular character.

pluripotent stem cell (**stem cell**)   Primitive blood cells found in the bone marrow, circulating bloodstream, and umbilical cord that are capable of reproducing and differentiating to make all varieties of mature blood cells (white blood cells, red blood cells, and platelets). This means that all blood cells originate from this single type of cell. Pluripotent stem cells are usually referred to as stem cells.

pneumatophore (**breathing roots**)   A specialized root structure that has numerous pores or lenticels over its surface, allowing gas exchange, and that grows up into the air, e.g., mangroves.

pocosin   A swamp on the coastal plain of the southeastern United States.

poikilotherm   An organism (e.g., fish or reptile) whose body temperature varies or fluctuates with the temperature of its surroundings; an ectotherm.

point mutation   When the base sequence of a codon is permanently changed. Four types exist:

Missense   A change in base sequence converts a codon for one amino acid to a codon for a different amino acid.

Nonsense   A codon for a specific amino acid is converted to a chain-terminating codon.

Silent   Conversion of a codon for an amino acid to another codon that specifies the same amino acid.

Frameshift   A nucleotide is deleted or added to the coding portion of a gene.

poison plants (**poison ivy; poison oak; poison sumac**)   The poison ivy plant is known as *Toxicodendron radicans* in the eastern United States and *T. rydbergii* in the midwestern United States. Historically it has been called *Rhus toxicodendron.* Western poison oak is known as *Toxicodendron diversilobum.*

These plants can cause a skin reaction. No reaction usually occurs the first time the skin is exposed to the plant. Subsequent contact with the plant or plant resin, however, can result in an allergic skin reaction that usually appears seven to 14 days after contact. Subsequent contact results in a more rapid reaction, usually within two to five days postcontact. The severity of the reaction is related to the amount of plant material that comes in contact with the skin, as well as to the degree of allergic sensitivity of the individual. The allergen (irritant from the plant) is often transferred from the hands or clothing to other parts of the body.

The poison ivy plant and its relatives are common throughout the United States. Poison ivy leaves are coated with a mixture of chemicals called urushiol. When people get urushiol on their skin, it causes allergic contact dermatitis. The body's immune system treats urushiol as foreign and attacks the complex of urushiol derivatives with skin proteins. The irony is that urushiol, in the absence of the immune attack, would be harmless.

Poison ivy can affect two out of three Americans, and of these, 15 percent may have severe allergic reactions that require medical treatment. Millions of Americans yearly seek remedies for the irritation caused by poison ivy, oak, and sumac.

polar covalent bond A type of chemical bond, based on electron affinity, where electrons shared by atoms spend a greater percentage of time closer to an oxygen nucleus rather than a hydrogen nucleus; bonds are polar, i.e., they have a partial electric charge across the molecule due to their geometry and the electronegativity difference between the two atoms (hydrogen is positive, oxygen is negative); in organisms, they can form weak hydrogen bonds. Water is an example, but peptide bonds and amines also form polar covalent bonds; these molecules can attract each other.

polar molecule A molecule that has both a positive and a negative end, such as water.

pollen Microscopic grains produced by plants in order to reproduce. Each plant has a pollinating period that can vary depending on the plant, climate, and region.

pollen allergy A hypersensitive reaction to pollen. While grass pollens are generally the most common cause of hay fever (seasonal allergic rhinitis), other pollen types are also important. These include tree pollens such as alder, hazel, birch, beech, cypress, pine, chestnut, and poplar, and weed pollens such as plantain, mugwort, and ragweed. The relative importance of the kinds of pollen that can cause hay fever varies between different climatic and vegetation zones. For example, ragweed pollen, although very common in North America, is present in Europe only in the French Rhône valley and some areas of Eastern Europe, while the pollen most associated with seasonal allergy in Mediterranean regions is the olive tree. A person allergic to one pollen is generally also allergic to members of the same group or family (e.g., Betulaceae). Pollen-induced reactions include extrinsic asthma, rhinitis, and bronchitis.

pollination The first step in plant reproduction. Occurs when the male germ cell of a plant, a pollen grain, reaches the female reproductive part, or stigma, of the same species of plant. This happens by wind transportation or by animal carriers, although 90 percent of flowering plants rely on animal delivery.

polyandry A rare mating system where one female mates with more than one male, although each male mates with only one female. Two types of polyandry exist: simultaneous polyandry, where each female maintains a large territory that contains smaller nesting territories of two or more males who care for the eggs and tend to the young; and sequential polyandry, where a female mates with a male, lays eggs, and terminates the relationship and leaves that male. While the male is left to incubate the eggs, she repeats the sequence with another male. The latter is more common. In human society, it is the practice of a woman to have more than one husband at a time.

polydentate *See* CHELATION; DONOR ATOM SYMBOL.

polygenic inheritance The interaction of several genes on a phenotype trait. A series of genes at multiple loci where each contributes a small additive effect on a plant's phenotype, for example, height in tobacco plants.

polygyny A mating system where one male mates with more than one female, while each female mates with only one male; believed to be the normal mating system in animals.

polyhedral symbol The polyhedral symbol indicates the geometrical arrangements of the coordinating atoms about the CENTRAL ATOM. It consists of one or more capital italic letters derived from common geometric terms (tetrahedron, square plane, octahedron, etc.), which denote the idealized geometry of the LIGANDs around the COORDINATION center, and an Arabic numeral that is the coordination number of the central atom. The polyhedral symbol is used as an affix, enclosed in parentheses, and separated from the name by a hyphen. Examples are *T-4*, *SP-4*, *TBPY-5*, *SPY-5*, *OC-6*, and *CU-8*.

polymer A macromolecule of high relative molecular mass composed of many similar or identical monomers linked together in chains. Plastics are polymers.

polymerase chain reaction (PCR) A laboratory technique used to rapidly amplify predetermined regions of double-stranded DNA. Generally involves the use of a heat-stable DNA polymerase.

polymorphic Refers to a phenotypic expression occurring in a number of forms appearing within an interbreeding population, such as fur coloration.

polymorphism Difference in DNA sequence among individuals in a plant or animal population expressed as two or more distinct forms of individuals in the same population. Polymorphisms can be inherited or environmentally created (polyphenism). Examples include sickle cell anemia and the caste system of bees.

polymorphonuclear leukocyte (PMN; granular leukocyte; granulocyte; inflammatory granulocyte; polymorphonuclear cell) A subgroup of leukocytes (white blood cells) filled with granules of toxic chemicals that enable them to digest microorganisms by phagocytosis. Examples of granulocytes are neutrophils, eosinophils, and basophils.

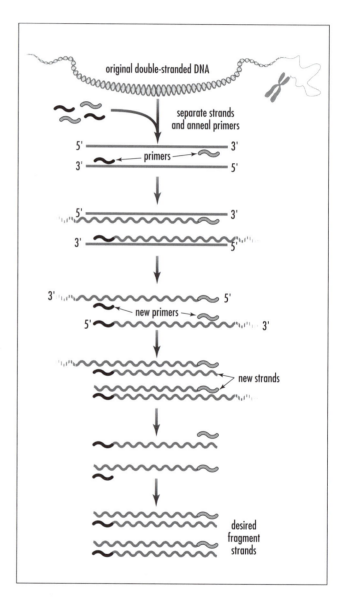

Polymerase chain reaction (PCR) is a fast, inexpensive technique for making an unlimited number of copies of any piece of DNA. Sometimes called "molecular photocopying," PCR has had an immense impact on biology and medicine, especially genetic research. *(Courtesy of Darryl Leja, NHGRI, National Institutes of Health)*

polyp   A cnidarian body form that is the sessile reproductive stage; the alternate is the mobile medusa stage. In medicine, a polyp is a small stalked benign growth or tumor protruding from a mucous membrane; can be precursors of cancer.

polypeptide   A polymer chain of amino acids linked by covalent peptide bonds. One or more polypeptides form proteins. Each polypeptide has two terminal ends; one, called the amino terminal or N-terminal, has a free amino group, while the other end is called the carboxyl terminal or C-terminal with a free carboxyl group.

polyphyletic   Refers to a group of organisms that might have some similarities but that do not include the most recent common ancestor of all the member organisms due to that ancestor lacking some or all characteristics of the group. Polyphyletic groups are not recognized in accepted taxonomies.

polyploidy   When the number of chromosomes in a cell gets doubled; two complete chromosome sets; a mutation. Polyploidy is very common in plants, where 30 percent to 70 percent of modern angiosperms are believed to be polyploids. Polyploidy is rare in animals but is found in some insects, amphibians, reptiles, and one mammal (a rat in Argentina).

polyribosome (polysome)   A cluster of ribosomes translating on one messenger RNA molecule.

polysaccharide   A carbohydrate (polymer) made by polymerizing any of more than 1,000 monosaccharides; a complex sugar.

population   Any group of interbreeding individuals of a particular species living in a specific geographic area.

population viability analysis (PVA)   A process to evaluate the likelihood of a population surviving and to identify threats facing the species. Used in endangered species recovery and management.

porins   A class of proteins that create water-filled channels across cell membranes.

porphyrin   A macrocyclic molecule that contains four pyrrole rings linked together by single carbon atom bridges between the alpha positions of the pyrrole rings. Porphyrins usually occur in their dianionic form coordinated to a metal ion.
See also COORDINATION.

positional information   Cells send signals indicating their locations relative to each other in the embryo development process; positional information is communicated through gap junctions, specific cell-cell adhesions, or diffusible signal molecules; part of the process of determining cell fate.

positive feedback   When a change occurs in a variable during homeostasis, the response is to reinforce the change in the variable. Examples include nerve impulse conduction, blood clotting, female ovarian cycles, labor and birth, and immune responses. A resulting NEGATIVE FEEDBACK may need to take over to halt the process.

postsynaptic membrane   The presynaptic membrane is at the tip of each nerve ending. A small cleft called the synaptic cleft separates the presynaptic membrane from the postsynaptic membrane, a special area on the receiving cell. When the presynaptic nerve ending receives nerve impulses, it sends neurotransmitters stored in synaptic vesicles into the synaptic cleft (the synapse), which in turn diffuse across and transmit the signal to the postsynaptic membrane receptor molecules, which are specialized molecules that open or close certain ion channels when activated by the correct neurotransmitter.

postzygotic barrier   A species-isolating mechanism preventing hybrids that are produced by two different species from developing into fertile and viable adults.
See also PREZYGOTIC BARRIER.

potency A comparative rather than an absolute expression of drug activity. Drug potency depends on both AFFINITY and EFFICACY. Thus, two AGONISTs can be equipotent but have different intrinsic efficacies, with compensating differences in affinity. Potency is the dose of DRUG required to produce a specific effect of given intensity as compared with a standard reference.

potential energy Stored energy that can be released or harnessed to do work.

power saturation A phenomenon used in ELECTRON PARAMAGNETIC RESONANCE SPECTROSCOPY to estimate the electron-spin relaxation times, thus providing information about distances between PARAMAGNETIC centers.

predator An animal that preys on or eats other animals.

predisposition, genetic A latent susceptibility to disease, at the genetic level, that can be activated under certain conditions.

prevalence The number of all new and old cases of a disease in a defined population at a particular point in time.

prezygotic barrier There are two types of reproductive isolation mechanisms: prezygotic (before gamete union) and postzygotic (after gamete union). A form of prezygotic barrier is mechanical isolation, which deals with the mechanics of the reproductive organs and physically prevents sexual intercourse between two different species. Temporal (two species reproduce at different times of day) and habitat isolation (two overlapping species in same range live in different habitats) are also prezygotic barriers.
See also POSTZYGOTIC BARRIER.

primary consumer Any animal that eats grass, algae, and other green plants in a food chain; a herbivore.

primary germ layers After gastrulation, development of the three germ layers—ectoderm, mesoderm, endoderm—occurs, and these eventually develop into all parts of the animal.

primary growth Growth initiated by the apical meristem that takes place relatively close to the tips of roots and stem and that involves extending the length of the plant.

primary immune response The immune response (cellular or humoral) to a first encounter with an antigen. The primary response is generally small, has a long induction phase or lag period, consists primarily of IgM antibodies, and generates immunologic memory.

primary producer An autotroph that acts as a food source for the next level up in the food chain. Green plants are primary producers.

primary productivity The rate at which new plant biomass is formed by photosynthesis. Gross primary productivity is the total rate of photosynthetic production of biomass; net primary productivity is gross primary productivity minus the respiration rate.

primary structure The amino acid SEQUENCE of a protein or the NUCLEOTIDE sequence of DNA or RNA.

primary succession Occurs when communities develop in a newly exposed habitat that had no former life (e.g., bare rock, newly deposited sand).

primate Mammals that include humans and other species that are closely related. There are two main groups: the anthropoids (humans, apes, monkeys) and prosimians (aye-ayes, galagos, lemurs, lorises, pottos, and tarsiers).

primer A short preexisting polynucleotide chain to which new deoxyribonucleotides can be added by DNA polymerase.

primordial germ cells Fetal cells that develop in the early fetus and that will develop into the gametes, the male sperm or female eggs.

principle of allocation Each organism has a limited, finite energy budget that is used for living out its life processes of growth, reproduction, obtaining nutrients, predator flight, and adjusting to environmental changes.

prion The smallest infectious particle known (though its existence is challenged by some scientists), consisting of hydrophobic protein without any nucleic acid. There are a number of prion-caused diseases. Scrapie, found in sheep and goats, causes those animals to lose coordination, and eventually they become incapacitated and develop an intense itch that leads them to scrape off their wool or hair (hence the name "scrapie"). Other diseases are transmissible mink encephalopathy, chronic wasting disease of mule deer and elk, feline spongiform encephalopathy, and bovine spongiform encephalopathy (mad cow disease). Kuru has been seen among the Fore highlanders tribesmen of Papua New Guinea, the result of cannibalism (but has almost disappeared). Creutzfeldt–Jakob disease affects one person in a million. The other two human prion diseases are Gerstmann–Straussler–Scheinker disease and fatal familial insomnia.

probability The statistical measure of an event's likelihood.

procambium A primary meristem tissue that differentiates into the vascular bundle and vascular tissues (xylem and phloem).

prodrug Any compound that undergoes BIOTRANS-FORMATION before exhibiting its pharmacological effects. Prodrugs can thus be viewed as DRUGs containing specialized nontoxic protective groups used in a transient manner to alter or to eliminate undesirable properties in the parent molecule.
See also DOUBLE PRODRUG.

prognosis Prediction of the future course of a disease.

prokaryote A unicellular organism characterized by the absence of a membrane-bound nucleus.
See also EUKARYOTES.

promoter The DNA region, usually upstream from the coding SEQUENCE of a GENE or OPERON, that binds and directs RNA polymerase to the correct transcriptional start site and thus permits TRANSCRIPTION at a specific initiation site. (In catalysis, a promoter is used differently: a cocatalyst usually present in much smaller amounts than the catalyst.)

prophage A phage (bacteriophage, a virus that infects bacteria) chromosome that has been inserted on a specific part of the DNA chromosome of a bacterium.

Comparison of Prokaryotic and Eukaryotic Cells

	Prokaryotic	Eukaryotic
Occurrence	bacteria	animals, plants, and fungi
Average diameter	1μm	20μm
Nuclear material	not separated from cytoplasm by membrane	bounded by nuclear membrane
DNA	circular and forming only one linkage group	linear and divided into a number of chromosomes
Nucleolus	-	+
Cell division	amitotic	usually by mitosis or meiosis
Cytoplasmic streaming	-	+
Vacuoles	-	+
Plastids	-	+
Ribosomes	smaller (70S)	larger (80S)
Endoplasmic reticulum	-	+
Golgi apparatus	-	+

+ indicates presence; - indicates absence

prophase   During prophase, the chromosomes are identical chromatids that are connected at the center by a centromere (x-shaped). The mitotic spindle, which is used to maneuver the chromosomes about the cell, is formed from excess parts of cytoskeleton and initially set up outside the nucleus.

The cell's centrioles are then duplicated to form two pairs of centrioles, with each pair becoming a part of the mitotic center forming the focus for the aster, an array of microtubules. The two asters lie side by side and close to the nuclear envelope while near the end of prophase; the asters then pull apart and the spindle is formed.

prophylaxis   Measures taken (treatment, drugs) to prevent the onset of a particular disease (primary prophylaxis) or recurrent symptoms in an existing infection that have been brought under control (secondary prophylaxis, maintenance therapy). Preventive treatment. Treatment intended to preserve health and prevent the occurrence or recurrence of a disease. Taking a drug to prevent yourself from getting an illness. Chemoprophylaxis is prevention of disease by chemical means.

proprioreceptor   Part of the sensory system. A sense organ that detects the relative position about body positions and movement.

prostaglandin (**PG**)   A lipid mediator, synthesized by the action of the enzyme phospholipase A2, which breaks down cell membrane components into arachidonic acid and then into prostaglandin by cyclooxygenase. Highly proinflammatory, bronchospastic, and vasodilatory.

prosthetic group   A tightly bound, specific nonpolypeptide unit in a protein that determines and is involved in its biological activity.
See also COFACTOR.

proteasome   A large, protein-complex protease found inside cells that degrades other proteins that

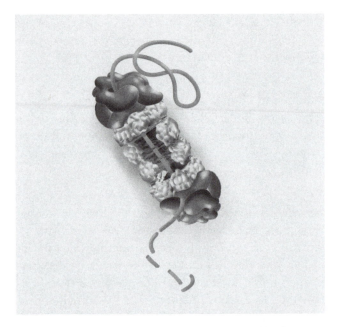

Breaking down unneeded proteins is accomplished by the orderly action of several multiprotein complexes. At the heart of this process is a multiprotein complex called the proteasome. These machines of destruction consist of a tunnellike core with a cap at either or both ends. The core is formed by four stacked rings surrounding a central channel that acts as a degradation chamber. The caps recognize and bind to proteins targeted by the cell for destruction, then use chemical energy to unfold the proteins and inject them into the central core, where they are broken into pieces. This is a fundamental kind of machine that has been highly conserved during evolution. Some form of it is found in organisms ranging from simple bacteria to humans. *(Courtesy of U.S. Department of Energy Genomes to Life program: www.DOEGenomesToLife.org)*

have been tagged for elimination by a smaller protein, ubiquitin, in a process called ubiquitination. Ubiquitin covalently attaches to lysines of the other proteins, tagging them for proteolysis (cleavage of proteins by proteases) within the proteasome.

protein   A molecule composed of many amino acids and with a complex structure. For example, immunoglobulin, casein.
See also AMINO ACID.

protein kinase   A kinase is an enzyme that transfers a phosphate group from ATP to some molecule. A pro-

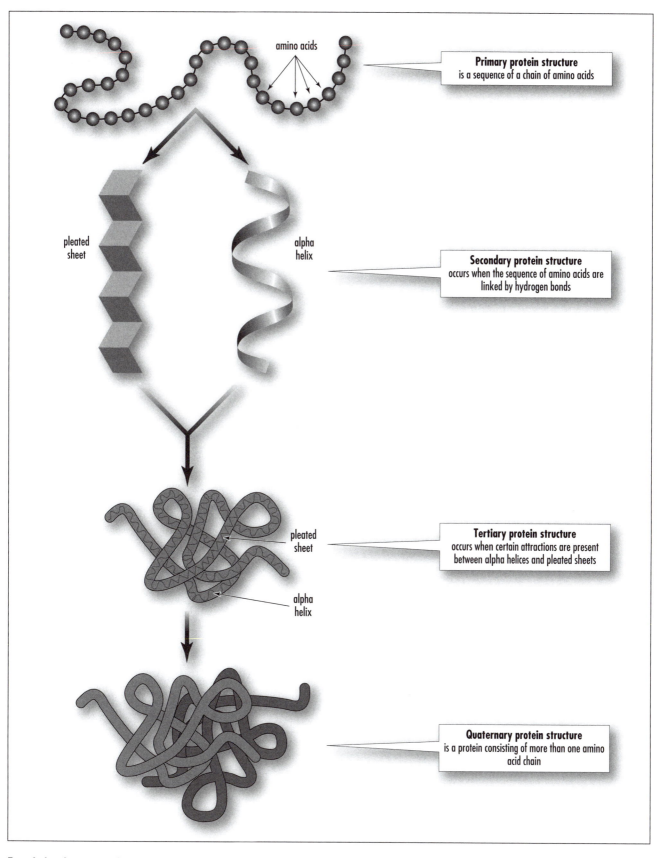

Primary protein structure
is a sequence of a chain of amino acids

amino acids

pleated sheet

alpha helix

Secondary protein structure
occurs when the sequence of amino acids are linked by hydrogen bonds

pleated sheet

alpha helix

Tertiary protein structure
occurs when certain attractions are present between alpha helices and pleated sheets

Quaternary protein structure
is a protein consisting of more than one amino acid chain

Protein is a large complex molecule made up of one or more chains of amino acids. Proteins perform a wide variety of activities in the cell. *(Courtesy of Darryl Leja, NHGRI, National Institutes of Health)*

tein kinase phosphorylates a protein, using ATP, and the phosphorylation modulates the shape and activity of proteins. Protein kinase A and protein kinase C are involved in intracellular signal transduction in most human cells.

See also PROTEIN PHOSPHATASE.

protein machines Genomes are "brought to life" by being read out or "expressed" according to a complex set of directions embedded in the DNA sequence. The products of expression are proteins that do essentially all the work of the cell: they build cellular structures, digest nutrients, execute other metabolic functions, and mediate much of the information flow within a cell and among cellular communities. To accomplish these tasks, proteins typically work together with other proteins or nucleic acids as multicomponent "molecular machines," structures that fit together and function in highly specific, lock-and-key ways.

protein phosphatase (phosphoprotein phosphatase) An enzyme that removes a phosphate group from a protein by the use of hydrolysis; opposite effect of a PROTEIN KINASE.

proteoglycan A type of glycoprotein with high carbohydrate content; component in the extracellular matrix of animal cells. Composed of one or more glycosaminoglycans, long polysaccharide chains covalently linked to protein cores.

protoderm One of the three primary meristem tissues differentiated from the apical meristem; the outermost tissue. The protoderm differentiates into the plant's epidermis of the roots and shoots.

proton An atomic particle found in the nuclei of atoms along with a neutron, but the proton has a positive electric charge.

protonephridium An example of primitive kidneys for osmoregulation and excretion among platyhelminthes.

Protonephridium consists of a network of two or more closed longitudinal branched tubules running the length of the body. The tubular system drains into excretory ducts that empty into the environment through nephridiopores, small external openings.

proton motive force Energy or force created by the transfer of protons (hydrogen ions) on one side only of a cell membrane and across the membrane during chemiosmosis; an electrochemical gradient that has potential energy. This force can be channeled to operate rotating flagella, generate ATP, and other needed activities.

proton pump Proton pumps are a type of active transport that use the energy of ATP hydrolysis to force the transport of protons out of the cell, thus creating a membrane potential.

proto-oncogene A normal gene that can become an active oncogene, one that is capable of causing cells to change into cancer cells, by mutation or insertion of viral DNA.

protoplasm The living material within cells.

protoplast Any cell from which the outer cell wall or membrane has been removed, leaving only the cell contents.

protoporphyrin IX The PORPHYRIN LIGAND of HEME b. Heme b is a Fe(II) porphyrin complex readily isolated from the hemoglobin of beef blood, but it is also found in other proteins, including other HEMOGLOBINS, MYOGLOBINS, CYTOCHROME P-450, CATALASEs, PEROXIDASEs, as well as b type CYTOCHROMEs. Protoporphyrin IX contains four methyl groups in positions 2, 7, 12, and 18, two vinyl groups in positions 3 and 8, and two propionic acid groups in positions 13 and 17.

protostome Any of the group of coelomates, such as annelids, mollusks, and arthropods, whose embryonic blastopore becomes a mouth and who have spiral and determinate cleavage.

protozoan   The simplest forms of animal life, consisting of aquatic single cells or colonies of single cells, such as amoeba and paramecium. The "protists" ingest food and live in both fresh and marine waters as well as inside animals, and most are motile. Estimates of the number of species of protozoa are between 12,000 to 19,000 species. Many protozoans are known to cause human disease, such as *Giardia lamblia*, a flagellated organism that can infect via water or be contracted via contaminated foods. It causes giardiasis, the most frequent cause of nonbacterial diarrhea in North America. *Cryptosporidium parvum* has a strong association between cases of cryptosporidiosis and immunodeficient individuals (such as those with AIDS). *Cyclospora cayetanensis* infection results in a disease with nonspecific symptoms: usually one day of malaise, low fever, diarrhea, fatigue, vomiting, and weight loss. *Cryptosporidium parvum* causes intestinal, tracheal, or pulmonary cryptosporidiosis. *Acanthamoeba* spp., *Naegleria fowleri*, and other amoebae are responsible for primary amoebic meningoencephalitis (PAM). *Naegleria fowleri* is associated with granulomatious amoebic encephalitis (GAE), acanthamoebic keratitis, and acanthamoebic uveitis. *Entamoeba histolytica* causes amoebiasis (or amebiasis), resulting in various gastrointestinal upsets, including colitis and diarrhea. In more severe cases, the gastrointestinal tract hemorrhages, resulting in dysentery.

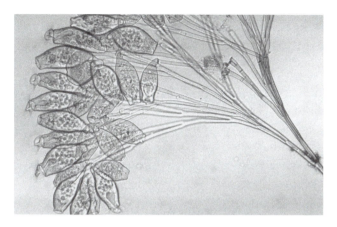

The *Vorticella* is a protist (protozoan) and belongs to the phyllum Ciliophora. A protist that turns as it moves. *(Courtesy of Hideki Horikami)*

Other protozoans such as *Naegleria fowleri* cause meningitis; *Trypanosoma cruzi* causes African sleeping sickness and Chagas disease; and *Plasmodium* spp. causes malaria.

provirus   Term given to a retrovirus DNA when it is integrated into the infected host cell genome. It can remain inactive for periods of time and can be passed to each of the infected cell's daughter cells.

proximate causation   Explains how organisms respond to their immediate environment (through behavior, physiology, and other methods) and the mechanics of those responses.

pseudocoelomate   Any invertebrate whose body cavity is not completely lined with mesoderm. The embryonic blastocoel persists as a body cavity and is not lined with mesodermal peritoneum (the lining of the coelom); therefore it is called a pseudocoel ("false cavity"). Examples include nematodes, rotifers, acanthocephalans, kinorhynchs, and nematomorphs.

pseudopodium   A dynamic protruding structure of the plasma membrane used by amoeboid-type cells used for locomotion and phagocytosis.

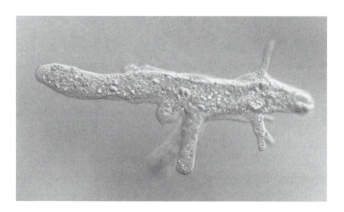

Amoebas are protozoans, which are the simplest form of animal life. Photo of *Amoeba proteus*. *(Courtesy of Hideki Horikami)*

pterin   2-amino-4-hydroxypteridine.
See also MOLYBDOPTERIN.

punctuated equilibrium   A part of evolutionary theory that states that evolution works by alternating periods of spurts of rapid change followed by long periods of stasis.

pupa   The stage preceding the adult in a holometabolous insect, for example, the stage in a butterfly or moth when it is encased in a chrysalis and undergoing metamorphosis.

quadruped   An animal that moves using four-footed locomotion; moves using all four limbs.

quantitative character (polygenic character; polygenic inheritance)   An inherited character or feature in a population whose phenotypes have continuous variation, such as height or weight, and whose distribution follows small discreet steps and can be numerically measured or evaluated; expressed often from one extreme to another. The effects are due to both environment and the additive effect of two or more genes.

quantum evolution   A rapid increase in the rate of evolution over a short period of time.

quarantine   A way to control or prevent importing, exporting, or transporting of plants, animals, agricultural products, and other items that may be able to spread disease or become a pest.

quasisocial   Refers to a situation where members of the same generation share a nest and care for the brood.

Quaternary period   The most recent geologic period of the Cenozoic era. It began 2 million years ago with the growth and movement of Northern Hemisphere continental glaciers and the Ice Age.

See also GEOLOGICAL TIME.

quaternary structure   There are four levels of structure found in polypeptides and proteins. The first or primary structure of a polypeptide or protein determines its secondary, tertiary, and quaternary structures. The primary structure is the amino acid sequence. This is followed by the secondary structure, how the amino acids adjacent to each other are organized in the structure. The tertiary or third structure is the folded three-dimensional protein structure that allows it to perform its role, and the fourth or quaternary structure is the total protein structure that is made when all the subunits are in place. Quaternary structure is used to describe proteins composed of multiple subunits or multiple polypeptide molecules, each called a monomer. The arrangement of the monomers in the three-dimensional protein is the quaternary structure. A considerable range of quaternary structure is found in proteins.

queen   A member of the reproductive caste in semisocial or eusocial insect species.

queenright   A colony that contains a functional queen. A monogynous colony in which one morphologically different female ant is the only reproducer.

queen substance   A pheromone secreted by queen bees and given to worker bees that controls them from producing more queens.

quiescent center   A region of cells behind the root cap that contains a population of mitotically inactive cells; precedes the organization of a root meristem and contains high levels of the enzyme ascorbic acid oxidase, which may suggest a mechanism that links gene expression with patterning. It is the region in the apical meristem that has little activity, and cells rarely divide or do so very slowly. However, cells can be encouraged to divide by wounding of the root. The quiescent center (QC) consists of a small group of cells at the tip called the initials. It is the initials that divide very slowly and create the quiescent center. Cells around the QC divide rapidly and form the majority of the cells of the root body.

raceme An inflorescence, a flower structure, in which stalked flowers are borne in succession along an elongate axis, with the youngest at the top and oldest at the base.

racemic Pertaining to a racemate, an equimolar mixture of a pair of ENANTIOMERS. It does not exhibit optical activity.

radial cleavage A form of embryonic development where cleavage planes are either parallel or perpendicular to the vertical axis of the embryo. Found in deuterostomes, which include echinoderms and chordates.

radial symmetry A body shape characterized by equal parts that radiate outward from the center like a pie. Found in cnidarians and echinoderms.

Radiata The animal phylum that includes cnidarians and ctenophores; radially symmetric.

radiation Released energy that travels through space or substances as particles or electromagnetic waves and includes visible and ultraviolet light, heat, X rays, and cosmic rays. Radiation can be nonionizing

(infrared, visible light, ultraviolet, electromagnetic) or ionizing (alpha, beta, gamma, and X rays). Ionizing radiation can have severe effects on human health, but it is also used in medical diagnostic equipment and can be used to provide a host of other economic benefits from electrical power generation to smoke detectors.

radical A molecular entity possessing one or more unpaired electrons, formerly often called "free radical." A radical can be charged positively (radical cation) or negatively (radical anion). Paramagnetic metal ions are not normally regarded as radicals.

radicle An embryonic plant root.

radioactive isotope An isotope is an element with a different amount of neutrons than protons. Isotopes are unstable and spin off energy and particles. There are radioactive and nonradioactive isotopes, and some elements have both, such as carbon. Each radioactive isotope has its own unique half-life, which is the time it takes for half of the parent radioactive element to decay to a daughter product. Some examples of radioactive elements, their stable daughters, and half-lives are: potassium 40–argon 40 (1.25 billion years); rubidium 87–strontium 87 (48.8 billion years); thorium 232–lead 208 (14 billion years); uranium 235–lead 207

(704 million years); uranium 238–lead 206 (4.47 billion years); carbon 14-nitrogen 14 (5,730 years).

See also ELEMENT.

radiocarbon dating By using radioactive isotopes, it is possible to qualitatively measure organic material over a period of time. Radioactive decay transforms an atom of the parent isotope to an atom of a different element (the daughter isotope) and ultimately leads to the formation of stable nuclei from the unstable nuclei. Archeologists and other scientists use radioactive carbon to date organic remains. The radioactive isotope of carbon, known as carbon-14, is produced in the upper atmosphere and absorbed in a known proportion by all plants and animals. Once the organism dies, the carbon-14 in it begins to decay at a steady, known rate. Measuring the amount of radiocarbon remaining in an organic sample provides an estimate of its age. The approximate half-life of carbon 14 is 5,730 plus or minus 30 years and is good for dating up to about 23,000 years. The carbon-14 method was developed by the American physicist Willard F. Libby in 1947.

radiometric dating The use of radioactive isotopes and their half-lives to give absolute dates to rock formations, artifacts, and fossils. Radioactive elements tend to concentrate in human-made artifacts, igneous rocks, the continental crust, and so the technique is not very useful for sedimentary rocks, although in some cases, when certain elements are found, it is possible to date them using this technique. Other radiometric dating techniques used are:

Electron Dating Spin Resonance
Electrons become trapped in the crystal lattice of minerals from adjacent radioactive material and alter the magnetic field of the mineral at a known rate. This nondestructive technique is used for dating bone and shell, since exposure to magnetic fields does not destroy the material (e.g., carbonates [calcium] in limestone, coral, egg shells, and teeth).

Fission Track Dating
This technique is used for dating glassy material like obsidian or any artifacts that contain uranium-bearing material, such as natural or human-made glass, ceramics, or stones that were used in hearths for food preparation. Narrow fission tracks from the release of high-energy charged alpha particles burn into the material as a result of the decay of uranium 238 to lead 206 (half-life of 4.47 billion years) or induced by the irradiation of uranium 235 to lead 207 (half-life of 704 million years). The number of tracks is proportional to the time since the material cooled from its original molten condition, i.e., fission tracks are created at a constant rate throughout time, so it is possible to determine the amount of time that has passed since the track accumulation began from the number of tracks present. This technique is good for dates from 20 million to 1 billion years ago. U-238 fission track techniques are from spontaneous fission, and induced-fission tracking from U-235 is a technique involving controlled irradiation of the artifact with thermal neutrons of U-235. Both techniques give a thermal age for the material in question. The spontaneous fission of uranium-238 was first discovered by the Russian scientists K. A. Petrzhak and G. N. Flerov in 1940.

Potassium-Argon Dating
This method has been used to date rocks as old as 4 billion years and is a popular dating technique for archeological material. Potassium-40, with a half-life of 1.3 billion years in volcanic rock, decays into argon-40 and calcium-40 at a known rate. Dates are determined by measuring the amount of argon-40 in a sample. Argon-40 and argon-39 ratios can also be used for dating in the same way. Potassium-argon dating is accurate from 4.3 billion years (the age of the Earth) to about 100,000 years before the present.

Radiocarbon Dating See RADIOCARBON DATING.

Thermoluminescence Dating
A technique used for dating ceramics, bricks, sediment layers, burnt flint, lava, and even cave structures like stalactites and stalagmites. It is based on the fact that some materials, when heated, give off a flash of light. The intensity of the light is used to date the specimen and is proportional to the quantity of radiation it has been exposed to and the time span since it was heated. Similar to the electron spin resonance (ESR)

technique. Good for dates between 10,000 and 230,000 years.

radionuclide A radioactive nuclide. The term *nuclide* implies an atom of specified atomic number and mass number. In the study of biochemical processes, radioactive isotopes are used for labeling compounds that subsequently are used to investigate various aspects of the reactivity or METABOLISM of proteins, carbohydrates, and lipids or as sources of radiation in IMAGING. The fate of the radionuclide in reactive products or metabolites is determined by following (counting) the emitted radiation. Prominent among the radionuclides used in biochemical research are : ^{3}H, ^{14}C, ^{32}P, ^{35}Ca, ^{99m}Tc, ^{125}I, and ^{131}I.

ragweed (**Ambrosia**) Ragweed refers to the group of approximately 15 species of weed plants, belonging to the Compositae family. Most ragweed species are native to North America, although they are also found in eastern Europe and the French Rhône valley. The ragweeds are annuals characterized by their rough, hairy stems and mostly lobed or divided leaves. The ragweed flowers are greenish and inconspicuously concealed in small heads on the leaves.

The ragweed species, whose copious pollen is the main cause of seasonal allergic rhinitis (hayfever) in eastern and middle North America, are the common ragweed (*A. artemisiifolia*) and the great, or giant, ragweed (*A. trifida*). The common ragweed grows to about 1 meter (3.5 feet), is common all across North America, and is also commonly referred to as Roman wormwood, hogweed, hogbrake, or bitterweed. The giant ragweed, meanwhile, can reach anywhere up to 5 meters (17 feet) in height and is native from Quebec to British Columbia in Canada and southward to Florida, Arkansas, and California in the United States. Due to the fact that ragweeds are annuals, they can be eradicated simply by being mowed before they release their pollen in late summer.

rain forest An evergreen forest of the tropics distinguished by a continuous, closed canopy of leafy trees of variable height and diverse flora and fauna, with an average rainfall of about 100 inches per year. Rain forests play an important role in the global environment, and destruction of tropical rain forests reduces the amount of carbon dioxide absorbed, causing increases in levels of carbon dioxide and other atmospheric gases. Cutting and burning of tropical forests contributes about 20 percent of the carbon dioxide added to the atmosphere each year. Rain forest destruction also means the loss of a wide spectrum of flora and fauna.

rash (**dermatitis**) An inflammation of the upper layers of the skin, causing rash, blisters, scabbing, redness, and swelling. There are many different types of dermatitis, including: acrodermatitis, allergic contact dermatitis, atopic dermatitis, contact dermatitis, diaper rash (diaper dermatitis), exfoliative dermatitis, herpetiformis dermatitis, irritant dermatitis, occupational dermatitis, perioral dermatitis, photoallergic dermatitis, phototoxic dermatitis, seborrheic dermatitis, and toxicodendron dermatitis.

Contact Dermatitis (allergic contact dermatitis, contact eczema, irritant contact dermatitis)
Contact dermatitis is a reaction that occurs when skin comes in contact with certain substances. There are two mechanisms by which substances can cause skin inflammation: irritation (irritant contact dermatitis) or allergic reaction (allergic contact dermatitis). Common irritants include soap, detergents, acids, alkalis, and organic solvents (as are present in nail polish remover). Contact dermatitis is most often seen around the hands or areas that touched or were exposed to the irritant/allergen. Contact dermatitis of the feet also exists, but it differs in that it is due to the warm, moist conditions in the shoes and socks.

An allergic reaction does not generally occur the first time one is exposed to a particular substance, but it can on subsequent exposures, which can cause dermatitis in 4 to 24 hours.

Treatment includes removal or avoidance of the substance causing the irritation, and cleansing the area with water and mild soap (to avoid infection). A recent recommendation for mild cases is to use a manganese sulfate solution to reduce the itching. Antihistamines are generally not very helpful. The most common treatment for severe contact dermatitis is with corticosteroid tablets, ointments, or creams, which diminish the immune attack and the resulting inflammation.

Toxicodendron Dermatitis

When people get urushiol—the oil present in poison ivy, poison oak, and poison sumac—on their skin, it causes another form of allergic contact dermatitis (see above). This is a T-cell-mediated immune response, also called delayed hypersensitivity, in which the body's immune system recognizes as foreign and attacks the complex of urushiol derivatives with skin proteins. The irony is that urushiol, in the absence of the immune attack, would be harmless.

Atopic Dermatitis

Atopic dermatitis is a chronic, itchy inflammation of the upper layers of the skin. Often develops in people who have hay fever or asthma or who have family members with these conditions. Most commonly displayed during infanthood, usually disappearing by the age of three or four. Recent medical studies suggest that *Staphylococcus aureus* (a bacteria) contributes to exacerbation of atopic dermatitis.

Treatment is similar to that of contact dermatitis.

Seborrheic Dermatitis

An inflammation of the upper layers of the skin where scales appear on the scalp, face, and sometimes in other areas. Usually more common in cold weather and often runs in families.

Stasis Dermatitis

A chronic redness, scaling, warmth, and swelling on the lower legs. Often results in dark brown skin due to a pooling of blood and fluid under the skin, thus usually displayed by those with varicose veins and edema.

rate-controlling step (**rate-determining step; rate-limiting step**) A rate-controlling step in a reaction occurring by a composite mechanism is an elementary reaction, the rate constant for which exerts a dominant effect—stronger than that of any other rate constant—on the overall rate.

Réaumur, René-Antoine Ferchault de (1683–1757) French *Philosopher, Naturalist* René Réaumur was born in La Rochelle, France, in 1683. After studying mathematics in Bourges he moved to Paris in 1703 at age 20 and under the eye of a relative. Like most scientists of the time, he made contributions in a number of areas, including meteorology. His work in mathematics allowed him entrance to the Academy of Sciences in 1708. Two years later, he was put in charge of compiling a description of the industrial and natural resources in France, and as a result he developed a broad-based view of the sciences. It also inspired him to invention, which led him into the annals of weather and climate and, ultimately, the invention of a thermometer and temperature scale.

In 1713 Réaumur made spun-glass fibers that were made of the same material as today's building blocks of Ethernet networking and fiber-optic cable. A few years later, in 1719, after observing wasps building nests, he suggested that paper could be made from wood in response to a critical shortage of papermaking materials (rags) at the time. He also was impressed by the geometrical perfection of the beehive's hexagonal cells and proposed that they be used as a unit of measurement.

He turned his interests from industrial resources such as steel to temperature, and in 1730 he presented to the Paris Academy his study "A Guide for the Production of Thermometers with Comparable Scales." He wanted to improve the reliability of thermometers based on the work of Guillaume Amontons, though he appears not to be familiar with Fahrenheit's earlier work.

His thermometer of 1731 used a mixture of alcohol (wine) and water instead of mercury, perhaps creating the first alcohol thermometer, and it was calibrated with a scale he created called the Réaumur scale. This scale had 0° for freezing and 80° for boiling points of water. The scale is no longer used today. However, most of Europe, with the exception of the British Isles and Scandinavia, adopted his thermometer and scale.

Unfortunately, his errors in the way he fixed his points were criticized by many in the scientific community at the time, and even with modifications in the scale, instrument makers favored making mercury-based thermometers. Réaumur's scale, however, lasted over a century, and in some places well into the late 20th century.

Between 1734 and 1742, Réaumur wrote six volumes of *Mémoires pour servir à l'histoire des insectes* (Memoirs serving as a natural history of insects). Although unfinished, this work was an important contribution to entomology. He also noticed that crayfish have the ability to regenerate lost limbs and demonstrated that corals were animals, not plants. In 1735 he introduced the concept of growing degree-days, later known as Réaumur's thermal constant of phenology. This idea led to the heat-unit system used today to study plant-temperature relationships.

In 1737 Réaumur became an honorary member of the Russian Academy of Sciences, and the following year he became a fellow of the Royal Society.

After studying the chemical composition of Chinese porcelain, in 1740 he formulated his own Réaumur porcelain. In 1750 while investigating the animal world, he designed an egg incubator. Two years later, in 1752, he discovered that digestion is a chemical process by isolating gastric juice and studied its role in food digestion by studying hawks and dogs.

Réaumur died in La Bermondière on October 18, 1757, and bequeathed to the Academy of Science his cabinet of natural history with his collections of minerals and plants.

recapitulation   The repetition of stages of evolution in the stages of development in the individual organism (ONTOGENY). The history of the individual development of an organism.

receptor   A molecule or a polymeric structure in or on a cell that specifically recognizes and binds a compound acting as a molecular messenger (neurotransmitter, HORMONE, lymphokine, lectin, DRUG, etc.).

receptor mapping   The technique used to describe the geometric and/or electronic features of a binding site when insufficient structural data for this RECEPTOR or ENZYME are available. Generally the active-site cavity is defined by comparing the superposition of active molecules with that of inactive molecules.

receptor-mediated endocytosis   Cells use RECEPTOR-mediated endocytosis—a method where specific molecules are ingested into the cell—for ingestion of nutrients, hormones, and growth factors. The specificity results from a receptor-LIGAND (a molecule or ion that can bind another molecule) interaction. Other ligands that can be ingested include toxins and lectins, viruses, and serum transport proteins and antibodies. A receptor, a specific binding protein such as clathrin, on the plasma membrane of the target tissue will specifically bind to ligands on the outside of the cell. An endocytotic process results, i.e., the cell folds inward with a portion of the plasma membrane, and the resulting clathrin-coated pit is pinched off to form a membrane-enclosed bubble or vesicle, called an endosome. After entering the cytoplasm, the endocytotic vesicle loses its clathrin coat, and the ligand (multiple ligands can enter the cell in the same coated pit) is ingested. The receptor can be recycled to the surface by vesicles that bud from the endosome targeting the plasma membrane. After these recycling vesicles fuse with the plasma membrane, the receptor is returned to the cell surface for binding and activity once more.

receptor potential (end-plate potential)   A change in a neuron's membrane potential (a change in voltage across the receptor membrane) caused by redistribution of ions responding to the strength of the stimulus. If the potential is high enough, an action potential will be fired in an afferent neuron. The more action potentials fired, the more neurotransmitters released, and stronger the signals reaching the brain.

recessive allele   An allele that is not expressed phenotypically in a heterozygote due to the presence of a dominant or masking allele. However, it will be expressed in a specific phenotype when a counterpart recessive gene is present. However, often when individuals inherit two mutant copies of a gene on one of the autosomes, they suffer from autosomal recessive disorders. Examples like hemophilia occur where the recessive gene associated with it lies on part of the X chromosome (X-linked recessive disorder). Most genetic mutations produce recessive alleles.

recessive disorder   A disorder associated with a recessive allele.

reciprocal altruism   The belief that if one acts kindly toward another unrelated individual, that

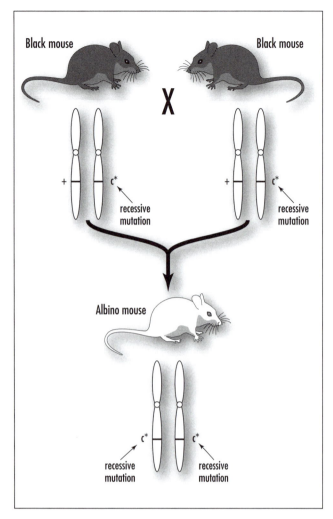

A recessive allele is expressed phenotypically only in offspring who have received two copies of the recessive gene, one from each parent. *(Courtesy of Darryl Leja, NHGRI, National Institutes of Health)*

individual will be inclined to perform the same act back or to another individual. The individual giving does not necessarily expect anything back, only that the kindly act will be repeated and benefit society as a whole.

recognition species concept   A concept of species as a set of individuals that recognize each other as potential mates; having a common or shared system of mate recognition, i.e., courtship displays, that ensures mating with members of the same population.

recombinant   An individual where the genotype is produced by the nonparental arrangement of alleles that results from independent assortment or crossing over (recombination); the phenotype individual is different from the parents; also, a term describing a new combination of genes that form DNA.

recombinant DNA   Altered DNA that has been joined, mostly by in vitro means, by genetic material from two different sources.

recombinant DNA technology   Refers to the modern techniques of gene cloning.

redox potential   Any oxidation–reduction (redox) reaction can be divided into two half reactions: one in which a chemical species undergoes oxidation, and one in which another chemical species undergoes reduction. If a half-reaction is written as a reduction, the driving force is the reduction potential. If the half-reaction is written as oxidation, the driving force is the oxidation potential related to the reduction potential by a sign change. So the redox potential is the reduction/oxidation potential of a compound measured under standard conditions against a standard reference half-cell. In biological systems, the standard redox potential is defined at pH = 7.0 versus the hydrogen electrode and partial pressure of hydrogen = 1 bar.

See also ELECTRODE POTENTIAL.

redox reaction   An abbreviated term for an oxidation-reduction reaction; a reaction that has both *red*uction (electrons are gained) and *oxi*dation (electrons are lost) occurring; one does not occur without the other.

reducing agent   The reactant that donates its electrons and in turn becomes oxidized when another substance is reduced.

See also OXIDATION; REDOX REACTION; REDUCTION.

reductase   *See* OXIDOREDUCTASE.

reduction The part of a redox reaction where the reactant has a net gain of electrons and in which a different reactant must oxidize (lose electrons).

See also OXIDATION.

reductive elimination The reverse of OXIDATIVE ADDITION.

reductive speciation Complete fusion of two previously independent evolutionary species—a hybridization/intergradation phenomenon.

reflex An active response to a stimulus that is usually involuntary and reproducible; many types exist from simple to complex. A common example is the kick response when a doctor hits a knee with a small hammer.

refractory period At the end of the action potential, the absolute refractory period (ARP) is the interval immediately following the discharge of a nerve impulse (action potential) during which the cell cannot be induced to fire again. A relative refractory period is the time following the ARP where a neuron can be induced to discharge again but only if there is a more intense stimulus than normal.

regulation Refers to control of activity of an ENZYME (system) or GENE EXPRESSION.

regulative development (indeterminate development) A development pattern in which cell fates are not determined until late in development; depends on interactions with neighboring cells. For example, in mice gastrulation, if the majority of embryonic cells are destroyed, a normal mouse can still develop from the remainder.

Reichstein, Tadeus (1897–1996) Swiss *Chemist* Tadeus Reichstein was born on July 20, 1897, at Wloclawek, Poland, to Isidor Reichstein and Gastava

Brockmann. He was educated at a boarding school at Jena after his family moved to Zurich in 1906 (where he was naturalized). He had a private tutor and then attended the Oberrealschule (technical school of junior college grade) and the Eidgenössische Technische Hochschule (E.T.H.) (state technical college).

In 1916 he began to study chemistry at the E.T.H. at Zurich and graduated in 1920. In 1922 he began research on the composition of the flavoring substances in roasted coffee, a project that lasted for nine years.

In 1931 he turned to other scientific research, and by 1938 he was professor in pharmaceutical chemistry and director of the pharmaceutical institute in the University of Basel. From 1946 to 1967 he was professor of organic chemistry at the University of Basel.

In 1933 he synthesized vitamin C (ascorbic acid) and worked on plant glycosides. From 1953 to 1954 he worked with several other scientists and was the first to isolate and explain the constitution of aldosterone, a hormone of the adrenal cortex. He also collaborated with E. C. KENDALL and P. S. HENCH in their work on the hormones of the adrenal cortex. For this work, Reichstein, Kendall, and Hench were jointly awarded the Nobel Prize in physiology or medicine in 1950.

After 1967 he worked on the study of ferns and published many papers on the subject. He died on August 1, 1996, in Basel.

relapsing fever A tropical disease associated with some 20 species of the bacteria genus *Borrelia* that is transmitted to humans by two vectors, soft ticks (*Ornithodoros* or *Argas*) or, in the case of *Borrelia recurrentis,* by lice. Louse-borne relapsing fever is more severe than the tick-borne variety.

relative configuration The CONFIGURATION of any stereogenic (asymmetric) center with respect to any other stereogenic center contained within the same molecular entity. A stereogenic unit is a grouping within a molecular entity that can be considered a focus of STEREOISOMERism.

relative fitness One genotype contributes to the next generation. If the population is stable, i.e., neither increasing nor decreasing, then the genotype with a relative fitness of more than one will increase in frequency, whereas if the genotype has a relative fitness of less than one, it will decrease.

relaxation If a system is disturbed from its state of equilibrium, it returns to that state, and the process is referred to as relaxation.

releaser A signal stimulus that functions as a communication signal between individuals of the same species and initiates a fixed action pattern (FAP), a stereotyped species-to-species behavior.

reniform Kidney shaped, such as a kidney bean.

repetitive DNA Repeated DNA sequences that may occur in the thousands of copies in the chromosomes of eukaryotes; represents much of the human genome. These sequences of variable length can be repeated up to 100,000 (middle repetitive) or over 100,000 (highly repetitive) copies per genome. Much of the DNA in eukaryotes is repetitive.

replication fork The Y-shaped portion of the replicating DNA where new strands are growing.

repressible enzyme An enzyme whose synthesis is inhibited or regulated by a regulatory molecule, a specific metabolite.

repressor A protein that prevents gene transcription; prevents RNA polymerase from commencing mRNA synthesis.

Reptilia The class of vertebrate animals that includes snakes, turtles, lizards, tuatara, crocodilians, and extinct fossil species. Reptiles have scales or modified scales, breathe air, are cold blooded, and usually lay eggs. The reptiles evolved from amphibians during the

A closeup of the head of a ribbonsnake, a member of the Reptilia class. *(Courtesy of Tim McCabe)*

late Carboniferous or early Permian periods. Today there are about 2,500 species of snakes, 3,000 of lizards, 250 of turtles and tortoises, and 21 species of crocodilians and distributed worldwide throughout temperate and tropical regions.

reservoir host The host, a vertebrate, that harbors a particular parasite and acts as a long-term source of infection of other vertebrates or vectors.

resistance The ability of an organism to resist microorganisms or toxins produced in disease.

resolving power A property of instruments, like microscopes and telescopes, that distinguish objects that are close to each other; the smaller the minimum distance at which two objects can be distinguished, the greater the resolving power.

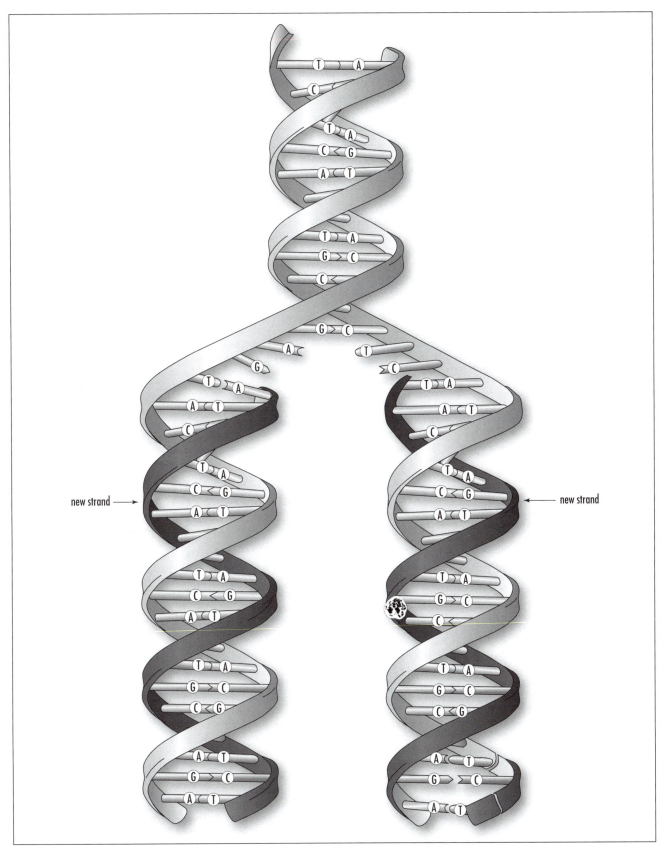

The process by which the DNA double helix unwinds and makes an exact copy of itself. *(Courtesy of Darryl Leja, NHGRI, National Institutes of Health)*

resonance Raman spectroscopy A spectroscopic technique increasingly used in bioinorganic chemistry for characterization and assignment of vibrations directly connected with a CHROMOPHORE, as well for the assignment of the chromophore. The excitation frequency is applied close to the absorption maximum of the chromophore. Particularly useful for deeply colored species.

resource partitioning The process whereby coexisting species living in the same ecosystem each find a separate niche so that resources can be divided up and used without having to compete. Often a few dominant species are able to exploit most of the resources, and the rest divide up the remainder.

respiration The process where mitochondria in the cells of plants and animals release chemical energy from sugar and other organic molecules through chemical oxidation.
See also KREBS CYCLE.

respiratory system The respiratory system is the system by which oxygen, essential for life, is taken into the body and the waste product, carbon dioxide, is expelled from the body. The respiratory system consists of the mouth and nose, airways, and lungs. Air enters through the mouth and nose and passes down the pharynx (throat) and through the larynx (voice box). Air then continues down through the trachea (windpipe), which branches into two bronchi (singular: bronchus) to each of the two lungs. The inflammation of the bronchus is called bronchitis. The bronchi branch many times until becoming much smaller airways called "bronchioles." At the end of each bronchiole are tiny air-filled cavities called alveoli. Each alveolus is surrounded by many blood capillaries, which allow oxygen to move into the bloodstream and carbon dioxide out. This exchange of substances is the primary function of the respiratory system.

resting potential The state of a neuron's charge, the gradient of electric potential across the membrane, when it is in a resting state and ready to receive a nerve impulse (the ACTION POTENTIAL); usually consists of a negative charge on the inside of the cell relative to the outside. At rest, the cell membrane electrical gradient maintains a negative interior charge of –70 mv.

restriction enzyme (**restriction endonuclease**) A DNA cutting protein that recognizes a specific nucleotide sequence in a DNA molecule and excises the DNA; found in bacteria. Some sites are common and occur every several hundred base pairs, while others are less common. Bacteria possess several hundred restriction enzymes that cut more than 100 different DNA sequences. Each restriction enzyme has a single, specific recognition sequence and cuts the DNA molecule at a specific site. Some restriction enzymes have been used in RECOMBINANT DNA TECHNOLOGY.

restriction fragment-length polymorphisms (**RFLPs**) A polymorphism is a genetic variant that appears in about 1 percent of the population, e.g., the human Rh factor. An RFLP is a variation between individuals in DNA fragment sizes cut by specific restriction enzymes. RFLPs usually are caused by mutation at a cutting site. The resulting polymorphic sequences in RFLPs are used as markers on both physical and genetic linkage maps. RFLPs are useful in screening human DNA for the presence of potentially deleterious genes, e.g., sickle cell anemia, and for DNA fingerprinting in forensic science.

restriction site The specific site on a DNA strand where a restriction enzyme cuts DNA.

reticulate Lacelike or netlike in texture, repeating intercrossing between lines; divided or marked to form a network. In evolution, the lateral transfer, back and forth, of genetic information; the complex pattern of species origin by hybridization.

retina The inner back surface, the posterior portion, of the eye; composed of neural tissue and photoreceptive cells that line the inner eye. The retina converts light energy into nerve impulses and sends

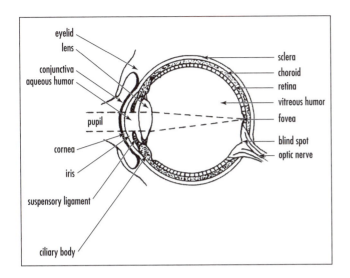

The retina converts light energy into nerve impulses and sends them to the optic nerve by using a set of nerve cells.

them to the optic nerve by using a set of nerve cells—horizontal, bipolar, amacine, and retinal ganglion cells—finally leading to the brain for interpretation. Two types of sensory cells, called photoreceptors, allow the visualization of color, contrast (sharp central vision), and night vision. Cone cells, which are more dense in the center of the retina, are responsible for seeing color and contrast during daylight, while the rods, which are more numerous around the peripheral areas of the retina, are thousands of times more sensitive than cones and allow vision in the dark and allow side or peripheral vision. The choroid is a layer of the retina filled with blood vessels that nourish it.

retinue The group of workers that surrounds a queen in insect societies; attendants or followers. Queen bees release a pheromone, queen mandibular pheromone (QMP), that is both a releaser and a primer. As a releaser it attracts worker bees to their queen, and as a primer it inhibits queen rearing and suppresses the ontogeny of worker foraging by diminishing the release of juvenile hormone in worker bees. QMP also stimulates the retinue response in honeybees by way of licking and antennating behavior that signals the presence of a dominant reproductive queen and establishes and stabilizes the social structure of the bee colony. QMP is spread throughout the colony by retinue behavior. When worker bees turn toward the queen and form a retinue around her, several bees concurrently touch the queen (using their antennae, forelegs, or mouthparts) and then proceed through the hive passing on the pheromone.

retrovirus An RNA virus that encodes an enzyme called REVERSE TRANSCRIPTASE to convert into DNA when inserted into a host's cell DNA, thereby reproducing the virus. HIV viruses are retroviruses, and they also are a class of cancer-causing viruses. *Retro* is short for REverse TRanscriptase Onko.

reverse transcriptase An enzyme that retroviruses use to encode RNA to convert to DNA.

revision In taxonomy, the reclassification of a group of species as the result of additional study.

Rh blood types A blood group, which involves 45 different antigens on the surface of red cells, that is

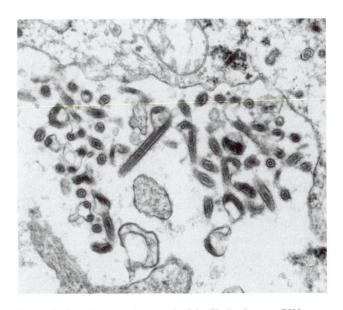

Transmission electron micrograph of the Ebola virus, an RNA virus that causes hemorrhagic fever. *(Courtesy of Centers for Disease Control and Prevention)*

controlled by two closely linked genes on chromosome 1. Can lead to serious medical complications between a mother and her developing fetus if they have different Rh blood types (+ or −). This can occur when the parents have different Rh blood types, e.g., the mother is Rh− and the father is Rh+.

See also BLOOD.

rhizome   A modified underground stem (has nodes and scalelike leaves) that grows horizontal along or below the ground and sends out roots and shoots.

rhodopsin   The red photosensitive pigment found in the retina's rod cells; contains the protein opsin linked to retinene (or retinal), a light-absorbing chemical derived from vitamin A and formed only in the dark. Light striking the rhodopsin molecule forces it to change shape and generate signals as the retinene splits from the opsin, which is then reattached in the dark to complete the visual cycle.

ribonucleic acid (RNA)   Linear polymer molecule composed of a chain of ribose units linked between positions 3 and 5 by phosphodiester groups. The bases adenine or guanine (via atom N-9) or uracil or cytosine (via atom N-1) are attached to ribose at its atom C-1 by ß-*N*-glycosidic bonds. The three most important types of RNAs in the cell are messenger RNA (mRNA), transfer RNA (tRNA), and ribosomal RNA (rRNA).

See also NUCLEOTIDES.

ribonucleotide reductases   ENZYMEs that catalyze the reduction of ribonucleotide diphosphates or triphosphates to the corresponding deoxyribonucleotides by a RADICAL-dependent reaction. The enzyme of animal, yeast, and AEROBIC *ESCHERICHIA COLI* cells contains an oxo-bridged DINUCLEAR iron center and a tyrosyl radical cation, and uses thioredoxin, a thiol-containing protein, as reductant. At least three other ribonucleotide reductases are known from bacteria, containing an IRON–SULFUR CLUSTER with a glycyl radical, adenosyl COBALAMIN, and a dinuclear manganese CLUSTER.

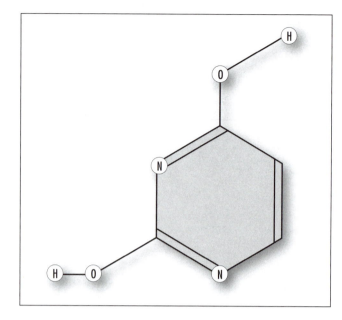

Uracil is one of the four bases in RNA. The others are adenine, guanine, and cytosine. Uracil replaces thymine, which is the fourth base in DNA. Like thymine, uracil always pairs with adenine. *(Courtesy of Darryl Leja, NHGRI, National Institutes of Health)*

ribose   The sugar component of RNA; a five-carbon (pentose) aldose.

ribosomal RNA (rRNA)   The most common form of RNA. When combined with certain proteins, it forms ribosomes that are responsible for translation of MESSENGER RNA (mRNA) into protein chains.

ribosome   A subcellular unit composed of specific rRNA and proteins that is responsible for the TRANSLATION of MESSENGER RNA (mRNA) into protein synthesis.

ribozyme   RNA with enzymatic or catalytic ability to specifically cleave (break down) or bind RNA molecules. Also known as autocatalytic or catalytic RNA.

ribulose-1,5-bisphosphate carboxylase/oxygenase (rubisco)   A magnesium-dependent ENZYME. The primary enzyme of carbon dioxide fixation in plants and

AUTOTROPHIC bacteria. It catalyzes the synthesis of 3-phospho-D-glycerate from ribulose bisphosphate and also the oxidation of ribulose bisphosphate by O_2 to 3-phospho-D-glycerate and 2-phosphoglycolate.

Richards, Dickinson Woodruff, Jr. (1895–1973)

American *Physiologist* Dickinson Woodruff Richards Jr. was born on October 30, 1895, in Orange, New Jersey, to Dickinson W. Richards, a New York lawyer, and Sally Lambert. He was educated at the Hotchkiss School in Connecticut, and in 1913 he went to Yale University to study English and Greek. In June 1917 he was awarded a B.A. After a period as instructor in artillery during 1917–18, he served as an artillery officer in France.

After the war, he attended Columbia University College of Physicians and Surgeons, receiving an M.A. in physiology in 1922 and an M.D. in 1923. From 1923 to 1927 he was on the staff of the Presbyterian Hospital, New York, followed by a year at the National Institute for Medical Research, London, under Sir Henry DALE, working on the control of the circulation in the liver.

In 1931 Richards collaborated with André COURNAND at Bellevue Hospital, New York, and in 1940 he developed a technique for catheterization of the heart, along with later studies of traumatic shock, the diagnosis of congenital heart diseases, the physiology of heart failure, measurement of the actions of cardiac drugs, and various forms of dysfunction in chronic cardiac and pulmonary diseases and their treatment. He was awarded, together with André Cournand and Werner FORSSMANN, the 1956 Nobel Prize in physiology or medicine for his work on catheterization.

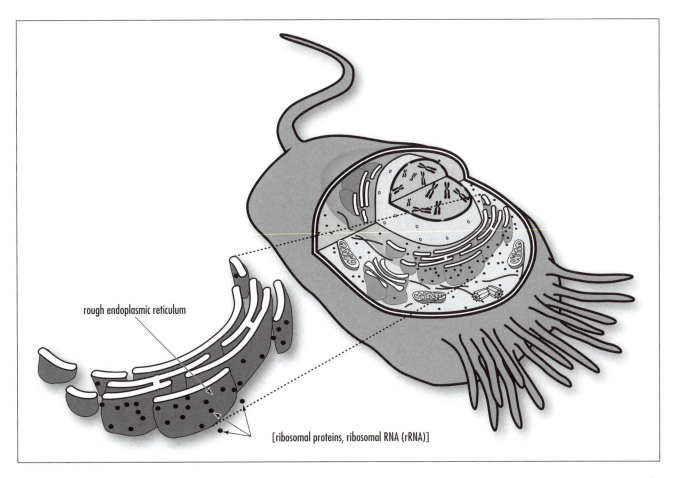

rough endoplasmic reticulum

[ribosomal proteins, ribosomal RNA (rRNA)]

Ribosomes are the cellular organelles that are the sites of protein synthesis. *(Courtesy of Darryl Leja, NHGRI, National Institutes of Health)*

In 1945 Richards was appointed professor of medicine at Columbia University and visiting physician and director of the First (Columbia) Division of the Bellevue Hospital, New York. In 1947 he became Lambert professor of medicine, and he retired in 1961. He served as editor of *The American Review of Tuberculosis* and was a member of the editorial board of *Medicine* and of *Circulation*. He died on February 23, 1973, in Lakeville, Connecticut.

Richet, Charles-Robert (1850–1935) French *Physiologist* Charles Richet was born on August 25, 1850, in Paris to Alfred Richet, a professor of clinical surgery in the Faculty of Medicine, Paris, and his wife Eugénie Renouard. He studied in Paris at the Faculty of Medicine, receiving degrees of doctor of medicine (M.D.) in 1869 and doctor of sciences in physiology in 1878, but he also wrote poetry and drama while in school to cope with the boredom. He became a professor of physiology in 1887 at the University of Paris, in the Sorbonne, where he stayed until 1927. In 1888 he showed that the blood of animals vaccinated against an infection protected the animals against this infection, and two years later he gave the first serotherapeutic injection in humans.

In 1902 Richet made contributions in the study of anaphylaxis, a term he used to describe a hypersensitive reaction (e.g., allergy) to injections of foreign proteins, e.g., serums, a phenomenon noted earlier by Theobald Smith. Anaphylaxis is a deadly reaction in a sensitized individual following a second injection of an antigen. It is called Richet's phenomenon in his honor. For his work on anaphylaxis he received the 1913 Nobel Prize in physiology or medicine.

He also worked on serum therapy, the nervous system, and role of animal heat in homeothermic animals, and he was interested in psychological research, working in hypnosis and coining the word *metaphysique* for research in the field of parapsychology.

From 1878 to 1902 he was editor of the *Revue Scientifique,* and beginning in 1917 he served as the coeditor of the *Journal de Physiologie et de Pathologie Générale.* He published numerous papers on physiology, physiological chemistry, experimental pathology, and normal and pathological psychology. Several of his important books were: *Suc Gastrique chez l'Homme et chez les Animaux* (Gastric juice in man and in animals),

1878; *Leçons sur les Muscles et les Nerfs* (Lectures on the muscles and nerves), 1881; *Leçons sur la Chaleur Animale* (Lectures on animal heat), 1884; *Essai de Psychologie Générale* (Essay on general psychology), 1884; *Souvenirs d'un Physiologiste* (Memoirs of a physiologist), 1933. He was also the editor of *Dictionnaire de Physiologie* (Dictionary of physiology), 1895–1912.

Aside from his medical research, he wrote poems, dramatic works, and even designed a self-powered airplane, one of the first. He died in Paris on December 4, 1935.

rickettsia Gram-negative bacteria (three genera, *Coxiella, Rickettsia,* and *Bartonell*) that infect mammals (hosts) and arthropods (parasites). *R. prowazekii* is the agent of typhus. Other associated diseases include trench fever, Rocky Mountain spotted fever, Q fever, Oriental spotted fever, Flinders Island spotted fever, boutonneuse fever, rickettsialpox, Siberian tick typhus, Queensland tick typhus, and scrub typhus (tsutsugamushi disease).

Rieske iron-sulfur protein An IRON-SULFUR PROTEIN of the MITOCHONDRIAL respiratory chain, in which the [2FE-2S] CLUSTER is coordinated to two sulfur LIGANDs from cysteine and two imidazole ligands from histidine. The term is also applied to similar proteins isolated from photosynthetic organisms and microorganisms and other proteins containing [2Fe-2S] clusters with similar coordination.

See also COORDINATION; PHOTOSYNTHESIS.

rigor mortis The stiffening of a dead body. A result of the depletion of adenosine triphosphate in the muscle fibers.

risk factor Anything in the environment, personal characteristics, or events that make it more or less likely that an individual might develop a certain disease or experience a change in its health status. The risk factors for breast disease are a first-degree relative with breast cancer, a high-fat diet, early menstruation, late menopause, first child after 30 years of age, or no children.

Enlarged view of mouthparts of an American dog tick, *Dermacentor variabilis,* magnified 779×. Ticks are of the class Arachnida, as are spiders and mites. *D. variabilis* is a known carrier of Rocky Mountain spotted fever caused by the bacterium *Rickettsia rickettsii. (Courtesy of Janice Carr, Centers for Disease Control and Prevention)*

RNA polymerase   The enzyme that directs the linking of ribonucleotides during transcription.

RNA   See RIBONUCLEIC ACID.

RNA processing   Synthesis of messenger RNA (mRNA) in a eukaryotic cell involves numerous successive steps: transcription initiation and elongation; RNA processing reactions, such as capping (modification of the 5'-ends of eukaryotic mRNAs), splicing (removing introns, "junk" DNA), and polyadenylation (modifica-

tion of the 3'-ends of eukaryotic mRNA); and finally termination of transcription.

RNA splicing   The process of removing introns, the noncoding regions of eukaryotic genes that are transcribed into mRNA.

See also RNA PROCESSING.

Robbins, Frederick Chapman (1916–) American *Pediatrician*   Frederick Chapman Robbins was born in Auburn, Alabama, on August 25, 1916, to William J.

Robbins, a plant physiologist who became director of the New York Botanical Gardens, and Christine (née Chapman).

He received a B.A. from the University of Missouri in 1936 and, two years later, a B.S. In 1940 he graduated from Harvard Medical School and became a resident physician in bacteriology at the Children's Hospital Medical Center in Boston, Massachusetts. He left briefly in 1942 to serve in the U.S. Army.

Robbins returned to the Children's Hospital Medical Center and completed his work in 1948. For the next two years he held a senior fellowship in viral diseases at the National Research Council, working with Dr. John F. ENDERS in the Research Division of Infectious Diseases at the Children's Hospital Medical Center. He was also a member of the faculty of the Harvard Medical School. While he was working with Enders, Robbins chiefly studied the cultivation of poliomyelitis virus in tissue culture. In 1954 he received the Nobel Prize in physiology or medicine, along with Enders and Thomas WELLER for successfully cultivating poliomyelitis virus in tissue cultures.

Robbins was an associate in pediatrics on the faculty of the Harvard Medical School; an associate in the research division of infectious diseases, associate physician, and associate director of the isolation service at the Children's Hospital Medical Center; a research fellow in pediatrics at the Boston Lying-in Hospital; and an assistant to the Children's Medical Service, Massachusetts General Hospital.

In May 1952 he became professor of pediatrics at Western Reserve University School of Medicine and director of the department of pediatrics and contagious diseases in the Cleveland Metropolitan General Hospital until 1966, and he served as a professor of pediatrics (1952–80) and dean (1966–80) at the Case Western Reserve University School of Medicine, Cleveland, Ohio. Over his long career he has been a member of many scientific organizations.

rod cell One of two photoreceptor cells in the retina, the other being cone cells. Rod cells are more numerous around the peripheral areas of the retina and are thousands of times more sensitive than cones. They allow vision in the dark as well as side or peripheral vision.

See also CONE CELL.

rodent A member (mammal) of the order Rodentia. Examples include rats and mice.

roentgen A measure of ionizing radiation named after Wilhelm Roentgen, a German scientist who discovered X rays in 1895. One roentgen is the amount of gamma rays or X rays needed to produce ions, resulting in a charge of 0.000258 C/kg of air under standard conditions.

Romaña's sign Various observations such as swelling of lymph glands and unilateral palpebral edema characteristic of Chagas' disease. Named for Argentine physician Cecilio Romaña.

root Part of the plant, found mostly underground, that withdraws water and other nutrient from the soil and sometimes accumulates reserves of nutrients.

root cap A cap-shaped structure found on the tips or ends of roots. They cover and protect the active growing region of the root known as the meristem.

root hair Thin, hairlike extensions behind the root tips in plants that are an extension of the root's epidermis. They function as a large surface area to absorb water and minerals.

root pressure Forced water pressure in roots that moves upward by the pumping of minerals into the xylem.

Ross, Ronald (1857–1932) English *Physician, Entomologist* Ronald Ross was born on May 13, 1857, in Almora, India, the son of Sir C. C. G. Ross, an English army general. He lived in India until the age of eight and then was sent to a boarding school in Southampton, England.

Ross began the study of medicine at St. Bartholomew's Hospital in London in 1875, passed his exams in 1879, and entered the Indian Medical

Service in 1881. Ross married Rosa Bessie Bloxam in 1889 and they had two sons and two daughters. He began the study of malaria in 1892 after being exposed to the sufferings of many while he lived in India. In 1894 he began to make an experimental investigation in India that mosquitoes were connected with transmitting the disease. After two and a half years of experimentation, on August 20, 1897, Ross finally succeeded in demonstrating the life cycle of the parasites of malaria in mosquitoes. He dubbed it "Mosquito Day." In 1898 he demonstrated the malarial parasite (plasmodium) in the stomach of the anopheles mosquito, and the following year, Ross joined the Liverpool School of Tropical Medicine under the direction of Sir Alfred Jones and was sent to West Africa to continue his investigations. There he found the species of mosquitoes that convey the deadly African fever.

In 1901 Ross was elected a fellow of the Royal College of Surgeons of England and also a fellow of the Royal Society (became vice president from 1911 to 1913). In 1902 His Majesty the King of Great Britain appointed him a companion of the Most Honorable Order of Bath. In 1911 he was elevated to the rank of knight commander of the same order. In Belgium, he was made an officer in the Order of Leopold II.

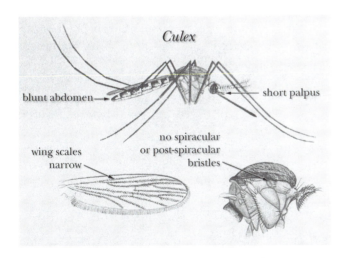

An illustration depicting morphological characteristics common to *Culex* mosquitoes. This image shows the common characteristics of *Culex* mosquitoes. In the United States, West Nile virus is transmitted by infected mosquitoes, primarily members of the genus *Culex*. *(Courtesy of Center for Disease Control)*

In 1902 the School of Tropical Medicine founded a chair of tropical medicine in University College. The Sir Alfred Jones' Chair of Tropical Medicine, as it was called, was given to Ross in 1902, which he retained until 1912. He left Liverpool and was appointed physician for tropical diseases at Kings College Hospital, London, a post that he held together with the chair of tropical sanitation in Liverpool. He stayed until 1917, when he was appointed consultant in malariology to the War Office. He was elevated to the rank of knight commander, St. Michael and St. George, in 1918 and was later appointed consultant in malaria to the Ministry of Pensions. In 1926 he assumed the post of director-in-chief of the Ross Institute and Hospital of Tropical Diseases and Hygiene. His admirers created the institute, and he remained there until his death. He was also a president of the Society of Tropical Medicine.

While his contributions were in the form of the discovery of the transmission of malaria by mosquitoes, he also found time to write poetry and plays and to paint. His poetic works gained him wide acclamation independent of his medical and mathematical standing.

He received many honors in addition to receiving the Nobel Prize in 1902 for his work in malaria, and was given honorary membership of learned societies of most countries of Europe and of many other continents. He received an honorary M.D. degree in Stockholm in 1910 at the centenary celebration of the Caroline Institute. He was knighted in 1911.

His wife died in 1931. Ross died the following year, after a long illness, at the Ross Institute, London, on September 16, 1932, and was buried next to his wife.

rough ER   Rough endoplasmic reticulum, the system of membranous interconnected folded sheets in the cytoplasm of cells and in which ribosomes are contained. It transports materials through the cell and produces cisternae (protein sacks) that are moved to the Golgi body or into the cell membrane. There is also a smooth ER that lacks ribosomes but contains enzymes and produces and digests lipids and membrane proteins; smooth ER buds off, forming vesicles from rough ER, and moves newly made proteins and lipids to the Golgi body and membranes.

See also CELL; SMOOTH ER.

roundworm *See* NEMATODE.

royal jelly A milky white creamy material, thick and nutritious, secreted by the hypopharyngeal glands of nurse bees and supplied to female larvae in royal cells. Needed for the transformation of larvae into queens. Queen bees live exclusively on royal jelly.

R plasmid A plasmid is a mobile, extrachromosomal, self-replicating DNA structure that is found in the cells of bacteria. The R plasmid carries genes that function to resist antibiotics. This is carried out by a resistance transfer factor (RTF), the component of the R plasmid that encodes the ability to conjugate and to transfer DNA.

r-selection Selection that shows a pattern of species reproduction that is fast, with many offspring that mature quickly; such species have short life spans and little parental involvement in rearing.

rubisco *See* RIBULOSE-1,5-BISPHOSPHATE CARBOXYLASE/OXYGENASE.

rubredoxin An IRON–SULFUR PROTEIN without acidlabile sulfur, in which an iron center is coordinated by four sulfur-containing LIGANDs, usually cysteine. The function, where known, is as an electron carrier.
  See also ACID-LABILE SULFIDE; COORDINATION.

rubrerythrin A protein assumed to contain both a RUBREDOXINlike iron center and a HEMERYTHRINlike dinuclear iron center.
  See also NUCLEARITY.

ruminant A polygastric (having more than one digestive cavity) herbivorous animal has a stomach divided into four chambers. Usually a hoofed, eventoed, usually horned mammal, such as sheep, goats, deer, elks, camels, antelope, giraffes, and cows, that chew a cud consisting of regurgitated, partially digested food. The four stomachs consist of the rumen, the first large compartment from which food is regurgitated after cellulose has been broken down by the action of the symbiotic bacterial, protozoal, and fungal populations in the rumen. The second stomach or reticulum (or honeycomb) consists of folds of the mucous membrane that form hexagonal cells and joins the omasum or third stomach by the reticulo-omasal orifice. The digested material in the rumen and the reticulum is exchanged about every 50 to 60 seconds by a rhythmic cycle of contraction. Some of the material from the reticulum also goes into the omasum through the reticulo-omasal orifice. The omasum, or third stomach, located between the reticulum and the abomasum, has muscular leaves that may serve an absorptive function. The abomasum, or fourth stomach, of a ruminant is the true digestive stomach. A ruminant can use cellulose as an energy source because of fermentation by bacteria in the rumen.

runner A modified horizontal, aboveground stem, e.g., strawberry.

rusts (Uredinales; Urediniomycetes) Rusts are obligate parasites, fungi, of vascular plants and cause economic problems with food crops. Rusts can complete their life cycle on one host (autoecious) or alternate between two different hosts (heteroecious). Some economically important rusts affect

barley stem rust—*Puccinia graminis* f. sp. *secalis*, *P. graminis* f. sp. *tritici*; leaf rust—*P. hordei*; stripe rust—*P. striiformis*
oats stem rust—*P. graminis* f. sp. AVENAE; crown rust—*P. coronata*
rye stem rust—*P. graminis* f. sp. AVENAE; crown rust—*P. coronata*
triticale stem rust—*P. graminis* f. sp. *tritici*; leaf rust—*P. recondite*
wheat stem rust—*P. graminis* f. sp. *tritici*; leaf rust—*P. recondite*; stripe rust—*P. striiformis*

There are about 7,000 species of rusts.

saltatory conduction The quick transmission of a nerve impulse along an axon by "jumping" from one node of Ranvier to another.

samara A dry, indehiscent winged fruit either one seeded (e.g., *Ulmus*) or two seeded (*Acer*). Dispersed by the wind.

SA node (sinoatrial node) A specialized cardiac tissue connected to the autonomic nervous system and found in the upper rear wall of the right atrium of the heart, near the superior vena cava. Acts as the heart's pacemaker by speeding or slowing down the heart rate when necessary. It generates a brief electrical impulse of low intensity around 72 times each minute in a resting adult, causing the contraction of two atria, which then pushes blood into the ventricles.

sap The liquid part of a plant, e.g., sap from the sugar maple.

saprobe Any organism that obtains its nutrients and energy from dead or decaying matter. Formerly called saprophyte.

saprophyte *See* SAPROBE.

sarcomere The basic unit, a myofibril, of striated muscle that causes it to contract. It is composed of overlapping myosin (thick) and actin (thin) filaments between two adjacent Z disks, (thin, dark disks that traversely cross through and bisect the clear zone of a striated muscle), giving it a characteristic striated look.

sarcophagous Flesh eating, e.g., tiger shark. Another term for carnivorous.

sarcoplasmic reticulum A cellular organelle, a modified endoplasmic reticulum, that is found only in striated muscle cells and which stores calcium ATPase used to trigger muscle contraction when stimulated.

saturated fatty acid Fatty acid with a hydrocarbon chain that contains no double bonds but has the maximum possible number of hydrogen atoms attached to every carbon atom; it is "saturated" with hydrogen atoms. A fatty acid is the building block of fat, and it exists mostly solid at room temperature. It raises the blood LDL cholesterol ("bad" cholesterol) level.

savanna A biome that is characterized by a flat extensive cover of tall grasses, often three to six feet tall, with scattered trees and shrubs and large herbivores. It is a transitional biome between forest dominates and grassland dominates. A midlatitude variant, the parkland, is located in the drier portions of the humid continental climate, such as the eastern United States. Associated with climates having seasonal precipitation, tropical wet and dry climate type, accompanied with a seasonal drought. Found close to the equator and tropical areas of Africa, Australia, and India.

scavenger   A substance that reacts with (or otherwise removes) a trace component or traps a reactive reaction intermediate (as in the scavenging of RADICALS or free electrons in radiation chemistry). Also an animal that feeds on dead animals.

See also TRAPPING.

Schiff bases   Imines bearing a hydrocarbyl group on the nitrogen atom: $R_2C=NR'$ ($R' \neq H$).

schizophrenia   A form of mental illness with associated problems such as psychotic disorder characterized by severe problems with a person's thoughts, feelings, or behavior; often includes delusions and/or hallucinations. Men usually are affected in late teens or early 20s, while females are affected between their 20s and 30s. Mental health disorders account for four of the top 10 causes of disability worldwide, and 2 million American adults are affected by schizophrenia each year.

school   A large group of similar fish swimming together.

Schwann cells   Schwann cells are the supporting cells of the peripheral nervous system (PNS) that wrap themselves around nerve axons to form a myelin sheath. A single Schwann cell makes up a single segment of an axon's myelin sheath. They also assist in cleaning up debris in the peripheral nervous system and guide regrowth of PNS axons.

scientific method   The steps taken by a scientist after he or she develops a hypothesis, tests its predictions through experimentation, and then changes or discards the hypothesis if the predictions are not supported by the results.

sclereid   A type of sclerenchyma cell. Sclereids are found throughout the plant body and are short, cubical cells that can be living or dead at maturity and form part of the shell and pits of fruits. Also called stone cells.

sclerenchyma cell   Cells that have a rigid secondary cell wall and serve to support and protect plants. Often dead at maturity and composed of lignin, they provide structural support. Sclerenchyma tissue can be composed of two different kinds of cells: sclereids and fibers.

screwworm   There are two, the Old World screwworm (*Chrysomia bezziana*) and the New World screwworm (*Cochliomyia hominivorax*). Larvae of screwworm flies are obligate parasites of mammals (including humans). The larvae of the flies cause lesions known as myiasis that can be fatal and causes serious livestock production losses. The New World species has been eliminated from the United States and several Central American countries.

scrub typhus   A form of typhus caused by a microorganism collectively called rickettsia (*Orientia tsutsugamushi* with multiple serologically distinct strains) and transmitted via chigger mites, by which they are transmitted to humans and other animals. Fatality is as high as 60 percent if not treated. Found in central, eastern, and southeastern Asia; from southeastern Siberia and northern Japan to northern Australia and Vanuatu, as far west as Pakistan, to as high as 10,000 feet above sea level in the Himalayan Mountains, and particularly prevalent in northern Thailand. Restricted mainly to adult workers who frequent scrub overgrown terrain or other mite-infested areas, such as forest clearings, reforested areas, new settlements, or even newly irrigated desert regions.

secondary compound   Compounds that are not directly a function in the process of growth and development in a plant but are parts of the normal part of a plant's metabolism. They often serve to discourage being eaten by making the plant taste bad or have a toxic effect. Examples of secondary compounds are nicotine and caffeine.

secondary consumer (carnivore) An organism in one trophic level that eats a primary consumer, such as a fox eating a hen.

secondary growth Secondary growth occurs in stems and roots and usually results in the thickening of the diameter by the addition of vascular tissue. Used also to describe an area of regrown forest.

secondary immune response The immune response that follows a second or subsequent encounter with a particular antigen. Usually the response is more severe than the PRIMARY IMMUNE RESPONSE.

secondary productivity The rate of new biomass production that is nutrient material synthesized by consumers over a specific time frame in an ecosystem.

secondary structure Level of structural organization in proteins described by the folding of the polypeptide chain into structural MOTIFs such as ALPHA HELIXes and BETA SHEETs, which involve hydrogen bonding of backbone atoms. Secondary structure is also formed in NUCLEIC ACIDS, especially in single-stranded RNAs by internal BASE PAIRING.

secondary succession Ecological succession in communities that have been disturbed, such as a forest turned to an agricultural field that then reverts back to forest; the natural recovery process from disturbance.

second law of thermodynamics States that entropy, a measure of disorder, increases in the universe and is spontaneous. Elements in a closed system will tend to seek their most probable distribution, and

entropy always increases. It is a measure of unusable energy and a gauge of randomness or chaos within a closed system. Also called the Law of Increased Entropy. Along with the first law of thermodynamics, these two laws serve as the fundamental principles of physics.

See also FIRST LAW OF THERMODYNAMICS.

second messenger An intracellular METABOLITE or ion increasing or decreasing as a response to the stimulation of RECEPTORs by AGONISTs, considered as the "first messenger." This generic term usually does not prejudge the rank order of intracellular biochemical events.

sedimentary rock A rock composed of sediment that accumulated over millions of years in the bottom of water bodies and is turned into rock by pressure through the process of lithification. Sedimentary rock identification is primarily based on composition and can be based on chemistry, fossil or organic composition, or clastics (small fragments of rock or mineral).

seed A structure produced by a terrestrial plant in which the young embryo is encased in a protective covering and contains stored food.

segmentation Being divided into segments, e.g., earthworms.

selection coefficient A measure of the relative strength of selection acting against a specific genotype. It is calculated by subtracting each fitness value from 1.0. It is the cost associated with a given mutation on an organism's fitness.

selective permeability Refers to the control a cell membrane has over what can pass through it. The membrane controls which specific molecules may enter or leave the cell by using either PASSIVE TRANSPORT or ACTIVE TRANSPORT or by way of a VESICLE.

self-incompatibility The process in which plants avoid inbreeding. It is the capacity for plants to block fertilization by pollen from members of its own species or closely related ones.

self-pollination The deposit of pollen from the anther of a flower to the stigma of the same flower or to different flowers on the same plant; must be complete flowers. Self-pollination is frequently prevented in plants because it would result in inbreeding, and some mechanisms have evolved to prevent it, such as imperfect flowers and dioecious plants. African violets and orchids are often self-pollinators.

semelparity The production of offspring where organisms produce all of their offspring over a short time in one single reproductive event.

semen The fluid that contains sperm and glandular secretions that is ejaculated from a male during orgasm.

semicircular canals Three small tubes in the inner ear that maintain equilibrium and allow an organism to detect changes in movement (acceleration) in a front-to-back, up-and-down, and side-to-side motion by using hair cells that detect movements of the fluid in the canals. The canals are connected to the auditory nerve.

semilunar valve Two valves, or connective tissue, around the origin of the aorta (aortic valve) and beginning of the pulmonary artery (pulmonary valve). Each valve is composed of three flaps of tissue that maintain the unidirectional blood flow and prevent blood from reentering the ventricles during diastole.

seminiferous tubules Highly convoluted channels in the testes, where male sperm is produced. The spermatogonia, stem cells for spermatogenesis, are nourished by Sertoli cells in the tubules, which also serve to phagocytose damaged germ cells; provide a barrier between the testis and blood; and produce mullerian

inhibition factor (MIF), inhibin, and androgen-binding protein (ABP).

Interstitial cells, the Leydig cells between the tubules, produce testosterone, the hormone responsible for secondary sexual characteristics (e.g., facial and pubic hair, male voice lowering at puberty). As the sperm cells undergo meiosis and age, they migrate to the center of the tubule and into the epididymus, located on top of each testicle in the scrotum. The scrotum keeps the testes about 2°C cooler than body temperature, increasing sperm production.

Inhibin, a polypeptide hormone produced in the ovary and testes, regulates the release of follicle-stimulating hormone (FSH) from the anterior pituitary and is released by the Sertoli cells in the testes and granulosa cells in the ovary. Androgens stimulate the release of inhibin.

Total length of a seminiferous tubule is around 250 meters in one testis.

semipermeable A membrane where some substances will pass through but others will not.

sensation The results of an impulse that is sent to the brain when receptors and sensory neurons are activated.

sensory neuron A specialized neuron that sends messages that it receives from external or internal stimuli such as light, sound, smell, and chemicals to the central nervous system.

sensory receptor A cell or organ that converts a stimulus from a form of sound, light, or thermal, chemical, or mechanical stimulation into a signal, or action potential, that can be transmitted through the organism. Sensory receptors are specific to the stimulus to which they are responding and fall into specific types such as chemoreceptors, photoreceptors, thermoreceptors, mechanoreceptors, and pain receptors, to name a few. Each changes the polarization of the cell that may eventually cause an action potential. Phasic receptors send action potentials quickly when first stimulated and then soon reduce the frequency of action potentials, even if the stimulus continues, e.g., odor or pres-

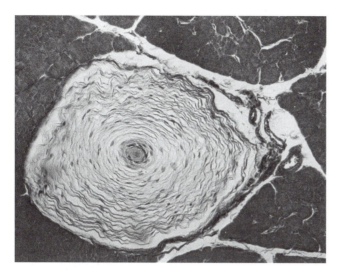

Light micrograph of sectioned Pacinian corpuscle (large circle, center), a sensory receptor for touch. The corpuscle consists of nerve endings surrounded by "onion-skin" membrane layers. The nerve cells are stimulated when the membranes are deformed by changes in pressure. Pacinian corpuscles are especially sensitive to vibration. They are found in skin and in some internal membranes. Magnification unknown. *(Courtesy ©CNRIoto Researchers)*

pinephrine); a neurotransmitter and an important vasoactive substance. A mediator of immediate hypersensitivity. Serotonin is one of many mediators released by circulating basophils and tissue mast cells. Found in blood platelets, in the gastrointestinal tract, and in certain regions of the brain.

It plays a major role in blood clotting, stimulating strong heart beats, initiating sleep, fighting depression, and causing migraine headaches.

serum *See* PLASMA.

sessile Stalkless, stationary, or attached to a substrate.

sex chromosomes The pair of chromosomes needed for determining the sex of an individual. In females,

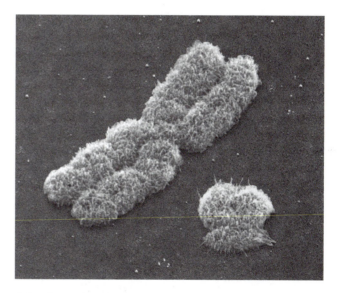

Colored scanning electron micrograph (SEM) of human X (center) and Y (lower right) sex chromosomes. Each chromosome has replicated to form two identical strands (chromatids). The area linking the chromatids is the centromere. The sex chromosomes inherited during fertilization determine a person's gender. Males have an X and a Y chromosome (as seen here), while females have two copies of the X chromosome. The Y chromosome carries instructions for the development of male characteristics. The sex chromosomes are one pair of the 23 pairs (present in most cells in the body) that contain the DNA necessary for growth and development. Magnification: ×7,150 at 6 × 7 cm size. *(Courtesy © Andrew Syred/Photo Researchers, Inc.)*

sure. They are useful for signaling sudden changes in the environment. Receptors that respond to light or mechanics are tonic receptors and produce a constant signal, after an initial amount of high-frequency action potentials, while the stimulus is being applied.

sepal A part of the calyx consisting of a whorl of modified leaves that encloses and protects the flower bud before it opens.

sequence The order of neighboring amino acids in a protein or the purine and pyrimidine bases in RNA or DNA.

See also PRIMARY STRUCTURE.

sequence-directed mutagenesis *See* MUTAGENESIS.

serotonin (5-hydroxytryptamine; 5-HT) A biogenic monoamine (like histamine, epinephrine, nore-

there are two X chromosomes. In males, there is one Y and one X. It is the presence of the Y chromosome that leads to the development into a male.

sex-linked gene   A gene located on one of the sex chromosomes. Some genetic diseases are located on a sex gene. For example, hemophilia A, which is a blood-clotting disorder caused by a mutant gene, is located on the X chromosome. Genes located on the X chromosome are called X-linked, while genes located on the Y chromosome are called Y-linked.

sexual dimorphism   The existence of noticeable physical differences between males and females.

sexual reproduction   The creation of a new individual following the union of two gametes (e.g., egg and sperm). In humans and most other eukaryotes, the two gametes differ in structure (anisogamy) and are contributed by different parents.

sexual selection   Sexual selection based on secondary sex characteristics and on the success of certain individuals over others in relation to the propagation of the species and the enhancement of sexual dimorphism.

Sherrington, Charles Scott (1857–1952) English *Neurophysiologist* Charles Scott Sherrington was born on November 27, 1857, in Islington, London, to James Norton Sherrington and Anne Brookes. He attended Ipswich Grammar School from 1870 to 1875 and in 1876 began medical studies at St. Thomas's Hospital. In 1879 he went to Cambridge, where he studied physiology under the "father of British physiology," Sir Michael Foster (1836–1907). He was professor of physiology at the Universities of Liverpool and London and at Oxford.

From 1891 to 1897 he conducted research on the nervous system, the efferent nerve (part of the peripheral nervous system, not in the brain or spinal cord) supply of muscles, and discovered that about one-third of the nerve fibers in a nerve supplying a muscle are efferent, the remainder being motor. He explained proprioception, the function of the nerve synapse, a word he created. He received the 1932 Nobel Prize in physiology or medicine with E. D. ADRIAN for their discoveries regarding the function of the neuron.

In 1906 he published his well-known book *The Integrative Action of the Nervous System*. His other books include *Mammalian Physiology: A Course of Practical Exercises*, 1919; *The Brain and Its Mechanism*, 1933; and *Man on His Nature*, 1940.

In 1913 he became the Waynfleet professor of physiology at Oxford, a post he held until his retirement in 1936. He died suddenly of heart failure at Eastbourne in 1952.

shock   A life-threatening condition where blood pressure is too low to sustain life. Occurs when a low blood volume (due to severe bleeding, excessive fluid loss, or inadequate fluid uptake), inadequate pumping action of the heart, or excessive dilation of the blood vessel walls (vasodilation) causes low blood pressure. This in turn results in inadequate blood supply to body cells, which can quickly die or be irreversibly damaged.

Anaphylactic shock is the severest form of allergy, which is a medical emergency; a Type I reaction according to the Gell and Coombs classification. An often severe and sometimes fatal systemic reaction in a susceptible individual upon exposure to a specific antigen (such as wasp venom or penicillin) following previous sensitization. Characterized especially by respiratory symptoms, fainting, itching, urticaria, swelling of the throat or other mucous membranes, and a sudden decline in blood pressure.

shoot system   The aboveground system of stems, leaves, and flowers.

short-day plant   A plant that flowers when the days are short; requires less than 12 hours of daylight to bloom or long periods of darkness before it can bloom, e.g., poinsettia.
See also PHOTOPERIODISM.

sib   Short for sibling, a brother or sister.

sibling species   Species that are closely related and are so similar morphologically that it is difficult to distinguish between them.

sickle cell   A genetically inherited disease where defective hemoglobin, the oxygen-carrying pigment in red blood cells, causes sickling (distortion) as well as becoming hard and sticky, resulting in loss of red blood cells and causing anemia. There are three types in the United States: hemoglobin SS or sickle-cell anemia, hemoglobin SC disease, and hemoglobin sickle beta-thalassemia. While anyone can get sickle-cell anemia, 8.5 percent of the African-American population carry the trait but do not have the disease. One out of 400 African Americans has sickle cell, and sickle cell affects eight out of every 100,000 people.

siderophore   Generic term for Fe(III)-complexing compounds released into the cell medium by bacteria for the purpose of scavenging iron.
See also SCAVENGER.

sieve-tube member   Living elongated cells arranged in chains that form sieve tubes in phloem. Transports sucrose, amino acids, other food materials, and hormones throughout the plant.

signal peptide   A sequence of amino acids that determines the location of a protein in a eukaryotic cell.

signal transductions pathway   Signal transduction refers to the movement of signals from outside the cell to the inside and is a mechanism connecting the stimulus to a cellular response.

sign stimulus   A stimulus or releaser such as an environmental cue that elicits a fixed action response. If the sign is not present, then the instinct will not occur.

sink habitat   A habitat where deaths are greater than births and where the population would cease to exist without immigration from more productive habitats.
See also SOURCE HABITAT.

sinus   *Sinus* means "cavity," and many structures of the human body are thus called sinuses. However, the term generally refers to the paranasal sinus. The sinuses (paranasal sinuses) are air cavities within the facial bones. They are lined by mucous membranes similar to those in other parts of the airways. The paranasal sinuses consist of the ethmoid sinus, frontal sinus, maxillary sinus, and sphenoid sinus.
An inflammation of the sinuses is called sinusitis.

siroheme   A HEMElike PROSTHETIC GROUP found in a class of ENZYMEs that catalyze the six-electron reduction of sulfite and nitrite to sulfide and ammonia, respectively.
See also NITRITE REDUCTASE; SULFITE REDUCTASE.

sister chromatids   Two identical copies of a single chromosome that are connected by a centromere. After a dividing cell has duplicated its chromosomes, it does not separate the copies until it is positive that duplication was a complete success and that the duplicated copies have been lined up correctly. While the chromosomes are duplicated in S phase, the two copies or sister chromatids are kept together until later, when they are pulled apart at the metaphase-to-anaphase transition during mitosis. The sister chromatids are pulled to opposite halves of the cell by microtubules coming from spindle poles located at opposite sides of the cell. Sister chromatids segregate away from each other because their kinetochores attach to microtubules emanating from opposite poles.

sister species   Species that have evolved from a common ancestral species and shared by no other species. The mushroom *Chalciporus piperatoides* is a sister species of *Chalciporus piperatus*.

site-directed mutagenesis   *See* MUTAGENESIS.

site-specific delivery   An approach to target a DRUG to a specific tissue using PRODRUGS or antibody recognition systems.

skeletal muscle   Striated muscle attached to the skeleton; contraction is under voluntary control. A single skeletal muscle is attached at its origin to a large area of bone, while at the other end the insertion tapers into a tendon.

skeleton   A set of bones or cartilage, ligaments, and tendons that acts as a frame and holds the body and its organs together while also allowing locomotion. Endoskeletons are within the body, while exoskeletons surround the outside of the body.

sliding filament model   An explanation as to why muscular contraction occurs. During muscular contraction, thin (actin) filaments within the sarcomere of a myofibril are pulled toward the center of the sarcomere (called the H zone) by the thick (myosin) filaments; the two myofilaments slide past each other, increasing overlap. During this process, the sarcomere's length shortens, the myofibril shortens, and the muscle contracts.

slugs   Snails without shells. Slugs are mollusks belonging to the class Gastropoda. Several species of slugs are frequently damaging to gardens and flowers, including the gray garden slug (*Peroceras reticulatum*), the banded slug (*Limax poirieri*), and the greenhouse slug (*Milax gagates*). Slugs are hermaphrodites, and there are at least 40 species of slugs in the United States.

small nuclear ribonucleoprotein (snRNP)   A combination of nuclear RNA and protein that is part of the spliceosome, the intron-removing structure in the nuclei of eukaryotes. Introns are DNA segments within a gene that interrupt the coding sequence of the gene. SnRNPs are not fully understood.

smooth ER   Smooth ER is part of the endoplasmic reticulum that does not contain ribosomes and is a

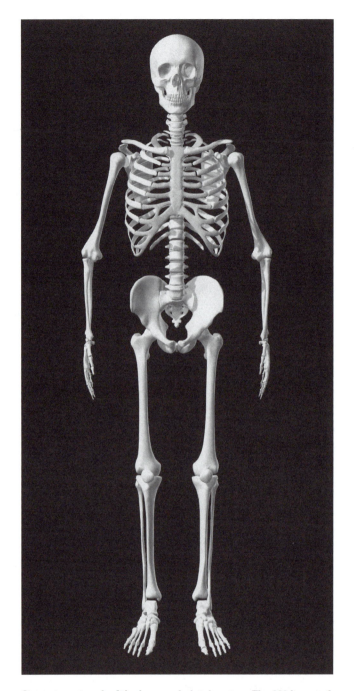

Computer artwork of the human skeletal system. The 206 bones of the skeleton provide protection and support, and their joints allow locomotion. The skull (at top) protects the brain. The ribs of the chest (at upper center) enclose the heart and lungs. The pelvis (at center) protects the lower abdominal organs. The flexible backbone runs from the skull to the pelvis and protects the spinal cord. The long leg and arm bones provide support. The humerus (upper-arm bone) articulates with the radius and ulna (lower-arm bones) at the elbow. The femur (thigh bone) articulates with the fibula and tibia (lower-leg bones) at the knee. (Courtesy © Roger Harris/Photo Researchers, Inc.)

system of membranous sacs and tubules (cisternae). Protein molecules move from the rough ER into the smooth ER, which forwards them as enclosed vesicles, usually to the Golgi complex. Smooth ER functions in metabolic processes: the synthesis of lipids, carbohydrate metabolism, and the detoxification of drugs and other toxins. Enzymes in the smooth ER deal with synthesis of fats, phospholipids, steroids, and other lipids, and they serve to detoxify drugs and other toxins in the liver cells so they can be discharged from the body. In muscle cells, the smooth ER stores calcium that is released upon stimulation to effect cell contraction.

See also ROUGH ER.

smooth muscle Smooth muscle is composed of single, spindle-shaped cells that have no visible striations. Each smooth muscle cell contains thick (myosin) and thin (actin) filaments that slide against each other to produce contraction of the cell and are anchored near the plasma membrane. Smooth muscle contraction is slower than that of striated muscle and can be sustained for long periods of time. Smooth muscle can be found in blood vessels, the gastrointestinal tract, bronchi of the lungs, the bladder, and the uterus. Smooth muscle is primarily under the control of the autonomic nervous system.

smuts Smuts are a group of organisms, along with rusts, belonging to the microfungi, class Ustomycetes. Microfungi also include molds and mildews. Smuts are parasites that appear on plants such as grain, corn, and other economically important plants as small galls that break open and spread spores. In the United States, about 13,000 species of microfungi on plants or plant products have been described.

society An interdependent system of organisms, plant or animal; in insects, two or more individuals that constitute a social unit (bee hive, ant hill).

sociobiology The study of the behavior of social animals, including humans, with a premise that genetics is the sole factor responsible for the behavior in humans and animals.

sodium-potassium pump An active transport mechanism of cell membranes to regulate pressure between the inside and outside of the cell and to pump potassium ions into the cell and keep sodium out, thereby preventing water retention and swelling within the cell. The sodium-potassium pump also maintains the electrical charge within each cell. ATP is used as the energy source for the pump.

See also ACTIVE TRANSPORT.

soft acid *See* HARD ACID.

soft base *See* HARD BASE.

soft drug A compound that is degraded in vivo to predictable nontoxic and inactive METABOLITEs after having achieved its therapeutic role.

solute Any dissolved substance in a solution.

solution Any liquid mixture of two or more substances that is homogeneous. A solution can be basic, that is, have more OH^- ions than H^+ ions with a pH greater than 7, or it can be acidic with more H^+ ions and have a pH lower than 7.

solvation Any stabilizing interaction of a solute (or solute moiety) and the solvent, or a similar interaction of solvent with groups of an insoluble material (e.g., the ionic groups of an ion-exchange resin). They generally involve electrostatic forces and Van der Waals forces, as well as more specific chemical effects such as hydrogen bond formation.

solvent Any liquid that dissolves another solute and forms a homogeneous solution. Several types of solvents exist such as organic solvents (acetone, ethanol) or hydrocarbon solvents (mineral spirits). Water is the most common solvent.

solvolysis Reaction with a solvent involving the rupture of one or more bonds in the reacting solute.

somatic cell Term for any cell in the body except gametes (sperm and egg) and their precursors.

somatic nervous system A branch of the peripheral nervous system containing both afferent and efferent nerves. It consists of peripheral nerve fibers that send sensory information to the central nervous system as well as motor nerve fibers that carry signals to skeletal muscle, which allows the movement of arms and legs.

sonar Short for sound navigation ranging. A system that transmits sound underwater and reads the reflected sound received back to detect and locate underwater objects or to measure distances. A form of echolocation.

soret band A very strong absorption band in the blue region of the optical absorption spectrum of a HEME protein.

source habitat A habitat that produces a surplus number of individuals, i.e., where reproduction exceeds mortality and excess individuals emigrate.
See also SINK HABITAT.

Southern blotting A gel technique invented by Edward M. Southern in 1975 to locate a particular DNA sequence within a complex mixture. DNA fragments are separated by electrophoresis in an agarose gel (involving in situ denaturation), transfer by capillary action to a nitrocellulose sheet, and hybridization to a labeled nucleic acid probe.

spawning Sexual reproduction in the form of microscopic eggs and sperm being discharged into water at the same location; common among fish. A spawning ground is the location where the female fish lays its eggs and the male fertilizes them.

speciation Refers to the chemical form or compound in which an element occurs in both nonliving and living systems. It can also refer to the quantitative distribution of an element. In biology, it refers to the origination of a new species.
See also BIOAVAILABILITY.

species A genomically coherent group of individuals sharing a large degree of similarity in independent features that interbreed and are reproductively isolated from other such groups. One of the levels of scientific classification in a taxonomic hierarchy that includes the genus, family, order, class, phylum, and kingdom. There are more than 1 million named species on Earth and millions more nameless.
See also TAXON.

species diversity Species in a biological community based on number and relative abundance.

species richness The quantitative number of species in a community, biome, or other defined region.

species selection Based on speciation and extinction rates, the observation that species living longer and producing more species will determine major evolutionary directions.

specific heat The amount of heat per unit mass required to move the temperature by one degree Celsius. Every substance has its own specific heat. Measured in joules per gram-degree Celsius (J/g°C), the specific heat of water is 2.02 J/g°C in gas phase, 4.184 in liquid phase, and 2.06 in solid phase.

specificity The property of antibodies that enables them to react with some antigenic determinants and not with others. Specificity is dependent on chemical composition, physical forces, and molecular structure at the binding site.

spectrophotometer An instrument that measures the intensity of light versus its wavelength.

Spemann, Hans (1869–1941) German *Embryologist* Hans Spemann was born on June 27, 1869, in Stuttgart to the publisher Wilhelm Spemann. From 1878 until 1888 he went to the Eberhard-Ludwig School at Stuttgart, left school in 1888, and spent a year in his father's publishing business.

In 1891 he entered the University of Heidelberg after spending his military service from 1889 to 1890. He studied at the University of Munich during the winter of 1893–94, and from the spring of 1894 to the end of 1908 he worked in the Zoological Institute at the University of Würzburg.

In 1895 he took his Ph.D. degree in zoology, botany, and physics, and three years later was qualified as a lecturer in zoology at the University of Würzburg. In 1908 he was asked to become professor of zoology and comparative anatomy at Rostock, and in 1914 he became associate director of the Kaiser Wilhelm Institute of Biology at Berlin-Dahlem. In 1919 he was appointed professor of zoology at the University of Freiburg-im-Breisgau, a post that he held until he retired in 1935, then becoming emeritus professor.

Spemann's work was in experimental embryology, focusing on the differentiation of embryo cells during an organism's development. He was an expert in micro-surgical technique. He worked on amphibian eggs and in 1924 discovered that a part of an embryo, when transplanted to other regions of the embryo, causes a change in the surrounding tissues. Those parts were named "organizer center" or "organizer" by him. The so-called organizer effect was the dominant topic in embryological research during the 1930s, and for this discovery he was awarded the Nobel Prize in 1935.

Spemann laid the foundations of the theory of embryonic induction by organizers, which led to biochemical studies of this process and the ultimate development of the modern science of experimental morphogenesis. He even conducted one of the first cloning experiments on a salamander. He described his researches in his book *Embryonic Development and Induction* (1938). Spemann died at Freiburg on September 9, 1941.

sperm The small, mobile, haploid gamete that is produced by sexually reproducing male eukaryotes.
See also SEMEN.

spermatogenesis The development of sperm cells in the tubules of the testes from spermatogonium.

S phase The time period when DNA is synthesized (S) in a cell so that new cells forming will have the correct amount of DNA. One part of a five-part process in cell division.
See also MITOSIS.

sphincter A ringlike muscle that surrounds a natural opening and closes it by contraction.

spiders Members of the class Arachnida, order Araneae, and one of the most feared organisms by humans. There are more than 30,000 species of spiders belonging to 105 families. All arachnids have four pairs of walking legs and fangs (chelicerae) adapted for liquid feeding. They lack jaws or other types of feeding structures.

All spiders have spinnerets, small fingerlike appendages at the end of their abdomen, that secrete silk during all stages of life. The silk is used for capturing prey, rearing young, mobility, making shelter, and for spinning webs. Spiders mate when the male transfers sperm to the female using pedipalps, specially modified appendages near the mouth. Some spiders are poisonous, and all are predators, eating many different kinds of insects.

spina bifida (myelomeningocele) A neural-tube defect, a birth defect, that results in the failure of the bones of the spine, usually at the bottom of the spinal cord, to close during the first month of pregnancy. Paralysis is the usual condition, along with no control of some other organs such as bowel and bladder. Mental retardation can also occur. It affects approximately one out of every 1,000 newborns in the United States and is the most frequent disabling birth defect.

spindle A group of microtubules originating from the centriole that move chromosomes during cell division in eukaryotic cells. Spindles attach at the kinetochore—a region on the centromere, the area that joins chromatids.

spin label A STABLE paramagnetic group that is attached to a part of another molecular entity whose microscopic environment is of interest and can be revealed by the ELECTRON PARAMAGNETIC RESONANCE spectrum of the spin label. When a simple paramagnetic molecular entity is used in this way without covalent attachment to the molecular entity of interest, it is frequently referred to as a spin probe.

spin-orbit coupling The interaction of the electron-spin magnetic moment with the magnetic moment due to the orbital motion of the electron.

spin probe *See* SPIN LABEL.

spin-spin coupling The interaction between the spin magnetic moments of different electrons and/or nuclei. In NMR spectroscopy it gives rise to multiple patterns and crosspeaks in two-dimensional NMR spectra. Between electron and nuclear spins, this is termed the nuclear HYPERFINE interaction. Between electron spins, it gives rise to relaxation effects and splitting of the electron paramagnetic resonance (EPR) spectrum.
See also NUCLEAR MAGNETIC RESONANCE SPECTROSCOPY.

spin trapping In certain solution reactions, a transient RADICAL will interact with a DIAMAGNETIC reagent to form a more "persistent" radical. The product radical accumulates to a concentration where detection and, frequently, identification are possible by

ELECTRON PARAMAGNETIC RESONANCE SPECTROSCOPY. The key reaction is usually one of attachment; the diamagnetic reagent is said to be a "spin trap," and the persistent product radical is then the "spin adduct."

spiral cleavage An embryonic development characteristic of protostomes. The developing embryo undergoes cell division (cleavage) from a four-cell embryo to an eight-cell embryo, with the cells dividing at slight angles to one another. None of the four cells in one plane of the eight-cell stage is directly over a cell in the other plane (oblique to polar axis).
See also RADIAL CLEAVAGE.

spirochetes Long, slender, coiled (looks like a telephone cord) bacteria that cause disease and are symbionts in the stomachs of ruminants. Spirochetes can be aerobic or anaerobic, free-living or parasitic. All spirochetes are chemoheterotrophs. Three genera exist, *Leptospira* (cause icterohemorrhagic fever), *Treponema* (cause syphilis), and *Borrelia* (cause Lyme disease).

spleen The spleen is located in the upper left quadrant of the abdomen. It has two main functions, acting as part of the immune system and as a filter. There are two distinct components of the spleen, the red pulp and the white pulp. It plays an important role in immune system activities as part of the lymphatic system.

spliceosome A complex of several snRNA molecules (small nuclear RNA) and proteins that remove introns (noncoding mRNA) and splice exons (remaining mRNA sequences).

splitter A taxonomist who prefers to create taxonomic categories that are narrowly defined, thus ending up with more genera than a lumper, who prefers to place closely related genera into a single genus.
See also LUMPER.

sporangium A tiny closed globe or capsule located on a sporangiophore in which sporangiospores and haploid spores (asexual) are produced by meiosis in fungi and other plants.

spore The reproductive organ in cryptogams (ferns, mosses, fungi, and algae), which functions like a seed but has no embryo. In alternation of generations, it germinates and produces a gametophyte in ferns and mosses. It is the reproductive unit encased in a protective coat of a fungus. Usually environmentally resistant and can lay dormant for many years.

See also SPOROPHYTE.

sporophyte The phrase *alternation of generations* refers to the way in which the diploid and haploid stages in a plant's life cycle take different and distinct forms. The diploid stage is called the sporophyte and the haploid stage the gametophyte. The diploid sporophyte phase results from a union of gametes that meiotically produces haploid spores. The haploid gametophyte phase develops following meiosis and develops mechanisms for fertilization by producing and accepting gametes. The haploid, gamete-producing vegetative phase is the gametophyte, and the diploid, spore-producing vegetative phase is the sporophyte.

See also GAMETOPHYTE.

sporopollenin An acetolysis resistant biopolymer that makes up most of the material of the exine, an outer layer of the wall of pollen grains and spores that is highly resistant to acids and bases. A granule of sporopollenin is called an orbicule.

spotted fever (Rocky Mountain spotted fever) A disease caused by *Rickettsia rickettsii* and transmitted by ticks. Confined to the Western Hemisphere and found in all states in the United States with the exception of Maine, Hawaii, and Alaska. *Dermacentor andersoni* (the wood tick) is the usual vector in the western United States, while *D. variabilis* (dog tick) is the vector in the eastern and southern United States. Without treatment, fatality is about 20 percent, but antibiotics reduces the rate if treated early.

See also RICKETTSIA.

square plane *See* COORDINATION.

stability constant An equilibrium constant that expresses the propensity of a species to form from its component parts. The larger the stability constant, the more STABLE the species. The stability constant (formation constant) is the reciprocal of the instability constant (dissociation constant).

stabilizing selection A form of natural selection that favors maintaining existing character traits in a population and eliminating those individuals who exhibit extreme variations of a trait. While it maintains the existing state of adaptation and exhibits the highest fitness, it results in an overall decrease in genetic diversity.

stable A term describing a system in a state of equilibrium corresponding to a local minimum of the appropriate thermodynamic potential for the specified constraints on the system. Stability cannot be defined in an absolute sense, but if several states are in principle accessible to the system under given conditions, then that with the lowest potential is called the stable state, while the other states are described as metastable. Unstable states are not at a local minimum. Transitions between metastable and stable states occur at rates that depend on the magnitude of the appropriate activation energy barriers separating them.

stamen The male organ of a pollen-bearing flower.

See also OVULE.

starch A polysaccharide containing glucose (long-chain polymer of amylose and amylopectin) that is the energy storage reserve in plants.

stasipatric speciation A form of speciation where new species are created as a result of chromosomal rearrangements or mutations (translocations, inversions, changes in chromosome numbers). Reproductive isolation occurs, and new species develop.

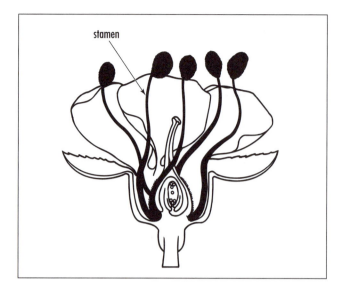

The male organ of a pollen-bearing flower.

statocyst   A vesicle, organ of balance, or mechanoreceptor in invertebrates that serves to regulate equilibrium by using statoliths, a granule of sand or limestone that responds to gravity by stimulating sensory cells when an animal is in motion.

statoliths   Equilibrium organs; starch grains in plant cells that act as a gravity sensor. Found in root-tip cells or tissues close to vascular bundles.

steady state   If, during the course of a chemical reaction, the concentration of an intermediate remains constant, the intermediate is said to be in a steady state. In a static system, a reaction intermediate reaches a steady state if the processes leading to its formation and those removing it are approximately in balance. The steady-state hypothesis leads to a great simplification in reaching an expression for the overall rate of a composite reaction in terms of the rate constants for the individual elementary steps. Care must be taken to apply the steady-state hypothesis only to appropriate reaction intermediates. An intermediate such as an atom or a free RADICAL, present at low concentrations, can usually be taken to obey the hypothesis during the main course of the reaction. In a flow system, a steady state can be established even for intermediates present at relatively high concentrations.

stele   The location of xylem and phloem in roots; central vascular cylinder.

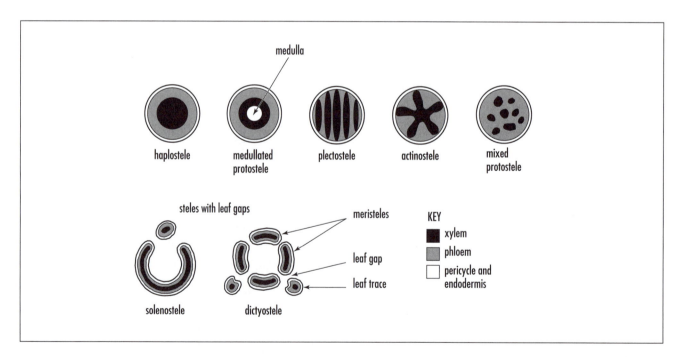

A schematic of different types of stele, where xylem and phloem are located in roots.

stellacyanin An ELECTRON-TRANSFER PROTEIN, containing a TYPE 1 COPPER site, isolated from exudates of the Japanese lacquer tree.

stem cell Primitive blood cells, found in the bone marrow, circulating bloodstream, and umbilical cord, that are capable of reproducing and differentiating to make all varieties of mature blood cells (white blood cells, red blood cells, and platelets). This means that all blood cells originate from this single type of cell. PLURIPOTENT STEM CELLS are also usually referred to as stem cells.

stereochemical Refers to the three-dimensional view of a molecule.

stereoisomer A special case of Isomerism created by a difference in the spatial arrangement of atoms without any difference in connectivity or bond multiplicity between the isomers.
   See also CONFIGURATION; CONFORMATION.

steroid hormone or drug A large family of structurally similar chemicals. Various steroids have sex-determining, anti-inflammatory, and growth-regulatory roles. Examples include corticosteroid and glucocorticoid.

Stevens-Johnson syndrome A severe allergic drug reaction characterized by blisters breaking out on the lining of the mouth, throat, anus, genital area, and eyes. A severe form of erythema multiform. Drugs that can cause this reaction include penicillin, antibiotics containing sulfa, barbiturates, and some drugs used to treat high blood pressure and diabetes.

stimulus The cause for making a nerve cell respond; can be in the form of light, sounds, taste, touch, smell, temperature, etc.

stochastic Mathematical probability; a stochastic model takes into account variations in outcome that are because of chance alone.

stock number *See* OXIDATION NUMBER.

stoma An opening to intercellular air spaces in the epidermis of leaves of plants, consisting of a microscopic pore surrounded by guard cells that open and close to regulate gas exchange and water loss.

strand The land bordering a body of water. The term is used in the Everglades for the isolated "islands" of trees and shrubs.

strict aerobe An organism that utilizes aerobic respiration and can survive only in an atmosphere of oxygen.

strict anaerobe An organism that cannot survive in an atmosphere containing any oxygen.

stroma The fluid matrix of the chloroplast in which the thylakoids are situated. Those specialized membrane structures where photosynthesis takes place.
   In human biology, the stroma is the thickest part of the cornea (450–600 microns, approximately 0.5 millimeters) and is located between both Bowman's and Decemet's membrane.

stromatolite Stromatolites are colonies of millions of single-cell cyanobacteria (formerly known as blue-green algae). Stromatolites are so far the oldest known fossils, dating back well over 3 billion years. They were the dominant life-form for 2 billion years and still persist today on the west coast of Australia. One of the largest exposures of fossilized stromatolites is the Petrified Sea Gardens in Saratoga Springs, New York. It is listed as one of America's National Historic Landmarks, designated in 1999 by the Secretary of the U.S. Department of the Interior. It is a 500-million-year-old ocean reef from a time when the land that is now Saratoga Springs was at the shore of a warm tropical sea.
   Stromatolites are also known as "stone cabbage," based on their fossil appearance.

This view of a stromatolite field ground by glaciers shows the concentric ring formation created over millions of years of deposition by blue-green alga. *(Courtesy of Joseph Deuel, Petrified Sea Gardens, Inc.)*

structural formula   The structural formula of a compound shows how the atoms and bonds are arranged in the molecule; indicates the arrangement of the atoms in space.

structure-activity relationship (SAR)   The relationship between chemical structure and pharmacological activity for a series of compounds.

structure-based design   A DRUG design strategy based on the three-dimensional structure of the target obtained by X ray or NMR.

structure-property correlations (SPC)   All statistical mathematical methods used to correlate any structural property to any other property (intrinsic, chemical, or biological); based on statistical regression and PATTERN RECOGNITION techniques.

subspecies   A geographical isolate of a species. The federally listed endangered Karner blue butterfly (*Lycaeides melissa samuelis*) is a subspecies of the Melissa blue butterfly (*Lycaeides melissa*).
    See also SPECIES.

substrate   A substrate can be a chemical species of particular interest, the reaction of which with some other chemical reagent is under observation (e.g., a compound that is transformed under the influence of a catalyst). It can also be the chemical entity whose conversion to a product or products is catalyzed by an ENZYME. Also a solution or dry mixture containing all ingredients that are necessary for the growth of a microbial culture or for product formation. Finally, it can be a component in the nutrient medium, supplying the organisms with carbon (C-substrate), nitrogen (N-substrate), etc.

substrate-level phosphorylation   The formation or synthesis of the energy source ATP (adenosine triphosphate) by transferring an inorganic phosphate group to ADP (adenosine diphosphate).

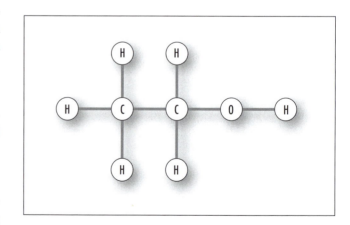

The structural formula of ethanol.

subunit An individual polypeptide chain in a protein containing more than one polypeptide chain. Different types of subunits are frequently designated by α, ß, γ, etc.

succession The process whereby species are replaced over time in an ecological community. Primary succession is when new communities develop on newly exposed habitat that has no previous life forms, such as on bare rock. Secondary succession is when communities take over an area that previously supported life, such as a recently burned forest or abandoned agricultural land. Succession usually occurs in one direction and often can be predicted. However, predictable succession can be altered, such as in fire disclimax communities like pine barrens, where fire periodically burns out species that would, if left unchecked, succeed to a new community type.

succulent A xeric-adapted plant that has fleshy stems, roots, or leaves that store water, thus preventing desiccation in harsh, dry environments, e.g., cactus, aloe.

sugar An organic compound that has the general chemical formula $(CH_2O)n$. All carbohydrates are sugars.

sulfite reductase ENZYMES that catalyze the reduction of sulfite to sulfide. All known enzymes of this type contain SIROHEME and IRON-SULFUR CLUSTERS.

superhyperfine *See* ELECTRON PARAMAGNETIC RESONANCE SPECTROSCOPY.

superior ovary (plants) *See* HYPOGYNOUS.

superoxide dismutases (SOD) ENZYMES that catalyze the dismutation reaction of superoxide anion to dihydrogen peroxide and dioxygen. The enzymes have ACTIVE SITES containing either copper and zinc (Cu/Zn-superoxide dismutase), or iron (Fe-superoxide dismutase), or manganese (Mn-superoxide dismutase).

See also DISPROPORTIONATION.

suppressor T cell (T8 cell; CD8 cell) The existence of these cells is a relatively recent discovery, and their function is still somewhat debated. The basic concept of suppressor T cells is a cell type that specifically suppresses the action of other cells in the immune system, notably B cells and T cells, thereby preventing the establishment of an immune response. How this is done is not known with certainty, but it seems that certain specific antigens can stimulate the activation of the suppressor T cells. This suppressor effect is thought to be mediated by some inhibitory factor secreted by suppressor T cells. The factor is not any of the known lymphokines. Another fact that renders the study of this cell type difficult is the lack of a specific surface marker. Most suppressor T cells are CD8-positive, as are cytotoxic T cells.

supramolecular chemistry This is defined as the chemistry of molecular assemblies and of the intermolecular bond—"chemistry beyond the molecule"—bearing on the organized entities of higher complexity that result from the association of two or more chemical species held together by intermolecular forces. Thus, supramolecular chemistry can be considered to represent a generalized COORDINATION chemistry extending beyond the coordination of TRANSITION ELEMENTS by organic and inorganic LIGANDs to the bonding of all kinds of SUBSTRATEs: cationic, anionic, and neutral species of either inorganic, organic, or biological nature.

surface tension The cohesive forces between liquid molecules. Surface tension is typically measured in dyne/cm, the force in dynes required to break a film of length 1 cm. Water at 20°C has a surface tension of 72.8 dyne/cm. Surfactants act to reduce the surface tension of a liquid.

survivorship curve Summarizes the pattern of survival in a population. There are basically three types of survivorship curves: Type I applies to species having a high survival rate of the young, where most of the individuals of the population survive to maturity and die in old age, such as humans. Type II species have a relatively constant death rate throughout their life span, such as fish and large birds. Type III curves are found in species that have many young, most of which die very early in their life, such as plants. Most type III offspring die before they reach reproductive age.

suspension Particles mixed with, but undissolved, in a fluid or solid.

suspension feeder (suspensivores) An organism that obtains food by capturing organic matter suspended in the water. Those that use a filter to capture food are called filter feeders. A mucous-bag suspension feeder uses a sheet or bag of mucous to trap particles, while a tentacle-tube-foot suspension feeder traps particles on distinct tentacles or tube feet.

sustainable A process that can continue indefinitely without overusing resources and causing damage to the environment.

sustainable agriculture Agricultural techniques and systems that, while economically viable to meet the need to provide safe and nutritious foods, utilize nonenvironmentally destructive methods such as the use of organic fertilizers, biological control of pests instead of pesticides, and minimized use of nonrenewable fuels.

sustainable development Development and economic growth that meets the requirements of the present generation without compromising the ability of future generations to meet their needs. Seeks a balance between development and conservation of natural resources.

Sutherland, Earl W., Jr. (1915–1974) American *Pharmacologist* Earl Sutherland was born on Novem-

ber 19, 1915, in Burlingame, Kansas. He received a B.S. from Washburn College in 1937 and an M.D. from Washington University School of Medicine in St. Louis in 1942.

Sutherland joined the faculty of Washington University, and in 1953 he became director of the department of medicine at Western Reserve University in Cleveland, Ohio. Here he discovered cyclic AMP (adenosine monophosphate) in 1956.

In 1963 he became professor of physiology at Vanderbilt University in Nashville, Tennessee. He was awarded the 1971 Nobel Prize in physiology or medicine for isolation of cyclic adenosine monophosphate (cyclic AMP) and for demonstrating its involvement in numerous animal metabolic processes. From 1973 until his death he was a member of the faculty of the University of Miami Medical School.

He died on March 9, 1974, in Miami, Florida. The Earl Sutherland Prize award is presented annually by the chancellor of Vanderbilt University to a Vanderbilt faculty member who has made a nationally recognized impact in a particular discipline.

swim bladder (air bladder) An organ—a gas-filled sac—that lies in the upper body cavity of many bony fishes; functions as a ballast tank by creating buoyancy; also used in respiration and as a sound organ.

symbiont One of the members of a symbiotic relationship.

symbiosis A mutually beneficial relationship between two or more different kinds of organisms in direct contact with each other.
See also MUTUALISM.

sympathetic division One of two divisions of the autonomic nervous system—the sympathetic and parasympathetic—that regulates involuntary functions. The sympathetic division prepares the body for survival response while under stress and increases energy use; autonomic motor neurons release the neurotransmitter norepinephrine, which accelerates heartbeats, constricts or dilates blood vessels, dilates the bronchi, and inhibits digestion. The sympathetic division opposes the actions of the parasympathetic division.

sympatric speciation   The creation of a new species within a geographical area caused by reproductive isolation among individuals coexisting in the area.

symplast   The interconnected protoplasm between cells in a plant due to the presence of pores between the cells that allow cytoplasmic flow (plasmadesmata).

symplesiomorph   A character state shared by two or more taxa thought to have originated as an evolutionary novelty in an earlier ancestor.

symptom   A subjectively perceived problem or complaint reported by an individual. For example, a rash is a symptom that the immune system is reacting to something such as dust.

syn   *See* ANTI.

synapomorphy   A derived or specialized character (apomorphy) that is shared by two or more groups that originated in their last common ancestor; used to infer common ancestry.

synapse   A gap or junction between the ends of two neurons in a neural pathway where nerve impulses pass from one to the other; an impulse causes the release of a neurotransmitter at the synapse that diffuses across the gap and triggers the next neuron's electrical impulse.
    See also NEURON.

synapsis   The alignment of chromosomes during prophase of the first division in meiosis so that each chromosome is next to its homologue; the time when crossing over occurs.

synaptic terminal (**synaptic vesicle**)   A membrane or bulb at the end of an axon terminal that stores and releases neurotransmitters.

syndrome   A group of symptoms and diseases that together are characteristic of a specific condition.

synergism   The combination of two or more substances in which the sum result of this combination is greater than the effects that would be exhibited from each of the individual substances alone.

syngamy   When male and female gametes meet and join. Same as sexual reproduction.

synonyms   One of several different scientific names that are spelled differently but refer to the same species. An example is *Epilobium canum* (E. Greene) Raven and *Zauschneria californica* K. Presl. The older name is the senior synonym and is how the species thenceforth is referred once the synonymy has been recognized. The more recent name is the junior synonym and its use is discontinued.

synthase   An ENZYME that catalyzes a reaction in which a particular molecule is synthesized, not necessarily by formation of a bond between two molecules (contrast with synthetase).
    See also LIGASE.

synthetase   *See* LIGASE.

system   Set of organs in the body with a common structure or function. Examples of body systems include the immune, gastrointestinal, respiratory, and lymphatic systems.

systematics   The study of diversity, evolutionary, and genetic relationships among organisms.

systemic   Relating to or affecting the whole body.

systemic acquired resistance (SAR)   A defensive response in infected plants characterized by an activation of a number of defense mechanisms. The mechanisms are activated at the site where the pathogen attacks and in tissues untouched by the pathogen. SAR provides resistance against a host of organisms such as fungi, bacteria, and viruses. Also called induced resistance.

systole   Contraction of the heart; the period when ventricles contract, pushing blood from the left ventricle into the aorta and from the right ventricle into the pulmonary artery. Systolic pressure is when the blood pressure is measured during contraction of the heart. In blood pressure readings, it is normally the higher of the two measurements.

Szent-Györgyi, Albert von (1893–1986) Hungarian *Biochemist*   Albert von Szent-Györgyi was born in Budapest on September 16, 1893, to Nicolaus von Szent-Györgyi, and Josefine, whose father, Joseph Lenhossék, and brother Michael were both professors of anatomy in the University of Budapest.

He took his medical degree at the University of Sciences in Budapest in 1917, and in 1920 became an assistant at the University Institute of Pharmacology in Leiden. From 1922 to 1926, he worked with H. J. Hamburger at the Physiology Institute, Groningen, Netherlands. In 1927, he went to Cambridge as a Rockefeller fellow, working under F. G. HOPKINS, and spent one year at the Mayo Foundation, Rochester, Minnesota, before returning to Cambridge. In 1930 he became chair of medical chemistry at the University of Szeged, and five years later he also was chair in organic chemistry. At the end of World War II, he was chair of medical chemistry at Budapest.

Szent-Györgyi's early researches concerned the chemistry of cell respiration. He pioneered the study of biological oxidation mechanisms and proved that hexuronic acid, which he isolated and renamed ascorbic acid, was identical to vitamin C and that it could be extracted from paprika. He won the 1937 Nobel Prize in physiology or medicine for his discoveries, especially of vitamin C.

In the late 1930s his work on muscle research quickly led him to discover the proteins actin and myosin and their complex. This led to the foundation of muscle research in the following decades. He also worked on cancer research in his later years. In 1947, he moved to the United States, where he became director of research, Institute of Muscle Research, Woods Hole, Massachusetts.

His publications include *Oxidation, Fermentation, Vitamins, Health and Disease* (1939), *Muscular Contraction* (1947), *The Nature of Life* (1947), *Contraction in Body and Heart Muscle* (1953), *Bioenergetics* (1957), and *The Crazy Ape* in 1970, which was a commentary on science and the future of the human race. He died on October 22, 1986.

T

taiga The far northern (50 to 60 degrees north latitude) open region of the boreal forest encompassing parts of Siberia, Eurasia, Scandinavia, and North America, just south of the tundra. Characterized by open coniferous woodlands composed of tamarack, black or white spruce, fir, pine and larch, and a floor of lichens. The boreal forests represent the largest terrestrial biome where seasons alternate between long, cold winters and short, cool summers. These forests have acidic, thin soils and a growing season of 130 days. The average temperature is below freezing for half of the year.

tapeworms One of the cestode worms (Cestoda), flatworms that are internal parasites.

tarantula A large hairy spider (arachnid) that lives in warm areas such as South America, southern North America, southern Europe, Africa, southern Asia, and Australia, with the greatest concentration of tarantulas in South America. There are about 300 species of tarantulas. *Aphonopelma* and *Cyrtopholis* are representative genera. The female can live over 30 years, but the life span of the male is considerably shorter.

tardigrade A microscopic animal that looks like a miniature bear, hence the name water bear. They have inflated round bodies with four pairs of stubby clawed legs used for walking. If the environment dries up, they have the ability to withstand hostile conditions by undergoing a process called anhydrobiosis (life without water), part of an adaptable technique of cryptobiosis.

They are not affected by changes in salinity, extreme vacuum, or lack of oxygen. They can survive freezing or thawing in temperatures as cold as −328°F or as hot as 304°F. When they turn into dried barrels called tuns, they are able to survive for long periods of time, and some have been brought back to life from 100-year-old mosses. Their typical environment is the thin film of water that coats mosses and lichens, but they have been found in a vast range of habitats. There may be up to 1,000 species.

tautonym A biological term where the genus and species are the same word, such as *Amoracia armoracia,* or *Rattus rattus* (the black rat). Triple tautonyms are common in biological nomenclature.

taxis The behavior exhibited when an organism reacts to an external stimulus by turning or moving away from the stimulus.

taxon A grouping of organisms by taxonomic rank, such as order, family, genus, etc.; also a group of organisms or other taxa sharing a single common ancestor.

taxonomy The name of the discipline given to the scientific study, naming, classification, and identification of organisms.

T cell (T-lymphocyte) A lymphocyte (white blood cell) that develops in the bone marrow, matures in the thymus, and expresses what appear to be antibody molecules on its surface. Unlike B cells, these molecules cannot be secreted. Also called a T-cell receptor (CD3 and CD4 or CD8). Works as part of the immune system in the body. Produces cytokine to help B lymphocytes produce immunoglobulin.

Several distinct T-cell subpopulations are recognized:

helper T cell (T4 cell, CD4 cell) A class of T cells that help trigger B cells to make antibodies against thymus-dependent antigens. Helper T cells also help generate cytotoxic T cells.

T killer cells (cytotoxic T cells) Cells that kill target cells bearing appropriate antigen within the groove of an MHC (major histocompatibility complex) class I molecule that is identical to that of the T cell.

suppressor T cell (T8 cell, CD8 cell) The existence of these cells is a relatively recent discovery, and their function is still somewhat debated. The basic concept of suppressor T cells is a cell type that specifically suppresses the action of other cells in the immune system, notably B cells and T cells, thereby preventing the establishment of an immune response. How this is done is not known with certainty, but it seems that certain specific antigens can stimulate the activation of the suppressor T cells. This suppressor effect is thought to be mediated by some inhibitory factor secreted by suppressor T cells. The factor is not any of the known lymphokines. Another fact that renders the study of this cell type difficult is the lack of a specific surface marker. Most suppressor T cells are CD8-positive, as are cytotoxic T cells.

telomerase (telomere terminal transferase) An enzyme, known as an immortalizing enzyme, that is composed of RNA and proteins that uses its RNA as a template to synthesize telomeric DNA onto the ends of chromosomes; acts as a reverse transcriptase, adding telomeres to the chromosome ends when activated.

Found in germ and inflammatory cells, fetal tissue, and tumor cells. When cells become cancerous, telomerase is activated, and the cells can replicate without limits in a process called immortalization.

telomere The protective and stabilizing ends of linear chromosomes in eukaryotes that are involved in the replication and stability of DNA molecules. Telomeres erode slightly with each cell division unless TELOMERASE is activated.

telophase The last stage in meiosis. Telophase I is the stage where the migration of the daughter chromosomes to the two poles is completed. In telophase II, the last stage of meiosis, a nuclear membrane forms around each set of chromosomes, and cytokinesis (division of the cytoplasm) takes place.

temperate deciduous forest Located in the midlatitude areas between the polar regions and the tropics. This biome has four seasons, the result of being exposed to warm and cold air masses. The temperature varies widely from season to season, with cold winters and hot, wet summers. Most of the trees are broadleaf trees such as oak, maple, beech, hickory, and chestnut. Animals have adapted to the seasons through a variety of strategies, including hibernating, migrating, or staying active.

temperate virus A virus that does not necessarily kill the host by lysis but that reproduces in synchrony with the host, called a lysogen. When this process is occurring the virus is called a prophage.

temperature A measure of the energy in a substance. The more heat energy in the substance, the higher the temperature. A number of temperature scales have evolved over time, but only three are used presently: Fahrenheit, Celsius, and Kelvin. The Fahrenheit temperature scale is a scale based on 32 for the freezing point of water and 212 for the boiling point of water, the interval between the two being divided into 180 parts. This scale is named after its inventor, the 18th-century German physicist Daniel Gabriel FAHRENHEIT. The Celsius or centigrade temperature scale is based on 0 for the freezing point of water and 100 for the boiling point of water. This scale was invented in 1742 by the Swedish astronomer Anders CELSIUS. It was once called the centigrade scale because of the 100-degree interval between the defined points. The Kelvin temperature scale is the base unit of thermodynamic temperature measurement and is defined as 1/273.16 of the triple point (equilibrium among the solid, liquid, and gaseous phases) of pure water. The Kelvin is also the base unit of the Kelvin scale, an absolute-temperature scale named for the British physicist William Thomson, Baron Kelvin. Such a scale has its zero point at absolute zero, the theoretical temperature at which the molecules of a substance have their lowest energy.

tendon (sinew) A fibrous connective tissue that attaches muscles to bones; a collagen-rich substance.

tendril A slender outgrowth of a plant stem that can clasp or wind around and help support climbing plants.

tepuis A unique ecological region in South America (Venezuela, Guyana, Suriname, and Brazil) of about 19,000 square miles, famous for isolated mountains, each called a tepui, with flat tops covered with species that may not exist elsewhere. Thirty three percent of the 2,300 vascular plants found there are endemic.

teratogen A substance that produces a malformation in a fetus.

territory An area within the living range of an individual animal that it will defend against intruders over food or protection of its young.

tertiary consumer Part of the food chain in which the tertiary consumer (e.g., hawk) eats the secondary consumer (e.g., snake), which eats the primary consumer (e.g., mouse), which eats a producer (e.g., grass). Carnivores that eat mainly other carnivores.

Tertiary period The first period of the Cenozoic era (after the Mesozoic era and before the Quaternary period), extending from 65 million to 1.8 million years ago.
See also GEOLOGICAL TIME.

tertiary structure The overall three-dimensional structure of a BIOPOLYMER. For proteins, this involves the side-chain interactions and packing of SECONDARY STRUCTURE motifs. For NUCLEIC ACIDS, this can be the packing of stem loops or supercoiling of DOUBLE HELIXES.

testcross A cross of an individual with a potentially ambiguous or unknown genotype to a homozygous recessive individual.

testis (testes; testicle) The male reproductive organ, gonad, where sperm is and hormones (testosterone) are produced; located inside the scrotum, behind and below the penis.
See also SPERM.

testosterone Male androgen hormone that regulates male sexual characteristics along with dihydrotestosterone. Testosterone is converted to dihydrotestosterone by the enzyme 5-alpha-reductase. Testosterone is converted to estradiol (the most potent estrogen, formed in both the ovary and the testes) by the enzyme 17-ketoreductase.

tetanus (lockjaw) A disease caused by the release of exotoxins by a bacterium (*clostridium tetani*) that causes muscular spasm, contraction, and convulsions; obtained through a contaminated wound, e.g., from stepping on a rusting nail. Also any tense muscle contraction by rapid and frequent action potentials (like electric shocks). Approximately 30 percent of reported cases of tetanus end in death. Tetanus kills 300,000 newborns and 30,000 birth mothers worldwide, from lack of immunization.

tetrahedron *See* COORDINATION.

tetrahydrofolate Reduced FOLATE derivative that contains additional hydrogen atoms in positions 5, 6, 7, and 8. Tetrahydrofolates are the carriers of activated one-carbon units and are important in the biosynthesis of amino acids and precursors needed for DNA synthesis.

See also FOLATE COENZYMES.

tetrapod Any vertebrate organism that has four limbs, or two sets of limbs, e.g., mammals, reptiles, birds, amphibians.

thalamus Consists of two egg-shaped masses at the base of the brain that form the dorsal subdivision of the diencephalon (forebrain) and that function together as a unit. The thalamus acts as switching relay, receiving sensory data from the nervous system and escorting it to the cerebral cortex and other parts of the brain. In botany, it is the receptacle or torus of a flower.

thalassemia A chronic inherited disease characterized by defective synthesis of HEMOGLOBIN. Defective synthesis of the α chain of hemoglobin is called α-thalassemia, and defective synthesis of the ß chain of hemoglobin is called ß-thalassemia. Thalassemias result in anemia that can be severe and are found more frequently in areas where malaria is endemic.

Theiler, Max (1899–1972) South African *Microbiologist* Max Theiler was born on January 30, 1899, in Pretoria, South Africa, to Sir Arnold Theiler, a well-known veterinary scientist, and Emma Theiler (née Jegge). He attended Rhodes University College, Grahamstown, and the University of Capetown Medical School (1916–18). He then went to England to study at St. Thomas's Hospital and at the London School of Tropical Medicine, receiving a medical degree in 1922.

In 1922 he joined the department of tropical medicine at the Harvard Medical School, Boston, Massachusetts. In 1930 he joined the staff of the International Health Division of the Rockefeller Foundation. In 1951 he became director of laboratories of the Rockefeller Foundation's Division of Medicine and Public Health, New York.

His early work dealt with amoebic dysentery and rat-bite fever, but he wanted to produce a vaccine for yellow fever, at the time a major disease in humans. By 1927 he and his colleagues had showed that the cause of yellow fever was not a bacterium but a virus and was easily transmitted to mice, making research on the subject cheaper. Theiler and his colleagues produced a vaccine, 17D, against the disease. He was awarded the 1951 Nobel Prize in physiology or medicine for this breakthrough.

He contributed to two books, *Viral and Rickettsial Infections of Man* (1948) and *Yellow Fever* (1951), and wrote many scientific papers. He also received numerous awards, including the Chalmer's Medal of the Royal Society of Tropical Medicine and Hygiene (London, 1939). He died on August 11, 1972, in New Haven, Connecticut.

Theorell, Axel Hugo Theodor (1903–1982) Swedish *Biochemist* Axel Hugo Theodor Theorell was born in Linköping, Sweden, on July 6, 1903, to Thure Theorell, surgeon-major to the First Life Grenadiers practicing medicine in Linköping, and his wife Armida Bill.

Theorell was educated at a state secondary school in Linköping and started studying medicine in 1921 at the Karolinska Institute. In 1924 he graduated with a bachelor of medicine and spent three months studying bacteriology at the Pasteur Institute in Paris. He received an M.D. in 1930 and became lecturer in physiological chemistry at the Karolinska Institute.

In 1924 he became part of the staff of the Medico-Chemical Institution as an associate assistant and temporary associate professor working on the influence of the lipids on the sedimentation of the blood corpuscles. In 1931 at Uppsala University, he studied the molecular weight of myoglobin. The following year he was appointed associate professor in medical and physiological chemistry at Uppsala University, and he continued and extended his work on myoglobin.

From 1933 until 1935 he held a Rockefeller fellowship and became interested in oxidation enzymes. He produced, for the first time, the oxidation enzyme called "the yellow ferment," and he succeeded in splitting it reversibly into a coenzyme part, which was found to be flavinmononucleotide, and a colorless protein part.

Returning to Sweden in 1935, he worked at the Karolinska Institute, and in 1936 he was appointed head of the newly established biochemical department of the Nobel Medical Institute that was opened in 1937.

He carried out research on various oxidation enzymes, contributing to the knowledge of cytochrome-c, peroxidases, catalases, flavoproteins, and pyridin-proteins, particularly the alcohol dehydrogenases. For his work on the nature and effects of oxidation enzymes, he was awarded the 1955 Nobel Prize in physiology or medicine.

He was a member of numerous scientific organizations and in 1954 was chief editor of the journal *Nordisk Medicin*. He died on August 15, 1982.

therapeutic index For a substance used to alleviate disease, pain, or injury, the therapeutic index is the ratio between toxic and therapeutic doses (the higher the ratio, the greater the safety of the therapeutic dose).

thermolysin A calcium- and zinc-containing neutral protease isolated from certain bacteria.

thermolysis An uncatalyzed bond cleavage resulting from exposure of a compound to a raised temperature.

thermoregulation The process of regulating body temperature. There are various types of thermoregulation. Ectothermic regulation is temperature based on an organism's behavior in response to the temperature of the environment, e.g., amphibians and reptiles. Heterothermic regulation is maintaining a constant body temperature but having the ability to fluctuate at other times. The body temperature can rise without hurting the animal, or it can drop significantly when the animal enters a state of torpor, with body temperature dropping almost to the level of the surroundings and with a concurrent reduction in metabolic rate, heart rate, respiration, and other functions. This can be in the form of hibernation during winter or estivation during summer.

thick filament A myosin filament and part of the myofibril in muscles. Contains myosin molecules.

thigmomorphogenesis The process in which a plant reacts to a mechanical disturbance, such as forming thicker stems in response to strong winds.

thigmotropism The sense of touch response, such as bending or turning of a plant when coming into contact with a solid surface.

thorium-lead dating A method to measure the age of rocks and other materials that contain thorium and lead through the use of the natural radioactive decay of ^{232}Th (half-life about 4.7 billion years) as it decays to ^{208}Pb. Because of the very long half-life, this dating technique is usually restricted to ages greater than 10 million years.

threatened species The classification provided to an animal or plant likely to become endangered within the foreseeable future throughout all or a significant portion of its range.

See also ENDANGERED SPECIES.

three-dimensional quantitative structure-activity relationship (3D-QSAR) The analysis of the quantitative relationship between the biological activity of a set of compounds and their spatial properties using statistical methods.

threshold potential The potential at which a neuron will fire and then reset to its reset potential; when a neuron receives inputs that increase its voltage from its resting voltage and reaches its threshold voltage, it sends out an action potential.

thylakoids Enclosed membrane structures inside CHLOROPLASTs and photosynthetic bacteria.

See also PHOTOSYNTHESIS.

thymus A major component of the lymphatic system. In the thymus, lymphoid cells undergo a process of maturation and education prior to release into circulation.

This process allows T cells to develop self-tolerance (distinguishing self from nonself). While developing in the thymus gland, any T cell that reacts to the thymus's major histocompatibility complex (MHC) is eliminated. T cells that tolerate the MHC learn to cooperate with cells expressing MHC molecules and are allowed to mature and leave the thymus.

The result is that surviving T lymphocytes tolerate the body's cells and cooperate with them when needed. However, some T lymphocytes lose this ability to differentiate self from nonself, which results in the autoimmune diseases such as systemic lupus erythematosus or multiple sclerosis.

thyroid gland   A two-lobed endocrine gland located at the base of the neck that secretes hormones such as triiodothyronine (T3), thyroxine (T4), and calcitonin, a hormone produced in the C cells that helps regulate blood calcium by slowing down the amount of calcium released from the bones; controls the rate of metabolism. An enlarged thyroid gland is called goiter and is treated with thyroid hormones.

thyroid-stimulating hormone (TSH)   TSH is a hormone secreted by the pituitary gland that stimulates the synthesis and secretion of T4 (thyroxine) and T3 (triiodothyronine) by the thyroid gland. TSH is itself stimulated by another hormone, thyroid-releasing hormone (TRH), which is released by the hypothalamus.

tick paralysis   Ticks produce a toxin that can cause paralysis in animals, especially if the bite is near the spine and causes ascending paralysis. It can be reversed if the offending tick is removed. Can be fatal if not removed.

tight junction   A cell-to-cell junction that forms a fluid-tight seal between cells, preventing leakage or passage of molecules from one side to the other.

Ti plasmid   Short for tumor-inducing plasmid. It is a plasmid of *Agrobacterium tumefaciens* that is responsible for inducing tumors in infected plants. These plasmids are also used as vectors to introduce foreign DNA into plant cells.

tissue   A grouping or layer of cells that, with their products and intercellular material, together perform specialized and specific functions. Examples include connective, epithelial, glandular, muscular, nervous, and skeletal tissue.

tobacco mosaic virus   This virus was the first recognized viral disease. Caused by a tobamovirus, the tobacco mosaic virus (TMV), it attacks the leaves of many plants, causing a mosaiclike pattern of discolorations in the plant.

tonoplast (vascular membrane)   A cytoplasmic membrane that surrounds the central vacuole separating the cell sap from the cytosol.

topless tree   An operational scheme for ANALOG design.

torpor   A physiological response resulting in a drop in body temperature almost to that of the surrounding environment and a large reduction in the metabolic rate, heart rate, respiration, and other functions. Hibernation is winter torpor and estivation is summer torpor.

totipotency   The ability of a single cell to develop unlike cells that will develop into a fully differentiated organism or part.

toxicity   The action of poisons (including XENOBIOTICS) on biochemical reactions or processes in living organisms or ecological systems. A study of this action is the subject matter of toxicology.

toxin   A poisonous material that can cause damage to living tissues.

trace elements Elements required for physiological functions in very small amounts that vary for different organisms. Included among the trace elements are Co, Cu, F, Fe, I, Mn, Mo, Ni, Se, V, W, and Zn. Excessive mineral intake can produce toxic symptoms.

trachea A linear portion of the respiratory tract that connects the larynx to the bronchial tubes, forming an inverted Y; part of it is in the throat, and the other is in the chest. Called the windpipe.

tracheae Tiny tubes that branch throughout the body cavity of insects that carry air and exchange gases.

tracheal system The part of the insect's respiratory system composed of the tracheae and tracheole (small end tubes formed within single cells of the tracheal epithelium); opens to the outside of the body through spiracles.

tracheid A thick-walled cell of the xylem tissue that is water conducting and is thickened and hardened by lignin; also helps in supporting the plant.

trans In inorganic nomenclature, a structural prefix designating two groups directly across a CENTRAL ATOM from each other (not generally recommended for precise nomenclature purposes of complicated systems).
See also CIS.

transcription The process by which the genetic information encoded in a linear SEQUENCE of NUCLEOTIDES in one strand of DNA is copied into an exactly complementary sequence of RNA.
See also GENETIC CODE.

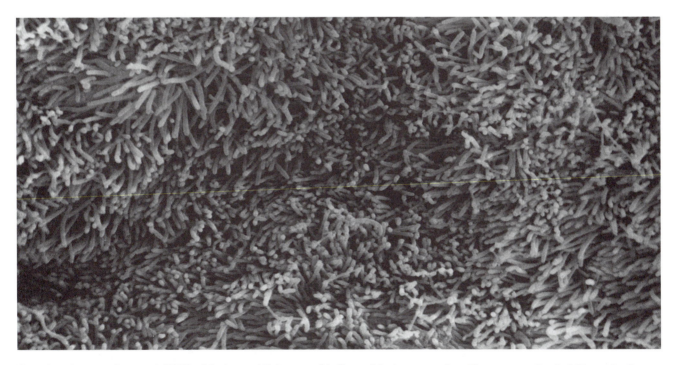

Scanning electron micrograph (SEM) of the internal lining, or epithelium, of the human trachea. The numerous fine hairlike projections, called cilia, protrude from the tops of specialized epithelial cells. Between the epithelial cells lie goblet cells, whose function is to release mucus onto the surface of the epithelium. Rhythmic movements of the cilia serve to move bacteria and other particles caught in the mucus away from the gas-exchanging parts of the lungs, up the trachea, and toward the throat, where they can be swallowed or coughed up. Magnification: ×4000 at 10 × 8 inch size. *(Courtesy © E. Gray/Photo Researchers, Inc.)*

transcription factor The regulatory protein that binds to DNA and stimulates the transcription of genes.

transduction (1) The transfer of genetic information from one bacterium to another by means of a transducing bacteriophage. When the phage is grown on the first host, a fragment of the host DNA can be incorporated into the phage particles. This foreign DNA can be transferred to the second host upon infection with progeny phage from the first experiment. (2) In cell biology, the transduction of a signal (mechanical signal, hormone, etc.) to cells or tissues summarizes the chain of events between the primary reception of the signal and the final response (change in growth and/or METABOLISM) of the target cells or tissues. Inorganic substances (e.g., calcium ions) are frequently involved in the transduction of signals.

See also GENETIC CODE.

transferase An ENZYME of EC class 2, which catalyzes the transfer of a group from one SUBSTRATE to another.

See also EC NOMENCLATURE FOR ENZYMES.

transferrin An iron-transport protein of blood PLASMA, that comprises two similar iron-binding DOMAINS with high affinity for Fe(III). Similar proteins are found in milk (lactoferrin) and eggs (ovotransferrin).

transfer RNA (tRNA) RNA that translates the nucleotide language of codons into the amino acid language of proteins. The tRNAs bond with amino acids and transfer them to the ribosomes, where the new proteins are assembled based on the genetic code carried by mRNA.

transformation The assimilation of external genetic material into a cell.

transition element A transition element is an element whose atom has an incomplete d-subshell, or that gives rise to a cation or cations with an incomplete

d-subshell. The first transition series of elements is Sc, Ti, V, Cr, Mn, Fe, Co, Ni, and Cu. The second and third transition series are similarly derived: these include the lanthanoids (lanthanides) and actinoids (actinides), respectively, which are designated inner (or **f**) transition elements of their respective periods in the periodic table.

transition-state analog A compound that mimics the transition state of a substrate bound to an ENZYME.

translation The unidirectional process that takes place on the RIBOSOMEs whereby the genetic information present in an mRNA is converted into a corresponding SEQUENCE of amino acids in a protein.

See also GENETIC CODE.

translocation A mutation that occurs and that moves a piece of chromosome to a different chromosome. A balanced translocation happens when two chromosomes break and exchange places, leaving no net loss of genetic material. While an individual with a balanced translocation will be unaffected, future children may be. A Robertsonian translocation occurs when the translocations of chromosomes involve an end-to-end fusion with the resulting loss of the short arms. While the carrier has 45 chromosomes and is normal, resulting children may be affected. A reciprocal translocation happens when single breaks in two nonhomologous chromosomes produce an exchange of chromosome sections between them. An insertion is a form of translocation where an interstitial segment from one chromosome that resulted from two breaks is inserted within a nonhomologous chromosome at a third break point.

Translocation is also the term used for the transport of food by way of the phloem.

transpiration The loss of water from a plant or tree by way of the stomata.

transposon (transposable element) A mobile piece of DNA that can move and be inserted into a new location (transposition). Formerly called jumping genes.

trapping The interception of a reactive molecule or reaction intermediate so that it is removed from the system or converted into a more STABLE form for study or identification.

tree A perennial plant that grows from the ground up, usually with a single, mostly tall, woody, self-supporting trunk or stem with bark, and an associated elevated crown of branches and foliage. May shed leaves or keep needles yearly.

trench fever A bacterial infection that causes repeated cycles of high fever. Common in World War I and II, due to bacteria that was passed among soldiers through contact with body lice, overcrowding, and bad hygiene conditions among soldiers in trenches. Two different bacteria can cause trench fever: *Bartonella quintana* and *Bartonella henselae*. *B. quintana* is carried by body lice; *B. henselae* is carried by ticks.

Triassic The earliest period of the Mesozoic era (248 million to 213 million years ago).
See also GEOLOGICAL TIME.

trichinosis A disease caused by the roundworm (nematode) *Trichinella spiralis*. Caused by eating unprepared or improperly cooked meats, particularly pork products.

trichocyst A structure in the outer cytoplasm of ciliates and dinoflaggelates that produces hairlike or dartlike fibers that can be discharged for defensive/offensive purposes in organisms, e.g., paramecium.

trilobite Ancient class (Trilobita) of hard-shelled, segmented creatures with jointed legs; one of the first arthropods, they existed over 300 million years ago in the Earth's ancient seas. There are nine orders, more than 150 families, 5,000 genera, and over 15,000 described species. Mostly detritivores, predators, or scavengers, the trilobites were one of the most diverse groups ever known.

triplet code A code in which a given amino acid is specified by a set of three nucleotides (codons).
See also CODON.

triploblastic Organisms formed with three cell layers, the endoderm, mesoderm, and ectoderm.

triploidy A rare chromosomal abnormality where three complete sets of the haploid genome exist instead of the normal two sets. In humans there are 69 chromosomes instead of the normal 46 chromosomes. Usually results in miscarriage or death in the first few hours or days after birth.

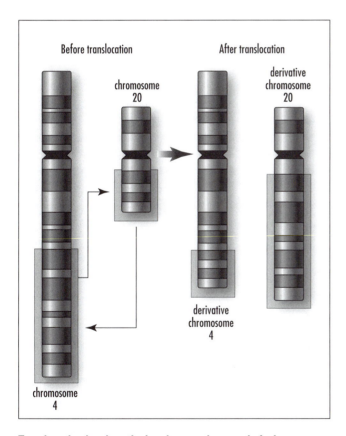

Translocation involves the breakage and removal of a large segment of DNA from one chromosome, followed by the segment's attachment to a different chromosome. *(Courtesy of Darryl Leja, NHGRI, National Institutes of Health)*

troop A group of lemurs, monkeys, apes, or other primates.

trophallaxis The mutual exchange of alimentary liquid among colony members, such as between adult and larvae of certain social insects, e.g., ants, bees, or wasps.

trophic level Serves as a way to delineate levels on the food chain from producers to consumers. Some organisms like humans are multitrophic.

trophic structure The distribution of the energy flow and its relationships through the various trophic levels.

trophoblast The outer layer of epithelium of the blastocyst that gives rise to the mammalian placenta and in which the embryo receives nourishment from the mother. The trophoblast differentiates into the syncytiotrophoblasts, the outer large cell layer that produces HCG (human chorionic gonadotropin) that stimulates the production of estrogen and progesterone within the ovary, and cytotrophoblast, the compacted cell inner layer next to the mesoderm. The trophoblast attaches itself to the uterus wall in a process called implantation and goes on to become the various life support systems (placenta, amniotic sac, umbilical cord), while the blastocyst develops into the embryo.

tropical rain forest A warm, moist terrestrial biome with a tree canopy. Two types of rain forest exist, the temperate and the tropical. Tropical rain forests are found close to the equator, while temperate rain forests are found near coastal areas, such as the American Pacific Northwest stretching from Oregon to Alaska for 1,200 miles. Characteristics of rain forests are trees that flare at the base, with very dense vegetation that is tall and green and rich in animal species. Both types receive lots of rain, with the tropical rain forest obtaining 400 inches a year, while the temperate rain forest receives about 100 inches per year. Tropical rain forests are warm and moist, but temperate rain forests are cool.

Tropical rain forests lie between the tropic of Cancer and the tropic of Capricorn and receive rainfall regularly throughout the year (80–400 inches per year) and are warm and frost free all year long (mean temperatures are between 70 and 85°F), with very little daily fluctuation. Tropical rain forests are found in Latin America (57 percent), while one-third of the world's tropical rain forests are in Brazil. Some 25 percent of the tropical rain forests can be found in Southeast Asia and the Pacific Islands, and 18 percent are found in West Africa.

Smaller temperate rain forests can be found on the southeast coast of Chile in South America and a few other coastal strips, including small areas in the United Kingdom, Norway, Japan, New Zealand, and southern Australia.

Half of the world's plant and animal species live in the tropical rain forests, even though they comprise only 20 percent of all forests. These richly diverse ecosystems are being destroyed at a rate of 30 acres of trees being cut every minute.

tropic hormone Hormones that cause secretion of other hormones.

tropism The movement of a plant toward (positive tropism) or away (negative tropism) from an environmental stimulus by elongating cells at different rates. Phototropism is induced by light; gravitropism is induced by gravitational pull; hydrotropism is a response to water gradients; and thigmotropism is a response to touch.

tsetse flies Vectors of African trypanosomiasis, causing nagana in livestock, a fatal disease of horses and cattle, and sleeping sickness in humans. Members of the Glossinidae; there are 23 species and eight subspecies. The simple proboscis of tsetse flies projects straight out front from the head, consisting only of labrum, hypopharynx, and labium. Found only in sub-Saharan Africa and two localities in the Arabian Peninsula.

tuber A short, swelled or enlarged, fleshy underground part of a stem or root used as food storage by the plant, e.g., potato.

tuberculosis (TB) A contagious bacterial infection caused by bacteria that are members of the *Mycobacterium* genus, usually by *Mycobacterium tuberculosis,* that usually affects the lungs or other body organs. Can be spread by person-to-person contact through watery airborne droplets of coughs and in the mucus that infected persons cough up from their throats. Some 10 to 20 million Americans have been infected, and one-third of the world's population. Can be treated with medicines, and a vaccine is available.

tularemia An infectious disease with high fever and other complications caused by the bacterium *Pasteurella tularensis,* which affects mostly people who handle infected wild rabbits, although it can be transmitted by ticks, contaminated food, water, and other animals. Mortality rate is about 6 percent.

tumor (neoplasm) A mass of abnormal tissue that arises from normal tissue; grows abnormally in rate and structure and serves no physiological function; abnormal regulation of cell growth.

Tumors can be benign, characterized by being slow-growing and harmless, depending on the location; malignant, characterized by being fast-growing, likely to spread, and capable of damaging other organs or systems; or intermediate, which is a mixture of benign and malignant cells.

Tumors can be caused by abnormalities of the immune system, radiation, genetic mutations, viruses, sunlight, tobacco, benzene and other mutagens, poisonous mushrooms, and aflatoxins.

Also called neoplasm, the preferred medical term.

tumor suppressor gene A normal gene whose purpose is to suppress cell growth or proliferation (cancer). However, if a mutation, inactivation, or deletion of part or all of one allele of a tumor suppressor gene occurs, it predisposes the individual to increased risk of tumor or cancer development.

tundra An area of flat or undulating treeless barrens found across the northern parts of Asia, Europe, and North America, covering over one-fifth of the earth's surface. Covered with lichens, sedges and grasses, mosses, and low shrubs (dwarf willows and birches), it is characteristic of the Arctic and alpine (high-altitude) regions. Below the tundra soil is permafrost, a permanently frozen layer of ground. During brief summers, the top section of the soil thaws long enough for plants and microorganisms to grow and reproduce. Animals found in the tundra include the Arctic fox, caribou, golden eagle, gray wolf, musk ox, Norway lemming, polar bear, red phalarope, ruddy turnstone, and Eskimos and their sled dogs. The average annual temperature is only 10 to 20°F (−12°C to −6°C), with temperatures sometimes dipping below −60°F (−51°C) and the annual precipitation usually less than 10 inches (25 centimeters).

turgid Swollen, rigid, or firm. Cells become turgid upon entry of water.

turgor pressure Like air pressure in a car tire, it is the outward pressure that is exerted against the inside surface of a plant cell wall under the conditions of water flowing into the cell by osmosis, and the resulting resistance by the cell wall to further expansion.

turtles A group of slow-moving armored reptiles (subclass Anapsida, order Testudines) with 11 families that include terrestrial, amphibious, freshwater, and marine species. They are oviparous, cold-blooded, have scaly skin, and lay eggs. A turtle's armor is the backbone fused to the shell, and the central disk of the upper shell, or carapace, is formed by a fusion of bony plates with the turtle's ribs. Their ribs lie outside the shoulder and pelvic bones, which allows them to retract their limbs and heads into the shells. They live worldwide except for Antarctica. Turtles are the oldest living group of reptiles, first appearing about 200 million years ago. Many of the terrestrial turtles are called tortoises. Many species today are endangered because of habitat destruction.

twins A pair of offspring that go through the same gestation period at the same time from the same mother. Two forms exist. Identical twins develop from a single fertilized egg that splits in two and creates two genetic replicas, while fraternal twins develop from separate eggs and sperm.

tympanic membrane (ear drum) A stretched, thin, semitransparent membrane attached to the ossicles of the middle ear and located at the end of the auditory canal. Sounds in the canal cause the tympanic membrane to vibrate. It is attached to the malleus (hammer) and its vibrations move the malleus, which in turn moves the incus (anvil).

type 1,2,3 copper Different classes of copper-BINDING SITES in proteins, classified by their spectroscopic properties as Cu(II). In type 1, or BLUE COPPER centers, the copper is coordinated to at least two imidazole nitrogens from histidine and one sulfur from cysteine. They are characterized by small copper HYPERFINE couplings and a strong visible absorption in the Cu(II) state. In type 2, or non-blue copper sites, the copper is mainly bound to imidazole nitrogens from histidine. Type 3 copper centers comprise two spin-coupled copper ions bound to imidazole nitrogens.

See also COORDINATION.

type species Refers to a species that was selected as the representative of a genus.

type specimen The actual specimen from which a new species is named. If the original author does not designate an individual, a single specimen from the original series can be selected as the lectotype.

typhus An infectious disease transmitted by the microorganism of genus *Rickettisia* (e.g., *Rickettsia prowazekii*) by lice, mites, ticks, and fleas. Also called typhus fever. Fatality is as high as 60 percent if not treated.

tyrosinase A copper protein containing an antiferromagnetically coupled dinuclear copper unit (TYPE 3-like site) that oxygenates the tyrosine group to catechol and further oxidizes this to the quinone.

See also FERROMAGNETIC; NUCLEARITY.

tyrosine kinase (PTKs) Protein enzymes that modulate a wide variety of cellular events, including differentiation, growth, metabolism, and apoptosis. Protein kinases add phosphate groups to proteins. Enzymes that add phosphate groups to tyrosine residues are called protein tyrosine kinases. These enzymes have important roles in signal transduction and regulation of cell growth, and their activity is regulated by a set of molecules called protein tyrosine phosphatases that remove the phosphate from the tyrosine residues. A tyrosine kinase is an enzyme that specifically phosphorylates (attach phosphate groups to) tyrosine residues in proteins and are critical in T- and B-cell activation.

tyrosine kinase receptor Proteins found in the plasma membrane of the cell that can phosphorylate (attach phosphate groups to) on a tyrosine residue in a protein. Insulin is an example of a hormone whose receptor is a tyrosine kinase. Following binding of the hormone, the receptor undergoes a conformational change, phosphorylates itself, then phosphorylates a variety of intracellular targets.

U

ubiquitous Existing or having presence everywhere at the same time. For example, ribonuclease, an enzyme that degrades RNA, is ubiquitous in living organisms.

ultimate causation An explanation for why a behavior has evolved and its context within the environment. Identifies the evolutionary factors responsible for the behavior and its purpose within a biological system.

ultrasound The use of sound with a frequency higher than 20,000 Hz. Used to obtain images for medical diagnostic purposes, especially during pregnancy.

umwelt The total sensory and perceptual world of an animal.

uniformitarianism A doctrine that states that current geological and biological processes, occurring at the same rates and in the same manner observed today, account for all of Earth's geological and biological features. "The present is the key to the past." The doctrine of uniformitarianism was advanced by James Hutton (1726–97) in his publication, *Theory of the Earth* (1785).

unsaturated fatty acids Fatty acids are the essential building blocks of all fats in our food supply and body,

but not all of them have beneficial results. There are five major fatty acids types: saturated (SAFA), unsaturated (UFA), monounsaturated (MUFA), polyunsaturated (PUFA), and essential (EFA).

An unsaturated fatty acid is a long-chain carboxylic acid that contains one or more carbon C=C double bonds. They occur when all of the carbons in a chemical chain are not saturated with hydrogen, so that the fat molecule contains one or more double bonds and produces a fat that is fluid at room temperature.

There are three types of unsaturated fatty acids: monounsaturates such as oleic acid found in olive and sesame oils that contain one double bond; polyunsaturated fats such as corn, soybean, and sunflower oils that contain more than one double bond; and essential fatty acids (EFA) that, while they cannot be created in the body, are important. EFAs include linoleic acid (LA) and alpha-linolenic acid (LNA).

The double bonds in a molecule of an unsaturated fatty acid can be found in two forms known as *cis* and *trans*.

Cis double bonds produce a kink, or a bend of about 30 degrees for each double bond into the backbone and can flip over to the *trans* form under high temperatures. *Trans* double bonds allow the molecule to lie in a straight line. However, the human body cannot convert the *trans* form into nutrients and thus cannot activate the metabolic activities required to convert to the active *cis* forms. This can lead to a deficiency in essential fatty acids. The more double bonds, and therefore the more kinks, the more beneficial it is to human health. The kinks, by completely changing the physical and chemical properties, allow it to form essential protein associations more easily, disperse more saturated fatty acids, and interact with water or blood.

See also SATURATED FATTY ACID.

urea   A nitrogen-containing waste product; a result of the normal breakdown of protein in the liver in mammals. It is created in liver cells from ammonia and carbon dioxide, and carried via the bloodstream to the kidneys, where it is excreted in the urine along with nitrogen. Urea accumulates in the body of people with renal failure. Urea is also a synthetic source of nitrogen made from natural gas.

urease   A nickel ENZYME, urea amidohydrolase, that catalyzes the HYDROLYSIS of urea to ammonia and carbon dioxide. The ACTIVE SITE comprises two Ni(II) ions bridged by a carbamate.

ureter   Tubes, about eight to 12 inches in length, that run from the kidney to the bladder on each side and carry urine from the kidneys to the bladder.

urethra   A narrow tube or canal that carries urine from the bladder, and semen from the prostrate and other sex glands, out via the tip of the penis to the outside of the body.

uric acid   The end results of urine breakdown, a product of protein metabolism, which is the major form of excreting metabolic nitrogen out of the body. Too much uric acid in the blood and its salts in joints lead to gout, which causes pain and swelling in the joints. When urine contains too much uric acid, "kidney" or uric acid stones can develop.

urochordate   A chordate (organism having a notochord at some stage of development) without a back-

Colored scanning electron micrograph (SEM) of crystals of calcium carbonate on the surface of an otolith, found in the utricle. An otolith or otoconium is a calcified stone that is found in the otolith organs of the inner ear. They are attached to sensory hairs, and, when the head tilts, the stones' movements cause nerve impulses that allow a sense of balance and orientation to be maintained. In humans, otoconia range from 3 to 30 microns (millionths of a meter) across. Magnification unknown. *(Courtesy © Susumu Nishinaga/Photo Researchers, Inc.)*

bone, also called a tunicate or sea squirt; a sessile marine animal composed of a sack with two siphons in which water enters and exits, filtered inside by the sack-shaped body.

urticating   The act of causing itching or burning sensation to the skin or eyes as a result of being inflicted by the entry of poison-filled spines (tips often break off) or setae that contain venom capable of causing pain and irritation. Found on the abdomen of tarantulas, some caterpillars, and the Urticaceae, plants of the nettle family.

uterus A pear-shaped hollow female reproductive organ in the pelvis area, at the top of the vagina, where eggs are fertilized and/or development of the young occurs.

utricle The larger of two membranous sacs (the other is the saccule) within the vestibule, the cavity at the entrance to the bony labyrinth of the inner ear. The utricle contains receptors for balance and a macula that is responsive to linear acceleration. The macula is a group of hair cells residing in either the saccule or the utricle and covered with gelatinous material containing otoliths, which are small granules of calcium carbonate. The ear's semicircular canals lead to and from the utricle.

vaccination   Injection of a killed or weakened infectious organism (virus, bacterium) in order to prevent the disease. Vaccinations are administered via needles, orally, or by aerosol spray.

vaccine   A preparation that stimulates an immune response in the body that can prevent an infection or create resistance to an infection or disease. Vaccines are administered via needles, orally, or by aerosol spray. There are several types of vaccines, ranging from monovalent types that contain only one antigen to combinational vaccines where several antigens are combined into one. Vaccines can be live or attenuated, i.e., a weakened strain, to induce the immune response while bypassing the severe effects of the disease. Common vaccines include measles, mumps, polio, and others. A live-vector vaccine uses a nondisease virus or bacterium to deliver a foreign substance to develop immunity. An inactive vaccine contains dead viruses or bacteria and cannot cause disease but will trigger a response.

There are acellular vaccines that contain only a partial amount of cellular material; DNA vaccines that inject genes with coding for a specific antigen protein; bacterial vaccines that use bacteria; intracellular vaccines within a cell; and conjugate vaccines that are made with polysaccharide (carbohydrate) antigens bound to proteins to improve the effect of immunity.

vacuole   A large membrane-bound, fluid-filled space within a cell. In plant cells, there usually is a single large vacuole filling most of the cell's volume, which helps maintain the shape of the cells. Vacuoles can contain food, gas, ingested bacteria, and other debris. In species such as the paramecium, two vacuoles are important: the contractile vacuole is used in osmoregulation, which is the removal of excess water; the food vacuole contains recently ingested food, where it eventually combines or fuses with the cell's lysosomes, which contain enzymes for digestion.

vagility   Free to move about, ability to move or migrate. A vagile species is one whose distribution can vary widely from year to year.

valence shell   Valence electrons are the electrons located in the outermost, highest energy orbits or "shell" of an atom. The "shell" is more of a field density and indicates the region where the electrons are located. The valence electrons determine the chemical properties of an element, since it is these valence electrons that are gained or lost during a chemical reaction.

van der Waals forces   Weak forces, such as seen in hydrogen bonding, that contribute to intermolecular bonding.

vascular cambium A secondary meristem, i.e., a thin layer of undifferentiated plant cells that divide indefinitely, which gives rise to secondary xylem and phloem, leading to an increase in stem girth. Tissue external to the vascular cambium is bark.

vascular plants The plant kingdom comprises algae, bryophytes, seedless vascular plants, and seed vascular plants (gymnosperms and angiosperms). Most of the familiar flora in the United States such as trees, shrubs, herbs, vines, grasses, ferns, and most other land plants belong to the Tracheophyta, or vascular plants.

These plants have systems for transporting water, sugars, and nutrients and are differentiated into stems, leaves, and roots. They have an elaborate system of conducting cells, consisting of xylem, where water and minerals are transported, and phloem, where carbohydrates are transported. This method of rigid internal support enables them to stand and grow erect and distribute nutrients against gravity.

There are about 17,000 species of vascular plants native to the United States, along with several thousand additional native subspecies, varieties, named natural hybrids, and about 5,000 exotic species known outside cultivation. More than 4,850 species (about 28 per-

cent) of the native U.S. vascular plants are considered globally rare.

Vascular plants first appear in the fossil record during the mid-Silurian period, about 410 million years ago. Rhyniophyta is the earliest known division of these plants, represented by several genera. *Vascular* is from Latin *vasculum*, meaning a vessel or duct.

Nonvascular plants, like mosses and liverworts, have poorly developed fluid transportation systems.

vascular system A network of specialized cells; the vascular tissue that transports water and nutrients from the roots throughout a plant's body. In animals, it is a specialized network of vessels—arteries, veins, and capillaries—for the circulation of fluids throughout the body tissue of an animal.

vascular tissue The collective name given to the xylem and phloem, both tissues that carry food, water, and minerals through the plant for the nourishment of the cells in vascular plants.

vas deferens (sperm duct) The excretory duct or tube that carries sperm from the epididymis, a long coiled tube in which spermatozoa are stored, to the

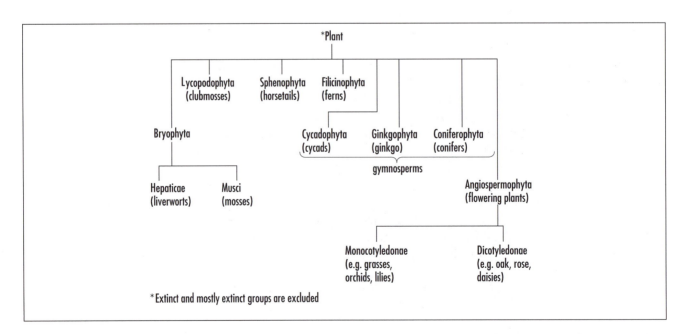

The plant kingdom comprises algae, bryophytes, seedless vascular plants, and seed vascular plants (gymnosperms and angiosperms).

ejaculatory duct and urethra. The tube connects the testes with the urethra and has thick muscles to move the sperm down the tract.

See also SPERM.

vector A vector can be a bacterium or virus that does not cause disease in humans. It is often used in genetically engineered vaccines to transport antigen-encoding genes into the body to initiate an immune response. Additionally, a vector can be an organism, usually an insect, that transmits an infectious agent to its alternate host. Examples are malaria, where the vector of the parasite is mosquitoes, and the hosts are humans.

vegetative reproduction Asexual reproduction in plants from vegetative parts (stems, leaves, or roots) or from modified stems (bulbs, tubers, rhizomes, and stolons). There is no exchange of gametes, and unless mutations occur, each new generation of plants is genetically identical to the parent plant.

vein A thin-walled blood vessel that carries blood to the heart. Smaller veins called venules connect veins to capillaries.

In plants, they are the vascular bundles in leaves that consist of xylem and phloem. The veins are large along the leaf midvein and petiole and get smaller as they radiate out into the leaf, becoming known as veinlets.

See also ARTERY.

ventilation Ventilation is the passage of air into and out of the respiratory tract. Ventilation exchange (VE) is the exchange of oxygen and carbon dioxide and other gases during the passage of air into and out of the respiratory passages.

ventricle The chamber of the heart that receives blood from the atrium (chamber that collects blood returning from the rest of the body) and contracts and pumps blood away from the heart. There are two ventricles in mammals and birds. The right ventricle pumps deoxygenated blood to the lungs via the pulmonary artery; the left ventricle pumps oxygenated blood to the body via the

aorta. The tricuspid valve separates the right atrium and the right ventricle and prevents backflow from the ventricle to the atrium. The mitral valve separates the left atrium and the left ventricle and prevents backflow from the ventricle to the atrium. The aortic valve prevents backflow of blood from the aorta into the left ventricle.

There are four brain ventricles. The right and left lateral ventricles lie within the cerebral cortex and connect with the very narrow third ventricle by way of openings called interventricular foramina. This third ventricle lies between the two halves of the thalamus and connects to the fourth ventricle via a long, thin tube called the cerebral aqueduct. The fourth ventricle lies in the hindbrain, under the cerebellum and dorsal to the pons and medulla.

vernalization A required chilling period leading to the breaking of dormancy in plants. Flowering can be accelerated or induced by exposure to a long period of near-freezing temperatures.

vertebra One of the segments of bone or cartilage composing the spinal column of vertebrates. In the spinal column of adult humans, there are 33 vertebrae: seven cervical vertebrae in the neck, 12 thoracic vertebrae that support the ribs, five lumbar vertebrae in the lower back, and other fused vertebrae forming the sacrum and the coccyx, beneath the sacrum.

vertebral bodies The oval segments of bone on the spine that support most of the axial load of the spine and consist mostly of the thoracic and lumbar vertebrae. The lumbar part of the spine—consisting of five lumbar vertebral bodies that sit on top of the sacrum—is above the coccyx, the tailbone. The lumbar spine supports most of the weight of the body, and its vertebral bodies are larger than the rest. The thoracic part of the spine consists of 12 bodies. The thoracic region supports the cervical spine (neck), which has seven bodies and supports the head.

vertebrates Chordate animals that have a spinal column or backbone (with the exception of hagfish and the rudimentary form of lampreys). The category

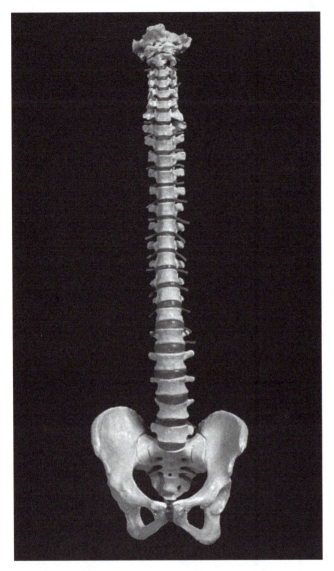

View of the spine and pelvis of a human skeleton. The spine is a column of 33 roughly cylindrical bones called vertebrae. Between each vertebra is a disc-shaped pad of cartilage. At the top of the image, seven cervical vertebrae support the skull. The 12 thoracic vertebrae are below this, and each of these supports an attached rib (not seen). The lumbar spine is five vertebrae that support the lower back, and these take most of the strain when lifting. At the bottom, the sacrum consists of 5 fused vertebrae that connect to the pelvic bone. The pelvic bone protects and supports abdominal organs. *(Courtesy © James Stevenson/Photo Researchers, Inc.)*

includes fish, amphibians, reptiles, birds, and mammals; a member of the phylum Chordata.

Three subphyla exist: Urochordata (sea squirts, larvaceans), Cephalochordata (amphioxus, lancelets), and Craniata (hagfish plus all the vertebrates). Another smaller phylum Hemichordates, the sawblades and acorns, are important in the study of vertebrate evolution. Although they contain only a few hundred species, they seem to share some of the chordate characteristics: brachial openings, or "gill slits," into the pharynx; a rudimentary structure called the stomochord that is similar to a notochord; and a dorsal nerve cord and a smaller ventral nerve cord. However, DNA studies are showing that hemichordates may be closer to echinoderms.

There are approximately 50,000 living species of vertebrates, with slightly fewer total fish vertebrates (25,988), the oldest group, than all others combined (4,500 mammals, 9,100 birds, 7,082 reptiles, 4,880 amphibians). They range in size from 0.1 gram to 100,000 kilograms. The Cephalochordata have the fewest number of living species with 45, followed by the Urochordata with 2,000 species, and the Craniata with 43,000 species.

Vertebrates are chordates with a distinct head and contain sense organs and a brain. Other characteristics include a segmented vertebral rod of cartilage/bone, a closed circulatory system, true coelom, and a bilaterally symmetrical body.

Vertebrates first appeared on Earth some 500 million years ago when continents were fragmented. Then the continents combined into the supercontinent Pangaea, which included most of the Earth's crust, some 300 million years ago, fragmenting again 100 million years ago into Laurasia (the northern continents) and Gondwanaland (the southern continents). Vertebrates live in almost every conceivable habitat on earth.

See also INVERTEBRATES.

vesicle   A small, membrane-enclosed sac, cyst, or bubble found in the cytosol, the semifluid inside the cell membrane of eukaryotic cells. Vesicles are used to transport proteins and lipids to various destinations in the cell, including the sites of glycolysis (the first stage of energy production by the cell), and fatty acid synthesis. Vesicles can import particles in a process called endocytosis or export waste in a process called exocytosis.

In botanical terms, a vesicle is a small bladderlike body in the substance of a vegetable or on the surface of a leaf. In medical terms, a vesicle is a small and more-or-less circular elevation of the cuticle, skin lesion, or blister containing a clear watery fluid; a cavity or sac, especially one filled with fluid. Anatomy refers to the umbilical vesicle, while zoology defines a vesicle as a small, convex, hollow prominence on the surface of a shell or a coral. In geological terms, a vesicle is a small cavity, nearly spherical in form and the size of a pea or smaller, common in some volcanic rocks and produced by the liberation of watery vapor in the molten mass.

vessel element Individual, short, wide, or fat cells arranged end to end, forming a system of tubes in the xylem. The cell walls are pitted and contain lignin, which gives them strength. They function to carry water and minerals upward in the stem and root.

vestigial organ Nonfunctional remains of organs that were previously functional and served a purpose in ancestral species, that no longer serve that purpose, but that remain part of the body and may still be functional in related species; e.g., the dewclaws of dogs, tails in human embryos, wisdom teeth in adults, wings of the ostrich, rudimentary legs in snakes, and whales with hip bones.

vetch *(Vicia sativa)* An annual forage legume, 40 to 80 cm in height. Chiefly pollinated by bees because it produces much nectar. Also, the maximum surface of a lake exposed to prevailing winds.

vicariance The separation of a group of organisms caused by a geographic barrier (e.g., mountain, lake, sea, etc.), resulting in the original group differentiating into new varieties or species.

viremia The presence of virus in the bloodstream. Often associated with malaise, fever, and aching in the back and extremities.

viroid (satellite RNA) Once thought to be a virus, a viroid is an infectious, pathogenic entity similar to a virus but having only one strand of nucleic acid without the protein coat that defines a virus; a naked unencapsulated strand of RNA. Viroids are known to cause plant diseases. For example, potato spindle tuber viroid (PSTV) can cause a destructive disease of potatoes.

virus A small microorganism that contains RNA or DNA and is surrounded by a protein coat. Viruses infect cells and then replicate new viruses after invasion using the protein of the infected cells to reproduce. Viruses can cause many human diseases, including chicken pox, measles, mumps, rubella, pertussis, and hepatitis, and they are not affected by common drugs such as antibiotics, which are effective against bacteria-based disease. Instead, vaccines are used to prevent or fight off virus attacks.

See also BACTERIA.

visceral (visceral muscle) Pertaining to the internal organs of the body, especially those within the abdominal cavity such as the digestive tract, bladder, and the heart; smooth muscle, as opposed to the two other muscle types, skeletal and cardiac.

visible light The portion of the electromagnetic spectrum that humans perceive. This excludes radio waves, microwaves, infrared light, ultraviolet light, X rays, and gamma rays. The human eye perceives the visible-light spectrum as a continuum of colors (red, orange, yellow, green, blue, indigo, and violet, along with various combinations and shades of these colors), with the perceived color depending on the wavelength. The spectrum of visible light ranges in wavelength from about 400 nm to about 700 nm. Visible light travels at the same speed as all other radiation, i.e., at 186,000 miles per second, and its wavelength is longer than ultraviolet light but shorter than X rays. Violet has the shortest wavelength, while red has the longest.

See also ENERGY.

vitalism Attributed to the teachings of Aristotle, the concept of vitalism, in its many forms, is the belief that life forms and natural phenomena cannot be explained by simple explanations of matter and processes. Vitalism posits that there must be another force, perhaps a

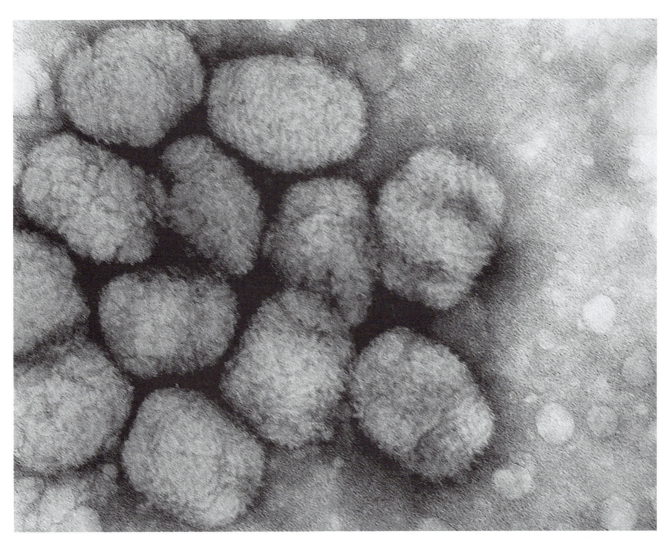

A transmission electron micrograph of smallpox viruses. Smallpox is a serious, highly contagious, and sometimes fatal infectious disease. There is no specific treatment for smallpox, and the only prevention is vaccination. *(Courtesy of Fred Murphy and Sylvia Whitfield, Centers for Disease Control and Prevention)*

soul or spirit, that must be added to the equation that brings us life as we know it. This was one of the central interests of scientists and scholars during the latter part of the 19th century, and many spent time investigating the relationship between human beings and nature by studying the physiology of perception, as opposed to the mechanistic theory that all living organisms are inanimate and mechanical. Most of the biologists of that era believed in the existence of a life force of some kind. It was opposed in the 19th century by meteorologist/scientist Hermann von Helmholtz (1821–94), who spent many years trying to prove it wrong.

vitamin   An essential organic nutrient that is needed in small amounts by an organism for metabolism and other processes. Organisms either synthesize them or obtain them in other ways. Examples of vitamins are vitamin C and vitamin E, both antioxidants. A vitamin usually functions as a coenzyme or a component of a coenzyme and is soluble in either water or organic solvent. The lack of certain vitamins can lead to disease such as in rickets (vitamin D), tooth decay (vitamin K), bone softening (vitamin D), or night blindness (vitamin A). Other vitamins include vitamin B_1 (thiamin), vitamin B_2 (riboflavin), niacinamide (niacin-vitamin B_3), vitamin B_6 (pyridoxine), vitamin B_{12} (cobalamin), pantothenic acid (vitamin B_5), pyridoxal (vitamin B_6), phylloquinone (vitamin K), biotin, folic acid, inositol, choline, and PABA (para amino benzoic acid). Vitamin supplements are a billion dollar per year industry.

vitamin B_{12} *See* COBALAMIN.

vitiligo A skin condition of unpigmented spots determined by a dominant gene that destroys special skin cells (melanocytes). These skin cells produce the pigment melanin in the skin along with tissues (mucus membranes) that line the inside of the mouth, nose, genital and rectal areas, and the retina of the eyes. About 40 to 50 million people worldwide have vitiligo, and in the United States alone, 2 to 5 million people have it.

viviparous Animals that are viviparous are born live after being nourished by blood from the placenta while in the uterus, and do not hatch from eggs.

voltage-gated channel Ion channels are pores in cell membranes that allow the passage of ions in and out of cells. There are two types, voltage-gated and chemically gated channels. The opening and closing of the voltage-dependent ion channels is regulated by voltage, the electrical charge or potential difference between the inside and outside of the membrane, while chemical stimuli are responsible for opening and closing the chemically gated channels. Neurons use these channels to pass sodium and potassium ions through them.

Wagner-Jauregg, Julius (1857–1940) Austrian *Neurologist, Psychiatrist* Julius Wagner was born on March 7, 1857, in Wels, Austria, to Adolf Johann Wagner. He attended the Schotten gymnasium in Vienna and in 1880 received his medical degree at the Institute of General and Experimental Pathology, where he stayed for two years.

In 1889 he was appointed extraordinary professor at the medical faculty of the University of Graz. Here he started his investigations on the connections between goiter and cretinism, and, based on his research, the government started selling salt laced with iodine in the areas most affected by goiter. From 1893 to 1928 he was professor at the University of Vienna.

Wagner's life work was to cure mental disease by inducing a fever, after observing that mental patients improve after surviving certain infections that have high fevers. In 1917, using malaria inoculation, he was able to cure syphilis patients of dementia paralytica, or paresis, caused by syphilis, bringing the disease under control. He attributed the success of the procedure to the induced malarial fever, and this discovery earned him the Nobel Prize in 1927. This is considered the first example of "shock therapy."

His main publication was a book titled *Verhütung und Behandlung der progressiven Paralyse durch Impfmalaria* (Prevention and treatment of progressive paralysis by malaria inoculation) in the memorial volume of the *Handbuch der experimentellen Therapie* (Handbook of experimental therapy) (1931). His other works include *Myxödem und Kretinismus* in the *Handbuch*

der Psychiatrie (1912) and *Lehrbuch der Organotherapie* (Textbook of organotherapy) with G. Bayer (1914). He published more than 80 papers after he retired in 1928.

Later in life, he devoted himself to research in forensic medicine and the legal aspects of insanity, and he assisted in formulating the law regarding certification of the insane in Austria. He died on September 27, 1940.

Waksman, Selman Abraham (1888–1973) American *Biochemist* Selman Abraham Waksman was born in Priluka, near Kiev, Russia, on July 22, 1888, to Jacob Waksman and Fradia London. He received his early education from private tutors and school training in Odessa in an evening school, also with private tutors.

In 1911 he entered Rutgers College, having won a state scholarship the previous spring, and received a B.S. in agriculture in 1915. He was appointed research assistant in soil bacteriology at the New Jersey Agricultural Experiment Station, and continued graduate work at Rutgers, obtaining an M.S. in 1916, the year he became a naturalized U.S. citizen. In 1918 he was appointed a research fellow at the University of California, where he received his Ph.D. in biochemistry the same year.

He was invited back to Rutgers, and by 1930 was a professor. When the Department of Microbiology was organized in 1940, he became professor of microbiology and head of the department, and nine years later he was appointed director of the Institute of Microbiology. He retired in 1958.

Waksman brought medicine from the soil. By studying soil-based acintomycetes, he was able to extract a number of antibiotics such as actinomycin (1940), clavacin, streptothricin (1942), streptomycin (1943), grisein (1946), neomycin (1948), fradicin, candicidin, candidin, and more. His discovery of streptomycin, which was the first effective treatment against tuberculosis, brought him the 1952 Nobel Prize in physiology or medicine.

He published more than 400 scientific papers and has written, alone or with others, 18 books, including *Principles of Soil Microbiology* (1927) and *My Life with the Microbes* (1954), an autobiography. He was a member of numerous scientific organizations. In 1950 he was made commander of the French Légion d'Honneur, and in 1952 was voted as one of the most outstanding 100 people in the world today. He died on August 16, 1973, in Hyannis, Massachusetts.

Wallace's line An imaginary line drawn by A. R. Wallace that passes between the Philippines and the Moluccas in the north and between Sulawesi and Borneo and between Lombok and Bali in the south (the Mariana Trench). It separates the Oriental and Australian biogeographical regions. It marks the limits of distribution for many major animal groups that appear on one side of the line but are absent on the other side.

Warburg, Otto Heinrich (1883–1970) German *Biochemist* Otto Heinrich Warburg was born on October 8, 1883, in Freiburg, Baden, to physicist Emil Warburg. He studied chemistry under Emil Fischer and received his doctor of chemistry from the University of Berlin in 1906, and a doctor of medicine from the University of Heidelberg in 1911.

In 1918 he was appointed professor at the Kaiser Wilhelm Institute for Biology, Berlin-Dahlem, and from 1931 to 1953 he was director of the Kaiser Wilhelm Institute for Cell Physiology (now Max Planck Institute) in Berlin.

He specialized in the investigation of metabolism in tumors and respiration of cells. He discovered that flavins and the nicotinamide were the active groups of the hydrogen-transferring enzymes, and early discovery of iron–oxygenase provided details of oxidation and reduction (redux reactions) in the living organisms. For his discovery of the nature and mode of action of the respiratory enzymes that enable cells to process oxygen, he was awarded the Nobel Prize in 1931. He was offered a second Nobel Prize in 1944 for his enzyme work, but he was not allowed to accept it, since he was living under the Hitler regime. He later discovered how the conversion of light energy to chemical energy is activated in photosynthesis. During the 1930s he showed the carcinogenic nature of food additives and cigarette smoke and demonstrated how cancer cells are destroyed by radiation.

Warburg is the author of *New Methods of Cell Physiology* (1962). He died on August 1, 1970.

warm-blooded Refers to organisms that maintain a constant body temperature. Also known as HOMEOTHERMIC.

water potential Direction of water flow based on solute concentration and pressure. An example is osmosis, which is the diffusion of water across a semipermeable barrier, such as a cell membrane, from high water potential to lower water potential.

Also a measure of the moisture stress in plants or soil, measured in megapascals. A more negative value indicates greater moisture stress. Soils with no moisture stress have a water potential of 0 to −1 mPa. The measurement of soil water potential involves the use of sensors that determine the energy status of the water in soil. The energy state describes the force that holds the water in the soil. Two methods of measuring soil water potential are the heat dissipation method or the electrical resistance method.

water table The level below the earth's surface at which the ground becomes saturated with water; usually mimics the surface contour and is set where hydrostatic pressure equals atmospheric pressure.

water vascular system An internally closed network of watery canals in echinoderms (e.g., starfish, sand dollars, sea urchins) that draws water from the surrounding sea and passes it through a perforated plate called the madreporite, which is used for locomotion and food gathering. Extensions of the water vascular system are called tube feet, which protrude from the body, usually ending in suckers, and are used for locomotion and for holding on to the sea bottom or prey.

wavelength   The physical distance between points of corresponding phase of two consecutive cycles of a wave.

Wegener, Alfred Lothar (1880–1930) German *Geophysicist, Meteorologist, Climatologist*   Alfred Wegener was born in Berlin on November 1, 1880, the son of a minister who ran an orphanage. He obtained his doctorate in planetary astronomy in 1904 at the University of Berlin. In 1905 Wegener took a job at the Royal Prussian Aeronautical Observatory near Berlin, studying the upper atmosphere with kites and balloons. Wegener was an expert balloonist, as proved the following year when he and his brother Kurt set a world record of 52 consecutive hours in an international balloon contest.

In 1911, at the age of 30, Wegener collected his meteorology lectures and published them as a book titled *The Thermodynamics of the Atmosphere*. It became a standard in Germany, and Wegener received acclaim. He also noticed the close fit between the coastlines of Africa and South America. He was formulating his theory of continental drift and began to search for paleontological, climatological, and geological evidence in support of his theory.

On January 6, 1912, at a meeting of the Geological Association in Frankfurt, he spoke about his ideas of "continental displacement" (continental drift), and presented his theory again days later at a meeting of the Society for the Advancement of Natural Science in Marburg.

In 1914 he was drafted into the German army, was wounded, and served out the war in the army weather-forecasting service. While recuperating in a military hospital, he further developed his theory of continental drift, which he published the following year as *Die Entstehung der Kontinente und Ozeane* (The origin of continents and oceans). Expanded versions of the book were published in 1920, 1922, and 1929. Wegener wrote that around 300 million years ago, the continents had formed a single mass, called Pangaea (Greek for "all the Earth"), which split apart, and its pieces had been moving away from each other ever since. While he was not the first to suggest that the continents had once been connected, he was the first to present the evidence, although he was wrong in thinking that the continents moved by "plowing" into each other through the ocean floor. His theory was soundly rejected, although a few scientists did agree with his premise.

In November 1930 he died while returning from a rescue expedition that brought food to a party of his colleagues camped in the middle of the Greenland ice cap. His body was not found until May 12, 1931, but his friends allowed him to rest forever in the area that he loved.

The theory of continental drift continued to be controversial for many years, but by the 1950s and 1960s, plate tectonics was all but an accepted fact and taught in schools. Today, we know that both continents and ocean floor float as solid plates on underlying rock that behaves like a viscous fluid due to being under such tremendous heat and pressure. Wegener never lived to see his theory proved. Had he lived, most scientists believe he would have been the champion of present-day plate tectonics.

Weller, Thomas Huckle (1915–) American *Microbiologist*   Thomas Huckle Weller was born in Ann Arbor, Michigan, on June 15, 1915, to Carl Vernon Weller, who was in the pathology department of the medical school at the University of Michigan. He attended this university in 1932 and received a B.A. in 1936 and an M.S. in 1937.

In 1936 he attended Harvard Medical School in Boston and worked in the facilities in the department of comparative medicine and tropical medicine. In 1939 he began research on viruses and tissue culture techniques to study infectious disease. He received an M.D. in 1940 and began training at the Children's Hospital in Boston. After a stint with the Army Medical Corps during the war, he returned to Boston and the Children's Hospital. In 1947 he joined John ENDERS in the organization of the new research division of infectious diseases at the Children's Medical Center. In 1949 he became assistant director of this division and later became an instructor in the department of comparative pathology and tropical medicine and an associate professor in the Harvard Medical School of Public Health. In July 1954 he was appointed Richard Pearson Strong Professor of Tropical Public Health and head of the department at the Harvard School of Public Health. From 1966 to 1981 he was director of Harvard's Center for the Prevention of Infectious Diseases.

He contributed a great deal of research on the helminthes parasites of humans, particular on the nematode *Trichirella spiralis* and schistosome trematodes, which cause schistosomiasis. He isolated the viruses of varicella and herpes zoster, showing that one caused both diseases. In 1955 he also isolated a virus that causes cytomegalic inclusion disease in infants. Together with J. F. Enders and F. C. ROBBINS, he was awarded the 1954 Nobel Prize in physiology or medicine for work in growing polio viruses in cultures of different tissues.

West Nile disease or fever   A mosquito-borne disease that can cause encephalitis, or inflammation of the brain, and caused by a flavivirus. West Nile fever is a disease that has occurred before in Egypt, Asia, Israel, South Africa, and parts of Europe, but it had never before been found in the Western Hemisphere until recently, appearing in America in 1999. In 2003, West Nile virus had killed 223 people in the United States, and there were 9,122 confirmed human cases of the disease worldwide. Closely related to the St. Louis encephalitis.

Whipple, George Hoyt (1878–1976) American *Pathologist* George Hoyt Whipple was born on August 28, 1878, in Ashland, New Hampshire, to Dr. Ashley Cooper Whipple and Frances Hoyt. Whipple was educated at Phillips Academy in Andover and received a B.A. at Yale University in 1900. He then completed course work at Johns Hopkins University and received his M.D. degree in 1905, when he was appointed assistant in pathology at the Johns Hopkins Medical School. In 1914 he was appointed professor of research medicine at the University of California Medical School and was named director of the Hooper Foundation for Medical Research at that university, serving as dean of the medical school during the years

1920 and 1921. In 1921 he was appointed professor of pathology and dean of the School of Medicine and Dentistry at the University of Rochester and became the founding dean of the university's School of Medicine (1921–53) and chair of the pathology department.

Whipple's main researches were concerned with anemia and the physiology and pathology of the liver. In 1908 he began a study of bile pigments that led to his interest in the body's manufacture of the oxygen-carrying hemoglobin, an important element in the production of bile pigments. His studies dealt with the effect of foods on the regeneration of blood cells and hemoglobin in 1918. Between 1923 and 1925, his experiments in artificial anemia were instrumental in determining that iron is the most potent inorganic factor in the formation of red blood cells.

For his work on liver research and treatment of anemia, he was awarded, together with GEORGE R. MINOT and WILLIAM P. MURPHY, the Nobel Prize in physiology or medicine in 1934. Whipple published many scientific papers in physiological journals.

He died on February 2, 1976, in Rochester, New York. His birthplace home on Pleasant Street in Ashland was listed on the National Register in 1978.

wild type   The normal form, genotype, or phenotype of an organism found or first seen in nature. It can refer to the particular whole organism or to a particular mutation. It is the most frequently encountered genotype in natural breeding populations.

Wilson's disease   An inherited condition in which copper fails to be excreted in the bile. Copper accumulates progressively in the liver, brain, kidney, and red blood cells. As the amount of copper accumulates, hemolytic anemia, chronic liver disease, and a neurological syndrome develop.

See also CHELATION THERAPY.

winterbottom's sign   Swelling of the posterior cervical lymph nodes at the base of the skull that is symptomatic of having African trypanosomiasis (African sleeping sickness). Caused by a parasite and transmitted by the bite of the tsetse fly. West African trypanosomiasis is caused by the parasite *Trypanosoma brucei gambiense*. East African trypanosomiasis is caused by *Trypanosoma brucei rhodesiense*. It is confined mainly to tropical Africa and is located between 15 degrees north and 20 degrees south latitude.

wobble The ability of certain bases at the third position of an anticodon in tRNA to form hydrogen bonds in various ways, causing alignment with several possible codons. The third base position within a codon is called the wobble position.

XANES (X-ray absorption near edge structure) *See* EXTENDED X-RAY ABSORPTION FINE STRUCTURE.

X chromosome The nuclei of human cells contain 22 autosomes (any chromosome that is not a sex chromosome) and two sex chromosomes. In females, the sex chromosomes are the two X chromosomes, and in males there is one X chromosome and one Y chromosome.

xenobiotic A xenobiotic (Greek *xenos* "foreign" *bios* "life") is a compound that is foreign to a living organism. Principal xenobiotics include drugs, carcinogens, and various compounds that have been introduced into the environment by artificial means.

xenodiagnosis A way to test, using laboratory-bred organisms, the diagnosis of certain parasitic diseases (like *Trypanosoma cruzi* [Chagas' disease] and *Trichinella spiralis* [trichinosis] infection) when it is not possible to identify the infecting organism.

xeric Dry and hot habitat, desertlike.

xerophytes Plants that grow under arid conditions with low levels of soil and water and have water-conserving features such as thick cuticles and sunken stomatal pits.

XO Sex is determined by genes, called sex-linked genes. An XO individual has only one chromosome, and the Y or X is missing. An XO human is viable and develops ovaries but fails to produce steroid hormones, and so the individual has no puberty and is infertile. Those with Turner's syndrome are missing or have a damaged X chromosome, a condition that affects about one in every 2,500 females. Certain organisms, such as caddis, butterflies and moths, birds, and some fish, have XO sex determination. A male is produced when two X chromosomes are present, and a female is produced when only one X chromosome is present. Thus XY produces a female and XX produces a male in Lepidoptera, the heterogametic sex being the female in this select group of organisms.

X-ray absorption near edge structure (XANES) *See* EXTENDED X-RAY ABSORPTION FINE STRUCTURE.

xylem A tissue found in a plant's vascular bundle (above the phloem) and in the center of the plant, where secondary growth occurs. Its purpose is to bring water and minerals (nutrients) from the roots upward throughout the plant.

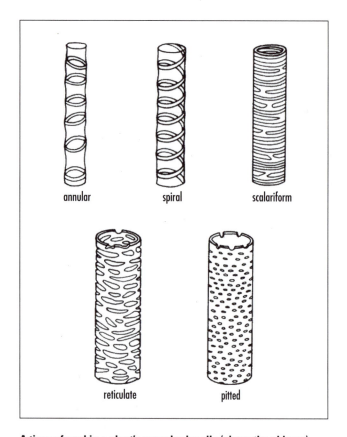

annular   spiral   scalariform

reticulate   pitted

A tissue found in a plant's vascular bundle (above the phloem) and in the center of the plant where secondary growth occurs.

There is a primary and secondary xylem. Xylem is made up of tracheids (long, dead cells that function as conductors), vessel elements (individual short, wide cells that make up the vessel, also responsible for water transport), fibers, and parenchyma cells (act as storage cells and help in assimilation or wound healing). Both tracheids and vessels have stiff lignin-containing secondary cell walls that give them rigidity.

Primary xylem comes from the plants procambium and contains protoxylem (annular and spiral vessels) and metaxylem (scalariform, reticulate, and pitted vessels). The secondary xylem comes from the vascular cambium in plants exhibiting secondary growth, such as wood.

See also PHLOEM.

Y

Y chromosome   The nuclei of human cells contain 22 autosomes (any chromosome that is not a sex chromosome) and two sex chromosomes. In females, the sex chromosomes are the two X chromosomes, and in males there is one X chromosome and one Y chromosome. The presence of the Y chromosome leads to a baby boy.

yeast   Considered to be the oldest "plant" cultivated by humans. Yeast is a unicellular fungus that belongs to the family Saccharomycetaceae. It lives in the soil, on plants, and is airborne, and it has been used in the production of bread, beer, and wine because it drives the process of fermentation. It produces carbon dioxide and alcohol when in the presence of sugar.

yellow fever   A tropical mosquito-borne viral hepatitis caused by an arbovirus (yellow fever virus) from the Flaviviridae family. It is transmitted by the mosquitoes *Aedes aegypti* and *Haemagogus capricorni,* among others.

yolk sac   An extraembryonic tissue (membranous sac) that is attached to an embryo and contains the nutritive food yolk. In fish larvae, the alevin-stage embryos swim with the yolk sac attached until it is absorbed. In placental mammals, it is vestigial (contains no nutrients). The primitive yolk sac divides into two parts: one portion becomes the fetal gastrointestinal tract, while the second portion connects to the fetal body via the vitelline duct and is located in the fluid space outside the amnion, the extraembryonic coelom. It produces the embryo's first blood cells and germ cells that develop into gonads; is the source of the mucous membrane of almost the entire intestinal tract; and deals with the transfer of nutritive fluid to the embryo from the trophoblast, primary mesenchyme, and extraembryonic coelom.

Z-DNA (zigzag DNA) A region of DNA that is "flipped" into a left-handed helix. DNA adopts the Z configuration when purines and pyrimidines alternate on a single strand. It may be important in regulating gene expression in eukaryotes.

zinc finger A DOMAIN, found in certain DNA-binding proteins, comprising a HELIX-loop structure in which a zinc ion is coordinated to two to four cysteine sulfurs, the remaining LIGANDS being histidines. In many proteins of this type the domain is repeated several times.
See also COORDINATION.

zoned reserve system An ecosystem preservation concept where certain lands are preserved and managed as wild or natural, while areas around it are used and altered by human activity. Several nations, such as Costa Rica, have adopted this strategy. Municipal zoning in the United States is another approach where undeveloped lands within a community are zoned for various uses such as industrial, residential, and conservation.

zones A series of life zones developed in 1889 by mammalogist Clinton Hart Merriam, who studied the distribution patterns of plants and animal. His developed the concept of a life zone. He recognized that similar zones of vegetation occurred with both increasing latitude and increasing elevation. He named them Arctic-Alpine, Hudsonian, Canadian, Transition, Upper Sonoran, and Lower Sonoran.

zoological nomenclature *See* BINOMIAL.

zooplankton Small floating or weakly swimming invertebrates that are transported by water currents and eat other plankton. Zooplankton are foraminifera and radiolarians, tiny crustaceans (*Daphnia*), as well as the larval or immature stages of animals like mollusks (snails and squid), crustaceans (crabs and lobsters), fish, jellyfish, and sea cucumbers.

zoospore An asexually produced flagellated, swimming spore.

zwitterionic compound A neutral compound having electrical charges of opposite sign, delocalized or not, on adjacent or nonadjacent atoms.

zygodactyl The trait of having two toes pointing forward and two toes pointing backward, found in

Viola cordata, an example of a zygomorphic flower from Virginia. *(Courtesy of Tim McCabe)*

climbing birds (e.g., parrots, woodpeckers). The first and fourth toes are directed backward, while it is the second and third that are forward.

zygomorphic   Flowers that are bilaterally symmetrical; can be bisected into similar halves in only one plane, thereby forming mirror images.

zygospore   A structure that forms in fungi from the diploid zygote. It forms sporangia, where meiosis takes place and forms spores.

zygote   When male (sperm) and female (ovum) gametes combine, they form a zygote that develops into an embryo. The zygote is a diploid cell created by the fusion of the mature male and female gametes, each of which has a haploid set of chromosomes (23 in human). Upon fusion, the zygote has two copies of chromosomes, one from each parent. It is the fertilized egg before it starts to divide and grow.

Appendix I

BIBLIOGRAPHY

Abbott, Derek, and Laszlo B. Kish. *In Unsolved Problems of Noise and Fluctuations: Upon'99,* Second International Conference, Adelaide, Australia. Melville, N.Y.: American Institute of Physics, 2000.

Aktories, K., and T. C. Wilkins. *Clostridium Difficile: Current Topics in Microbiology and Immunology.* New York: Springer, 2000.

Andow, D. A., R. J. Baker, and C. P. Lane, eds. *Karner Blue Butterfly: A Symbol of a Vanishing Landscape.* Miscellaneous Publication 84-1994, [viii]. St. Paul: Minnesota Agricultural Experiment Station, 1994.

Baddeley, Roland, Peter J. B. Hancock, and Peter Földiák. *Information Theory and the Brain.* New York: Cambridge University Press, 2000.

Balkwill, Frances R. *The Cytokine Network.* Frontiers in Molecular Biology. Vol. 25. New York: Oxford University Press, 2000.

Barash, David P., and I. Barash. *The Mammal in the Mirror: Understanding Our Place in the Animal World.* New York: W.H. Freeman, 2000.

Barlow, Connie C. *The Ghosts of Evolution: Nonsensical Fruit, Missing Partners, and Other Ecological Anachronisms.* New York: Basic Books, 2000.

Barrows, Edward M. *Animal Behavior Desk Reference: A Dictionary of Animal Behavior, Ecology, and Evolution.* 2d ed. Boca Raton, Fla.: CRC Press, 2001.

Beckerle, Mary C. *Cell Adhesion.* Frontiers in Molecular Biology, Vol. 39. New York: Oxford University Press, 2001.

Benedek, George Bernard, and Felix Villars. *Physics, with Illustrated Examples from Medicine and Biology.* 2d ed. Biological Physics Series. New York: AIP Press, 2000.

Beurton, Peter J., Raphael Falk, and Hans-Jörg Rheinberger. *The Concept of the Gene in Development and Evolution: Historical and Epistemological Perspectives, Cambridge Studies in Philosophy and Biology.* New York: Cambridge University Press, 2000.

Bonner, John Tyler. *First Signals: The Evolution of Multicellular Development.* Princeton, N.J.: Princeton University Press, 2000.

Boorman, Kathleen E., Barbara E. Dodd, and P. J. Lincoln. *Blood Group Serology Theory, Techniques, Practical, Applications.* New York: Churchill Livingston, 1977.

Boulter, Michael Charles. *Extinction: Evolution and the End of Man.* New York: Columbia University Press, 2002.

Brier, B. *Ancient Egyptian Magic.* New York: Perennial, 2001.

———. *Egyptian Mummies: Unraveling the Secrets of an Ancient Art.* New York: Quill, William Morrow, 1994.

Brinkmann, Bernd. "Overview of PCR-Based Systems in Identity Testing," *Methods in Molecular Biology* 98: 105–117.

Brock, James P. *The Evolution of Adaptive Systems.* San Diego, Calif.: Academic Press, 2000.

Brooks, D. R., and Deborah A. McLennan. *The Nature of Diversity: An Evolutionary Voyage of Discovery.* Chicago: University of Chicago Press, 2002.

Brown, A. *Genesis of the United States.* Vol. 1. New York: Russell & Russell, 1890.

Bryant, J. A., John M. Garland, and S. G. Hughes. *Programmed Cell Death in Animals and Plants.* Oxford, U.K.: BIOS Scientific, 2000.

Budowle, Bruce. *DNA Typing Protocols: Molecular Biology and Forensic Analysis.* Natick, Mass.: Eaton Pub., 2000.

Butler, John M. *Forensic DNA Typing: Biology & Technology behind STR Markers.* San Diego, Calif.: Academic Press, 2001.

Chiueh, Chuang C., and Daniel L. Gilbert. *Reactive Oxygen Species: From Radiation to Molecular Biology. A Festschrift in Honor of Daniel L. Gilbert.* Annals of the New York Academy of Sciences. Vol. 899. New York: New York Academy of Sciences, 2000.

Clark, Robert P. *Global Life Systems: Population, Food, and Disease in the Process of Globalization.* Lanham, Md.: Rowman & Littlefield Publishers, 2000.

Clote, Peter, and Rolf Backofen. *Computational Molecular Biology: An Introduction.* Wiley Series in Mathematical and Computational Biology. New York: John Wiley & Sons, 2000.

Cockell, Charles, and Andrew R. Blaustein. *Ecosystems, Evolution, and Ultraviolet Radiation.* New York: Springer, 2001.

Coleman, William B., and Gregory J. Tsongalis. *The Molecular Basis of Human Cancer.* Totowa, N.J.: Humana Press, 2002.

Collins, M. M., and R. D. Weast. *Wild Silk Moths of the United States.* Cedar Rapids, Iowa: Collins Radio Company, 1961.

Covell, C. V. *A Field Guide to the Moths of Eastern North America.* Boston: Houghton Mifflin Co., 1984.

Cowlishaw, Guy, and R. I. M. Dunbar. *Primate Conservation Biology.* Chicago: University of Chicago Press, 2000.

Crellin, J. K., and J. Philpott. *Herbal Medicine Past and Present.* Durham, N.C.: Duke University Press, 1990.

Cronk, J. K., and M. Siobhan Fennessy. *Wetland Plants: Biology and Ecology.* Boca Raton, Fla.: Lewis Publishers, 2001.

Crotch, W. *A Silkmoth Rearer's Handbook.* London: The Amateur Entomologists Society, 1956.

Czech, Brian, Paul R. Krausman, and Center for American Places. *The Endangered Species Act: History, Conservation Biology, and Public Policy.* Baltimore, Md.: Johns Hopkins University Press, 2001.

David, Rosalie, and Rick Archbold. *Conversations with Mummies.* New York: HarperCollins, 2000.

Davies, Geoffrey. *Forensic Science.* Rev ed. ACS Symposium Series. Washington, D.C.: American Chemical Society, 1986.

DeForest, Peter R., R. E. Gaensslen, and Henry C. Lee. *Forensic Science: An Introduction to Criminalistics.* New York: McGraw-Hill Book Co., 1983.

DePamphilis, Melvin L. *Gene Expression at the Beginning of Animal Development.* Advances in Developmental Biology and Biochemistry. Vol. 12. New York: Elsevier, 2002.

DeWan, Amielle A. "The Ecological Effects of Carnivores on Small Mammals and Seed Predation in the Albany Pine Bush." M. S. thesis, SUNY at Albany, N.Y., 2002.

Dirig, R. "Karner Blue, Sing Your Purple Song." *American Butterflies* 5, 1 (spring 1997): 14–20.

———. "Theme in Blue: Vladimir Nabokov's Endangered Butterfly." In *Nabokov at Cornell.* G. Shapiro, ed. Ithaca, N.Y.: Cornell University Press, 2003.

Dirig, R., and J. F. Cryan. "The Karner Blue Project: January 1973 to December 1976." *Atala* 4 (1976): 22–26.

Doerfler, Walter. *Foreign DNA in Mammalian Systems.* New York: Jossey-Bass, 2001.

Dunand, Françoise, Roger Lichtenberg, and Ruth Sharman. *Mummies: A Voyage through Eternity.* New York: Abrams, 1994.

Elgin, Sarah C. R., and Jerry L. Workman. *Chromatin Structure and Gene Expression.* 2d ed. Frontiers in Molecular Biology. Vol. 35. New York: Oxford University Press, 2000.

Eliot, I. M., and C. G. Soule. *Caterpillars and Their Moths.* New York: The Century Co., 1902.

Emini, Emilio A. *The Human Immunodeficiency Virus: Biology, Immunology, and Therapy.* Princeton, N.J.: Princeton University Press, 2002.

Entwistle, Abigail, and N. Dunstone. *Priorities for the Conservation of Mammalian Diversity: Has the Panda Had Its Day?* New York: Cambridge University Press, 2000.

Evans, Peter G. H., and Juan Antonio Raga. *Marine Mammals: Biology and Conservation.* New York: Kluwer Academic/Plenum Publishers, 2001.

Ferson, S., and Mark A. Burgman. *Quantitative Methods for Conservation Biology.* New York: Springer, 2000.

Fini, M. Elizabeth. *Vertebrate Eye Development, Results and Problems in Cell Differentiation.* Vol. 31. New York: Springer, 2000.

Flint, S. Jane. *Principles of Virology: Molecular Biology, Pathogenesis, and Control.* Washington, D.C.: ASM Press, 2000.

Forbes, W. T. M. *The Lepidoptera of New York and Neighboring States.* Memoirs 68. Ithaca, N.Y.: Cornell University Experimental Station, 1923.

Fox, Charles W., Derek A. Roff, and Daphne J. Fairbairn. *Evolutionary Ecology: Concepts and Case Studies.* New York: Oxford University Press, 2001.

Freinkel, Ruth K., and David Woodley. *The Biology of the Skin.* New York: Parthenon Pub. Group, 2001.

French, L. G. "The Sassafras Tree and Designer Drugs." *Journal of Chemical Education* 72, 6 (1995): 484–491.

Freund, Jan A., and Thorsten Pöschel. *Stochastic Processes in Physics, Chemistry, and Biology, Lecture Notes in Physics.* Vol. 557. New York: Springer, 2000.

Gaensslen, R. E. *Sourcebook in Forensic Serology, Immunology, and Biochemistry.* Washington, D.C.: National Institute of Justice, U.S. Department of Justice, 1983.

Gavin, Ray H. *Cytoskeleton Methods and Protocols.* Totowa, N.J.: Humana Press, 2000.

Gibbs, R. D. *Chemotaxonomie of Flowering Plants.* Montreal: McGill-Queen's University Press, 1974.

Gordon-Weeks, Phillip R. *Neuronal Growth Cones.* Developmental and Cell Biology Series. Vol. 37. New York: Cambridge University Press, 2000.

Gould, Stephen Jay. *I Have Landed: The End of a Beginning in Natural History.* New York: Harmony Books, 2002.

————. *The Structure of Evolutionary Theory.* Cambridge, Mass.: Belknap Press of Harvard University Press, 2002.

Graham, Colin A., and Alison J. M. Hill. *DNA Sequencing Protocols.* 2d ed. Totowa, N.J.: Humana Press, 2001.

Grant, Gregory A. *Synthetic Peptides: A User's Guide.* 2d ed. Advances in Molecular Biology. New York: Oxford University Press, 2002.

Gray, J. *Lyme Borreliosis: Biology, Epidemiology, and Control.* New York: CABI Pub., 2002.

Gray, P. H. K. "Radiography of Ancient Egyptian Mummies." *Med. Radiogr. Photogr.* 43 (1967): 34–44.

Guthrie, Christine, and Gerald R. Fink. "Guide to Yeast Genetics and Molecular and Cell Biology." In *Methods in Enzymology.* Vol. 351. San Diego, Calif.: Academic Press, 2002.

————. "Guide to Yeast Genetics and Molecular and Cell Biology." In *Methods in Enzymology.* Vol. 350. San Diego, Calif.: Academic Press, 2002.

Habib, Nagy A. *Cancer Gene Therapy: Past Achievements and Future Challenges.* Advances in Experimental Medicine and Biology. Vol. 465. New York: Kluwer Academic/Plenum Publishers, 2000.

Hall, A. Gtpases. *Frontiers in Molecular Biology.* Vol. 24. New York: Oxford University Press, 2000.

Harold, Franklin M. *The Way of the Cell: Molecules, Organisms and the Order of Life.* New York: Oxford University Press, 2001.

Harris, Harry. *The Principles of Human Biochemical Genetics.* Amsterdam: North-Holland Publishing Co., 1970.

Harris, J. E., and E. F. Wente. *Mummification in Ancient Egypt: Development, History and Techniques.* Chicago: University of Chicago Press, 1980.

Harris, J. E., and K. R. Weeks. *X-Raying the Pharaohs.* New York: Scribner, 1973.

Hart, John P., and John Terrell. *Darwin and Archaeology: A Handbook of Key Concepts.* Westport, Conn.: Bergin & Garvey, 2002.

Harwood-Nash, D. C. F. "Computed Tomography of Ancient Mummies," *J. Comput. Assist. Tomogr.* 3 (1979): 768–773.

Hazon, N., and G. Flik. *Osmoregulation and Drinking in Vertebrates.* Oxford, U.K.: BIOS, 2002.

Held, Lewis I. *Imaginal Discs: The Genetic and Cellular Logic of Pattern Formation.* Developmental and Cell Biology Series. Vol. 39. New York: Cambridge University Press, 2002.

Herrera, Carlos M., and Olle Pellmyr. *Plant-Animal Interactions: An Evolutionary Approach.* Malden, Mass.: Blackwell Science, 2002.

Herring, Peter J. *The Biology of the Deep Ocean, Biology of Habitats.* New York: Oxford University Press, 2002.

Hochberg, Michael E., and Anthony R. Ives. *Parasitoid Population Biology.* Princeton, N.J.: Princeton University Press, 2000.

Hodgkin, Jonathan. "Seven Types of Pleiotropy," *Int. J. Dev. Bio.* 42 (1958): 501–505.

Hoelzel, A. Rus. *Marine Mammal Biology: An Evolutionary Approach.* Malden, Mass.: Blackwell Science, 2002.

Hoffman, H., W. E. Torres, and R. D. Ernest. "Paleoradiology: Advanced CT in the Evaluation of Nine Egyptian Mummies." *RadioGraphics* 22 (2002): 377–385.

Hofrichter, Robert. *Amphibians: The World of Frogs, Toads, Salamanders and Newts.* Buffalo, N.Y.: Firefly Books, 2000.

Holland, W. J. *Moths: A Popular Guide to a Knowledge of the Moths of North America.* Garden City, N.Y.: Doubleday, Page & Co., 1920.

Hubbell, Stephen P. "The Unified Neutral Theory of Biodiversity and Biogeography." *Monographs in Population Biology* 32 (2001).

Hunt, Susan M. *Investigation of Serological Evidence.* Springfield, Ill.: Charles C. Thomas, 1984.

Kahl, Günter. *The Dictionary of Gene Technology: Genomics, Transcriptomics, Proteomics.* 2d ed. New York: Wiley-VCH, 2001.

Kaiser, Jamil. *Bioindicators and Biomarkers of Environmental Pollution and Risk Assessment.* Enfield, N.H.: Science Publishers, 2001.

Kaplanis, J. N., M. J. Thompson, W. E. Robbins, and B. M. Bryce. "Insect Hormones: Alpha Ecdysone and 20-Hydroxyecdyson in Bracken Fern." *Science* 157 (1967): 1436–1438.

Kareiva, Peter M., Simon A. Levin, and Robert T. Paine. *The Importance of Species: Perspectives on Expendability and Triage.* Princeton, N.J.: Princeton University Press, 2003.

Keddy, Paul A. *Competition.* 2d ed. Boston: Kluwer Academic, 2001.

Kessin, Richard H. *Dictyostelium: Evolution, Cell Biology, and the Development of Multicellularity.* Developmental and Cell Biology Series. Vol. 38. New York: Cambridge University Press, 2001.

Kinoshita, Roy K. "The Evolution of Skeletons: What Modern Foraminiferans Are Telling Us." Microform. Ph.D thesis, SUNY at Albany, N.Y., 2001.

Kirby, Lorne T. *DNA Fingerprinting: An Introduction.* New York: Stockton Press, 1990.

Kitano, Hiroaki. *Foundations of Systems Biology.* Cambridge, Mass.: MIT Press, 2001.

Koeberl, Christian, and Kenneth G. MacLeod. *Catastrophic Events and Mass Extinctions: Impacts and Beyond.* Boulder, Colo.: Geological Society of America, 2002.

Kot, Mark. *Elements of Mathematical Ecology.* New York: Cambridge University Press, 2001.

Krieger, R. I., P. P. Feeny, and C. F. Wilkinson. "Detoxication Enzymes in the Guts of Caterpillars: An Evolutionary Answer to Plant Defenses?" *Science* 172 (1971): 579–581.

Lawrence, Eleanor. *Henderson's Dictionary of Biological Terms.* 12th ed. New York: Prentice Hall, 2000.

Lederhouse, R. C., M. P. Ayres, J. K. Nitao, and J. M. Scriber. "Differential Use of Lauraceous Hosts by Swallowtail Butterflies, *Papilio troilus* and *P. palamedes* Papilionidae." *OIKOS* 63 (1992): 244–252.

Lennox, James G. "Aristotle's Philosophy of Biology: Studies in the Origins of Life Sciences." In *Cambridge Studies in Philosophy and Biology.* New York: Cambridge University Press, 2001.

Lesk, Arthur M. *Introduction to Protein Architecture: The Structural Biology of Proteins.* New York: Oxford University Press, 2000.

Liebler, Daniel C. *Introduction to Proteomic: Tools for the New Biology.* Totowa, N.J.: Humana Press, 2002.

Livingstone, Margaret. *Vision and Art: The Biology of Seeing.* New York: Harry N. Abrams, 2002.

Lutz, Peter L. *The Rise of Experimental Biology: An Illustrated History.* Totowa, N.J.: Humana Press, 2002.

MacArthur, Robert H., and Edward Osborne Wilson. "The Theory of Island Biogeography." In *Princeton Landmarks in Biology.* Princeton, N.J.: Princeton University Press, 2001.

Macbeth, Helen M., and Paul Collinson. *Human Population Dynamics: Cross-Disciplinary Perspectives.* Biosocial Society Symposium Series. Vol. 14. New York: Cambridge University Press, 2002.

Macieira-Coelho, Alvaro. *Biology of Aging.* Progress in Molecular and Subcellular Biology. Vol. 30. New York: Springer, 2003.

Magner, Lois N. *A History of the Life Sciences.* 3rd ed. New York: Marcel Dekker, 2002.

Maldonado, Rafael. *Molecular Biology of Drug Addiction.* Totowa, N.J.: Humana Press, 2003.

Margulis, Lynn, and Dorion Sagan. *Acquiring Genomes: A Theory of the Origins of Species.* New York: Basic Books, 2002.

Marsh, Mark. *Endocytosis.* Frontiers in Molecular Biology. Vol. 36. New York: Oxford University Press, 2001.

Martin, E. A., and Robert Hine. *A Dictionary of Biology.* 4th ed. New York: Oxford University Press, 2000.

Marx, M., and S. D'Auris. "Examination of Eleven Egyptian Mummies." *RadioGraphics* (1986): 321–330.

Mason, C. F. *Biology of Freshwater Pollution.* 4th ed. New York: Prentice Hall, 2002.

Matera, Lina, and Robert Rapaport. *Growth and Lactogenic Hormones.* Neuroimmune Biology. Vol. 2. New York: Elsevier, 2002.

Matis, James H., and Thomas Kiffe. *Stochastic Population Models: A Compartmental Perspective.* New York: Springer, 2000.

Mayr, Ernst. *What Evolution Is.* New York: Basic Books, 2001.

McCabe, T. L. "A Revision of the Genus Hypsoropha Hubner (Lepidoptera: Noctuidae: Ophiderinae)." *Journal of the New York Entomological Society* 100 (1992): 273–285.

McCallum, Hamish. *Population Parameters: Estimation for Ecological Models.* Malden, Mass.: Blackwell Science, 2000.

McGillicuddy-De Lisi, Ann V., and Richard De Lisi. *Biology, Society, and Behavior: The Development of Sex Differences in Cognition.* Westport, Conn.: Ablex Pub., 2002.

Mertz, Leslie A. *Recent Advances and Issues in Biology.* Oryx Frontiers of Science Series. Phoenix, Ariz.: Oryx Press, 2000.

Minugh-Purvis, Nancy, and Ken McNamara. *Human Evolution through Developmental Change.* Baltimore, Md.: Johns Hopkins University Press, 2002.

Mishra, N. C. *Nucleases: Molecular Biology and Applications.* Hoboken, N.J.: Wiley-Interscience, 2002.

Moodie, R. L. *Roentgenologic Studies of Egyptian and Peruvian Mummies.* Chicago: Field Museum of Natural History, 1931.

Moore, John Alexander. *From Genesis to Genetics: The Case of Evolution and Creationism.* Berkeley: University of California Press, 2002.

Mullis, Kary. "The Unusual Origin of the Polymerase Chain Reaction." *Scientific American* (April 1990): 56–65.

Nault, L. R., and W. E. Styer. "Effects of Sinigrin on Host Selection by Aphids." *Entomol. Exp. Appl.* 15 (1972): 423–437.

Nerlich, A. G., A. Zink, U. Szeimies, and H. G. Hagedorn. "Ancient Egyptian Prosthesis of the Big Toe." *Lancet* 356 (2000): 2176–79.

Neuweiler, Gerhard. *The Biology of Bats.* New York: Oxford University Press, 2000.

Nielsen, Claus. *Animal Evolution: Interrelationships of the Living Phyla.* 2d ed. New York: Oxford University Press, 2001.

Norris, Ken, and Deborah J. Pain. *Conserving Bird Biodiversity: General Principles and Their Application.* New York: Cambridge University Press, 2002.

Novacek, Michael J., and American Museum of Natural History. *The Biodiversity Crisis: Losing What Counts.* New York: New Press, 2001.

O'Neill, Kevin M. *Solitary Wasps: Behavior and Natural History.* Cornell Series in Arthropod Biology. Ithaca, N.Y.: Cornell University Press, 2001.

Pagel, Mark D. *Encyclopedia of Evolution.* New York: Oxford University Press, 2002.

Pinhey, E. *Emperor Moths of South and South Central Africa.* Cape Town, South Africa: C. Struik (PTY) Ltd., 1972.

Porter, G. S. *Moths of the Limberlost.* Garden City, N.Y.: Doubleday, Page & Co., 1916.

Priami, Corrado. *Computational Methods in Systems Biology: First International Workshop.* Proceedings, Lecture Notes in Computer Science, No. 2602. New York: Springer, 2003.

Puttaswamaiah, K. *Cost-Benefit Analysis: Environmental and Ecological Perspectives.* New Brunswick, N.J.: Transaction Publishers, 2002.

Reading, Richard P., and Brian Miller. *Endangered Animals: A Reference Guide to Conflicting Issues.* Westport, Conn.: Greenwood Press, 2000.

Richter, Dietmar. *Cell Polarity and Subcellular RNA Localization.* Results and Problems in Cell Differentiation. Vol. 34. New York: Springer, 2001.

Rittner, D., ed. *Pine Bush: Albany's Last Frontier.* Albany, N.Y.: Pine Bush Historic Preservation Project, 1976.

Robinson, Richard, and NetLibrary, Inc. *Biology.* The Macmillan Science Library. New York: Macmillan Reference USA, 2002.

Rockson, Stanley G. "The Lymphatic Continuum: Lymphatic Biology and Disease." In *Annals of the New York Academy of Sciences.* Vol. 979. New York: New York Academy of Sciences, 2002.

Ruse, Michael. *The Evolution Wars: A Guide to the Debates, Controversies in Science.* Santa Barbara, Calif.: ABC-CLIO, 2000.

Saferstein, Richard. *Criminalistics: An Introduction to Forensic Science.* New York: Prentice Hall, 1990.

Scarce, Rik. *Fishy Business: Salmon, Biology, and the Social Construction of Nature, Animals, Culture, and Society.* Philadelphia, Pa.: Temple University Press, 2000.

Schluter, Dolph. *The Ecology of Adaptive Radiation.* Oxford Series in Ecology and Evolution. New York: Oxford University Press, 2000.

Sen, Chandan K., and Lester Packer. "Redox Cell Biology and Genetics." In *Methods in Enzymology.* Boston: Academic Press, 2002.

Seutin, Gilles, Catherine Jeanne Potvin, and Margaret Kraenzel. *Protecting Biological Diversity: Roles and Responsibilities.* Montreal: McGill-Queen's University Press, 2001.

Sewald, Norbert, and Hans-Dieter Jakubke. *Peptides: Chemistry and Biology.* New York: Wiley & Sons, 2002.

Shaw, A. Jonathan, and Bernard Goffinet. *Bryophyte Biology.* New York: Cambridge University Press, 2000.

Shi, Yun-Bo. *Amphibian Metamorphosis: From Morphology to Molecular Biology.* New York: Wiley-Liss, 2000.

Singleton, Paul, and Diana Sainsbury. *Dictionary of Microbiology and Molecular Biology.* 3d ed. New York: Wiley, 2001.

Smith, Malcolm T. *Human Biology and History.* New York: Taylor & Francis, 2002.

Sommer, Ulrich. "Worm. Competition and Coexistence." In *Ecological Studies.* Vol. 161. New York: Springer-Verlag, 2002.

Sternberg, Robert J., and James C. Kaufman. *The Evolution of Intelligence.* Mahwah, N.J.: L. Erlbaum Associates, 2002.

Stuefer, J. F. "Ecology and Evolutionary Biology of Clonal Plants." In *Proceedings of Clone-2000.* International Workshop held in Obergurgl, Austria. Boston: Kluwer Academic Publishers, 2002.

Swain, T. "Secondary Plant Compounds as Protective Agents." *Annual Review of Plant Physiology* 42 (1977): 255–302.

Templer, Richard H., and Robin Leatherbarrow. *Biophysical Chemistry: Membranes and Proteins.* Cambridge, U.K.: Royal Society of Chemistry, 2002.

Tietz, H. M. *An Index to the Described Life Histories, Early Stages and Hosts of the Macrolepidoptera of the Continental United States and Canada.* Sarasota, Fla.: Allyn Mus. Entomol., 1972.

Turner, I. M. *The Ecology of Trees in the Tropical Rain Forest.* Cambridge Tropical Biology Series. New York: Cambridge University Press, 2001.

Tuskes, P. M., J. P. Tuttle, and M. M. Collins. *The Wild Silk Moths of North America: A Natural History of the Saturniidae of the United States and Canada.* Ithaca, N.Y.: Cornell University Press, 1996.

Tyrrell, Andy M., Pauline C. Haddow, and Jim Torresen. "Evolvable Systems: From Biology to Hardware." In *5th Ices International Conference Proceedings.* Lecture notes in computer science, No. 2606. New York: Springer-Verlag, 2003.

Villiard, P. *Moths and How to Rear Them.* New York: Dover Publications, 1975.

Wagle, W. A. "Toe Prosthesis in an Egyptian Human Mummy." *American Journal of Roentgenology* (April 1994): 999–1000.

Ward, Peter Douglas. *Rivers in Time: The Search for Clues to Earth's Mass Extinctions.* New York: Columbia University Press, 2000.

Weissmann, Gerald. *The Year of the Genome: A Diary of the Biological Revolution.* New York: Times Books, 2002.

Wheeler, Alfred George. *Biology of the Plant Bugs (Hemiptera: Miridae): Pests, Predators, Opportunists.* Ithaca, N.Y.: Comstock Pub. Associates, 2001.

White, T. D., and Pieter A. Folkens. *Human Osteology.* 2d ed. San Diego, Calif.: Academic Press, 2000.

Whitehouse, W. M. "Radiologic Findings in the Royal Mummies." In *X-Ray Atlas of the Royal Mummies.* J. E. Harris and E. F. Wente, eds. Chicago: University of Chicago Press, 1980.

Wolpert, L. *Principles of Development.* 2d ed. New York: Oxford University Press, 2002.

Zelditch, Miriam. *Beyond Heterochrony: The Evolution of Development.* New York: Wiley-Liss, 2001.

Zug, George R., Laurie J. Vitt, and Janalee P. Caldwell. *Herpetology: An Introductory Biology of Amphibians and Reptiles.* 2d ed. San Diego, Calif.: Academic Press, 2001.

APPENDIX II

Ask a Biologist. Available on-line. URL:
http://ls.la.asu.edu/askabiologist/index.html. Accessed May 20, 2003.
Anyone is welcome to use Ask a Biologist, intended primarily for an audience of grades K–12, and for teachers as a resource. Arizona State University faculty and staff maintain the site. Experiments and reading materials are also included.

Biointeractive. Available on-line. URL:
http://www.biointeractive.org/. Accessed May 20, 2003.
A site from the Howard Hughes Medical Institute that features virtual labs, animations, Web-based video, and other interactive features relating to biology.

Biology On-line. Available on-line. URL:
http://biology-online.org/. Accessed May 20, 2003
Many on-line tutorials on various biology-related subjects, as well as many other links to biology and environment Websites.

Biology Website References for Students and Teachers. Available on-line. URL:
http://www.hoflink.com/~house/genbkgd.html. Accessed May 20, 2003.
Find out where you can find a career in biology, order biological equipment or specimens, find biological classifications of animals and plants, and locate other general biology material.

Botanical Society of American Online Images. Available on-line. URL:
http://images.botany.org/. Accessed May 20, 2003.
A collection of approximately 800 images available for instructional use in 14 collections of images: plant geography, plant morphology, phloem development, xylem development, floral ontogeny, lichens, economic botany, carnivorous plants, organography, pollen, paleobotany, plant defense mechanisms, plant anatomy, and cellular communication channels.

Ecology www Pages. Available on-line. URL:
http://www.people.fas.harvard.edu/~brach/Ecology-WWW.html. Accessed May 20, 2003.
An alphabetical listing of Websites that focuses on the field of ecology.

Internet Directory of Botany. Available on-line. URL:
http://www.botany.net/IDB/. Accessed May 20, 2003.
A database of subject areas relating to botany. You can find listings of arboreta, software, universities, courses, plant checklists, downloadable database sets, and just about everything you can think of that relates to botany. You can search alphabetically or by categories. Contains hundreds of listings.

Library of Biological Resources via the Internet. Available on-line. URL:
http://vm.cfsan.fda.gov/~frf/biologic.html. Accessed May 20, 2003.

A great source for biological collections and databases accessible via the Web. Toxicology and risk assessment, biotechnology, and biological journals are also listed.

National Human Genome Research Institute. Available on-line. URL:
http://www.genome.gov/. Accessed May 20, 2003.
The Human Genome Project (HGP) is an international research effort to determine the DNA sequence of the entire human genome. You can find many resources here on the progress of this endeavor, including fact sheets, educational curricula, and teaching modules.

U.S. Fish and Wildlife Service Threatened and Endangered Animals and Plants. Available on-line. URL:
http://endangered.fws.gov/wildlife.html. Accessed May 20, 2003.
This Federal Website lists the laws governing endangered and threatened species in the United States and contains listings of all species that are listed or proposed for listing in the United States. It also contains world-related species information and a wealth of information on a species level.

Appendix III

BIOLOGY SOFTWARE AND ANIMATIONS SOURCES

Biocatalog of Software

http://www.ebi.ac.uk/biocat/biocat.html
The Biocatalog is a software directory of general interest in molecular biology and genetics. Includes domains in:

DNA
Proteins
Alignments
Genomes
Genetic
Mapping
Molecular evolution
Molecular graphics
Databases
Servers
Miscellaneous

Biology-Related Simulations

ELECTROCARDIOGRAM
http://www.nobel.se/medicine/educational/ecg/index.html
On-line game that demonstrates how an electrocardiogram works.

PAVLOV'S DOG
http://www.nobel.se/medicine/educational/pavlov/index.html

On-line game demonstrating Pavlov's famous dog experiments.

CONTROL OF THE CELL CYCLE
http://www.nobel.se/medicine/educational/2001/index.html
On-line game that demonstrates the different phases in the cell cycle.

BLOOD TYPING
http://www.nobel.se/medicine/educational/landsteiner/index.html
On-line game that demonstrates various blood types and transfusion.

THE DISCOVERY OF PENICILLIN
http://www.nobel.se/medicine/educational/penicillin/index.html
On-line game that lets you discover penicillin like Sir Fleming.

MICROSCOPES
http://www.nobel.se/physics/educational/microscopes/1.html
On-line game that teaches how various types of microscopes work.

THE BIOCHEMISTRY LAB

http://www.nobel.se/chemistry/educational/vbl/index.html
On-line site that lets you perform 12 biochemistry experiments.

EBI

http://www.ebi.ac.uk/
or ftp://ftp.ebi.ac.uk/pub/software/
The European Bioinformatics Institute, home to EMBL databank and others, and a home to a large molecular biology software archive.

EXPASY MOLECULAR BIOLOGY SERVER

http://ca.expasy.org/
The ExPASy (Expert Protein Analysis System) proteomics server of the Swiss Institute of Bioinformatics (SIB) is dedicated to the analysis of protein sequences and structures.

GENAMICS SOFTWARESEEK

http://genamics.com/software/
Genamics SoftwareSeek is a repository and database of freely distributable and commercial tools for use in molecular biology and biochemistry. Windows, MS-DOS, Mac, Unix, and Linux platforms. The database presently contains more than 1,200 entries.

IU-BIO-ARCHIVE

http://iubio.bio.indiana.edu/
This is an Internet archive of biology data and software, established in 1989, to promote public access to freely available information, primarily in the field of molecular biology. There are hundreds of free and shareware programs available for downloading for almost all computer platforms from PC to Mac.

MOLECULAR BIOLOGY SOFTWARE

http://www.yk.rim.or.jp/~aisoai/soft.html
A Website that links to many other sites that contain biology software. Direct downloads of features programs from modeling to RNA analysis are available directly from this site as well.

NATIONAL CENTER FOR BIOTECHNOLOGY INFORMATION

http://www.ncbi.nlm.nih.gov/
Established in 1988 as a national resource for molecular biology information, NCBI creates public databases, conducts research in computational biology, develops software tools for analyzing genome data, and disseminates biomedical information. A great deal of software available for downloading.

Appendix IV

NOBEL LAUREATES RELATING TO BIOLOGY

Chemistry Laureates

1901   Jacobus Henricus van 't Hoff
"in recognition of the extraordinary services he has rendered by the discovery of the laws of chemical dynamics and osmotic pressure in solutions"

1902   Hermann Emil Fischer
"in recognition of the extraordinary services he has rendered by his work on sugar and purine syntheses"

1903   Svante August Arrhenius
"in recognition of the extraordinary services he has rendered to the advancement of chemistry by his electrolytic theory of dissociation"

1907   Eduard Buchner
"for his biochemical researches and his discovery of cell-free fermentation"

1912   Paul Sabatier
"for his method of hydrogenating organic compounds in the presence of finely disintegrated metals whereby the progress of organic chemistry has been greatly advanced in recent years"

1912   Victor Grignard
"for the discovery of the so-called Grignard reagent, which in recent years has greatly advanced the progress of organic chemistry"

1913   Alfred Werner
"in recognition of his work on the linkage of atoms in molecules by which he has thrown new light on earlier investigations and opened up new fields of research especially in inorganic chemistry"

1915   Richard Martin Willstätter
"for his researches on plant pigments, especially chlorophyll"

1918   Fritz Haber
"for the synthesis of ammonia from its elements"

1921   Frederick Soddy
"for his contributions to our knowledge of the chemistry of radioactive substances, and his investigations into the origin and nature of isotopes"

1923   Fritz Pregl
"for his invention of the method of micro-analysis of organic substances"

1925   Richard Adolf Zsigmondy
"for his demonstration of the hetero-

geneous nature of colloid solutions and for the methods he used, which have since become fundamental in modern colloid chemistry"

1927   Heinrich Otto Wieland
"for his investigations of the constitution of the bile acids and related substances"

1928   Adolf Otto Reinhold Windaus
"for the services rendered through his research into the constitution of the sterols and their connection with the vitamins"

1929   Arthur Harden, Hans Karl August Simon von Euler-Chelpin
"for their investigations on the fermentation of sugar and fermentative enzymes"

1930   Hans Fischer
"for his researches into the constitution of haemin and chlorophyll and especially for his synthesis of haemin"

1932   Irving Langmuir
"for his discoveries and investigations in surface chemistry"

1937   Paul Karrer
"for his investigations on carotenoids, flavins, and vitamins A and B2"

1937   Walter Norman Haworth
"for his investigations on carbohydrates and vitamin C"

"for his investigations on carotenoids, flavins and vitamins A and B2"

1938   Richard Kuhn
"for his work on carotenoids and vitamins"

1939   Leopold Ruzicka
"for his work on polymethylenes and higher terpenes"

1939   Adolf Friedrich Johann Butenandt
"for his work on sex hormones"

1945   Artturi Ilmari Virtanen
"for his research and inventions in agricultural and nutrition chemistry, especially for his fodder preservation method"

1946   John Howard Northrop, Wendell Meredith Stanley
"for his research and inventions in agricultural and nutrition chemistry, especially for his fodder preservation method"

"for their preparation of enzymes and virus proteins in a pure form"

1946   James Batcheller Sumner
"for their preparation of enzymes and virus proteins in a pure form"

1947   Sir Robert Robinson
"for his investigations on plant products of biological importance, especially the alkaloids"

1948   Arne Wilhelm Kaurin Tiselius
"for his research on electrophoresis and adsorption analysis, especially for his discoveries concerning the complex nature of the serum proteins"

1950   Otto Paul Hermann Diels, Kurt Alder
"for their discovery and development of the diene synthesis"

1952   Archer John Porter Martin, Richard Laurence Millington Synge
"for their invention of partition chromatography"

1953   Hermann Staudinger
"for his discoveries in the field of macromolecular chemistry"

1954   Linus Carl Pauling
"for his research into the nature of the chemical bond and its application to the elucidation of the structure of complex substances"

1955   Vincent Du Vigneaud
"for his work on biochemically impor-

tant sulfur compounds, especially for the first synthesis of a polypeptide hormone"

1957   Lord (Alexander R.) Todd
"for his work on nucleotides and nucleotide co-enzymes"

1958   Frederick Sanger
"for his work on the structure of proteins, especially that of insulin"

1960   Willard Frank Libby
"for his method to use carbon-14 for age determination in archaeology, geology, geophysics, and other branches of science"

1961   Melvin Calvin
"for his research on the carbon dioxide assimilation in plants"

1962   Max Ferdinand Perutz, John Cowdery Kendrew
"for their studies of the structures of globular proteins"

1963   Karl Ziegler, Giulio Natta
"for their discoveries in the field of the chemistry and technology of high polymers"

1964   Dorothy Crowfoot Hodgkin
"for her determinations by X-ray techniques of the structures of important biochemical substances"

1965   Robert Burns Woodward
"for his outstanding achievements in the art of organic synthesis"

1969   Derek H. R. Barton, Odd Hassel
"for their contributions to the development of the concept of conformation and its application in chemistry"

1970   Luis F. Leloir
"for his discovery of sugar nucleotides and their role in the biosynthesis of carbohydrates"

1971   Gerhard Herzberg
"for his contributions to the knowledge of electronic structure and geometry of molecules, particularly free radicals"

1972   Stanford Moore, William H. Stein
"for their contribution to the understanding of the connection between chemical structure and catalytic activity of the active centre of the ribonuclease molecule"

1972   Stanford Moore
"for his contributions to the knowledge of electronic structure and geometry of molecules, particularly free radicals"

1972   Christian B. Anfinsen
"for his work on ribonuclease, especially concerning the connection between the amino acid sequence and the biologically active conformation"

1973   Ernst Otto Fischer, Geoffrey Wilkinson
"for their pioneering work, performed independently, on the chemistry of the organometallic, so-called sandwich compounds"

1974   Paul J. Flory
"for his fundamental achievements, both theoretical and experimental, in the physical chemistry of the macromolecules"

1975   Vladimir Prelog
"for his research into the stereochemistry of organic molecules and reactions"

1975   John Warcup Cornforth
"for his work on the stereochemistry of enzyme-catalyzed reactions"

1978   Peter D. Mitchell
"for his contribution to the understanding of biological energy transfer through the formulation of the chemiosmotic theory"

1979   Herbert C. Brown, Georg Wittig
"for their development of the use of

boron- and phosphorus-containing compounds, respectively, into important reagents in organic synthesis"

1980   Walter Gilbert, Frederick Sanger
"for their contributions concerning the determination of base sequences in nucleic acids"

1980   Paul Berg
"for his fundamental studies of the biochemistry of nucleic acids, with particular regard to recombinant-DNA"

1982   Aaron Klug
"for his development of crystallographic electron microscopy and his structural elucidation of biologically important nuclei acid-protein complexes"

1986   Dudley R. Herschbach, Yuan T. Lee, John C. Polanyi
"for their contributions concerning the dynamics of chemical elementary processes"

1987   Donald J. Cram, Jean-Marie Lehn, Charles J. Pedersen
"for their development and use of molecules with structure-specific interactions of high selectivity"

1988   Johann Deisenhofer, Robert Huber, Hartmut Michel
"for the determination of the three-dimensional structure of a photosynthetic reaction centre"

1989   Sidney Altman, Thomas R. Cech
"for their discovery of catalytic properties of RNA"

1990   Elias James Corey
"for his development of the theory and methodology of organic synthesis"

1991   Richard R. Ernst
"for his contributions to the development of the methodology of high resolution nuclear magnetic resonance (NMR) spectroscopy"

1992   Rudolph A. Marcus
"for his contributions to the theory of electron transfer reactions in chemical systems"

1993   Michael Smith
"for contributions to the developments of methods within DNA-based chemistry"

"for his fundamental contributions to the establishment of oligonucleotide-based, site-directed mutagenesis and its development for protein studies"

1993   Kary B. Mullis
"for contributions to the developments of methods within DNA-based chemistry"

"for his invention of the polymerase chain reaction (PCR) method"

1994   George A. Olah
"for his contribution to carbocation chemistry"

1995   Paul J. Crutzen, Mario J. Molina, F. Sherwood Rowland
"for their work in atmospheric chemistry, particularly concerning the formation and decomposition of ozone"

1997   Jens C. Skou
"for the first discovery of an ion-transporting enzyme, Na+, K+ -ATPase"

1997   Paul D. Boyer, John E. Walker
"for their elucidation of the enzymatic mechanism underlying the synthesis of adenosine triphosphate (ATP)"

2002   Kurt Wüthrich
"for his development of nuclear magnetic resonance spectroscopy for determining the three-dimensional structure of biological macromolecules in solution"

2002   John B. Fenn, Koichi Tanaka
"for the development of methods for

identification and structure analyses of biological macromolecules"

"for their development of soft desorption ionization methods for mass spectrometric analyses of biological macromolecules"

2003   Peter Agre, Roderick MacKinnon
"for discoveries concerning channels in cell membranes"

Physiology or Medicine Laureates

1901   Emil von Behring
"for his work on serum therapy, especially its application against diphtheria, by which he has opened a new road in the domain of medical science and thereby placed in the hands of the physician a victorious weapon against illness and deaths"

1902   Ronald Ross
"for his work on malaria, by which he has shown how it enters the organism and thereby has laid the foundation for successful research on this disease and methods of combating it"

1903   Niels Ryberg Finsen
"in recognition of his contribution to the treatment of diseases, especially lupus vulgaris, with concentrated light radiation, whereby he has opened a new avenue for medical science"

1904   Ivan Pavlov
"in recognition of his work on the physiology of digestion, through which knowledge on vital aspects of the subject has been transformed and enlarged"

1905   Robert Koch
"for his investigations and discoveries in relation to tuberculosis"

1906   Camillo Golgi, Santiago Ramón y Cajal
"in recognition of their work on the structure of the nervous system"

1907   Alphonse Laveran
"in recognition of his work on the role played by protozoa in causing diseases"

1908   Ilya Mechnikov, Paul Ehrlich
"in recognition of their work on immunity"

1909   Theodor Kocher
"for his work on the physiology, pathology and surgery of the thyroid gland"

1910   Albrecht Kossel
"in recognition of the contributions to our knowledge of cell chemistry made through his work on proteins, including the nucleic substances"

1911   Allvar Gullstrand
"for his work on the dioptrics of the eye"

1912   Alexis Carrel
"in recognition of his work on vascular suture and the transplantation of blood vessels and organs"

1913   Charles Richet
"in recognition of his work on anaphylaxis"

1914   Robert Bárány
"for his work on the physiology and pathology of the vestibular apparatus"

1919   Jules Bordet
"for his discoveries relating to immunity"

1920   August Krogh
"for his discovery of the capillary motor regulating mechanism"

1922   Archibald V. Hill, Otto Meyerhof
"for his discovery relating to the production of heat in the muscle"

1923   Frederick G. Banting, John Macleod
"for the discovery of insulin"

1924   Willem Einthoven
"for his discovery of the mechanism of the electrocardiogram"

1926   Johannes Fibiger
"for his discovery of the Spiroptera carcinoma"

1927   Julius Wagner-Jauregg
"for his discovery of the therapeutic value of malaria inoculation in the treatment of dementia paralytica"

1928   Charles Nicolle
"for his work on typhus"

1929   Christiaan Eijkman, Sir Frederick Hopkins
"for his discovery of the antineuritic vitamin"

"for his discovery of the growth-stimulating vitamins"

1930   Karl Landsteiner
"for his discovery of human blood groups"

1931   Otto Warburg
"for his discovery of the nature and mode of action of the respiratory enzyme"

1932   Sir Charles Sherrington, Edgar Adrian
"for their discoveries regarding the functions of neurons"

1933   Thomas H. Morgan
"for his discoveries concerning the role played by the chromosome in heredity"

1934   George H. Whipple, George R. Minot, William P. Murphy
"for their discoveries concerning liver therapy in cases of anemia"

1935   Hans Spemann
"for his discovery of the organizer effect in embryonic development"

1936   Sir Henry Dale, Otto Loewi
"for their discoveries relating to chemical transmission of nerve impulses"

1937   Albert Szent-Györgyi
"for his discoveries in connection with the biological combustion processes, with special reference to vitamin C and the catalysis of fumaric acid"

1938   Corneille Heymans
"for the discovery of the role played by the sinus and aortic mechanisms in the regulation of respiration"

1939   Gerhard Domagk
"for the discovery of the antibacterial effects of prontosil"

1943   Edward A. Doisy
"for his discovery of the chemical nature of vitamin K"

1943   Henrik Dam
"for his discovery of vitamin K"

1944   Joseph Erlanger, Herbert S. Gasser
"for their discoveries relating to the highly differentiated functions of single nerve fibers"

1945   Sir Alexander Fleming, Ernst B. Chain, Sir Howard Florey
"for the discovery of penicillin and its curative effect in various infectious diseases"

1946   Hermann J. Muller
"for the discovery of the production of mutations by means of X-ray irradiation"

1947   Bernardo Houssay
"for his discovery of the part played by the hormone of the anterior pituitary lobe in the metabolism of sugar"

1947   Carl Cori, Gerty Cori
"for their discovery of the course of the catalytic conversion of glycogen"

1948   Paul Müller
"for his discovery of the high efficiency of DDT as a contact poison against several arthropods"

1949   Egas Moniz
"for his discovery of the therapeutic

value of leucotomy in certain psychoses"

1949   Walter Hess
"for his discovery of the functional organization of the interbrain as a coordinator of the activities of the internal organs"

1950   Edward C. Kendall, Tadeus Reichstein, Philip S. Hench
"for their discoveries relating to the hormones of the adrenal cortex, their structure and biological effects"

1951   Max Theiler
"for his discoveries concerning yellow fever and how to combat it"

1952   Selman A. Waksman
"for his discovery of streptomycin, the first antibiotic effective against tuberculosis"

1953   Hans Krebs, Fritz Lipmann
"for his discovery of co-enzyme A and its importance for intermediary metabolism"

1953   Hans Krebs
"for his discovery of the citric acid cycle"

1954   John F. Enders, Thomas H. Weller, Frederick C. Robbins
"for their discovery of the ability of poliomyelitis viruses to grow in cultures of various types of tissue"

1955   Hugo Theorell
"for his discoveries concerning the nature and mode of action of oxidation enzymes"

1956   André F. Cournand, Werner Forssmann, Dickinson W. Richards
"for their discoveries concerning heart catheterization and pathological changes in the circulatory system"

1957   Daniel Bovet
"for his discoveries relating to synthetic compounds that inhibit the action of

certain body substances, and especially their action on the vascular system and the skeletal muscles"

1958   Joshua Lederberg
"for his discoveries concerning genetic recombination and the organization of the genetic material of bacteria"

1958   George Beadle, Edward Tatum
"for their discovery that genes act by regulating definite chemical events"

1959   Severo Ochoa, Arthur Kornberg
"for their discovery of the mechanisms in the biological synthesis of ribonucleic acid and deoxyribonucleic acid"

1960   Sir Frank Macfarlane Burnet, Peter Medawar
"for discovery of acquired immunological tolerance"

1961   Georg von Békésy
"for his discoveries of the physical mechanism of stimulation within the cochlea"

1962   Francis Crick, James Watson, Maurice Wilkins
"for their discoveries concerning the molecular structure of nucleic acids and its significance for information transfer in living material"

1963   Sir John Eccles, Alan L. Hodgkin, Andrew F. Huxley
"for their discoveries concerning the ionic mechanisms involved in excitation and inhibition in the peripheral and central portions of the nerve cell membrane"

1964   Konrad Bloch, Feodor Lynen
"for their discoveries concerning the mechanism and regulation of the cholesterol and fatty acid metabolism"

1965   François Jacob, André Lwoff, Jacques Monod
"for their discoveries concerning genetic control of enzyme and virus synthesis"

1966   Peyton Rous, Charles B. Huggins
"for his discovery of tumor-inducing viruses"

1967   Ragnar Granit, Haldan K. Hartline, George Wald
"for their discoveries concerning the primary physiological and chemical visual processes in the eye"

1968   Robert W. Holley, H. Gobind Khorana, Marshall W. Nirenberg
"for their interpretation of the genetic code and its function in protein synthesis"

1969   Max Delbrück, Alfred D. Hershey, Salvador E. Luria
"for their discoveries concerning the replication mechanism and the genetic structure of viruses"

1970   Sir Bernard Katz, Ulf von Euler, Julius Axelrod
"for their discoveries concerning the humoral transmittors in the nerve terminals and the mechanism for their storage, release and inactivation"

1971   Earl W. Sutherland Jr.
"for his discoveries concerning the mechanisms of the action of hormones"

1972   Gerald M. Edelman, Rodney R. Porter
"for their discoveries concerning the chemical structure of antibodies"

1973   Karl von Frisch, Konrad Lorenz, Nikolaas Tinbergen
"for their discoveries concerning organization and elicitation of individual and social behavior patterns"

1974   Albert Claude, Christian de Duve, George E. Palade
"for their discoveries concerning the structural and functional organization of the cell"

1975   David Baltimore, Renato Dulbecco, Howard M. Temin
"for their discoveries concerning the interaction between tumor viruses and the genetic material of the cell"

1976   Baruch S. Blumberg, D. Carleton Gajdusek
"for their discoveries concerning new mechanisms for the origin and dissemination of infectious diseases"

1977   Rosalyn Yalow
"for the development of radioimmunoassays of peptide hormones"

1977   Roger Guillemin, Andrew V. Schally
"for their discoveries concerning the peptide hormone production of the brain"

1978   Werner Arber, Daniel Nathans, Hamilton O. Smith
"for the discovery of restriction enzymes and their application to problems of molecular genetics"

1979   Allan M. Cormack, Godfrey N. Hounsfield
"for the development of computer assisted tomography"

1980   Baruj Benacerraf, Jean Dausset, George D. Snell
"for their discoveries concerning genetically determined structures on the cell surface that regulate immunological reactions"

1981   David H. Hubel, Torsten N. Wiesel
"for their discoveries concerning information processing in the visual system"

1981   Roger W. Sperry
"for his discoveries concerning the functional specialization of the cerebral hemispheres"

1982   Sune K. Bergström, Bengt I. Samuelsson, John R. Vane
"for their discoveries concerning prostaglandins and related biologically active substances"

1983   Barbara McClintock
"for her discovery of mobile genetic elements"

1984   Niels K. Jerne, Georges J. F. Köhler, César Milstein
"for theories concerning the specificity in development and control of the immune system and the discovery of the principle for production of monoclonal antibodies"

1985   Michael S. Brown, Joseph L. Goldstein
"for their discoveries concerning the regulation of cholesterol metabolism"

1986   Stanley Cohen, Rita Levi-Montalcini
"for their discoveries of growth factors"

1987   Susumu Tonegawa
"for his discovery of the genetic principle for generation of antibody diversity"

1988   Sir James W. Black, Gertrude B. Elion, George H. Hitchings
"for their discoveries of important principles for drug treatment"

1989   J. Michael Bishop, Harold E. Varmus
"for their discovery of the cellular origin of retroviral oncogenes"

1990   Joseph E. Murray, E. Donnall Thomas
"for their discoveries concerning organ and cell transplantation in the treatment of human disease"

1991   Erwin Neher, Bert Sakmann
"for their discoveries concerning the function of single ion channels in cells"

1992   Edmond H. Fischer, Edwin G. Krebs
"for their discoveries concerning reversible protein phosphorylation as a biological regulatory mechanism"

1993   Richard J. Roberts, Phillip A. Sharp
"for their discoveries of split genes"

1994   Alfred G. Gilman, Martin Rodbell
"for their discovery of G-proteins and the role of these proteins in signal transduction in cells"

1995   Edward B. Lewis, Christiane Nüsslein-Volhard, Eric F. Wieschaus
"for their discoveries concerning the genetic control of early embryonic development"

1996   Peter C. Doherty, Rolf M. Zinkernagel
"for their discoveries concerning the specificity of the cell mediated immune defense"

1997   Stanley B. Prusiner
"for his discovery of Prions—a new biological principle of infection"

1998   Robert F. Furchgott, Louis J. Ignarro, Ferid Murad
"for their discoveries concerning nitric oxide as a signaling molecule in the cardiovascular system"

1999   Günter Blobel
"for the discovery that proteins have intrinsic signals that govern their transport and localization in the cell"

2000   Arvid Carlsson, Paul Greengard, Eric R. Kandel
"for their discoveries concerning signal transduction in the nervous system"

2001   Leland H. Hartwell, Tim Hunt, Sir Paul Nurse
"for their discoveries of key regulators of the cell cycle"

2002   Sydney Brenner, H. Robert Horvitz, John E. Sulston
"for their discoveries concerning 'genetic regulation of organ development and programmed cell death'"

2003   Paul C. Lauterbur, Sir Peter Mansfield
"for their discoveries concerning magnetic resonance imaging"

Appendix V

Periodic Table of the Elements

Key:
1
H
1.008

1 — atomic number
H — symbol
1.008 — atomic weight

1 H 1.008																		2 He 4.003
3 Li 6.941	4 Be 9.012											5 B 10.81	6 C 12.01	7 N 14.01	8 O 16.00	9 F 19.00		10 Ne 20.18
11 Na 22.99	12 Mg 24.31											13 Al 26.98	14 Si 28.09	15 P 30.97	16 S 32.07	17 Cl 35.45		18 Ar 39.95
19 K 39.10	20 Ca 40.08	21 Sc 44.96	22 Ti 47.88	23 V 50.94	24 Cr 52.00	25 Mn 54.94	26 Fe 55.85	27 Co 58.93	28 Ni 58.69	29 Cu 63.55	30 Zn 65.39	31 Ga 69.72	32 Ge 72.59	33 As 74.92	34 Se 78.96	35 Br 79.90		36 Kr 83.80
37 Rb 85.47	38 Sr 87.62	39 Y 88.91	40 Zr 91.22	41 Nb 92.91	42 Mo 95.94	43 Tc (98)	44 Ru 101.1	45 Rh 102.9	46 Pd 106.4	47 Ag 107.9	48 Cd 112.4	49 In 114.8	50 Sn 118.7	51 Sb 121.8	52 Te 127.6	53 I 126.9		54 Xe 131.3
55 Cs 132.9	56 Ba 137.3	57–71* 	72 Hf 178.5	73 Ta 180.9	74 W 183.9	75 Re 186.2	76 Os 190.2	77 Ir 192.2	78 Pt 195.1	79 Au 197.0	80 Hg 200.6	81 Tl 204.4	82 Pb 207.2	83 Bi 209.0	84 Po (210)	85 At (210)		86 Rn (222)
87 Fr (223)	88 Ra (226)	89–103‡ 	104 Rf (261)	105 Db (262)	106 Sg (263)	107 Bh (262)	108 Hs (265)	109 Mt (266)	110 Ds (271)									

Numbers in parentheses are atomic mass numbers of radioactive isotopes.

*lanthanide series

57 La 138.9	58 Ce 140.1	59 Pr 140.9	60 Nd 144.2	61 Pm (145)	62 Sm 150.4	63 Eu 152.0	64 Gd 157.3	65 Tb 158.9	66 Dy 162.5	67 Ho 164.9	68 Er 167.3	69 Tm 168.9	70 Yb 173.0	71 Lu 175.0

‡actinide series

89 Ac (227)	90 Th 232.0	91 Pa 231.0	92 U 238.0	93 Np (237)	94 Pu (244)	95 Am (243)	96 Cm (247)	97 Bk (247)	98 Cf (251)	99 Es (252)	100 Fm (257)	101 Md (258)	102 No (259)	103 Lr (260)

The Chemical Elements

element	symbol	a.n.	element	symbol	a.n.	element	symbol	a.n.	element	symbol	a.n.
actinium	Ac	89	einsteinium	Es	99	mendelevium	Md	101	samarium	Sm	62
aluminum	Al	13	erbium	Er	68	mercury	Hg	80	scandium	Sc	21
americium	Am	95	europium	Eu	63	molybdenum	Mo	42	seaborgium	Sg	106
antimony	Sb	51	fermium	Fm	100	neodymium	Nd	60	selenium	Se	34
argon	Ar	18	fluorine	F	9	neon	Ne	10	silicon	Si	14
arsenic	As	33	francium	Fr	87	neptunium	Np	93	silver	Ag	47
astatine	At	85	gadolinium	Gd	64	nickel	Ni	28	sodium	Na	11
barium	Ba	56	gallium	Ga	31	niobium	Nb	41	strontium	Sr	38
berkelium	Bk	97	germanium	Ge	32	nitrogen	N	7	sulfur	S	16
beryllium	Be	4	gold	Au	79	nobelium	No	102	tantalum	Ta	73
bismuth	Bi	83	hafnium	Hf	72	osmium	Os	76	technetium	Tc	43
bohrium	Bh	107	hassium	Hs	108	oxygen	O	8	tellurium	Te	52
boron	B	5	helium	He	2	palladium	Pd	46	terbium	Tb	65
bromine	Br	35	holmium	Ho	67	phosphorus	P	15	thallium	Tl	81
cadmium	Cd	48	hydrogen	H	1	platinum	Pt	78	thorium	Th	90
calcium	Ca	20	indium	In	49	plutonium	Pu	94	thulium	Tm	69
californium	Cf	98	iodine	I	53	polonium	Po	84	tin	Sn	50
carbon	C	6	iridium	Ir	77	potassium	K	19	titanium	Ti	22
cerium	Ce	58	iron	Fe	26	praseodymium	Pr	59	tungsten	W	74
cesium	Cs	55	krypton	Kr	36	promethium	Pm	61	uranium	U	92
chlorine	Cl	17	lanthanum	La	57	protactinium	Pa	91	vanadium	V	23
chromium	Cr	24	lawrencium	Lr	103	radium	Ra	88	xenon	Xe	54
cobalt	Co	27	lead	Pb	82	radon	Rn	86	ytterbium	Yb	70
copper	Cu	29	lithium	Li	3	rhenium	Re	75	yttrium	Y	39
curium	Cm	96	lutetium	Lu	71	rhodium	Rh	45	zinc	Zn	30
darmstadtium	Ds	110	magnesium	Mg	12	rubidium	Rb	37	zirconium	Zr	40
dubnium	Db	105	manganese	Mn	25	ruthenium	Ru	44			
dysprosium	Dy	66	meitnerium	Mt	109	rutherfordium	Rf	104	a.n. = atomic number		

APPENDIX VI

BIOCHEMICAL CYCLES

Oxygen cycle Oxygen plays a vital part in the respiration of animals and plants.

1 Oxygen in air
2 Oxygen breathed in by animals
3 Carbon dioxide (a carbon-oxygen compound) breathed out by living things as waste
4 Carbon dioxide absorbed by plants and used in photosynthesis to make carbohydrate foods
5 Surplus oxygen released into the air by plants as waste

Carbon cycle Plant material is a valuable source of carbon. Oxidizing carbon compounds provide energy for animals and plants.

1 Carbon dioxide (a carbon oxygen compound) in air
2 Carbon dioxide absorbed by plants for making food
3 Plants eaten by animals
4 Carbon dioxide waste breathed out by animals and plants
5 Dead organisms broken down by bacteria
6 These give off carbon dioxide waste
7 Remains of long-dead plants and microscopic organisms forming hydrocarbon fossil fuels: coal, oil, and gas
8 Carbon dioxide released back into the air by burning fossil fuels

Nitrogen cycle As an ingredient in proteins and nucleic acids, nitrogen is vital to all living things.

1 Nitrogen in air
2 Atmospheric nitrogen trapped by some plants' roots
3 Nitrogen used by plants for making proteins
4 Plant proteins eaten by animals
5 Proteins in dead organisms and body wastes converted to ammonia by bacteria and fungi
6 Ammonia converted to nitrate by other bacteria
7 Nitrate taken up by plant roots

(continues)

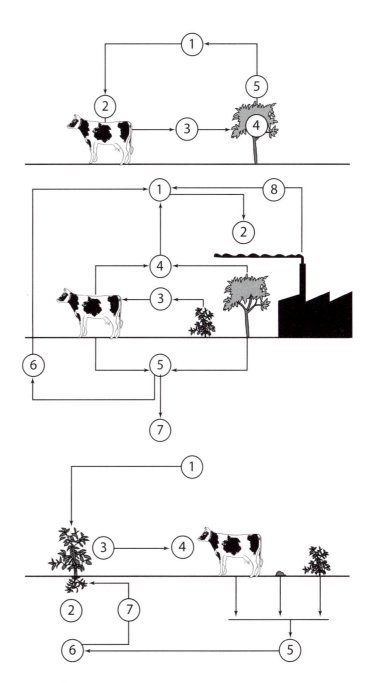

BIOCHEMICAL CYCLES

(continued)

Sulfur cycle Sulfur is in two of the 20 amino acids which are used by the body to make proteins.

1 Sulfates (sulfur–oxygen compounds) absorbed by plant roots
2 The oxygen in the sulfate is replaced by hydrogen in a plant process that produces certain amino acids
3 Plants eaten by animals
4 Sulfur-containing amino acids of dead plants and animals broken down to hydrogen sulfide (which gives off a rotten egg odor) by decomposer microorganisms
5 Sulfur extracted from sulfides by bacteria
6 Other bacteria combine sulfur with oxygen, producing sulfates

Phosphorus cycle Phosphorus is a vital ingredient of proteins, nucleic acids, and some other compounds found in living things.

1 Phosphates (compounds of phosphorus, hydrogen, and oxygen) absorbed by plant roots
2 Phosphates used by plants in making organic phosphorus compounds
3 Plants eaten by animals
4 Compounds in dead plants and animals broken down to phosphates by microorganisms

Krebs cycle The Krebs or citric acid cycle is the second stage of aerobic respiration in which living things produce energy from foods. It requires oxygen; enzymes (proteins that promote but are not used up in chemical changes) create successive compounds, thus transforming pyruvate to carbon dioxide and water and releasing energy.

1 Acetic acid combines with...
2 Oxaloacetic acid to form...
3 Citric acid. Later changes produce...
4 Aconitic acid
5 Isocitric acid
6 Ketoglutaric acid
7 Succinic acid, carbon dioxide, and energy-rich ATP (adenosine triphosphate)
8 Fumaric acid
9 Malic acid

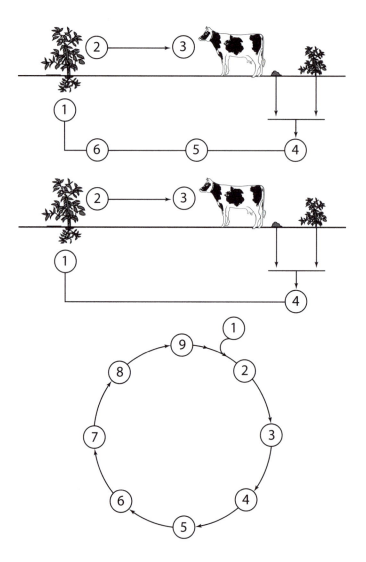

APPENDIX VII
The "Tree of Life"

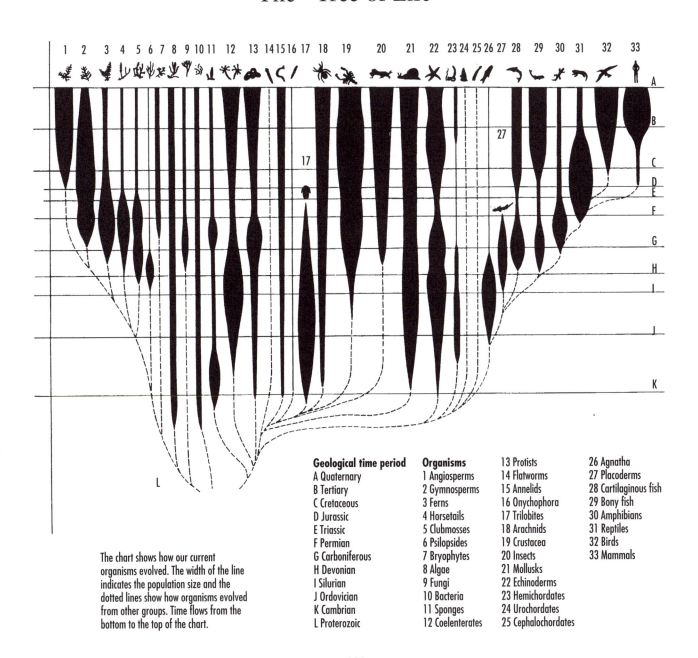

Geological time period
A Quaternary
B Tertiary
C Cretaceous
D Jurassic
E Triassic
F Permian
G Carboniferous
H Devonian
I Silurian
J Ordovician
K Cambrian
L Proterozoic

Organisms
1 Angiosperms
2 Gymnosperms
3 Ferns
4 Horsetails
5 Clubmosses
6 Psilopsides
7 Bryophytes
8 Algae
9 Fungi
10 Bacteria
11 Sponges
12 Coelenterates

13 Protists
14 Flatworms
15 Annelids
16 Onychophora
17 Trilobites
18 Arachnids
19 Crustacea
20 Insects
21 Mollusks
22 Echinoderms
23 Hemichordates
24 Urochordates
25 Cephalochordates

26 Agnatha
27 Placoderms
28 Cartilaginous fish
29 Bony fish
30 Amphibians
31 Reptiles
32 Birds
33 Mammals

The chart shows how our current organisms evolved. The width of the line indicates the population size and the dotted lines show how organisms evolved from other groups. Time flows from the bottom to the top of the chart.

381

INDEX

Note: Page numbers in **boldface** indicate main entries; page numbers in *italic* indicate illustrations.